BIOLOGY OF BRYOPHYTES

R.N. CHOPRA
*Professor
Department of Botany
University of Delhi,
Delhi-110007, INDIA*

P.K. KUMRA
*Department of Botany
Hans Raj College,
University of Delhi,
Delhi-110007, INDIA*

JOHN WILEY & SONS
NEW YORK CHICHESTER BRISBANE TORONTO SINGAPORE

First Published in 1988
Reprint 1989

WILEY EASTERN LIMITED
4835/24 Ansari Road, Daryaganj
New Delhi 110 002, India

Distributers:

Australia and New Zealand
JACARANDA WILEY LTD., JACARANDA PRESS
JOHN WILEY & SONS, INC.
GPO Box 859, Brisbane, Queensland 4001, Australia

Canada:
JOHN WILEY & SONS CANADA LIMITED
22 Worcester Road, Rexdale, Ontario, Canada

Europe and Africa:
JOHN WILEY & SONS LIMITED
Baffins, Chichester, West Sussex, England

South East Asia
JOHN WILEY & SONS, INC
05-05, Blick B, Union Industrial Building
37 Jalan Pemimpin, Singapore 2057

Africa and South Asia:
WILEY EASTERN LIMITED
4835/24 Ansari Road, Daryaganj
New Delhi 110 002, India

North and South America and rest of the world:
JOHN WILEY & SONS, INC
605 Third Avenue, New York, NY 10158 USA

Copyright © 1988, WILEY EASTERN LIMITED
New Delhi, India

Library of Congress of Cataloging-in-Publication Data

ISBN 0-470-21359-0 John Wiley & Sons, Inc.
ISBN 0 85226 240 X Wiley Eastern Limited

Printed in India at Ram Printograph, New Delhi

Dedicated to
PANCHANAN MAHESHWARI

Foreword

In the last few years several books about nearly all aspects of bryophytes have appeared ranging in sophistication from the introductory level of *The Biology of Mosses* of Richardson to the very broad and in some way overflowing edition of the *New Manual of Bryology* by Schuster. Just few months ago a comprehensive book by Demkiv and Sytnik in Russian has appeared. Therefore one may question whether a new book is necessary?

"Necessary" may be not the correct word to establish that a new book should be written. Sufficient justification may be to sum up the many accumulated data regarding a certain topic under one particular point of view so as to give a new or at least different insight into the fullness of data scattered throughout the literature.

The senior author, R.N. Chopra has been known for many years as the leader of group of scholars who have treated with consequence and in an industrious manner the problem of moss and liverwort development under the influence of external factors and chemical treatments. From the experimental work of Georg Klebs at the beginning of this century we know, that an exact determination of external factors can reveal the key to the understanding of the internal regulation of developmental processes. If we consider the main chapters of the book, we can easily recognize that this subject forms the theme of this volume. Each crucial step of development, namely germination of spores, formation and growth of gemmae, protonema differentiation, regeneration, gametangia production, pathways in the life cycle including sporophyte development, is treated in the similar way, which gives a realistic framework for understanding how the different processes of development are integrated into the whole life cycle of a bryophyte. One chapter is dedicated to the photomorphogenesis, emphasizing the special importance of this particular aspect, from which we know the internal response better than for all other external factors. Independent of this, the consequent treatment of bryophyte development under identical aspect distinguishes this book from other comparable publications.

Altogether, we can see that bryophytes are a fascinating group of beautiful plants, of which not only a greater knowledge is desired, but which also offer a tool for further research into the understanding of the mechanisms used for the regulation of growth and development of plants in general.

The book is written in a clear language avoiding complicated argumentation and will be useful to all students interested in this field and those who need a comprehensive summary for their own research. The collection of the vi *Foreword*

many data, offered in an accessible form, serves as a base for future research, in which the main direction is to go into depth of understanding rather than in the pure accumulation of more and more details. We thank the authors that they have grounded this basis, and I wish both of them a good response for their work and the book a wide distribution.

Heidelberg   MARTIN BOPP

Preface

In recent years there has been an increasing emphasis on investigations concerning the ultrastructure, reproductive biology, ecology, morphogenesis, physiology, biochemistry and related aspects of bryophytes. These themes have also rightfully found their place in the syllabi at all levels in most universities in India and abroad. However, the writing of texts in this area has lagged behind. Since the literature is scattered and at times not easy to reach, there is an urgent need for a book which deals with the modern topics of bryology. This volume is intended to fill this gap. It is primarily meant for postgraduate students, but can also be profitably consulted by undergraduate students and research scholars. The authors have tried to make the compilation of the literature as up-to-date as possible. The references cited in the text have been listed at the end of each chapter for those interested in more details.

The authors are thankful to the bryologists and publishers whose figures are reproduced and contributions have been included in this book. A work of this magnitude can not be free of errors and omissions. We will greatly appreciate receiving constructive suggestions from any quarter.

R. N. Chopra
P. K. Kumra

Acknowledgements

Professor Dr. Martin Bopp, Heidelberg, has very kindly given useful suggestions, and has also written the Foreword to this volume. Dr. (Mrs.) Suman Kumra and Dr. B.D. Vashistha rendered valuable help in various ways, and both of them deserve special thanks.

The senior author is indebted to his wife, Sudesh Chopra, for her willing cooperation and encouragement.

The authors are grateful to the University Grants Commission, New Delhi, for providing financial assistance for the preparation of the manuscript.

R. N. Chopra
P. K. Kumra

Contents

Foreword, by Professor Dr. Martin Bopp　　v
Preface　　vii
Acknowledgements　　viii

1. EXPERIMENTS ON SPORES AND GEMMAE　　1-38

Spore Germination in Liverworts—Jungermanniales, Marchantiales, Anthocerotales, Sphaerocarpales; Types of Spore Germination in Mosses—Sphagnales, Andreaeales, Tetraphidales, Bryales; Factors Affecting Spore Germination—Light, Temperature, Sugars, Minerals, Growth Regulators, Hydrogen-ion concentration (pH), Other Factors; Mechanism of Spore Germination; Experiments on Gemmae—Light, Temperature, Humidity and Other Physical Factors, Growth Regulators, Nitrogenous Substances, Hydrogen-ion concentration (pH), Other Chemical Factors; References.

2. PROTONEMAL DIFFERENTIATION AND BUD FORMATION IN MOSSES　　39-53

Protonemal Differentiation; Bud Formation; Factors Affecting Bud Formation—Light, Temperature, Auxins, Cytokinins, Gibberellins, Adenosine 3', 5'-cyclic monophosphate, Adenine and Amino Acids, Minerals and Chelates, Vitamins, Abscisic Acid, Sugars, pH, Influence of Other Organisms; References.

3. REGENERATION　　54-91

Potentialities of Various Organs for Regeneration—Regeneration from Leaves, Regeneration from Setae; Morphology of Regenerants; Factors Affecting Regeneration—Light, Radiation, pH, Season, Humidity, Wounding, Temperature, Size of the Fragment, Reserve Food Material, Location in the Plant, Age, Correlative Inhibition, Polarity and Apical Dominance; Changes Occurring in Regenerating Cells; References.

4. REPRODUCTIVE BIOLOGY　　92-106

Factors Affecting Gametangial Induction—Light Duration, Light Level, Light Quality, Temperature, Temperature-Photoperiod Interaction, Humidity, Hydration, Carbohydrates, Nitrogenous Substances, Growth Regulators, Chelating Agents, pH and Other Factors; References.

5. ALTERNATIVE PATHWAYS IN LIFE CYCLE   107-130

Apogamy—Occurrence of Apogamy in Diplophase and Haplophase, Spore Production in Apogamous Sporophytes, Differentiation of Apogamous Sporophytes from Callus, Factors Controlling Differentiation of Apogamous Sporophytes: Exogenous Factors, Endogenous Factors, Differentiation of Sporophyte and Gametophyte, Role of Calyptra in Sporogon Development; Apospory; Callus Formation and its Differentiation—Formation of Callus, Differentiation in Callus; Controls in Differentiation; Alternation of Generations; References.

6. PHOTOMORPHOGENESIS   131-147

Spore Germination—Liverworts, Mosses; Growth—Liverworts, Mosses; Vegetative Propagation—Liverworts, Mosses; Metabolism—Liverworts, Mosses; Senescence; Bud Induction in Mosses; Tropic Responses—Liverworts, Mosses; References.

7. ULTRASTRUCTURAL STUDIES   148-174

Spore; Protonema; Stem; Leaf; Gametangia—Antheridium; Gametogenesis—Spermatogenesis, Oogenesis; Sporogenesis—Spore Sac or Tapetum. Fluctuation in Plastid Number, Structural Changes in Plastids, Cytoplasm, and Other Organelles, Meiosis, Spore Wall Formation; Sporophyte-Gametophyte Junction; Seta; Histoenzymological Studies—Localization of Enzymes in the Haustorial Foot; References.

8. CHEMICAL CONSTITUENTS OF BRYOPHYTES   175-248

Antibiotics; Growth Substances—Specified Growth Substances, Non-specified Growth Substances; Lipids—Alkanes, Fatty Acids, Cuticular Components; Terpenoids—Monoterpenoids, Sesquiterpenoids, Diterpenoids, Triterpenoids and Sterols; Flavonoids—Flavones, Isoflavones, Flavonols, Dihydroflavonoids and Biflavonyls, Aurones and Chalcones, Acylated Flavonoids, Anthocyanins and Proanthocyanidins, Sphagnorubins; Lignins; Other Constitutents—Carotenoids, Carbohydrates, Organic Acids, Dihydrostilbenes, Enzymes, Amino Acids and Quinones, Inorganic Compounds, Miscellaneous, Antitumour Activities, Allergenic Activities; References.

9. BRYOPHYTES AS INDICATORS OF POLLUTION   249-267

Heavy Metals—Lead, Cadmium, Zinc, Mercury, Arsenates, Chromium, Nickel, Vanadium; Stability Pattern of Metal Ions; Metal Tolerance—Copper Mosses, Peat Mosses; Gaseous Pollutants—Sulphur dioxide, Fluorides, Ozone; Radionuclides (Radio Isotopes)—Cesium, Strontium, Uranium; Radiations; References.

10. PROTOPLAST CULTURE   268-284

Isolation of Protoplasts—Mechanical Method, Enzymatic Method, Source Material for Protoplasts, Factors Affecting Protoplast Isolation; Culture of Protoplasts—Liquid Culture; Agar Plating of Protoplasts; Regeneration of Protoplasts—Factors Affecting Regeneration; Protoplast Fusion and Somatic Hybrids; Induction and Isolation of Mutants; References.

11. CONDUCTION IN BRYOPHYTES   285-307

External Conduction—Gametophyte, Sporophyte, Significance of External Conduction; Internal Conduction—Cells Involved in Conduction, Anatomy of the Horizontal Axis, Midribs and Leaf Traces, Conducting Strand in Seta and Capsule, Development and Structure of Mature Conducting Tissues, The Interphase, Internal Conduction of Water, Conduction of Organic Compounds; Evolutionary Trends in the Conducting Strands; References.

12. WATER RELATIONS   308-317

Absorption and Conduction of Water—Endohydric, Ectohydric, Myxohydric; Water Holding Capacity and Growth Rate; Desiccation and Rehydration; Mechanism of Damage; Growth-Forms; References.

SUBJECT INDEX   319-330

PLANT INDEX   331-339

AUTHOR INDEX   340-350

1 Experiments on Spores and Gemmae

HEDWIG (1782, 1784) was the first to study spore germination in bryophytes. Since then extensive work has been done on the mode of spore germination and effect of various factors on this process.

SPORE GERMINATION IN LIVERWORTS

JUNGERMANNIALES

Hofmeister (1851, 1862) described early stages of spore germination and recognized three methods: formation of globose mass of cells, as in *Frullania*; of filament, as in *Jungermannia* and disc, as in *Radula*. Fulford (1956) reviewed the work done on the young stages of acrogynous Jungermanniales and recognized 10 types of sporelings,* which she divided into two main groups. Group A, with two types (*Nardia & Cephalozia*) in which the protonema develops outside the exospore. The remaining eight types (*Radula, Frullania, Lopholejeunea, Leucolejeunea, Lejeunea, Stictolejeunea, Ceratolejeunea,* and an unnamed type) were put in Group B, in which the protonema develops at least in part within the stretched exospore. Nehira (1966) recognized 24 sporeling types in Jungermanniales, 17 of the acrogynous and seven of the anacrogynous. In the former group, ten are those listed by Fulford, and of the seven new types four were taken out of the *Nardia* type. Among the characters used for classification within a major type are: (1) presence or absence of germ tube, (2) shape of protonema, (3) number of cells in protonema, (4) time and place of rhizoid formation, (5) growth by apical cell or by intercalary divisions, (6) number of cutting faces in the apical cell, (7) cell structure is firm or flexible, and (8) only primary protonema is present or secondary protonema also develops.

In acrogynous Jungermanniales six types of protonemata develop outside the exospore:

    (1) *Calobryum type:* Protonema is globose and without rhizoids. Germ tube is absent (Fig. 1.1 A,B). In the sporeling pattern *Calobryum* is most closely related to Anthocerotales.

    (2) *Cephalozia type:* A simple or branched filamentous protonema

---

* The type of sporeling produced in liverworts and mosses is likely to be influenced by the environmental factors.

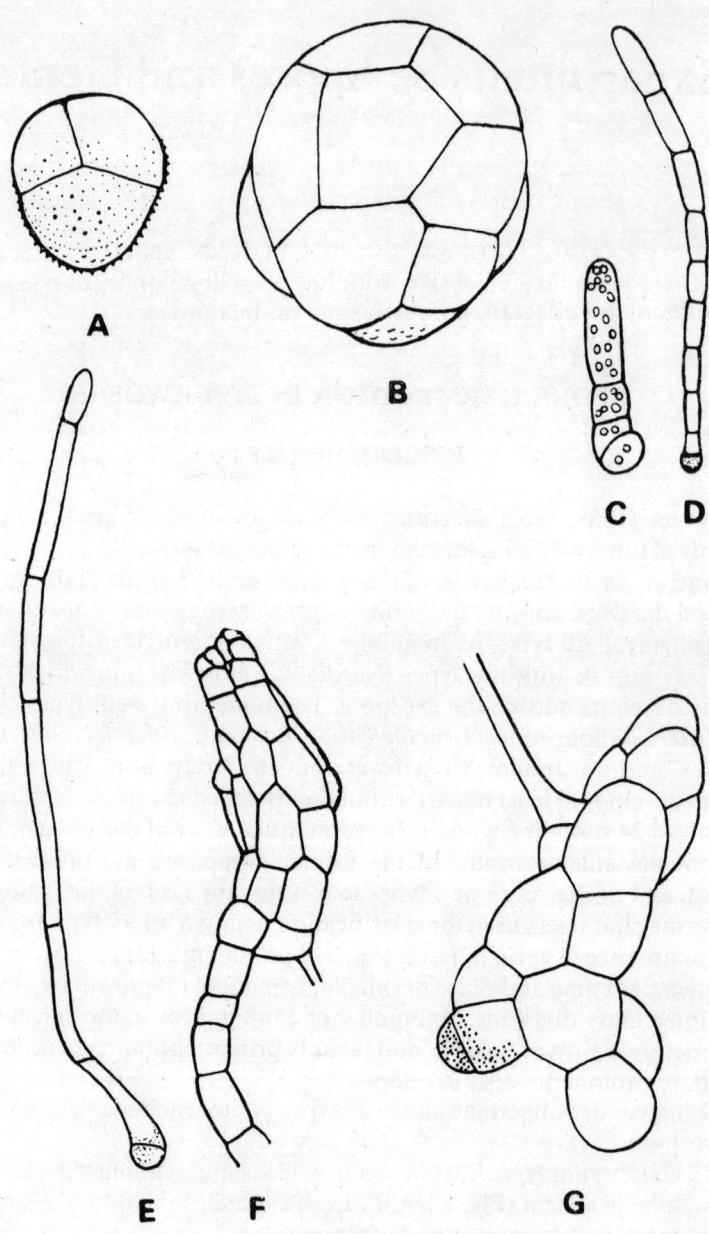

Fig. 1.1. A-G. Sporeling types in acrogynous Jungermanniales. A, B. *Calobryum rotundifolium*. A. Three-celled stage. B. Globose protonema. C, D. *Cephalozia hamatiloba*. Stages in development of protonema. E, F. *Nardia sieboldii*. E. Filamentous protonema F. Leafy shoot, G. *Calypogeia tosana*. Germinated spore.

arises from the germ tube (Fig. 1.1 C,D). Rhizoids may be present. An apical cell with three cutting faces develops and this gives rise to a leafy shoot.
(3) *Nardia type:* A germ tube-like cell appears and gives rise to the filamentous protonema. Eventually a massive, cylindrical protonema with a few rhizoids is formed (Fig. 1.1 E,F), and this develops into the leafy shoot with juvenile leaves. At times the germ tube does not appear and a massive globose protonema is directly formed. The *Nardia* type is also characterized by flexible protonemal cell structure.
(4) *Calypogeia type:* It differs from *Nardia* type in that the germ tube-like cell is absent, but it resembles it in having flexible cells (Fig. 1.1 G).
(5) *Scapania type:* The germ tube-like cell is usually absent as in the *Calypogeia* type, but the protonemal cells are firm. The typical protonema of the *Scapania* type is multicellular and massive (Fig. 1.2 A,B).
(6) *Bazzania type:* The sporeling is cylindrical, and its apical cell develops into the leafy shoot with juvenile leaves and underleaves (Fig. 1.2 C,D). Several rhizoids emerge from the protonema and on the leafy shoot, especially at the base of the primary underleaves.

In the remaining 11 types the development of protonema starts inside the stretched exospore:
(7) *Trichocoleopsis type:* The sporeling is radially symmetrical, whereas the adult plant is dorsiventral. One to four rhizoids are formed on the spore circumference (Fig. 1.2 E,F).
(8) *Jubula type:* While within the capsule, the spores become one to four-celled. After dispersal the protonema develops up to six-celled stage within the stretched exospore. The primary protonema gives rise to the secondary protonema which is unistratose, three to four cells broad and variable in length. The secondary protonema bears rhizoids (Fig. 1.2 G,H) and grows by an apical cell with three cutting faces.
(9) *Radula type:* Usually a unistratose, multicellular, disciform protonema is formed through intercalary divisions, and it bears rhizoids. The leafy shoot develops from a marginal cell of the protonema (Fig. 1.3 A,B).
(10) *Frullania type:* A massive, globose protonema of 20 or more cells develops. One or more rhizoids are usually present. The leafy shoot is formed by the activity of a cell at one end (Fig. 1.3 C,D). The primary leaves are ovate to oblong, and the juvenile leaves are saccate. The underleaves are narrowly ovate and entire in contrast to the bifid leaves in the adult plant.
(11) *Lopholejeunea type:* This is similar to the *Frullania* type except that there are fewer cells in the protonema. Further, the protonema of *Lopholejeunea* is oval and has a distinct developmental axis which is

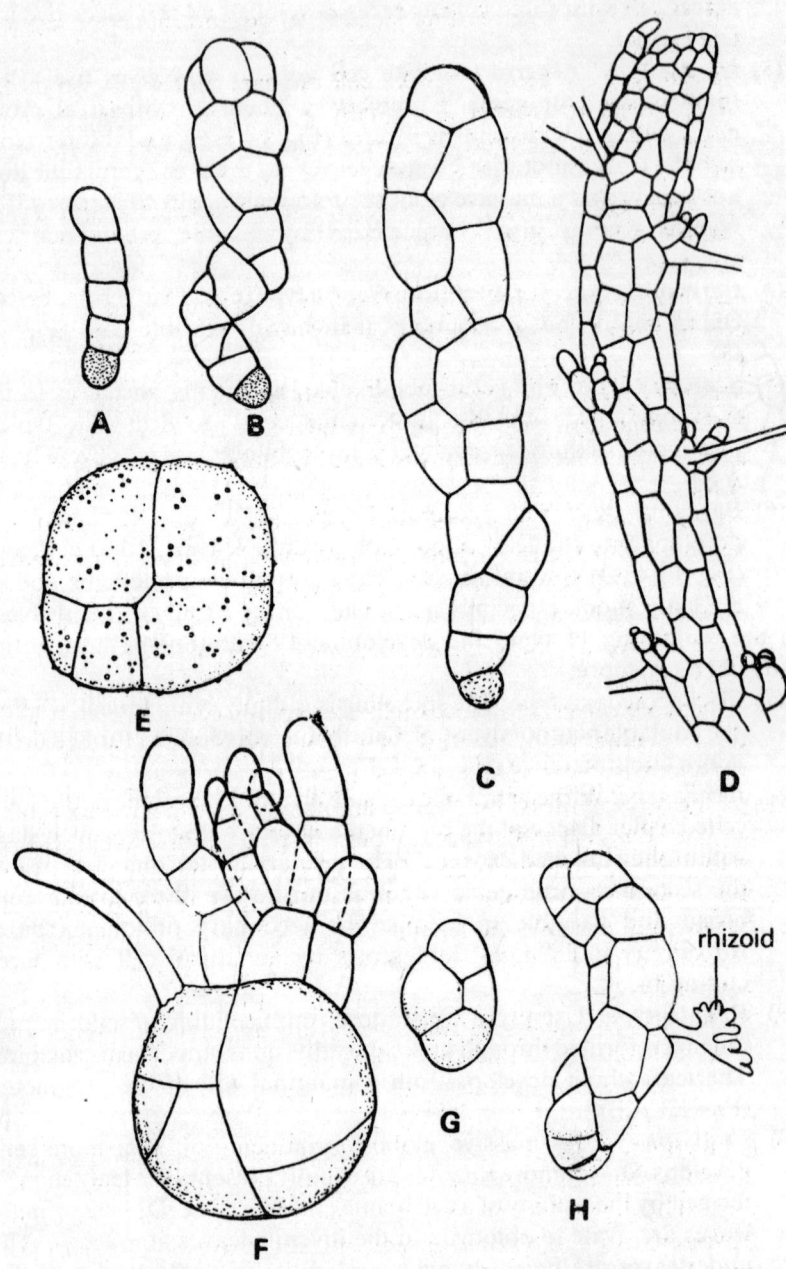

Fig. 1.2 A-H. Sporeling types in acrogynous Jungermanniales. A, B. *Scapania ligulata*. Stages in development of protonema. C, D. *Lepidozia* sp. C. Germinated spore. D. Sporeling. E, F. *Trichocoleopsis sacculata*. E. Globose protonema. F. Sporeling with a leafy shoot. G, H. *Jubula javanica*. Developmental stages of protonema.

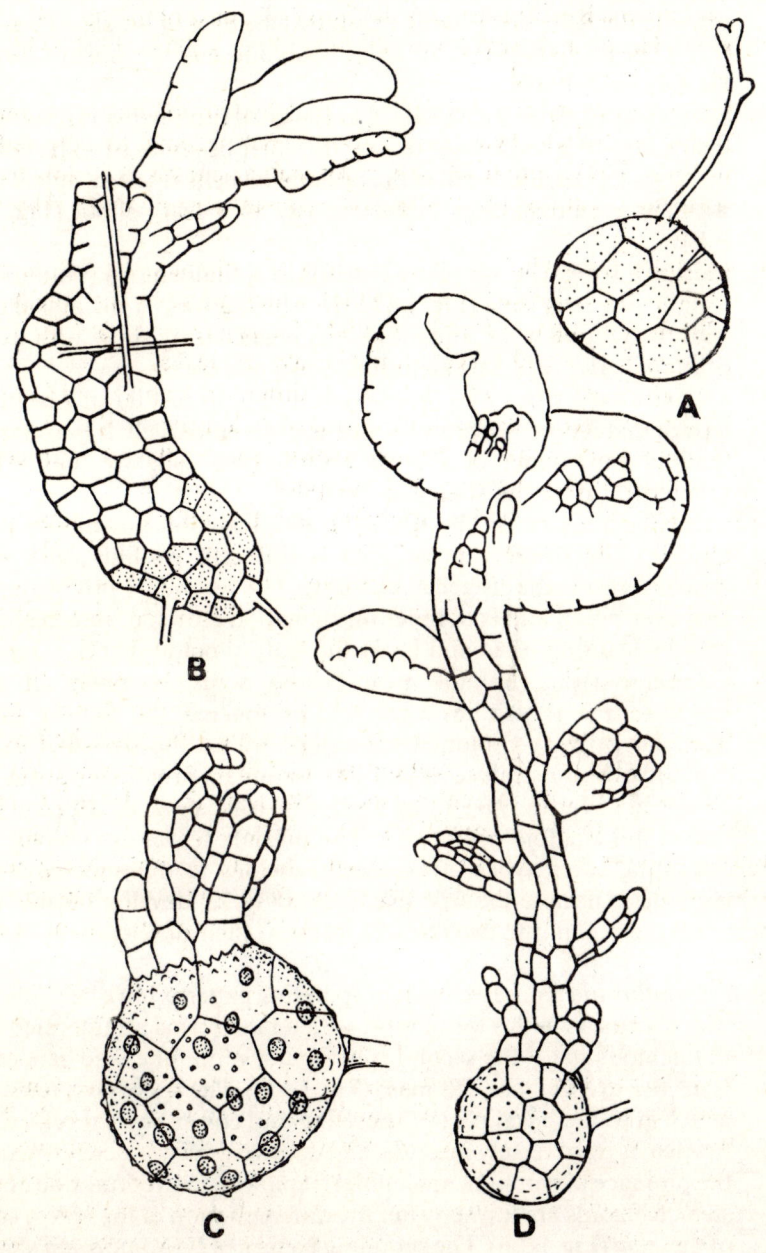

Fig. 1.3 A-D. Sporeling types in acrogynous Jungermanniales. A, B. *Radula* sp. A. Disc-protonema. B. Young plant. C, D. *Frullania truncatifolia*. C. Globose protonema. D. Young plant.

absent in the *Frullania* type (Fig. 1.4 A,B).

(12) *Leucolejeunea type:* This is characterized by an elongate, cylindrical protonema. Rhizoids usually develop at the base of the stem (Fig. 1.4 C,D). The primary leaves are ovate to oblong, and the juvenile leaves are saccate-inflated.

(13) *Cololejeunea type:* A biseriate, 8 to 10-celled protonema is produced inside the stretched exospore, and this later becomes two-layered or massive. The terminal cell of the protonema behaves as an apical cell with three cutting faces and gives rise to a leafy shoot (Fig. 1.4 E,F).

(14) *Lejeunea type:* The sporeling consists of a filamentous protonema, usually two cells wide (Fig. 1.4 G,H), which grows by the apical cell with two cutting faces, and eventually bears a leafy shoot with ovate primary leaves and saccate-inflated juvenile leaves.

(15) *Stictolejeunea type:* The protonema differs from that of *Lejeunea* type in that it does not grow by means of an apical cell. It is a simple, thalloid protonema of limited growth (two cells by four cells) developed inside the stretched exospore.

(16) *Ceratolejeunea type:* The sporeling is a two-fold, unistratose protonema. The primary protonema is two cells by four cells, and develops inside the stretched exospore. The secondary protonema is four cells broad and is of indefinite length. It grows by an apical cell with two cutting faces and bears the leafy shoot at its tip.

(17) *Unnamed type:* To this type belong some members of the Lejeuneaceae. It also has a two-fold protonema; the globose, multicellular primary protonema develops within the stretched exospore; and the unistratose ribbon-like secondary protonema grows by an apical cell with two cutting faces. The leafy shoot develops at the end of this thalloid protonema. The primary leaves are oblong.

The remaining seven types of protonema belong to the anacrogynous Jungermanniales. Of these the first five (Nos 18 to 22) develop outside the exospore and the remaining two (Nos 23, 24) start their development inside the spore:

(18) *Fossombronia type:* Two types of sporeling patterns have been observed: the first is with a germ tube, which later forms an intermediate filamentous stage; the second is without a germ tube, and generally gives rise to a globose cell mass (Fig. 1.5 A). The spore coat remains attached to the basal part of sporeling and covers several cells. The first leaf is transversely inserted, and dorsiventrality is determined by the presence of a minute unicellular papilla on the ventral side of the stem. Rhizoids are formed after the differentiation of the apical cells of the stem (Fig. 1.5 B). The sporeling pattern of *Fossombronia* shows similarities with those of Marchantiales.

(19) *Pallavicinia type:* In this type a strap-shaped or cylindrical protonema is formed, which becomes thalloid by further segmentations (Fig. 1.5 C,D). A few rhizoids appear quite early on the posterior side

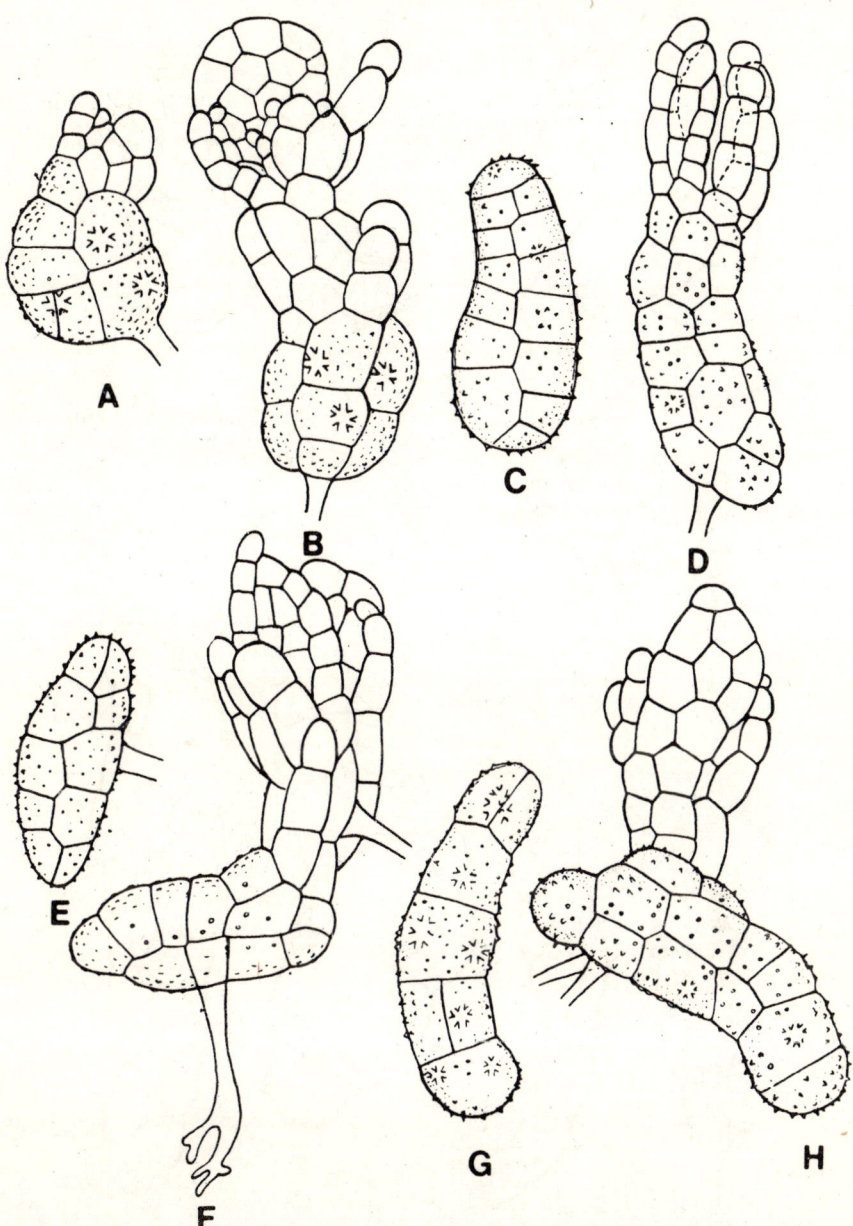

Fig. 1.4 A-H. Sporeling types in acrogynous Jungermanniales. A, B, *Ptychanthus striatus*. Developmental stages of sporeling. C, D. *Cheilolejeunea imbricata*. C. Protonema developed inside a stretched exospore. D. Sporeling. E, F. *Drepanolejeunea japonica*. Developmental stages of sporeling. G, H. *Lejeunea vaginata*. G. Cylindrical protonema developed inside a stretched exospore. H. Sporeling.

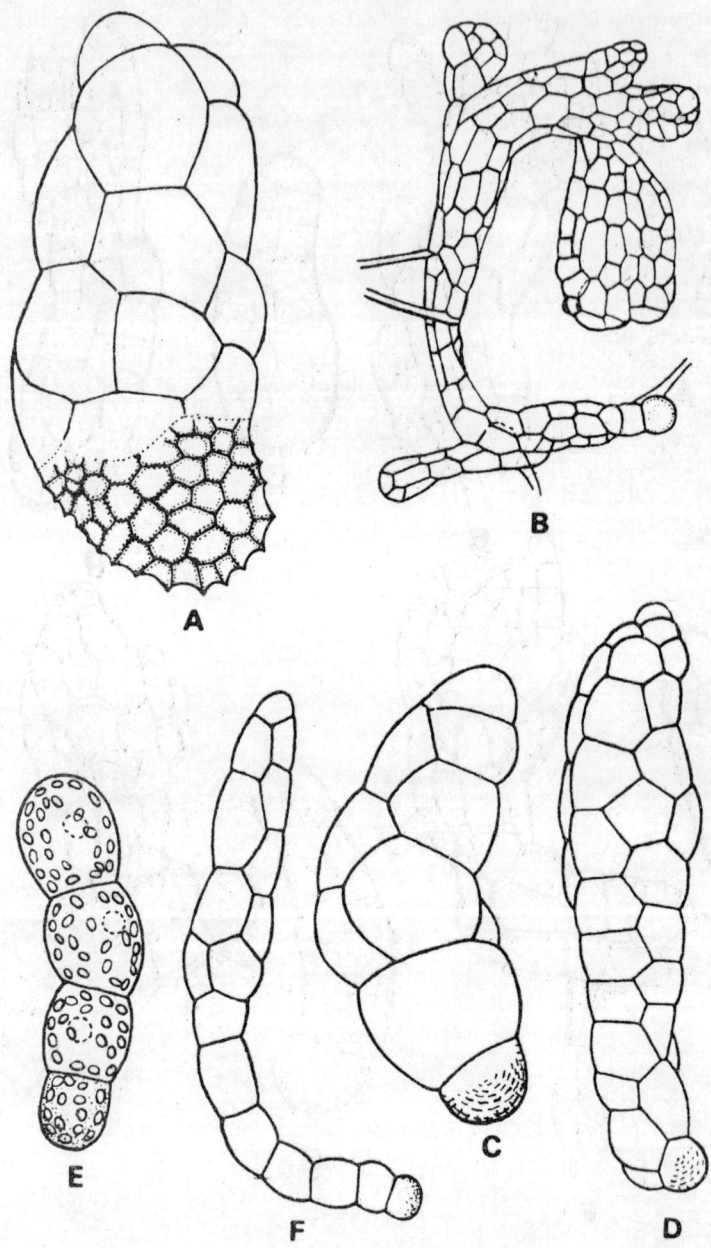

Fig. 1.5 A-F. Sporeling types in anacrogynous Jungermanniales. A, B. *Fossombronia japonica*. A. Globose cell mass. B. Juvenile plant. C, D. *Pallavicinia longispina*. C. Germinated spore. D. Young thalloid protonema. E, F. *Riccardia miyakeana*. E. Primary filamentous protonema. F. Secondary thalloid protonema developed from the primary protonema.

of the sporeling.

(20) *Riccardia type:* A filamentous protonema is followed by a thalloid one (Fig. 1.5 E,F) produced by the activity of either intercalary meristem or an apical cell. A few rhizoids are observed.

(21) *Metzgeria type:* The filamentous protonema later becomes thalloid and unistratose. A two-cell wide secondary protonema is usually formed near the base of the primary protonema. Rhizoids usually appear from the primary filamentous or thalloid protonema (Fig. 1.6 A,B).

(22) *Makinoa type:* The first cell division takes place after the exospore ruptures. The daughter cell is divided by a transverse wall or occasionally by a longitudinal one. The basal cell does not grow or divide further (Fig. 1.6 C,D). An apical cell with two cutting faces is formed at the 10-celled stage, and this gives rise to the thallus. The first rhizoid is generally purplish-brown as in the adult thallus.

(23) *Cavicularia type:* The 6 to 20-celled, globose, massive protonema is formed inside the exospore (Fig. 1.7 A,B). The leafy shoot usually

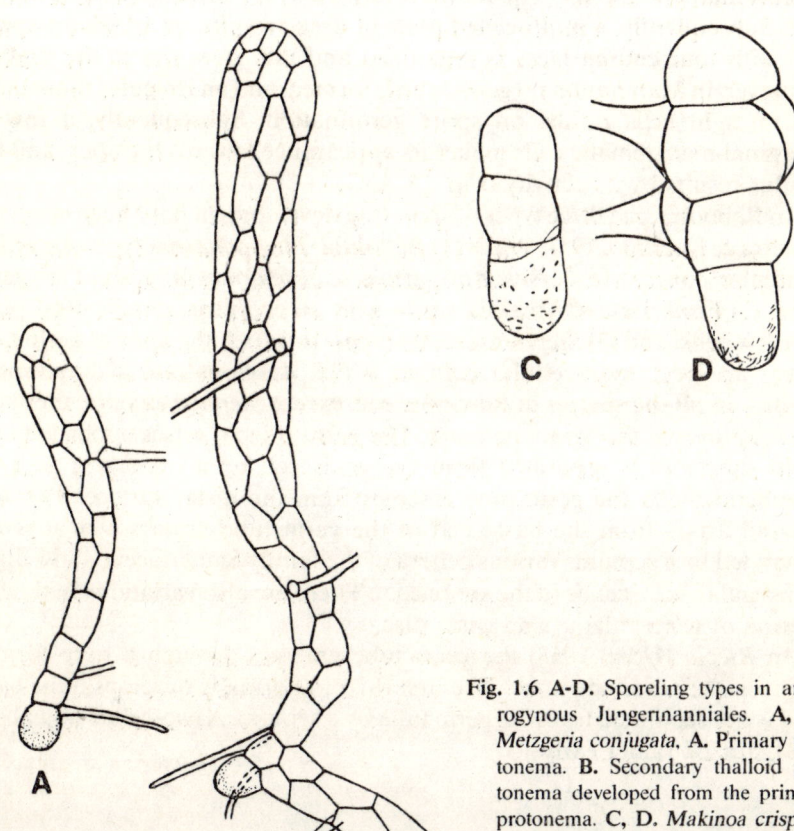

Fig. 1.6 A-D. Sporeling types in anacrogynous Jungermanniales. A, B. *Metzgeria conjugata.* A. Primary protonema. B. Secondary thalloid protonema developed from the primary protonema. C, D. *Makinoa crispata.* C. Three-celled stage. D. Sporeling with rhizoids.

develops by the activity of a superficial cell of the protonema. After differentiation of the apical meristematic cells on the juvenile stem, the juvenile leaves develop laterally (Fig. 1.7C).

(24) *Pellia type:* The spores germinate before dispersal. A row of four cells is formed by transverse divisions: a basal cell, two central cells, and one apical cell. Later, a few vertical divisions occur in the central cells, and a protrusion appears in the basal cell to form the first rhizoid (Fig. 1.7 D,E). The second rhizoid may also arise from the basal cell or from one of those adjacent to it.

MARCHANTIALES

In the majority of Marchantiales a germ papilla appears after the spore coat ruptures. The germ cell usually divides to form a germ rhizoid and a germ tube. The pattern of spore coat dehiscence and germ rhizoid formation is characteristic for a given taxon. The germ tube elongates and divides to form a short filament of cells. A quadrant is formed by the division of the terminal cell. Subsequently, a multi-celled plate of tissue results, on which an apical cell with four cutting faces is organized and this gives rise to the thallus. However, in *Marchantia* no germ tube is formed, and an irregular filament of six to eight cells results on spore germination. Subsequently, a row of marginal meristematic cells makes its appearance towards the apex, and the thallus results by its activity (Fig. 1.8 A-G).

In Rebouliaceae three types of sporeling development have been observed (Mehra & Kachroo, 1951, 1952): (1) *Reboulia-Plagiochasma* type—in which a circular, concentric, dorsiventral germ disc develops at the apex of the germ tube, (2) *Fimbriaria-Grimaldia* type—with an eccentric, dorsiventral plate of a few cells, and (3) *Stephensoniella* type—in which the apex of germ tube forms an erect, multicellular column which develops into a dorsiventral thallus. In all the species of Rebouliaceae, except *Stephensoniella*, the spore coat ruptures at the triradiate mark. The germ rhizoid arises opposite to the germ tube and is separated from the germ cell by a transverse wall. In *Stephensoniella* the germ tube emerges from the distal surface. The first rhizoid arises from the basal cell of the germ tube, from which it is not separated by a septum. Various genera of the family Sauteriaceae show distal or irregular dehiscence of the spore coat. There are also variations in the formation of germ rhizoid and germ disc.

In *Riccia* (Udar, 1958) the germ tube emerges through a pore formed opposite the triradiate mark. The first rhizoid normally develops quite early and is not separated from the germ tube by a septum. A two-celled apical cell is organized rather early.

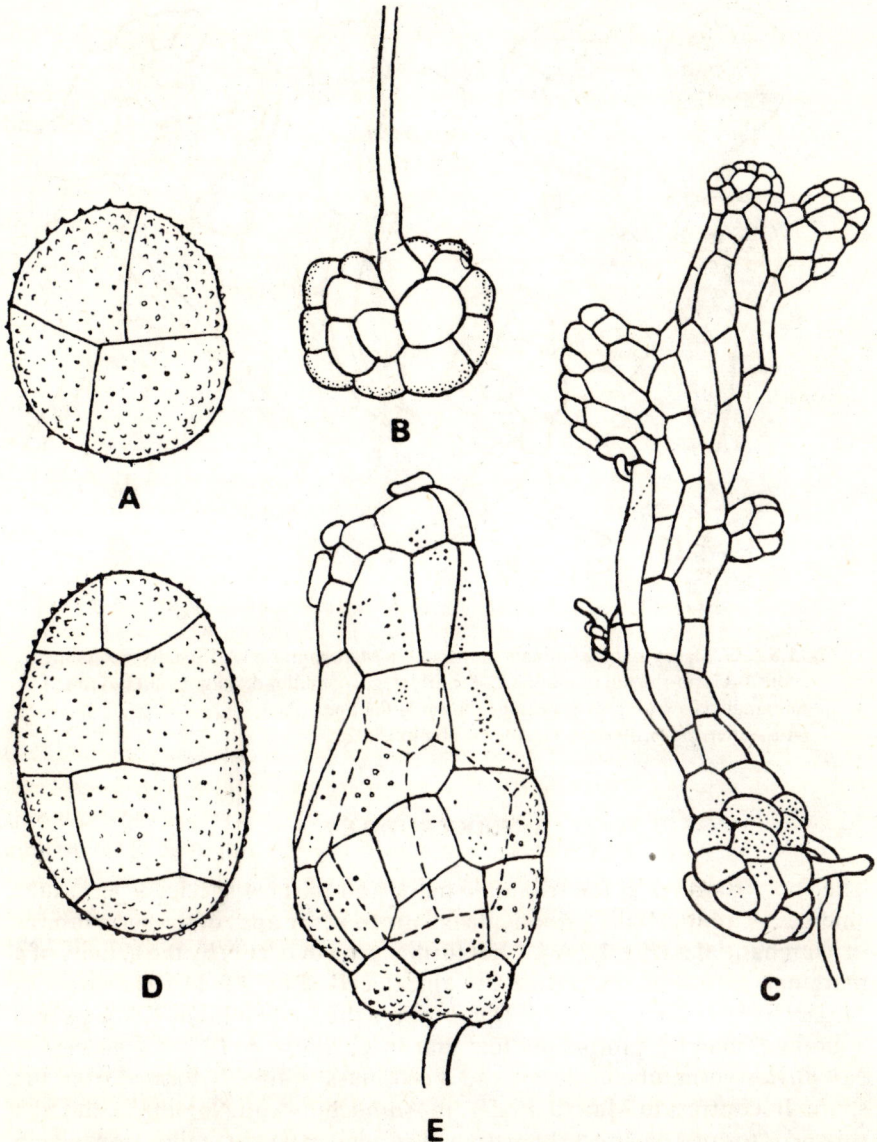

Fig. 1.7 A-E. Sporeling types in anacrogynous Jungermanniales. **A-C.** *Cavicularia densa.* **A.** Globose protonema consisting of four cells. **B.** Globose protonema consisting of many cells. **C.** Young thallus. **D, E.** *Pellia neesiana.* **D.** Multi-celled spore. **E.** Initial stage of development of thallus.
(Figs 1.1 to 1.7. After Nehira, 1966)

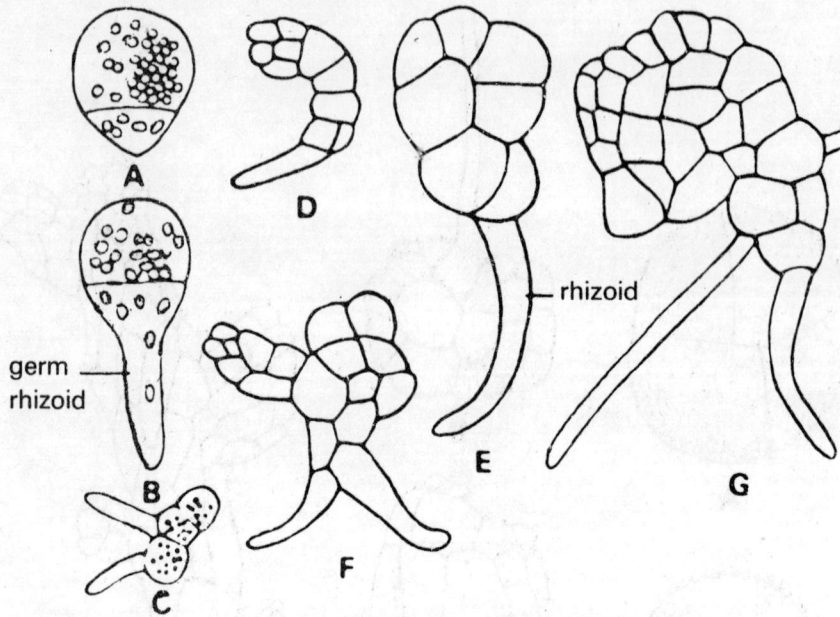

Fig. 1.8 A-G. Stages in the germination of spore in *Marchantia* sp. A. First division of the spore. B. Germ-rhizoid formation. C-F. Early stages of thallus development. G. A row of marginal cells makes its appearance towards the apex.
(A-E, After Inoue, 1960; F, G. After O'Hanlon, 1926).

ANTHOCEROTALES

The spore ruptures at the triradiate mark and in most species a germ tube emerges. A multicellular protonema is formed at the apex of the germ tube as in Marchantiales (Fig. 1.9 A-C), but further growth occurs by the activity of a marginal meristem instead of an apical cell (Fig. 1.9 D). However, in *Megaceros* and some species of *Dendroceros* the spore initially develops into a globose, massive protonema like that in *Calobryum*. In some species of *Notothylas* germ tube is absent and a cell mass is directly formed from the spore. In contrast to Marchantiales, in *Anthoceros* and *Notothylas* the first rhizoid is formed by direct elongation of any cell of young thallus from which it is not separated by a wall. In *Dendroceros javanicus* spore germination starts within the stretched exospore as in *Conocephalum* and some members of Jungermanniales.

SPHAEROCARPALES

A slender germ tube emerges either from the distal surface (as in *Sphaerocarpos* and *Riella*) or from the proximal surface as in *Geothallus*. The germ tube

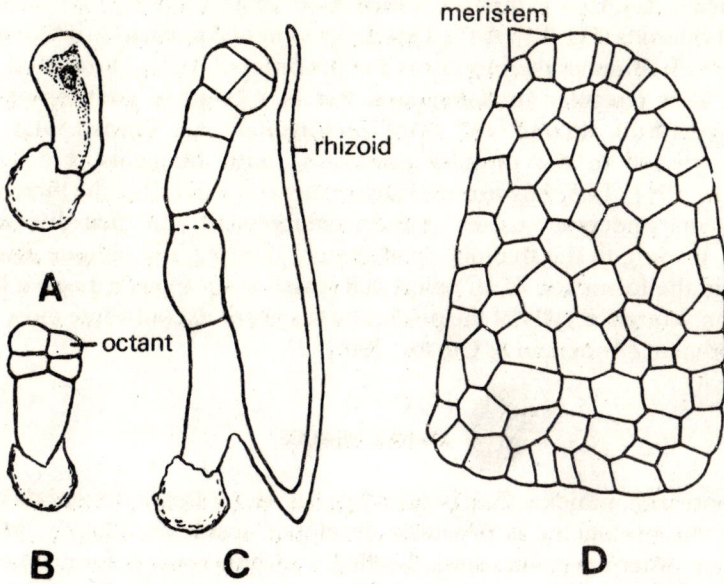

Fig. 1.9 A-D. Stages in the germination of spore in *Anthoceros* sp. A. Germ tube formation. B. Octant stage. C. Rhizoid formation. D. Organization of multicellular marginal meristem.
(After Mehra & Kachroo, 1962).

divides by a transverse wall to form a terminal cell and a basal cell. The latter does not divide further and gives rise to the first rhizoid. The terminal cell develops into a filament of variable cells. Further divisions vary in different genera: (1) In *Sphaerocarpos* subsequent divisions in the cells of the filament result in the formation of a germinal disc at right angles to a short multiseriate filament. The adult plant arises as a lateral outgrowth of the germinal disc. (2) In *Geothallus* an erect, unistratose flap of determinate growth is formed from the filament. Only one adult plant arises as a lateral outgrowth from the base of each juvenile plant. (3) In *Riella* also an erect, unistratose flap of determinate growth is formed from the filament, but it is larger than that of *Geothallus*, and occasionally two adult thalli are formed from each juvenile plant.

TYPES OF SPORE GERMINATION IN MOSSES

SPHAGNALES

The spore coat ruptures and a protuberance is formed usually at the proximal side of the spore. It develops into a sort of germ tube which gets transversely divided into two cells. Subsequently the upper cell divides transversely resulting in a three-celled protonema. At the apical part of this short filamentous protonema develops a thallose, unistratose, multicellular protonema with several rhizoids (Fig. 1.10 A-C). This developmental pattern is similar to that in the prothallium of pteridophytes, but the order of rhizoid formation is different: it is irregular in *Sphagnum*, but is regular in pteridophytes. In *S. girgensohnii*, second and third protonemata are formed, and these are connected to the primary protonema with filaments (Fig. 1.10 D; Noguchi, 1958). In *Sphagnum meridense* the terminal cell of the three-celled protonema undergoes rather uniform enlargement and then divides vertically, leading to the thalloid development. Subsequent oblique divisions result in the formation of an apical cell which is soon pushed into a lateral position. Marginal cells of the prothallium appear to contribute most to the development (Anderson & Crosby, 1965).

ANDREAEALES

The spore wall is thicker than in most Bryales. In *Andreaea petrophila* and *A. fauriei* the protonema is primarily developed inside the slightly enlarged exospore. After it becomes several-celled, a protuberance is formed and this develops into a filamentous or thallose protonema which subsequently bears gametophytic buds (Fig. 1.11 A-C; Nehira, 1963; Nishida, 1971).

TETRAPHIDALES

On germination the exospore ruptures and a germ tube emerges through the cleavage. The germ tube elongates and develops into a well-branched filamentous chloronema. Some of the chloronemal branches are transformed into terminal rhizoids. The initial cell of the thallose protonema appears on the basal regions of chloronemal branches and of the leafy gametophyte. Eventually the thallose protonema becomes spathulate and is unistratose (Fig. 1.12A). The caulonema is formed from the chloronema on marginal cells of the basal region of thallose protonema. The sporeling consists of chloronema, spathulate-thallose protonema, and terminal rhizoids.

BRYALES

On spore germination a protuberance is formed, which elongates and divides transversely into two cells. Subsequently, a filamentous branched protonema is formed (Fig. 1.12 B-D). In *Schistostega pennata*, the luminous moss, a flexible filamentous protonema develops (Kanda, 1971). After branching and stretching of the protonema, some lens-shaped cells are formed by budding (Fig. 1.12 E,F), especially in lower light levels. The protonemal cells are so shaped as to focus the incoming light upon the chloroplasts at the inner side of the cell cavity (Fig. 1.12G). The light is reflected back, causing a greenish-

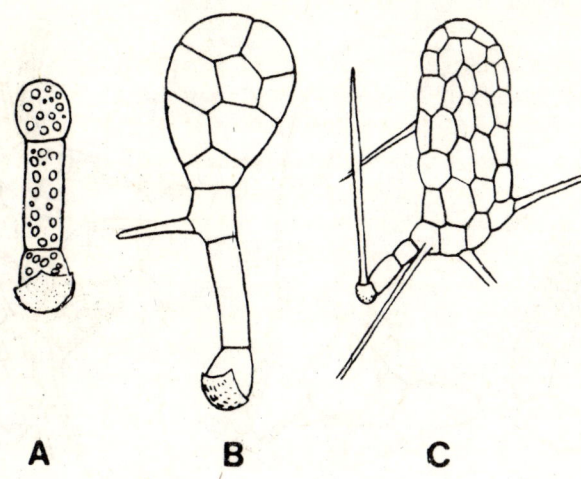
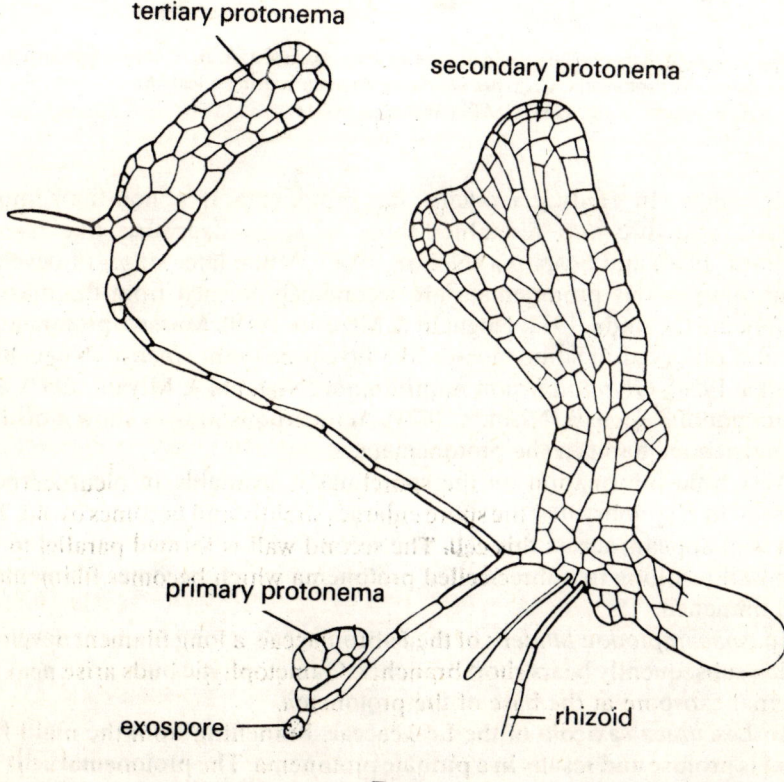

Fig. 1.10 A-D. Stages in the germination of spore in *Sphagnum*. A-C. *S. imbricatum*. A. Spore germination, B, C. Thallose protonemata. D. *S. girgensohnii*. Formation of primary, secondary, and tertiary protonemata.
(A-C. After Nehira, 1963; D. After Noguchi, 1958).

16 Biology of Bryophytes

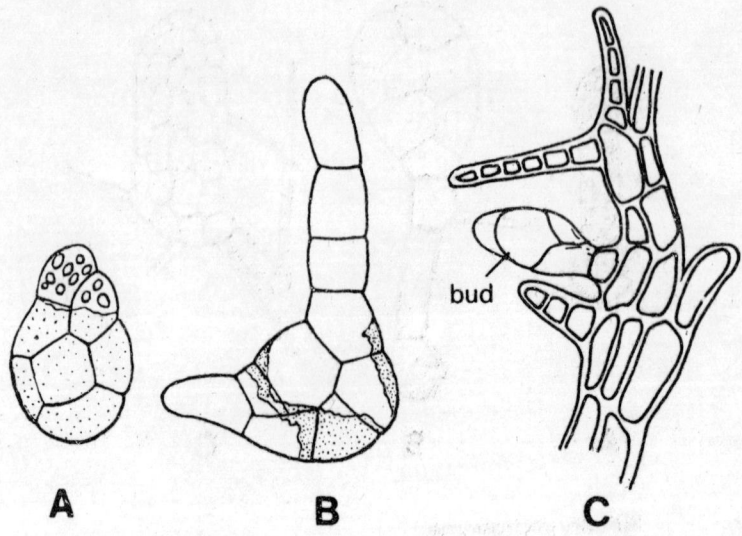

Fig. 1.11 A-C. Stages in the germination of spore in *Andreaea*. A, B. *A. fauriei*, germinating spore and protonema. C. *A. petrophila*, protonema bearing a leafy bud.
(A, B. After Nehira, 1963; C. After Waldner, 1887)

golden glow. In *Hedwigia ciliata*, the protonema is a non-filamentous, massive structure and resembles those of some liverworts like *Nardia* (Fulford, 1956) and *Scapania* (Nehira, 1966). At the later stages of development, filamentous protonemata are secondarily formed from the massive protonema (Nishida, 1972; Noguchi & Mizuno, 1959). Massive protonemata are also observed in some mosses like *Ptychomitrium sinense* (Noguchi & Otsuka, 1954), *Glyphomitrium humillimum* (Noguchi & Miyata, 1957), and *Drummondia sinensis* (Nishida, 1973). Acrocarpous mosses show a distinct heterotrichous habit in the protonemata.

Very little information on the sporelings is available in pleurocarpous mosses. In Hypnobryales, the spore enlarges slightly and becomes ovoid. The first wall appears across this cell. The second wall is formed parallel to the first wall resulting in a three-celled protonema which becomes filamentous and branched.

In *Anacamptodon latidens* of the Fabroniaceae, a long filament develops, which subsequently bears short branches. Gametophytic buds arise near the original exospore at the base of the protonema.

In *Lescuraea saxicola* of the Leskeaceae, branching from the main filament is profuse and results in a pinnate protonema. The protonemal cells are comparatively short, and thick-walled. The commonest position of a pro-bud cell is on the middle part of the prostrate system.

In *Thuidium bipinnatulum* of the Thuidiaceae, branching is much activated even during early stages. Ovoid or globose cells are sometimes formed near the exospore (Nehira, 1976).

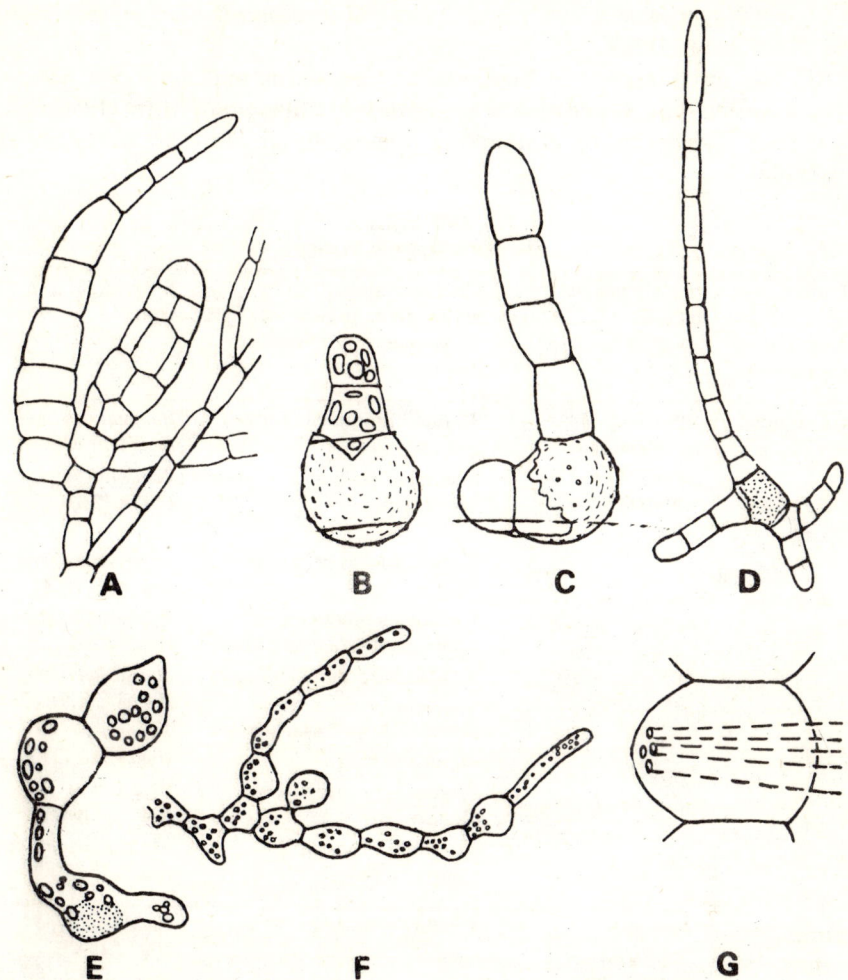

Fig. 1.12 A-G. Sporeling types in Musci. A. *Tetraphis pellucida*: Spathulate-thallose protonema developing from chloronemal branches. B-D. *Macromitrium gymnostomum*. B. Spore germination. C, D. Filamentous protonemata. E-G. *Schistostega pennata* (the luminous moss). E. Spore germination. F. A part of protonema. G. One cell of protonema, to show focussing of incoming light upon the chloroplasts at the inner side of the cell. This light is reflected back, causing a greenish-golden glow.
(A. After Nishida, 1978; B-D. After Nehira, 1965; E-G. After Kanda, 1971).

Spore germination in mosses can also be broadly classified under two heads:
(1) Both photopositive chloronema and photonegative rhizoid develop as in *Funaria hygrometrica* (Allsopp & Mitra, 1958; Bauer, 1942; Fitting, 1950; Heitz, 1942; Müller, 1874), *Physcomitrium* spp (Goebel, 1889; Kachroo, 1954; Meyer, 1947), *Polytrichum commune*

(Wigglesworth, 1947), and *Atrichum undulatum* (Becquerrel, 1906; Fitting, 1950).

(2) Only chloronemal filaments are formed as in *Dicranella heteromalla* and *Brachythecium rutabulum* (Allsopp & Mitra, 1958).

Nishida (1978) described 14 sporeling types in Musci. These are summarized in Table 1.1.

TABLE 1.1
Sporeling types in Musci.

Exosporous or endosporous germination	Protonema at earlier stage	Characteristics of spore germination, protonema development, or sporeling development	Sporeling type
Exosporous	Primarily filamentous protonema present	Thalloid protonema formed on apex of short filament	*Sphagnum* type (Fig. 1.10 A-D)
		Spathulate thalloid protonema	*Tetraphis* type (Fig. 1.12A)
		Funnel-shaped protonema	*Diphyscium* type (Fig. 1.13A)
		Sex organs producing on apex of chloronema branch	*Buxbaumia* type (Fig. 1.13B)
		Vesiculated protonemal cells	*Schistostega* type (Fig. 1.12E,F)
		Primary rhizoid developing from spore	*Funaria* type (Fig. 1.13C,D)
		Protonema consisting of longer cells	*Bryum* type (Fig. 1.12B-D)
		Protonema consisting of shorter cells	*Macromitrium* type (Fig. 1.13E,F)
	Primarily massive protonema present	Primary protonema consisting of a few globose cells	*Encalypta* type (Fig. 1.14A,B)
		Primary protonema consisting of 10-20 globose cells	*Ptychomitrium* type (Fig. 1.14C,D)
		Caulonema and rhizoid developing on massive protonema	*Hedwigia* type (Fig. 1.14E-G)
Endosporous	Massive protonema developed inside stretched exospore	Thalloid and filamentous protonemata developing from massive protonema	*Andreaea* type (Fig. 1.11A-C)
		Chloronema and caulonema developing from endosporous protonema	*Glyphomitrium* type (Fig. 1.15A,B)
		Only rhizoid developing from endosporous protonema	*Drummondia* type (Fig. 1.15C,D)

Source. Nehira, 1983.

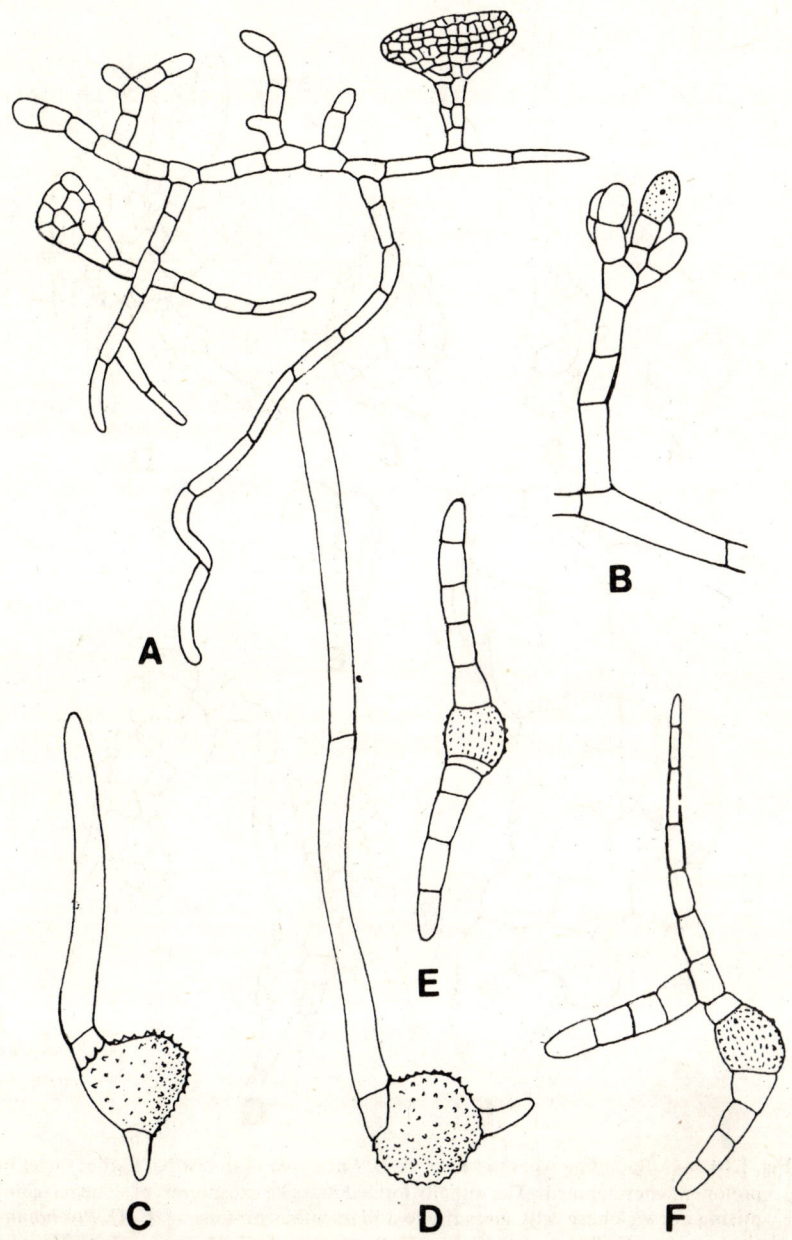

Fig. 1.13 A-F. Sporeling types in Musci. **A.** *Diphyscium fulvifolium*. Sporeling with chloronema, funnel-shaped protonema, and terminal rhizoids. **B.** *Buxbaumia aphylla*. Formation of male inflorescence from the apical cell of a chloronemal branch. C, D. *Aphanorrhegma serratum*. Earlier stages of spore germination, showing chloronema and primary rhizoid. E, F. *Aulacopilum piliferum*. Earlier stages of protonema development. Protonema is filamentous and comprises short, cylindrical cells.
(**A, B.** After Nishida, 1978; **C, D.** After Nehira, 1983; **E, F.** After Nehira, 1964.)

20 *Biology of Bryophytes*

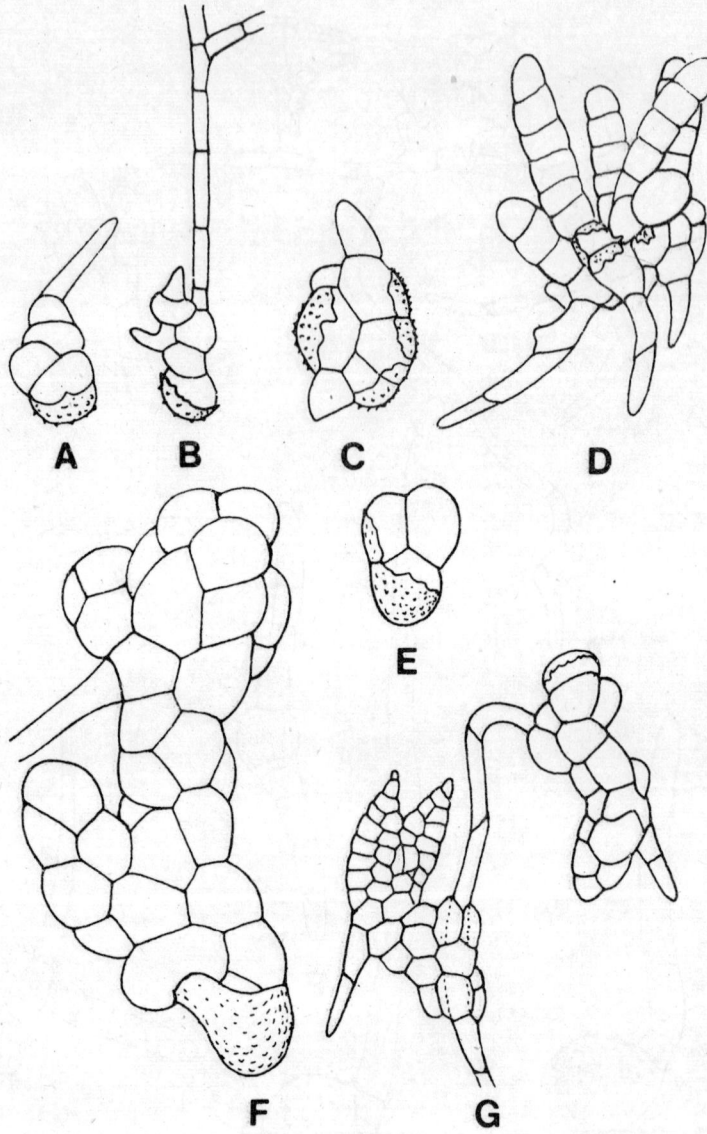

Fig. 1.14 A-G. Sporeling types in Musci. **A, B.** *Encalypta rhaptocarpa*. Earlier stages of protonema development. The initially formed massive exosporous protonema, comprising a few globose cells, gives rise to a filamentous protonema. **C, D.** *Ptychomitrium sinense* **C.** Spore germination. **D.** Protonema. **E-G.** *Hedwigia ciliata*. Massive protonemata developed outside exospore.

(A, B. After Nishida & Iwatsuki, 1980; E, F. After Nehira, 1983; C, D, G. After Nishida, 1978).

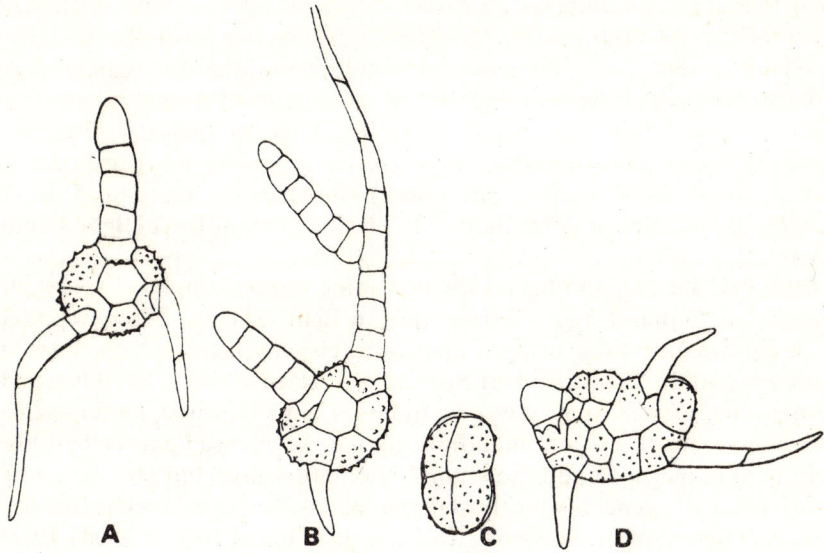

Fig. 1.15 A-D. Sporeling types in Musci. A, B. *Glyphomitrium humillimum*. Sporelings. C, - D. *Drummondia sinensis*. C. Massive protonema developed inside exospore. D. Rhizoid formation from massive protonema.
(After Nishida, 1978).

FACTORS AFFECTING SPORE GERMINATION

LIGHT

The fact that spores of bryophytes generally do not germinate in dark has been well known since the time of Borodin (1868). Leitgeb (1876; *cited in* Heald, 1898) and Heald (1898) reported that *Marchantia* spores failed to germinate in dark, and that normal sporelings and thalli developed only in light of high level.

Inoue (1960) subjected the germinating spores of five liverworts to three light levels: 1300, 120, and 20 lux. For *Plagiochasma* and *Targionia* light of 1300 lux is optimum, and at this level cent percent germination occurred. In *Reboulia* maximum germination (94.8%) takes place at 120 lux; higher and lower levels are not so favourable. In *Marchantia* all the spores germinate both at 1300 and 120 lux, but at 20 lux the percentage of germinating spores is only 14. In *Mannia* all the three levels of light are equally effective, and 100 percent germination is observed. Germ rhizoid differentiation occurs in cent percent spores at the three levels of light in *Mannia* and *Plagiochasma*. In the other three liverworts such a response is noticed only at 1300 lux, and with decrease in light level the percentage decreases. In *Marchantia* and *Targionia* none of the germinating spores develops a germ rhizoid at 20 lux. Higher levels retard spore germination in *Cryptothallus mirabilis* (Benson-Evans,

1960). In *Riccia crystallina* germ rhizoids develop in 2000 to 2500 lux, whereas in lower (50 to 500 lux) or higher (4000 lux) light levels no germ rhizoid is produced. The extent of development of germ tube also varies with light level. At 50 to 500 lux it may be as long as 1184 um, whereas in 4000 lux it is only 75 to 118 μm long (Chopra & Sood, 1973). In *Asterella angusta, Targionia hypophylla*, and *Plagiochasma intermedium* spores do not germinate in dark. In light the percentage of spore germination is fairly high. In *T. hypophylla* maximum germination is observed in diffused light (Kaul, 1974).

Meyer (1948) stated in his review that some workers observed spore germination in *Funaria hygrometrica* only in light, whereas others reported spore germination both in light and dark. He observed that the form of filaments and the nature of their tropism is affected by light. In 20 lux each germinating spore forms only one filament, and this is a photopositive chloronema. In 60 lux, germination is bipolar. In intense light the first filament is photonegative and the second is photopositive, but both resemble chloronema. It is generally the third filament, arising either from the first filament or directly from the spore, which is the rhizoid (Kofler et al., 1963). Chopra and Gupta (1967) reported that the dark-grown spores of *F. hygrometrica* merely swelled in the absence of sucrose and kinetin. Addition of sucrose, with or without kinetin, resulted in spore germination. On a medium without sucrose but with kinetin, spore germination was delayed and very few spores germinated. In *Anoectangium thomsonii* (Rashid, 1968), *Pogonatum aloides, Entodon myurus* (Sood, 1972, 1975), and *Bryum argenteum* (Bhatla, 1983) the number of filaments emerging from a spore increases with increase in light level. All the filaments are of the chloronema type in *A. thomsonii* and no rhizoid could be identified (Rashid, 1968). In *Bryum coronatum* and *Dicranella coarctata* also no rhizoidal filaments have been observed. In these two mosses the time taken for spore germination decreases and the percentage of germination increases with increasing light level (100 to 6000 lux). In the former the pattern of germination is also influenced by light level. In dark and at lower levels germination is unipolar, whereas at higher levels it is uni-, bi-, and tripolar (Kumra, 1981).

Photoperiod is also known to affect spore germination in Hepaticae (Ishikawa & Ohusa, 1954). In *Targionia* a light duration of more than 2 h is essential for appreciable spore germination, and with increase in light period, percentage of germination improves (Kaul, 1974). Valanne (1966) observed that a light period of six hours is necessary for spore germination in *Ceratodon* and *Funaria*. Sood (1972) noticed that in *Entodon* percentage of germination increases with increase in photoperiod. In 16- and 24-h photoperiods cent percent germination is observed. Egunyomi (1978) reported that for germination of spores in *Octoblepharum* these have to be exposed to a light level of 800 lux for 48 h.

In general, moss spores grown in yellow light produce longer and thinner protonema than those grown in white light. Yellow light also increases the time taken for germination, reduces the percentage of germination and slows down development (Mueller & Mueller, 1969). Red light promotes spore germination in *Plagiochasma intermedium* (Ahmad et al., 1977; Kaul, 1974), *P. appendiculatum* (Shukla et al., 1983), and *Asterella angusta* (Singhivi, 1979).

In *Funaria hygrometrica* phytochrome seems to be involved, since spore germination is enhanced by red light, and its effect is reversed by far-red light (Bauer & Mohr, 1959). The polarity of spores is also light-determined, through mediation by phytochrome. However, in *Physcomitrella patens* there is no evidence that the position of protonemal emergence is oriented with respect to light (Cove et al., 1978). Inoue (1960), Steiner (1963, 1964), and Doyle (1963) reported the importance of blue light in spore germination of liverworts (*see* Table 6.1).

TEMPERATURE

Higher temperatures retard germination of spores in *Cryptothallus mirabilis* (Benson-Evans, 1960). On the other hand, for cent percent germination of spores *Sphaerocarpos donnellii* requires a long pre-treatment with relatively high temperature (25 C) (Steiner, 1964). Unlike the spores of *Sphaerocarpos*, those of *Riccia crystallina* require a low temperature treatment (8 to 15 C) for germination. At 25 C the spores germinate after three to four months in only four to five percent cultures. They germinate within 15 days at 8 to 15 C, but if subjected to low (8 C) and high (25 to 35 C) temperature alternatingly germination is delayed. Cold treatment shortens the dormancy period and also increases the percentage of germination. Even in continuous cold (8 to 15 C) the spores do not germinate in the absence of light (Sood, 1972).

In *Asterella angusta* and *Plagiochasma intermedium* germination of spores is observed at 30 C, but diffused day light is essential. At 35 C spores do not germinate even in the presence of diffused light. In *Targionia hypophylla* germination does not occur at 8 and 12 C; at 16 C a small percentage of spores germinates, and at 27 C the percentage is improved.

SUGARS

Heald (1898) noticed that germination did not occur in the absence of light unless peptone or glucose was added to the medium. Later, some workers observed that addition of certain organic substances like sugars (glucose, fructose, maltose, or sucrose) to distilled water or to the inorganic media stimulates spore germination in dark in some systems (Hoffman, 1964; Richards, 1932; Treboux, 1905; Valanne, 1966). In *Bartramia pomiformis* effect of different sugars such as raffinose, sucrose, lactose, maltose, glucose, galactose, mannose, fructose, and sorbose has been studied on spore germination. In light of 100 and 30 lux, disaccharides such as sucrose, lactose, and maltose; and aldohexoses such as glucose, galactose, and mannose are effective. Raffinose, a trisaccharide, the ketohexoses, fructose, and sorbose have no appreciable effect on spore germination (Takao, 1977). Sugar is effective in *Bartramia* only in weak light (100 & 30 lux), whereas it is helpful in *Funaria* and *Ceratodon* in dark. It is, therefore, postulated that accumulation of sugar is essential for germination of moss spores, and the degree of accumulation differs in different species. A small amount of sugar is enough for germination of spores in *Funaria* (Treboux, 1905) and *Ceratodon* (Valanne, 1966). In *Bartramia* a little influence of sugar on germination has been observed at 1.0 to 0.1 percent, but at less than 0.1 percent fructose and sorbose have slightly better effect (Takao, 1977). Glucose at 0.02 percent is effective in *Funaria* (Treboux, 1905), and in *Ceratodon* sucrose gives good results at 0.1 percent (Valanne, 1966).

Sucrose and maltose are more effective in *Funaria* as compared to glucose and fructose. Germination percentages are greater in light than in dark, although there is less inhibition in dark in stronger sugar solutions (Hoffman, 1964). In all those systems in which sugars stimulate germination, the protonemata swell to large sizes and accumulate starch (Richards, 1932). In *Bryum coronatum* and *Dicranella coarctata* the percentage of cultures showing spore germination in light increases with increase in sucrose concentration up to two percent. There is no germination in control (without sucrose) or at four percent sucrose. Sucrose can in part replace the effect of light on spore germination, since a small percentage of spores do germinate in dark with sucrose. However, sucrose does not alter the germination pattern in light or dark (Kumra, 1981). Rahbar (1981) reported that in *Bartramidula bartramioides* the spores germinate in dark irrespective of the presence or absence of sucrose, whereas spores of *Hyophila involuta* germinate in dark only in the presence of sucrose. However, in *Octoblepharum albidum* the sugars tested do not substitute the effect of light in spore germination (Egunyomi, 1978). Similarly in *Physcomitrella patens* no germination occurs in darkness even when an exogenous carbon source is provided (Cove et al., 1978; Schild, 1981).

MINERALS

Very little is known about mineral requirement during spore germination (Inoue, 1960; Voth, 1943; Voth & Hamner, 1940). Rashid (1970) observed that on nitrogen deficient medium spores divided to form globular cell masses. In *Asterella angusta*, *Targionia hypophylla*, and *Plagiochasma intermedium*, 1/2 Knop's and 1/2 Hoagland's solutions prove better than their full strength (Kaul, 1974). Higher concentrations of KNO_3 inhibit spore germination in *Asterella*, which germinate only between 0.05 to 0.1 percent. Spores show better tolerance for $CaCl_2$ than for KNO_3. In *Targionia* and *Plagiochasma* germination percentage is quite high in the nutrient medium comprising KH_2PO_4, $CaSO_4$, $MgSO_4$, and KNO_3 (Kaul, 1974). In *Marchantia polymorpha* Fe and $Ca(NO_3)_2$ induce germination, but optimal germination and growth have been observed with $Ca(NO_3)_2$, KNO_3, and $MgSO_4$

(Gemmrich, 1976). The spores of *Plagiochasma appendiculatum* can germinate even in water, but inorganic salts influence germination up to a certain extent. Ca, K, and Na do not have any beneficial effect, whereas Mg plays a significant role. Except $Ca(NO_3)_2$, higher concentrations (0.5%) of salts in solution have a lethal effect (Shukla et al., 1981).

GROWTH REGULATORS

Application of 2,4-D and IAA stimulates germination of mature and immature spores in some bryophytes (Benson-Evans, 1953; Fulford, 1956; Heitz, 1942; Kofler, 1959). Potassium gibberellate also stimulates development in the multi-celled spores of *Pellia* (Asprey et al., 1958). Gibberellic acid promotes spore germination in some moss species: *Dicranum scoparium*, *D. undulatum*, *Dicranoweisia crispula*, and *Pogonatum urnigerum* (Vaarama & Tarén, 1959). However, higher levels of IAA and gibberellins interrupt spore germination in some systems. These factors induce the spores to swell at an abnormal rate, and result in their death. NAA inhibits spore germination in *Marchantia polymorpha* (Rousseau, 1952). Günther (1960) observed that IAA strongly increases the size of cellulose fibrils which affect the internal polarity in germinating spores.

In *Cryptothallus mirabilis* application of 2,4-D and IAA at 10 mg/l to spores pretreated with low temperature results in germination together with a slight increase in cell size in the sporelings. The untreated spores also germinate, but development does not progress beyond the two-celled stage. Potassium gibberellate induces early stages of germination in spores still arranged in tetrads, but no subsequent development occurs (Benson-Evans, 1960). In *Funaria hygrometrica*, GA_3 and kinetin induce germination in dark, whereas IAA does not do so, although in light it is as effective as kinetin and gibberellic acid (Valanne, 1966). In *Entodon myurus* GA_3, IAA, and kinetin do not affect spore germination in dark, but promote it in light. IAA is most effective followed by GA_3 and kinetin (Sood, 1972). In *Targionia hypophylla* gibberellic acid retards spore germination above 25 mg/l, and at 150 mg/l germination was totally stopped (Kaul, 1974). In *Marchantia polymorpha*, GA_3 is ineffective in inducing spore germination in dark (Gemmrich, 1976).

HYDROGEN-ION CONCENTRATION (pH)

In *Mnium hornum* spores germinate best in neutral or slightly acidic conditions; pH 5.0 being the most favourable (Benson-Evans, 1953). Inoue (1960) studied the effect of different pH values (5.0, 5.6, 6.2, 6.8, 7.4, 8.0, & 8.8) on spore germination in five liverworts. Spores of *Marchantia polymorpha* show 100 percent germination in pH ranging from 5.0 to 7.4, whereas in *Reboulia hemisphaerica* such a response is observed up to pH 6.8. Maximum percentage of germination is observed at pH 6.2 in *Plagiochasma intermedium*, *Mannia fragrans*, and *Asterella odora*. Spores of none of the above liverworts germinate at pH 8.8. In *Targionia hypophylla* and *Plagiochasma intermedium*, pH range towards acidic side is quite suitable for germination. At pH 8.0 spores do not germinate (Kaul, 1974). Thus, in general, pH beyond seven is not favourable for spore germination in the investigated bryophytes.

OTHER FACTORS

With chloral hydrate the polarity of spores is suppressed, and irregular cell masses of restricted growth result. Colchicine also interrupts spore germination (von Wettstein, 1953). Treatment of spores with H_2O_2 induces germination in dark in *Funaria* (Valanne, 1966).

Addition of agar (0.3%) to mineral sucrose medium promotes germination and growth of sporelings in *Frullania asagrayana* (Iverson, 1957).

Spore germination is also stimulated by the addition of mold fungi like *Aspergillus flavus*, *Penicillium martensii*, *Mucor racemosus*, and *Fusarium scripi*, and the yeast *Rhodotorula*, in the cultures of *Tetraphis pellucida* and *Pogonatum urnigerum* (Vaarama & Tarén, 1959).

MECHANISM OF SPORE GERMINATION

Whether protein synthesis is essential for the onset of spore germination is still not clear. Application of the protein inhibitor, cycloheximide, retards spore germination and prevents normal growth of germ tubes. If dormant spores are treated with an enzyme solution (β-glucuronidase or arylsulfatase) and then planted on a medium containing cycloheximide they germinate earlier than the non-enzyme treated dormant spores (Kurz, 1976).

EXPERIMENTS ON GEMMAE

Gemmae of mosses may be unicellular, bicellular, or multicellular, and are of varied shapes (globose, discoid, stellate, or plate-like). They are borne on rhizoids, leaves, leaf axils, shoot apices, or in gemma receptacles. Goebel (1905) opined that gemmae are arrested buds, and designated them as 'broodbuds' or 'restricted-buds'. According to Worsdell (1915) gemmae are miniature shoots which have become fixed abnormalities. Protonemal gemmae have been reported in several mosses (Table 1.2).

TABLE 1.2.
Mosses forming protonemal gemmae under cultural conditions.

Taxa	Investigator/s
Anisothecium molliculum	Kumra and Chopra (1985)
Anisothecium spirale	Vashistha (1985)

(Table contd.)

Experiments on Spores and Gemmae 27

Amblystegium serpens	Selkirk (1981)
Aulacomnium androgynum	Whitehouse (1980)
Barbula sp.	Maheu (1908)
Barbula arcuata	Whitehouse (1980)
Barbula glauca	—do—
Barbula gregaria	Kumra (1981)
Barbula revoluta	Müller (1874)
Barbula rigidula	Whitehouse (1980)
Barbula trifaria	—do—
Bryum coronatum	Chopra and Rawat (1977)
Bryum klinggraeffii	—do—
Bryum micro-erythrocarpum	Whitehouse (1980)
Bryum radiculosum	—do—
Bryum riparium	Whitehouse (1963)
Bryum rubens	Whitehouse (1980)
Bryum sauteri	—do—
Bryum tenuisetum	—do—
Bryum violaceum	—do—
Buxbaumia aphylla	Hancock and Brassard (1974)
Desmatodon bogosicus	Whitehouse (1980)
Dicranella jamesonii	—do—
Dicranella staphylina	—do—
Dicranella varia	—do—
Didymodon michiganensis	Lal and Parihar (1980)
Didymodon recurvus	Vashistha and Chopra (1984)
Ditrichum cornubicum	Whitehouse (1980)
Ditrichum cylindricum	—do—
Ditrichum pusillum	—do—
Eucladium verticillatum	Saito (1972); Whitehouse (1980)
Funaria hygrometrica	Berkley (1941)
Gyroweisia tenuis	Whitehouse (1980)
Hyophila tortula syn	Correns (1899); Andrews and Redfearn (1965);
Hyophila involuta	Rahbar and Chopra (1982); Sharma and Chopra (1986)
Leptobryum pyriforme	Pringsheim (1921); Chopra and Rawat (1977); Whitehouse (1980)
Octoblepharum albidum	Egunyomi (1978)
Pohlia lescuriana	Whitehouse (1980)
Schistostega pennata	Edwards (1978)
Semibarbula orientalis	Dhingra-Babbar and Chopra (1985)
Tetraphis geniculata	Muraoka and Noguchi (1961)
Tetraphis pellucida	Narayanaswami (1956) Muraoka and Noguchi (1961); Schneider and Sharp (1962)
Tortula amplexa	Whitehouse (1980)
Tortula bolanderi	—do—
Tortula desertorum	Lazarenko (1959)
Tortula latifolia	Whitehouse (1980)
Tortula muralis	Müller (1874); Maheu (1908)
Tortula papillosa	Maheu (1908)
Tortula ruralis	Müller (1874); Maheu (1908)
Tortula stanfordensis	Whitehouse (1961)
Trematodon brevicalyx	Dhingra and Chopra (1983)

Andrews and Redfearn (1965) reported the development of secondary gemmae on the protonema derived from gemmae of *Hyophila tortula*. They considered that gemmae were formed due to unfavourable conditions. Later, Chopra and Rawat (1973, 1977) studied the production and behaviour of protonemal gemmae in three gemmiferous mosses of the Bryaceae: *Bryum coronatum, B. klinggraeffii,* and *Leptobryum pyriforme.* In *Bryum klinggraeffii* the protonema obtained from rhizoid-borne gemmae produced numerous secondary gemmae (Plate 1.1A) instead of gametophytic buds in cultures maintained at 20 to 25 C, both in light and dark. In the beginning the gemmae were solitary. After seven to eight weeks, gemmae appeared in groups of two to four (composite gemmae). These were either interconnected by one to three-celled filaments or developed directly on a gemma (Plate 1.1B). In *B. coronatum* gemmae developed at 30 C in light as well as in dark, whereas in *Leptobryum* they were formed on protonemata and gametophores in dark. In dark, the protonemata of *Bryum* spp. produce pro-buds which for want of light/kinetin mature into gemmae. However, the dark-induced gemmae of *Leptobryum* appear to be basically different from shoot buds, since they do not develop into shoots under conditions favourable for shoot formation.

That the protonemal gemmae of *B. klinggraeffii* are not basically different from shoot buds is indicated by the following observations:

(1) There is a relationship between the size/age of protonema and gemma formation, comparable to the critical size of protonema required for bud formation.
(2) Factors responsible for formation of buds and gemmae are common, and so are the inhibitory factors.
(3) During gemma formation a gemma stimulatory factor resembling the bud stimulatory factor 'H' is produced.
(4) Kinetin produces moruloid buds, but with gradual decrease in its concentration the size of buds decreases, so much so that at 0.01 mg/l they are indistinguishable from gemmae.
(5) A young, green gemma primordium when subjected to low temperature and high light level, or kinetin treatment at low temperature, develops into a leafy shoot.
(6) Gemmae of *B. klinggraeffii* are homologous to the gemma buds of *B. coronatum* which are undoubtedly restricted buds.

In *Bryum klinggraeffii,* darkness, low light level, temperatures higher than 25 C, sugar-free medium, higher concentrations of agar (2 to 4%), pH below 5.0 or above 8.9, auxins (0.001 to 1.0 mg/l), TIBA (1 to 10 mg/l), GA_3 (10 to 20 mg/l), adenine (5.0 mg/l), ABA (0.1 to 10 mg/l), morphactin (2 to 10 mg/l), and coconut milk (5 to 50%) retard protonemal growth. Consequently, gemma formation is correspondingly delayed and gemmae appear in smaller numbers. On the other hand, increased light level, optimal temperature (20 C), sucrose, casein hydrolysate (up to 1000 mg/l), and yeast extract (up to 100 mg/l) promote protonemal growth and consequently increase the number of gemmae. (Rawat, 1976).

During the initiation of gemmae on the protonema of *B. klinggraeffii*, a morphoregulatory substance is released in the substrate. This substance causes induction of gemmae on a fresh protonema six to eight days earlier, while it suppresses protonemal growth. A gemmiferous protonema produces effective amounts of the diffusate in 24 h, whereas a younger protonema, which had not yet started bearing gemmae, does not produce any such factor (Rawat & Chopra, 1976).

Fresh protonemal gemmae of *B. klinggraeffii* germinate readily (Plate 1.1C), but on storage they lose their viability within a month. Similarly, rhizoid-borne gemmae germinate in the growing season, but on storage they become non-viable.

In *Bryum coronatum*, smaller, round, and leafless structures (resembling the rhizoid-borne gemmae) are initiated in dark, irrespective of temperature. In the presence of light the pathway of differentiation depends on temperature; at high temperature (30 C) leafless structures are formed, whereas at 15 C, buds are produced and these grow into stunted shoots.

During winter, the protonemata of *Bryum* spp. develop normal shoots under natural conditions of light and temperature. When transferred to culture room, numerous gemmae develop not only on the protonemata, but also on the shoots in *B. klinggraeffii* (Plate 1.1 D). When a young, germinating gemma of *B. klinggraeffii* is planted on fresh nutrient medium and the culture kept under field conditions, it grows into a normal shoot. However, the dark-induced gemmae of *Leptobryum* do not grow into shoots under conditions favourable for shoot formation (Rawat, 1976).

Under unfavourable cultural conditions protonemal gemmae have also been observed in *Barbula gregaria*. In addition to the normal gametophytic buds, gemmae develop on the protonema of *Barbula* grown on media with 2.4 and 3.2 percent agar. Among other factors which induce the formation of gemmae are: very low and high light levels, temperature of 12 and 18 C, and autoclaved GA_3. The gemmae are initially green but later turn brown (Kumra, 1981). Kumra and Chopra (1985) reported that at lower temperatures (12 & 18 C) gemmae are formed on the protonema and shoots of *Anisothecium molliculum*, in addition to buds. The protonema of *Trematodon brevicalyx* bears brown, multicellular gemmae in chains at 25 C (Dhingra & Chopra, 1983). In *Didymodon recurvus* low irradiance (below 11.55 Wm^{-2}) and high temperature (25 C) favour production of protonemal gemmae (Vashistha & Chopra, 1984). On the contrary in *Hyophila involuta* gemmae are formed under high light level and not in dark or low light level (Sharma & Chopra, 1986). Cytokinins alone or along with IAA also induce gemmae in *H. involuta* (Rahbar & Chopra, 1982).

The behaviour of gemmae in liverworts has also been studied under cultural conditions. Fitting and his co-workers (1936-1950) did some interesting experiments on the gemmae of *Marchantia* and *Lunularia*. The gemmae have two more or less identical convex surfaces. Under natural conditions, whichever of these faces upwards becomes the dorsal surface of the mature thallus with air chambers. Ordinarily a gemma starts to grow from both the lateral notches and soon forms two ribbon-shaped plants with rhizoids growing downwards from the lower surface. The form, growth rate, and general behaviour of a developing gemma, depend partly on its genetic make up and partly on the environment in which its genes operate. Among the environmental factors which determine the polarity of a gemma are the chemical and physical characteristics of the substratum; the level, quality, and angle of incidence of light; temperature and gravity. By using a perfectly transparent agar jelly Fitting was able to make the light level equal on both sides of the young plants. By filling the petri dish and its lid with jelly, he was able to grow the gemmae between the layers of jelly, like the filling of a sandwich, and thus when necessary, observe the growth when in contact with a 'substratum' both above and below. By placing such cultures on a klinostat, the usual one-sided influence of gravity could also be eliminated. Thus, at any given temperature the action of light, gravity, and the substratum could be controlled and all these factors could be altered in every possible way. He observed that when the gemmae developed in a completely uniform environment, i.e. when illumination was equal above and below, when there was contact with the substratum on both surfaces, and when the effects of gravity had been eliminated, the young plant became a kind of Siamese twin, the two growing points divided horizontally to become four and they appeared like two thalli joined face to face by their dorsal surfaces, the rhizoids in both arising towards the substrate. It is then described as 'isolateral' (Fig. 1.16 A-C).

His further experiments revealed that the interplay of these four factors is highly complex. At 17 to 18 C the effect of light on development is greater than that of gravity, but at 28 to 29 C gravity is relatively more important. Furthermore, the behaviour of a gemma depends not only on the conditions acting on it at a given moment, but also on its previous history. In other words there is some kind of an 'induction' effect.

Several other investigators have also studied the effect of various physical and chemical factors on gemma germination, production, and growth under cultural conditions, and a brief account follows:

LIGHT

Voth and Hamner (1940) observed that plants of *Marchantia polymorpha* grown under short photoperiods (9 h of light) produced more gemma cups as compared to those maintained under long photoperiods (18 h of light). Schwabe and Nachmony-Bascomb (1963) reported that in *Lunularia cruciata* continuous light prevented growth of gemmae, whereas vigorous growth occurred in short days (10 h of light). Production of rhizoids on gemmae also depended on light. In total dark rhizoids were not formed on gemmalings. Red light promoted and far-red light inhibited rhizoid formation (Valio & Schwabe, 1969).

In *Marchantia nepalensis* diameter of gemma cups and number of gemmae varied with light level. In high light level (6000 lux) the width of thalli and the production of gemma cups and gemmae increased. In 60 percent of the cups on plants grown in high light level all the gemmae became adherent to

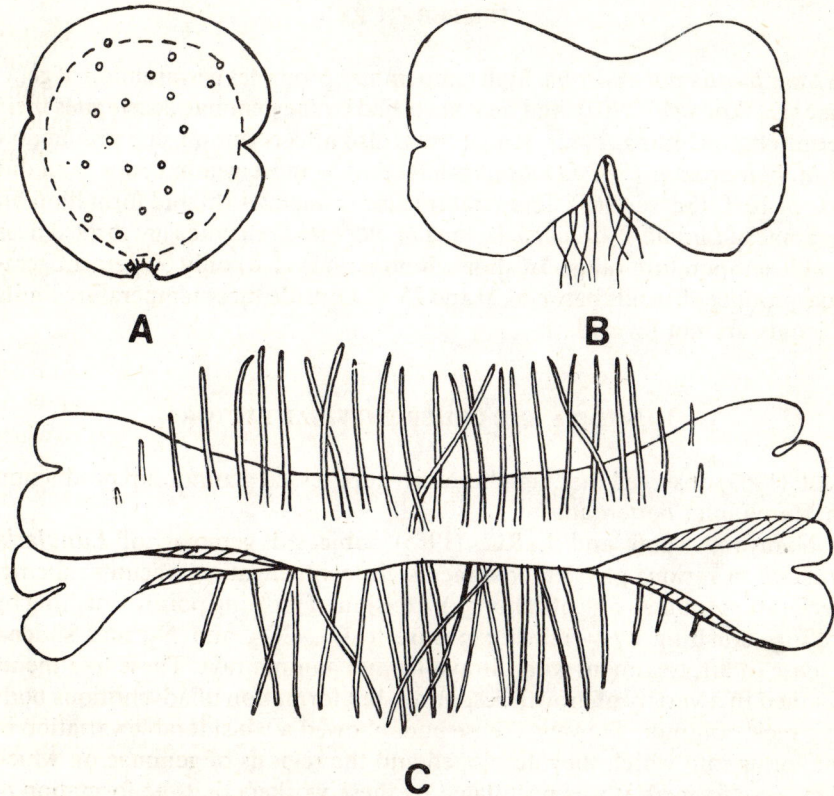

Fig. 1.16 A-C. Experiment on gemmae of *Marchantia polymorpha*. A. A detached gemma. B. Young thalli growing from gemma showing rhizoids and the two apical notches. C. Siamese twin growing from a gemma mounted on klinostat with equal illumination on both sides. For details see text.
(After Fitting, 1936a)

one another to form what may be called a 'gemma column. On the other hand, low light level (50 to 100 lux) inhibited production of gemma cups (Chopra & Sood, 1970).

Otto and Halbsguth (1976) reported that rhizoid formation on gemmae of *Marchantia polymorpha* is dependent on wavelength. Irradiation below 550 nm and above 670 nm was ineffective. Under conditions of unilateral gravitation about 20 percent of the gemmae formed rhizoids. Under conditions of alternating gravitational direction, and absence of irradiation almost no rhizoids developed. The rhizoids were formed in the direction of the gravitational pull. The direction from which the gemmae were irradiated influenced the emergence of rhizoids only when the orientation of rhizoids to the gravitational pull was not constant. The formation of rhizoids on the light-exposed surface was stimulated to a lesser degree than on the other surface (Otto, 1976). Egunyomi (1978) observed that in *Octoblepharum albidum* light is absolutely necessary for gemma germination.

TEMPERATURE

In *Marchantia polymorpha*, high temperature promotes germination of gemmae (Dacknowski, 1907), and heat absorbed by the gemmae accelerates their germination (Fitting, 1942). Temperature also affects gemma cup production in *M. berteroana*; 15 C favoured production of more gemma cups than did 5 C or 10 C (Scott, 1963). Temperature also influences rhizoid formation on gemmae of *Lunularia cruciata*. In long photoperiods rhizoids are formed over a wide temperature range. In short photoperiods (2 h) only 50 percent gemmae produce rhizoids between 20 and 25 C. Outside these temperature limits rhizoids are not formed.

HUMIDITY AND OTHER PHYSICAL FACTORS

Scott (1963) observed that high humidity suppressed gemma cup production in *Marchantia berteroana*.

Narayanaswami and LaRue (1955) subjected gemmae of *Lunularia cruciata* to various physical treatments viz. centrifugation; vacuum; slicing; perforation (puncturing of gemmae); excision of growing points; crushing by rolling; crushing by weights; exposure to O_2, CO_2, and N_2; and sudden release of air, treatment with ultraviolet and gamma rays. These treatments resulted in a variety of growth responses, but formation of adventitious buds was most common. However, these buds showed a considerable variation in the forms into which they developed and the regions of gemmae on which they were formed. It was postulated by these workers that the formation of adventitious buds was due to the removal of apical inhibitors.

Dickson (1932) reported that gemmae of *Marchantia polymorpha* produced adventitious growing points or buds, when subjected to treatments such as UV-light, X-rays, drying, and plasmolysing solutions.

In *Marchantia nepalensis* more gemma cups and gemmae were produced in liquid cultures as compared to solid ones. Germinating gemmae are frequently found within gemma cups, and these are present on solid but not in corresponding liquid cultures. Germinating gemmae without rhizoids is a common feature in liquid cultures, whereas the opposite is true for solid cultures. Only smooth-walled rhizoids are formed (Kaul et al., 1962).

GROWTH REGULATORS

Fitting (1939) reported that IAA stimulated rhizoid formation in dark-grown gemmae of *Marchantia polymorpha*. Narayanaswami (1956) studied the effect of some hormones and alkaloids on the behaviour of gemmae in *Tetraphis pellucida*. In control cultures germination of gemmae followed the normal pattern. Certain specialized peripheral cells of gemmae, called nematogones, gave rise to protonemal filaments. These filaments developed thallose appendages in whose axils arose the leafy gametophores (Fig. 1.17 A-

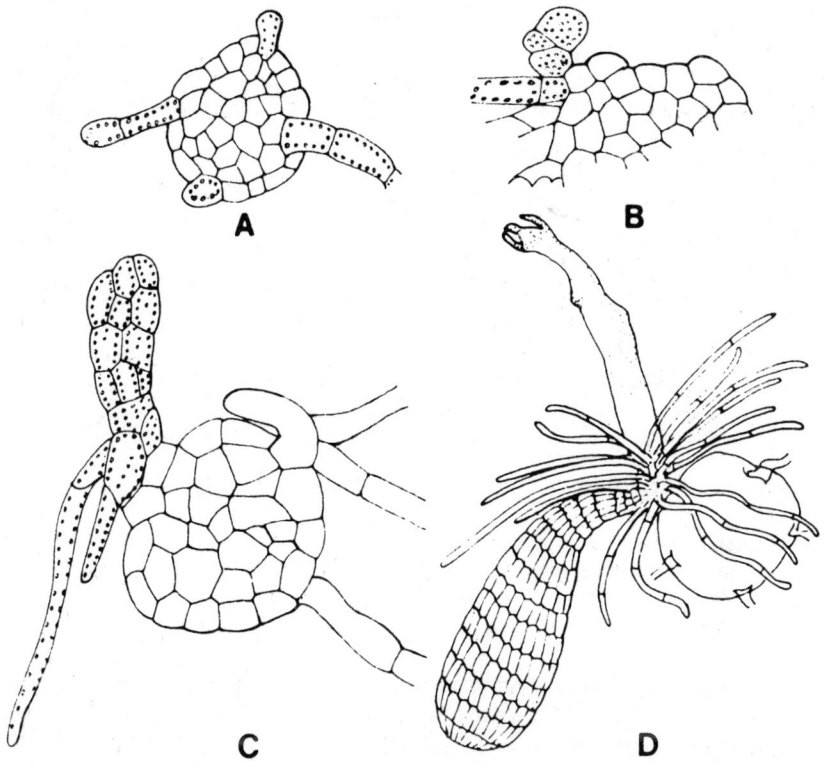

Fig. 1.17 A-D. *Tetraphis pellucida*. Germination of gemmae. A. Gemma developing filaments from massive nematogones (special cells on the periphery). B, C. Developmental stages of thallose protonema. D. Erect leafy shoot arising from the axil of a thallose appendage.
(After Narayanaswami, 1956).

D). When gemmae were grown directly on the IAA solution their germination was inhibited. However, when they were allowed to germinate on White's solution, and were cultured on IAA-supplemented medium (10^{-3}M) prior to the formation of the thallose protonemata, varied responses were observed. Two responses are described here. Frequently a small callus-like mass of tissue developed by divisions of a protonemal cell and this mass gave rise to a cluster of stalked gemmae (Fig. 1.18A). In still others, the individual cells of a filamentous protonema rounded up and formed a chain of single-celled bodies (Fig. 1.18B) or irregular masses of loosely aggregated brood cells. LaRue and Narayanaswami (1957) observed that in *Lunularia cruciata* removal of distal portion of thalli above the region of gemma cup caused in situ germination of gemmae, whereas application of growth hormones, especially some auxins (IAA, IBA, NAA), following excision prevented germination. Narayanaswami (1957) noted that higher concentrations of auxins caused inhibition of normal growth of thalli from the growing points of gem-

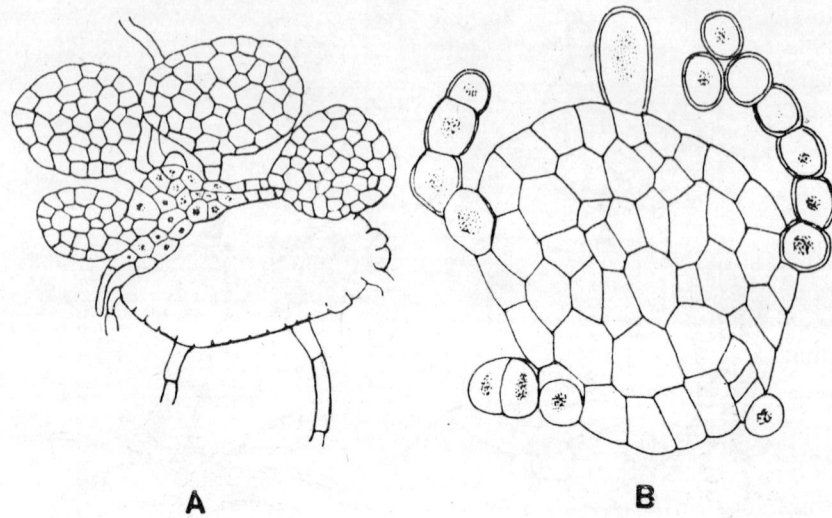

Fig. 1.18 A,B. Responses of germinating gemmae of *Tetraphis pellucida* on transfer to IAA- supplemented medium. A. Formation of secondary gemmae on parent gemma. B. Formation of short chains of rounded brood cells.
(After Narayanaswami, 1956).

mae. At lower concentrations gemma cups grew into bowls, and their margins were fringed and reflexed as a flap. Numerous rhizoids were formed on gemmae while within the cups. Tarén (1958) reported that initiation of growth and production of rhizoids on the gemmae contained in gemma cups of *M. polymorpha* could be altered by the application of IAA. In *Marchantia nepalensis*, gemma cups were formed in liquid but not on solid cultures (containing NOA (0.1 mg/l), IPA (1.0 mg/l), and TIBA (1.0 mg/l) (Kaul et al., 1962). Maravolo and Voth (1966) and Allsopp et al. (1968) reported that anti-auxins (MH, TIBA, & Coumarin) inhibited gemma growth in *M. polymorpha*. Otto and Halbsguth (1976) observed that IAA significantly promoted rhizoid formation on the gemmae of *M. polymorpha*.

Allsopp et al. (1968) reported that kinetin had little effect on growth and development of *Marchantia* gemmae, but in the presence of GA_3 growth was slow. Prior and Brown (1970) observed that GA_3 delayed rhizoid development in *Marchantia* gemmae, but had no effect on germination and initial intercalary growth. Binns and Maravolo (1972) noticed that in *M. polymorpha* BAP suppressed normal gemma germination and resulted in the development of nodular callus-like growth.

NITROGENOUS SUBSTANCES

Voth and Hamner (1940), and Voth (1941) observed that plants of *Marchantia polymorpha* possessed very few gemma cups on solutions lacking nitrate.
Basile (1965) studied the effect of three amino acids (arginine, histidine, &

glutamic acid) on gemma formation in the leafy liverwort *Scapania nemorosa*. All these amino acids stimulated production of gemmae, and the degree to which they did so depended on the ontogenetic stage of the plant.

HYDROGEN-ION CONCENTRATION (pH)

Narayanaswami and LaRue (1955) observed that when gemmae of *Lunularia cruciata* were grown in media with pH ranging from 3 to 11, there was not much difference in growth, but pH values below 3 and above 11 were lethal.

OTHER CHEMICAL FACTORS

Chopra and Sood (1970) observed that addition of sucrose to the medium brings about in situ germination of gemmae in *Marchantia nepalensis*.

Coumarin (Rousseau, 1954) and parasorbic acid (Moewus & Schader, 1951) inhibited the development of rhizoids and growth of gemmae in *Marchantia polymorpha*.

LaRue and Narayanaswami (1955) investigated the morphogenetic effects of various chemicals (benzoguanimine, formalin, ether, naphthalene, colchicine, caffeine, sodium cacodylate, acenaphthene, phenol, actidione, chromic acid, & parathion) on the gemmae of *Lunularia* and observed many malformations, such as inhibition of one growing point or displacement of growing point to one side. *Marchantia* gemmae showed profound disturbances in normal morphogenesis under the influence of azaguanine and thiouracil (Allsopp et al., 1968).

Nehira (1981) observed that rhizoid formation in *Marchantia* gemmae was prevented by the chelating agent GEDTA.

REFERENCES

AHMAD. S.M., N. SULTAN, and J.C. DAGAR, *Biologia*, 23, 137 (1977).
ALLSOPP, A. and G.C. MITRA, *Ann. Bot.*, 22, 95 (1958).
ALLSOPP, A., C. PEARMAN, and A.N. RAO, *Phytomorphology*, 18, 84 (1968).
ANDERSON, L.E. and M.R. CROSBY, *Bryologist*, 68, 47 (1965).
ANDREWS, S. and P.L. REDFEARN, *Bryologist*, 68, 345 (1965).
ASPREY, G.F., KATHRYN BENSON-EVANS, and A.G. LYON, *Nature*, 181, 1351 (1958).
BASILE, D.V., *Am. J. Bot.*, 52, 443 (1965).
BAUER, L., *Flora*, 36, 30 (1942).
BAUER, L. and H. MOHR, *Planta*, 54, 68 (1959).
BECQUERREL, P., *Rev. gén. Bot.*, 18, 49 (1906).
BENSON-EVANS, KATHRYN, *Trans. Br. bryol. Soc.*, 2, 291 (1953).
BENSON-EVANS, KATHRYN, *Trans. Br. bryol. Soc.*, 3, 729 (1960).
BERKLEY, E.E., *Trans. Ill. State Acad. Sci.*, 24, 102 (1941).
BHATLA, S.C., *Beitr. Biol. Pfl.*, 57, 157 (1983).
BINNS, A.N. and N.C. MARAVOLO, *Am. J. Bot.*, 59, 691 (1972).
BORODIN, J., *Bull. Acad. Imp. Sci. de St. Petersbourg*, 12, 432 (1868).

CHOPRA, R.N. and URMILLA GUPTA, *Bryologist*, **70**, 102 (1967).
CHOPRA, R.N. and SNEH SOOD, *Bryologist*, **73**, 592 (1970).
CHOPRA, R.N. and SNEH SOOD, *Bryologist*, **76**, 278 (1973).
CHOPRA, R.N. and M.S. RAWAT, *Bryologist*, **76**, 183 (1973).
CHOPRA, R.N. and M.S. RAWAT, *Bryologist*, **80**, 655 (1977).
CORRENS, C., *Untersuchungen über die Vermehrung der Laubmoose durch Brutorgane und Stecklings*, Verl. G. Fisher, Jena 1899.
COVE, D.J., A. SCHILD, N.W. ASHTON, and E. HARTMANN, *Photochem. Photobiol.*, **27**, 249 (1978).
DACKNOWSKI, A., *Jb. wiss. Bot.*, **44**, 244 (1907).
DHINGRA, SADHANA and R.N. CHOPRA, *J. Bryol.*, **12**, 571 (1983).
DHINGRA-BABBAR, SADHANA and R.N. CHOPRA, *J. Bryol.*, **13**, 533 (1985).
DICKSON, H., *Ann. Bot.*, **46**, 683 (1932).
DOYLE, W.T., *Bryologist*, **66**, 238 (1963).
EDWARDS, S.R., *J. Bryol.*, **10**, 69 (1978).
EGUNYOMI, A., *J. Hattori bot. Lab.* No. **44**, 25 (1978).
FITTING, H., *Jb. wiss. Bot.*, **82**, 333 (1936a).
FITTING, H., *Jb. wiss. Bot.*, **82**, 696 (1936b).
FITTING, H., *Jb. wiss. Bot.*, **85**, 169 (1937a).
FITTING, H., *Jb. wiss. Bot.*, **85**, 243 (1937b).
FITTING, H., *Jb. wiss. Bot.*, **86**, 107 (1938).
FITTING, H., *Jb. wiss. Bot.*, **88**, 633 (1939).
FITTING, H., *Biol. Zbl.*, **62**, 336 (1942).
FITTING, H., *Planta*, **37**, 635 (1950).
FULFORD, M., *Phytomorphology*, **6**, 199 (1956).
GEMMRICH, A.R., *Flora*, **165**, 479 (1976).
GOEBEL, K., *Flora*, **72**, 1 (1889).
GOEBEL, K., *Organography of Plants*, Part II, Hafner Publishing Company, London, 1905.
GÜNTHER, I., *J. Ultrastruct. Res.*, **4**, 304 (1960).
HANCOCK, J.A. and G.R. BRASSARD, *Bryologist*, **77**, 501 (1974).
HEALD, F.D.F., *Bot. Gaz.*, **26**, 169 (1898).
HEDWIG, J., *Fundamentum Historiae Naturalis Muscorum Frondosum*, Bd. 2, Leipzig, 1782.
HEDWIG, J., *Theoria Generations et Fructifications Plantarum Cryptogamicarum Petropoli*, 1784.
HEITZ, E., *Ber. dt. bot. Ges.*, **60**, 17 (1942).
HOFFMAN, G.R., *Bryologist*, **67**, 321 (1964).
HOFMEISTER, W., *Vergleichende Untersuchungen der Keimung, Entfaltung und Fruchtbildung höherer Kryptogamen, und der Samenbildung der Coniferen*, Leipzig, 1851.
HOFMEISTER, W., *On the Germination, Development, and Fructification of the Higher Cryptogamia*, London, 1862.
INOUE, H., *J. Hattori bot. Lab.* No. 23, 148 (1960)
ISHIKAWA, S. and T. OHUSA, *Bot. Mag. Tokyo*, **67**, 193 (1954).
IVERSON, G.B., *Bryologist*, **60**, 348 (1957).
KACHROO, P., *J. Indian bot. Soc.*, **33**, 263 (1954).
KANDA, H., *Hikobia*, **6**, 60 (1971).
KAUL, A., *J. Hattori bot. Lab.* No. 38, 283 (1974).
KAUL, K.N., G.C. MITRA, and B.K. TRIPATHI, *Ann. Bot.*, **26**, 447 (1962).
KOFLER, L., *Rev. Bryol. Lichénol.*, **27**, 1 (1959).
KOFLER, L., J. DUTEL, and F. NURIT, *J. Linn. Soc. (Bot.)*, **58**, 311 (1963).
KUMRA, P.K., "Morphogenetic and physiological studies on some mosses", Ph.D. Thesis, Univ. Delhi, India, 1981.
KUMRA, SUMAN and R.N. CHOPRA, *Phytomorphology*, **35**, 223 (1985).
KURZ, E.H., *Z. PflPhysiol.*, **78**, 58 (1976).

LAL, J. and N.S. PARIHAR, *J. Indian bot. Soc.*, **59** (Suppl.), 36 (1980).
LaRue, C.D. and S. NARAYANASWAMI, *Bull. Torrey bot. Club*, **82**, 198 (1955).
LaRue, C.D. and S. NARAYANASWAMI, *New Phytol.*, **56**, 61 (1957).
LAZARENKO, A.S., *Ukr. bot. Zh.*, **16**, 55 (1959).
MAHEU, J., *Bull. Soc. bot. Fr.*, **55**, 445 (1908).
MARAVOLO, N.C. and P.D. VOTH, *Bot. Gaz.*, **127**, 79 (1966).
MEHRA, P.N. and P. KACHROO, *Bryologist*, **54**, 1 (1951).
MEHRA, P.N. and P. KACHROO, *Bryologist*, **55**, 59 (1952).
MEHRA, P.N. and P. KACHROO, *J. Hattori bot. Lab.* No. 25, 145 (1962).
MEYER, S.L., *Bryologist*, **50**, 403 (1947).
MEYER, S.L., *Bryologist*, **51**, 213 (1948).
MOEWUS, F. and E. SCHADER, *Z. Naturforsch.*, **6b**, 112 (1951).
MUELLER, T.A. and M.K. MUELLER, *Radiat. Bot.*, **9**, 283 (1969).
MÜLLER, H., *Die Sporenvorkeime und Zweigvorkeime der Laubmoose*, Leipzig, 1874.
MURAOKA, S. and A. NOGUCHI, *Misc. Bryol. Lichenol.*, **2**, 83 (1961).
NARAYANASWAMI, S., *Phytomorphology*, **6**, 323 (1956).
NARAYANASWAMI, S., *J. Indian bot. Soc.*, **36**, 180 (1957).
NARAYANASWAMI, S. and C.D. LaRue, *Phytomorphology*, **5**, 356 (1955).
NEHIRA, K., *Hikobia*, **3**, 288 (1963).
NEHIRA, K., *Hikobia*, **4**, 43 (1964).
NEHIRA, K., *Misc. Bryol. Lichenol.*, **3**, 135 (1965).
NEHIRA, K., *J. Sci. Hiroshima Univ. Ser. B, Div. 2 (Bot.)*, **11**, 1 (1966).
NEHIRA, K., *J. Hattori bot. Lab.* No. 41, 157 (1976).
NEHIRA, K., "Rhizoid formation in *Marchantia* gemmae (Abstr.)," *The Proceedings of XIII International Botanical Congress*, Sydney, Australia, 1981, Pp. 291.
NEHIRA, K., "Spore germination, protonema development, and sporeling development", In R.M. Schuster, Ed., *New Manual of Bryology*, 1, The Hattori Botanical Laboratory, Nichinan, Miyazaki, Japan, 1983, Pp. 343-385.
NISHIDA, Y., *Bot. Mag. Tokyo*, **84**, 187 (1971).
NISHIDA, Y., *Misc. Bryol. Lichenol.*, **6**, 52 (1972).
NISHIDA, Y., *Bot. Mag. Tokyo*, **86**, 35 (1973).
NISHIDA, Y., *J. Hattori bot. Lab.* No. **44**, 371 (1978).
NISHIDA, Y. and Z. IWATSUKI, *Proc. Bryol. Soc. Jap.*, **2**, 137 (1980).
NOGUCHI, A., *J. Hattori bot. Lab.* No. 19, 71 (1958).
NOGUCHI, A. and I. MIYATA, *Kumamoto Journ. Sci. ser. b, sect. 2*, **3**, 1 (1957).
NOGUCHI, A. and T. MIZUNO, *Kumamoto Journ. Sci. ser. b, sect. 2*, **4**, 113 (1959).
NOGUCHI, A. and H. OTSUKA, *Res. Bull. Fac. Lib. Arts. Oita Univ.*, **3**, 38 (1954).
O'HANLON, S.M.E., *Bot. Gaz.*, **82**, 215 (1926).
OTTO, K.-R., *Z. PflPhysiol.*, **80**, 189 (1976).
OTTO, K.-R. and W. HALBSGUTH, *Z. PflPhysiol.*, **80**, 197 (1976).
PRINGSHEIM, E.G., *Jb. wiss. Bot.*, **60**, 499 (1921).
PRIOR, P.V. and P.R. BROWN, *Bryologist*, **73**, 687 (1970).
RAHBAR, KAVITA and R.N. CHOPRA, *New Phytol.*, **91**, 501 (1982).
RASHID, A., "Morphogenetic studies on some Indian bryophytes", Ph.D. Thesis, Univ. Delhi, India, 1968.
RASHID, A., *Phytomorphology*, **20**, 49 (1970).
RAWAT, M.S., "Morphogenic studies on some Bryaceae", Ph.D. Thesis, Univ. Delhi, India, 1976.
RAWAT, M.S. and R.N. CHOPRA, *Z. PflPhysiol.*, **78**, 372 (1976).
RICHARDS, P.W., "Ecology", In Fr. Verdoorn, Ed., *Manual of Bryology*, Amsterdam, 1932. Pp. 367-395.
ROUSSEAU, J., *Bull. Soc. bot. France*, **99**, 308 (1952).
ROUSSEAU, J., *C.r. hebd. Séanc. Acad. Sci., Paris*, **239**, 1420 (1954).
SAITO, K., *Misc. Bryol. Lichenol.*, **6**, 41 (1972).
SCHILD, A., "Untersuchungen zur Sporenkeimung und Protonemaentwicklung bei dem Laubmoos *Physcomitrella patens*," Ph.D. Thesis, Univ. Mainz, West Germany, 1981.

SCHNEIDER, M.J. and A.J. SHARP, *Bryologist*, **65**, 154 (1962).
SCHWABE, W.W. and S. NACHMONY-BASCOMB, *J. exp. Bot.*, **14**, 353 (1963).
SCOTT, G.A.M., *Nature*, **200**, 1123 (1963).
SELKIRK, P.M., *J. Bryol.*, **11**, 719 (1981).
SHARMA, POONAM and R.N. CHOPRA, *J. Hattori bot. Lab.* No. 60, 137 (1986).
SHUKLA, R.M., R.R. DAS, and A. KAUL, *Geobios*, **8**, 116 (1981).
SHUKLA, R.M., K.C. PATIDAR, D. JAIN, and A. KAUL, *Misc. Bryol. Lichenol.*, **9**, 153 (1983).
SINGHIVI, U., "Studies on reproductive biology of certain species of *Asterella*," Ph.D. Thesis, Univ. Udaipur, India, 1979.
SOOD, SNEH, "Experimental studies on some Indian bryophytes", Ph.D. Thesis, Univ. Delhi, India, 1972.
SOOD, SNEH, *Beitr. Biol. Pfl.*, **51**, 99 (1975).
STEINER, A.M., *Z. Bot.*, **51**, 399 (1963).
STEINER, A.M., *Z. Bot.*, **52**, 245 (1964).
TAKAO, A., *The Journal of the Faculty of General Education, Univ. Tottori*, **10**, 63 (1977).
TAREN, NIINA, *Bryologist*, **61**, 191 (1958).
TREBOUX, O., *Ber. dt. bot. Ges.*, **23**, 397 (1905).
UDAR, R., *J. Indian bot. Soc.*, **37**, 300 (1958).
VAARAMA, A., and N. TAREN, *Bot. Notiser*, **112**, 481 (1959).
VALANNE, N., *Ann. Bot. Fenn.*, **3**, 1 (1966).
VALIO, I.F.M. and W.W. SCHWABE, *J. exp. Bot.*, **20**, 615 (1969).
VASHISTHA, B.D., "In vitro investigations on some Indian bryophytes." Ph.D. Thesis, Univ. Delhi, India, 1985.
VASHISTHA, B.D. and R.N. CHOPRA, *New Phytol.*, **97**, 83 (1984).
VOTH, P.D., *Bot. Gaz.*, **103**, 310 (1941).
VOTH, P.D., *Bot. Gaz.*, **104**, 591 (1943).
VOTH, P.D. and K.C. HAMNER, *Bot. Gaz.*, **102**, 169 (1940).
WALDNER, M., *Die Entwicklung der Sporogone von Andreaea und Sphagnum*, Leipzig, 1887
VON WETTSTEIN, D., *Z. Bot.*, **41**, 199 (1953).
WHITEHOUSE, H.L.K., *Trans. Br. bryol. Soc.*, **4**, 84 (1961).
WHITEHOUSE, H.L.K., *Trans. Br. bryol. Soc.*, **6**, 389 (1963).
WHITEHOUSE, H.L.K., *J. Bryol.*, **11**, 133 (1980).
WIGGLESWORTH, G., *Trans. Br. bryol. Soc.*, **1**, 4 (1947).
WORSDELL, W.C., *The Principles of Plant-Teratology*, Vol. I., Ray Society, London, 1915.

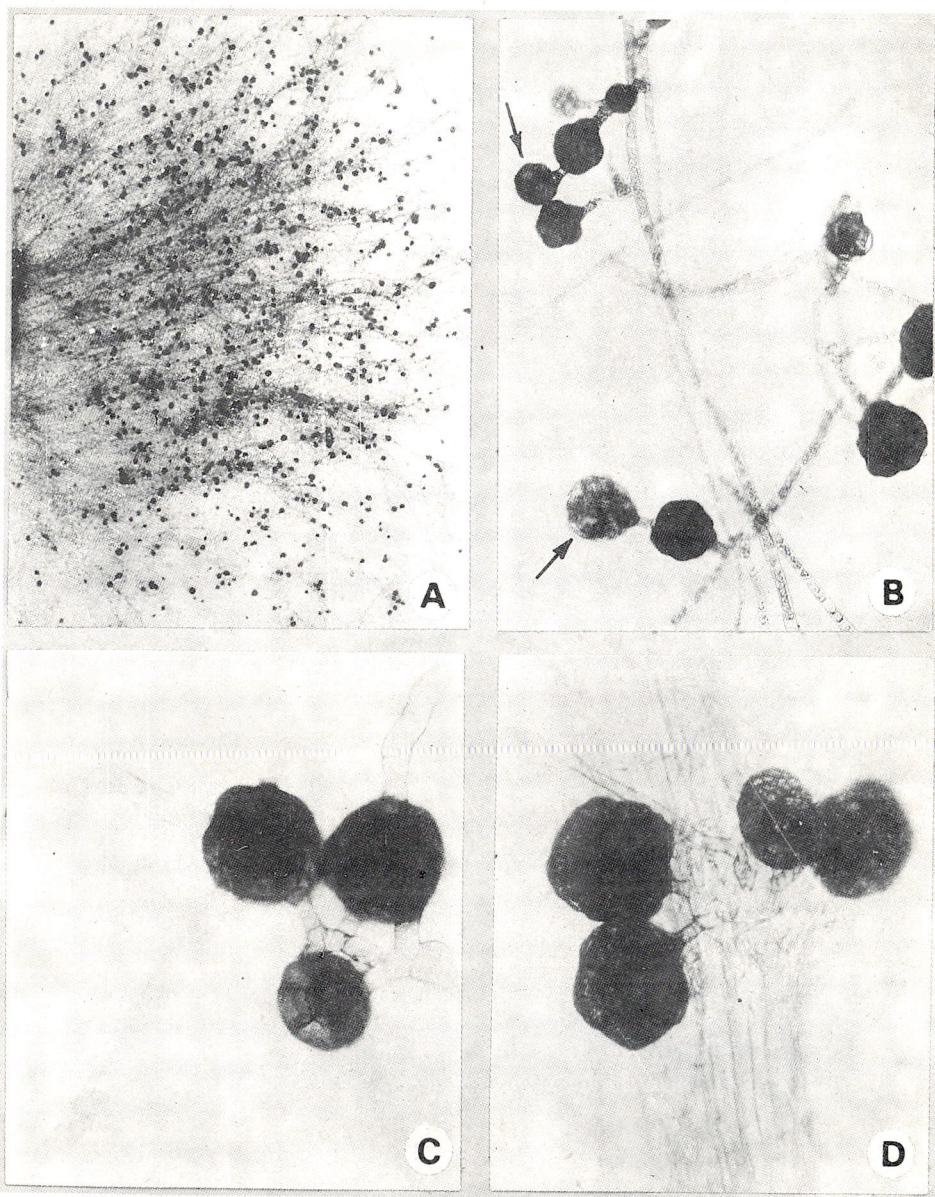

Plate 1.1 A–D. Gemmae in *Bryum klinggraeffii*. **A.** Protonema grown on basal medium bearing numerous gemmae. **B.** Gemmae in groups of 2-4 (arrow-marked), interconnected by short filaments. **C.** Gemmae germinating in situ. **D.** Gemmae borne at the tips of short filaments arising from a shoot.
(After Rawat. 1976).

2 Protonemal Differentiation and Bud Formation in Mosses

On spore germination the first-formed protonema is chloronema. It is irregularly branched, has hyaline cell walls, transversely placed cross walls, and numerous rounded chloroplasts per cell (Plate 2.1A, B). It gradually matures into caulonema which has regular branching (Plate 2.1C). Its horizontal filaments show unlimited growth, have brown cell walls, obliquely placed cross walls, and fewer spindle-shaped chloroplasts per cell. The vertical branches of caulonema have limited growth and mostly resemble the chloronemal filaments (Plate 2.1D). In addition to these morphological distinctions the two stages of protonema also differ in physiological characteristics, which have been described later in this chapter.

In vivo, germination of spores, differentiation of protonema into chloronema and caulonema, and transformation of the uni-dimensional growth of filamentous protonema into the three-dimensional growth of leafy shoot occur in a smooth sequence. However, attempts to copy these processes in vitro are not always successful (see Chopra, 1981). Besides external factors, protonemal differentiation and bud induction are regulated by internal factors of hormonal nature transported from one cell to another, and also by substances leached into the substratum by some cells of the system.

PROTONEMAL DIFFERENTIATION

Depending upon their origin, the protonemata of mosses are termed primary or secondary. Primary protonema arises from spores, and secondary protonema arises from any part of gametophyte or sporophyte. Whether a culture is started with a single spore, a clump of spores, or by regeneration of any part of the plant, the resulting protonema grows into a circular patch. The majority of mosses have freely branched filamentous protonema, but in some mosses like *Sphagnum* and *Andreaea* it is thalloid.

As regards the development of caulonema from chloronema, Sironval (1947) reported that in *Funaria hygrometrica* after about 20 days of spore germination all cells of chloronemal filaments excepting the apical cells degenerate, and caulonema arises from the persisting apical cells. Subsequent researchers are of the view that the development of caulonema from chloronema is a continuous process (Allsopp & Mitra, 1956, 1958; Bopp, 1954; Kofler, 1956, 1959).

Formation of caulonema depends largely on cultural conditions and composition of the nutrient media. Low temperature, inadequate light levels, short days, low concentration of agar, and hydrolysed agar are unfavourable for caulonema differentiation. The two forms of protonema are interconvertible (Bopp & Bohrs, 1965). Caulonema de-differentiates into chloronema on isolation and subculture. Some workers have not been able to observe clear cut distinction between chloronema and caulonema in certain mosses (Nehira, 1976; Van Andel, 1952).

Caulonema is reported to have a heterotrichous habit (Allsopp & Mitra 1958; Bhatla, 1982; Chopra & Rekhi, 1979; Dua, 1983; Kumra, 1981; Rawat, 1976; Servettaz, 1913; Sood, 1975; Vashistha, 1985). The degree of development of erect and prostrate systems depends upon the species, and is also influenced by environmental conditions.

In recent years chemical control of protonemal differentiation has been demonstrated. Auxins favour caulonema differentiation and inhibit growth of chloronema (Bopp, 1963, 1968; Hurel-Py, 1948; Johri & Desai, 1973; Szweykowska, 1962). Application of cAMP to the protonema of *Funaria hygrometrica* results in the differentiation of chloronema from caulonema. A compound indistinguishable from cAMP is reported to be present endogenously in this moss, and its level is four to seven-fold higher in the chloronema cells than in the caulonema cells. A specific ratio of cAMP/IAA (>20) leads to chloronema production. So, the differentiation of chloronema to caulonema depends on the relative levels of cAMP (which promotes the chloronema stage) and auxin (which enhances caulonema formation) in the system (Handa & Johri, 1976, 1977, 1979). Some enzymes like cyclic nucleotide phosphodiesterase, IAA oxidase, and peroxidase regulate the endogenous levels of cAMP and IAA, which in turn regulate the differentiation of specific cell types (Sharma et al., 1979).

Recently, Bhatla and Bopp (1985) isolated four auxin-resistant mutants from *Funaria hygrometrica*. These mutants differ in the extent of their inability to form caulonema on Knop's medium (minimal medium without auxin). Among these mutants, NAR-2 remains totally in chloronema state on minimal medium. Their studies also supported the hypothesis that the morphogenetic step of caulonema formation is regulated by the internal auxin level.

It has been demonstrated that endogenous morphoregulatory substances are involved in protonemal differentiation and bud formation in mosses (Bopp, 1963; Bopp & Klein, 1963; Gorton & Eakin, 1957; Hatanaka-Ernst, 1966; Jahn, 1964; Klein, 1967; Mitra & Allsopp, 1959a; Nebel & Naylor, 1968; Rawat & Chopra, 1976). The chloronemal cells of *Funaria hygrometrica* produce a heat labile factor 'F' (first letter of the German word Förderung, meaning stimulation), which promotes the differentiation of caulonema from chloronema, but inhibits bud initiation. The caulonema cells synthesize a heat stable factor 'H' (from Hemmung, meaning hinderance), which at higher concentrations promotes bud initiation and inhibits caulonemal growth, but at lower concentrations it exclusively promotes bud formation (Bopp, 1963;

Bopp & Klein, 1963; Klein, 1967). This factor influences protonemal differentiation not only in the species generating it but also in other species (Plate 2.2). It has been reported that factor 'H' contains two active ingredients which are of the nature of steroids and terpenes (Eltz, 1975). In *Bryum klinggraeffii* a morphoregulatory substance is produced by secondary protonema during the initiation of protonemal gemmae. This substance inhibits protonemal growth but promotes gemma formation. Like factor 'H', its gemma-stimulatory effect can not be destroyed by heat (Rawat & Chopra, 1976).

Caulonema is normally a pre-requisite for bud formation. This concept is supported by the studies on the X-ray induced mutants of *Funaria hygrometrica*, which do not form caulonema and remain bud-free (Oehlkers, 1956). However, buds arise even on chloronemal cells, as reported in *Leucodon* (Lal, 1961; Rashid, 1968), *Entodon* (Sood & Chopra, 1973), *Pogonatum* (Sood, 1975), *Microdus* (Nair & Raghavan, 1976), and *Pylaisiella* (Spiess, 1979). The time and manner of bud formation is generally species specific under a given set of cultural conditions (Allsopp & Mitra, 1958). In vitro the manner of bud formation on the protonemata of mosses is of two principal types: (1) the *Funaria* type, in which buds are formed in a ring at a certain distance from the centre, and (2) the *Polytrichum* type, in which buds arise in the centre of the fully developed protonema (Bopp, 1965).

There is normally a correlation between protonemal growth and bud formation. In some systems protonema should reach a 'critical size' before bud formation. In some others protonemal age rather than its size appears to be important. Bopp and Brandes (1964) postulated that protonema has to attain a critical size before bud formation takes place. Similar observation was made by Chopra and Rekhi (1979) in respect of bud induction in *Timmiella anomala*. However, in *Pylaisiella selwynii* (Spiess et al., 1971), *Barbula gregaria*, *Bryum coronatum* (Kumra, 1981), *Bryum argenteum* (Bhatla, 1982), *Bartramidula bartramioides* (Chopra & Rahbar, 1982), and *Anisothecium molliculum* (Kumra & Chopra, 1985) protonemal age rather than its size appears to be the critical factor. Nebel and Naylor (1968) noticed that in *Physcomitrium turbinatum* bud initiation is independent of colonial growth as a whole, and buds are initiated only after the caulonemal filaments had attained a critical length. In *Bryum klinggraeffii* a specific size of protonemal patch as well as the age of protonema, are required for gemma initiation (Rawat & Chopra, 1976). In *Physcomitrium sphaericum* formation of buds does not depend on the age or size of protonema but is determined by the number of cells from the tip of the filament. Caulonema cells at positions less than 15 cells from a spore do not form buds (Yoshida & Yamamoto, 1982). In this moss the position of first bud is not affected even by application of cytokinin (Yoshida & Yamamoto, 1985).

BUD FORMATION

The bud primordia normally arise on caulonemal filaments. These give rise to leafy shoots which develop their rhizoidal systems (Plate 2.3 A-D).

Parameters in the studies on bud induction include size of protonemal patch, time of appearance of buds, percentage response of cultures, number of buds per culture, place of origin of buds, and morphology of the gametophores.

FACTORS AFFECTING BUD FORMATION

Among the physical factors which affect bud induction are light (level and quality) and temperature. The chemical factors include auxins, cytokinins, gibberellins, adenosine 3':5'- cyclic monophosphate, adenine, amino acids, minerals, chelates, vitamins, abscisic acid, sugars, and pH. Other organisms growing along with mosses also give out chemical substances which affect morphogenesis.

Light

It was Klebs who in 1893 emphasized that adequate light level is required for bud initiation, since cultures of *Funaria* growing in weak light remained indefinitely bud-free. Subsequently, there were some unsuccessful attempts to induce buds in dark (Goebel, 1896; Pringsheim & Pringsheim, 1935; Robbins, 1918; Servettaz, 1913; Ubisch, 1913). These were followed by a few reports of bud formation in the absence of light, but since the methodology was defective, the results are not dependable (Belkengren, 1962; Fries, 1945; Keil, 1949; Naef & Simon, 1980). There have been some authentic reports of bud induction in dark in *Funaria hygrometrica* (Plate 2.4A-E) and *Ceratodon purpureus* on media supplemented with kinetin (Chopra & Gupta, 1967; Szweykowska, 1963). However, in some mosses like *Timmiella* it is not possible to induce buds in dark even with the addition of kinetin (Chopra & Rekhi, 1979).

Light quality is also important in protonemal development and bud initiation. Red light stimulates protonemal growth as well as bud formation in *Funaria hygrometrica* (Jahn, 1964; Pringsheim & Pringsheim, 1935) and *Pohlia nutans* (Mitra et al., 1959). Blue light inhibits bud formation in both the mosses. According to Jahn (1964) an inhibitor of bud formation is produced in blue light. In green light buds are not formed in *Pohlia*, and even protonemal growth is poor (Mitra et al., 1959). The inhibitory effect of blue light can be overcome by kinetin or by red light. A critical balance between red and blue light is required for the development of buds into leafy shoots. Red light followed by blue light is stimulatory, whereas blue light followed by red light is inhibitory (Mitra et al., 1965). For purposes of bud induction, red light can replace day light, and kinetin in turn replaces the effect of red light (Mitra & Allsopp, 1959b). Therefore, it can be presumed that red light brings about the synthesis of kinetin-like substances in the system. Light also controls mitotic rhythm, protonemal cell-length, and ramification.

Temperature

Buds are formed on protonema in a certain temperature range (12-30C), and usually low temperature is inhibitory. In some systems bud initiation is dependent on the interaction between light and temperature. In *Physcomitrium turbinatum* an increase in light level at 16 C did not reduce the time taken for bud initiation, but did so at 26 C (Nebel & Naylor, 1968).

Auxins

The effect of auxins on bud formation depends on their concentration. At lower levels auxins stimulate bud induction, whereas at high concentrations they not only cause complete inhibition of bud formation, but even at times bring about de-differentiation of the bud primordia and cause development of rhizoids on buds. Auxins also induce buds in some systems like *Ceratodon purpureus* (Szweykowska, 1962), *Anoectangium thomsonii* (Chopra & Rashid, 1969a), *Dicranella coarctata* (Kumra, 1981), *Anoectangium stracheyanum*, and *Bryum atrovirens* (Vashistha, 1985), which remain bud-free on basal medium.

Auxins influence bud formation through caulonema differentiation. In an X-ray mutant of *Funaria hygrometrica*, which had lost its capacity to synthesize auxin, buds could be produced only with the addition of IAA, while the wild type produced buds on basal medium (Hatanaka-Ernst, 1966). Studies on mutants of *Physcomitrella patens* (Ashton & Cove, 1977; Ashton et al., 1979a,b; Cove et al., 1979) have also indicated involvement of auxins in caulonema differentiation and bud formation. Bopp (1982), and Lehnert and Bopp (1983) reported that applied auxins and their precursors induce caulonema formation.

The inhibitory effect of high auxin concentration can be overcome by the addition of cytokinins, as in *Tortella caespitosa* (Gorton & Eakin, 1957) and *Entodon myurus* (Sood & Chopra, 1973). Sood and Hackenberg (1979) reported that antiauxins like PCIB cause de-differentiation of caulonema to chloronema and inhibit the bud induction response to IAA in *Funaria*. They also reported that IAA, 2,4-D, and tryptophan increase bud formation on intact protonema, but in isolated filaments such enhancement is possible only with a combination of an auxin with cytokinin. In the same moss, Bopp (1979, 1980) observed that with the application of PCIB growth speed of the protonema is reduced but not the frequency of cell division, which results in shorter cells. Cells become smaller in diameter and the newly formed cell walls are perpendicular, as in chloronema.

Cytokinins

The term cytokinin is a generic name for substances which promote cell division and have growth regulatory functions. A striking example of the morphogenetic effect of cytokinins is the induction of buds on moss protonemata (Bopp, 1963; Bopp & Klein, 1963; Gorton & Eakin, 1957; Jahn, 1964; Mitra & Allsopp, 1959b,c). In contrast to auxins, cytokinins are effective over a very wide concentration range. They bring about bud initiation under noninductive conditions like sub-optimal light level, blue light (Jahn, 1964), and even in dark; cause marked reduction in the time required for bud initiation; and considerably enhance their number. In *Anoectangium thomsonii* (Rashid, 1970), *Microdus miquelianus* (Nair & Raghavan, 1976), *Trematodon brevicalyx* (Chopra & Dhingra-Babbar, 1984), *Bryum pallescens* (Sarla & Chopra, 1985), *Campylopus richardii* (Sharma & Chopra, 1985), *Anoectangium stracheyanum*, and *Bryum atrovirens* (Vashistha, 1985) buds are induced by cytokinins. These mosses remain bud-free on basal medium. In some systems callus masses are formed in response to cytokinins; in still others abnormal or moruloid buds are induced, and these fail to develop into gametophores (Gorton & Eakin, 1957; Iwasa, 1965; Spiess et al., 1972).

In several mosses kinetin interacts with IAA during bud formation: *Splachnum ampullaceum* (von Maltzahn, 1959). *Ceratodon purpureus* (Szweykowska, 1962), *Anoectangium thomsonii* (Chopra & Rashid, 1969a), *Pogonatum aloides* (Sood, 1975), *Bryum klinggraeffii* (Chopra & Kumra, 1978), *Timmiella anomala* (Chopra & Rekhi, 1979), and *Anisothecium molliculum* (Kumra, 1985). Ashton et al. (1979a) demonstrated that in *Physcomitrella patens* both endogenous auxin and cytokinin play interdependent roles in several steps of gametophytic development. In *Dicranella coarctata* normal gametophores are formed only with an interaction of kinetin and IAA (Kumra, 1981). In *Anoectangium* the addition of IAA together with kinetin has a three-fold favourable effect: the time for bud initiation is reduced further; a still larger number of buds appear; and more importantly, the buds develop into normal gametophores (Plate 2.5A,B; Chopra & Rashid, 1969a). Likewise, in *Hyophila involuta* buds are induced only by the interaction of IAA with kinetin or DMAAP, and these develop into normal gametophores (Rahbar & Chopra, 1982). Kinetin also acts synergistically with the endogenous growth regulators in some systems (Chopra & Kumra, 1978). Normally, caulonema is a pre-requisite for bud formation. However, as stated earlier, in some mosses, it has been possible to set aside the requirement of 'critical size', and also bypass the caulonemal stage under the influence of cytokinins. On BAP-supplemented medium buds arose directly from 35 to 40 percent spores in *Entodon myurus* (Plate 2.5C,D; Sood & Chopra, 1973).

The participation of kinins in the process of bud induction is confirmed by the report of endogenous cytokinin 6-(γ, γ-dimethylallylamino) purine, also known as bryokinin or 2iP, from callus cells of the hybrid sporophyte *Funaria hygrometrica* × *Physcomitrium pyriforme* (Beutelmann & Bauer, 1977), and further by the production of mutants in *Physcomitrella* which had lost the capacity to synthesize kinins and could produce buds only with the addition of kinetin to the medium (Ashton & Cove, 1977). Presence of 2iP has also been reported in the protonemata of *Funaria hygrometrica* (see Bopp,

1982) and *Physcomitrella patens* (Wang et al., 1980, 1981 a,b).

It has been demonstrated by several workers that the cytokinin-induced buds can be reverted to protonemal phase if the cytokinin is withdrawn by washing during the process of bud induction. If the protonemata are then transferred to nutrient agar containing cytokinins, bud formation continues normally. This proves that it is the cytokinin itself which is removed from the cells during washing and not some other factor(s) required for bud formation. Furthermore, that cytokinin has to be present for a critical period of time during which differentiation is 'stabilized'. When filaments of *Funaria* are treated with ^{14}C-benzyladenine, the autoradiographs show that the label is accumulated by the 'target cells'. This label is loosely bound in the protonemal cells, since rinsing with water removes the hormone. Thus, the presence of binding sites may be the biochemical basis for the difference between responding and non-responding cells (Brandes & Kende, 1968). Bopp (1982) regards that auxins create target cells and cytokinins act on these cells to produce buds. Kinetin also influences the pattern of distribution of glucose in the protonema. Under the influence of kinetin the labelled sugar accumulates at the apical tips of side branches, which eventually swell and develop into buds (Bopp & Fell, 1976). According to Bopp (1984) the first visible effect of kinetin treatment is an arrest in the lengthwise growth of the side branches of caulonema.

The sensitivity of moss protonemata to cytokinins, and the specific bud induction response has led to the development of a bioassay for some cytokinins using *Funaria hygrometrica* (Hahn & Bopp, 1968; Szweykowska et al., 1970).

Regarding the internal changes accompanying bud induction, it is on record that the total RNA content of bud primordia induced in response to kinetin treatment, or appearing spontaneously, is 15 times (22 μμg/cell) of that in protonemal cells (1.6 μμg/cell) (Schneider et al., 1969). Activities of RNase and DNase decrease during the first 24 h of kinetin treatment, and this might be related to the increase in RNA accompanying bud induction (Schneider & Szweykowska, 1974). It is also reported that kinetin does not increase decomposition of RNase by proteases, but inhibits the de novo synthesis of these enzymes (Spychala et al., 1976). If the cultures are treated with actinomycin-D, an inhibitor of RNA synthesis, there is a complete inhibition of spontaneous and kinetin-induced bud formation (Brandes & Bopp, 1965). In addition to this quantitative increase there is a qualitative change in the composition of RNA. The adenine (A) to guanine (G) ratio of RNA from bud cells is much lower (1.0) than that from the protonemal cells (1.7). Thus, there is an evidence for the activation of a specific part of genetic information during bud initiation (Schneider et al., 1969).

It has also been shown that the absolute values of dry weight and protein content are higher in cultures supplied with kinetin. There is an increase in protein concentration during spontaneous or kinetin-induced bud formation in mosses (Szweykowska & Handszu, 1965). By microgel electrophoretic separation it was initially demonstrated that caulonema has some supplementary proteins which are absent in chloronema, and these were designated as 'caulonema specific proteins' (CSP). It was postulated that CSP are involved in the competence of caulonema cells to respond to cytokinins (Bopp et al., 1978; Erichsen et al., 1978; see also Sood et al., 1978; Sood & Hackenberg, 1979).

However, subsequently the role and even the presence of such proteins has been doubted. Bopp (1980) opined that caulonema specific proteins may be important but are not solely responsible for the target character of the cells. Gardner et al. (1978) failed to notice any such clearly defined proteins in the caulonema cells. In recent experiments Bopp (pers. comm.) has not observed any proteins specific to caulonema (see also Reski & Abel, 1985).

Several cytokinins and cytokinin-like adenine derivatives have been tested for their effectiveness in inducing buds on moss protonemata: 6-benzylaminopurine, mercaptopurine, methylaminopurine, 6-phenylureidopurine, 6-(3-methylbut-2-enylamino) purine, and zeatin. All these induce buds in *Funaria*, but 6-phenylureidopurine is most effective (Szweykowska et al., 1971; Valadon & Mummery, 1971; Szweykowska et al., 1972). The activities of benzyladenine, 6-(3-methylbut-2-enylamino) purine, and zeatin were compared with their corresponding ribosides in promoting bud formation in *F. hygrometrica*. Free cytokinins are highly active in promoting bud formation, whereas the ribosides of the free bases are relatively inactive (Whitaker & Kende, 1974; Spiess, 1975; Gardner et al., 1978). Order of effectiveness of some cytokinins in increasing bud number has been summarized in Table 2.1.

TABLE 2.1.
The order of effectiveness of some cytokinins for bud formation on protonemata of mosses.

Taxa	Order of effectiveness	Investigator/s
Anisothecium molliculum	2iP > Kinetin > BAP > Zeatin	Kumra (1985)
A. spirale	2iP > Zeatin > Kinetin > BAP	Vashistha (1985)
Anoectangium stracheyanum	BAP > 2iP > Kinetin > Zeatin	Vashistha (1985)
A. thomsonii	SD 8339 > 2iP > Kinetin > BAP	Rashid (1968)
Barbula gregaria	Kinetin > BAP	Kumra (1981)
Bartramidula bartramioides	Kinetin > 2iP	Chopra and Rahbar (1982)
Bryum atrovirens	BAP > 2iP > Kinetin > Zeatin	Vashistha (1985)
B. coronatum	BAP > Kinetin	Kumra (1981)
B. pallescens	BAP > 2iP > Kinetin	Sarla and Chopra (1985)
Entodon myurus	2iP > SD 8339 > Kinetin > BAP > Tricanthine	Sood and Chopra (1973)
Pohlia elongata	Kinetin > BAP > Zeatin > 2iP	Vashistha (1985)

Gibberellins

Gibberellins increase the number of buds in some mosses: *Splachnum ampullaceum* (von Maltzahn & MacQuarrie, 1958); *Pohlia nutans* (Mitra & Allsopp, 1959b,c); *Funaria hygrometrica* (Prusińska et al., 1969); *Bryum klinggraeffii* (Chopra & Kumra, 1978); *Barbula gregaria*, *Bryum coronatum* (Kumra, 1981); and *Anisothecium molliculum* (Dua, 1983). Gibberellic acid delays bud initiation in *Funaria hygrometrica* (Jahn, 1964) and *Bryum coronatum* (Kumra, 1981). However, it reduces the time required for bud induction in *Pohlia nutans* (Mitra & Allsopp, 1959b,c). When IAA and GA_3 are supplied together there is an increase in the number of buds in *Anoectangium thomsonii* (Rashid, 1970). Sood (1972) observed that in *Pogonatum aloides*, GA_3 failed to induce buds. However, the addition of GA_3 to kinetin + IAA medium enhanced bud number. In *Anoectangium thomsonii* (Rashid, 1970) and *Bryum atrovirens* (Vashistha, 1985) GA_3 induces buds. Some endogenous gibberellin-like substances have also been reported in *Polytrichum commune* (Muromtsev et al., 1964).

Adenosine 3':5'-Cyclic Monophosphate

The occurrence and regulatory role of cAMP in animals and microorganisms is well documented, but in bryophytes its endogenous presence is still debated (Kaul & Sachar, 1982). Exogenously supplied cAMP is either ineffective or proves inhibitory for bud induction. In some systems, a synergism is noticed between kinetin and cAMP in respect of increase in the number of buds. In *Funaria hygrometrica* and *Ceratodon purpureus* this effect seems to be non-specific, since it can be obtained with 5'-AMP or 5'-GMP as well (Schneider et al., 1975). In *Timmiella anomala* cAMP inhibits bud formation, but stimulates growth of gametophores. Cyclic AMP in combination with kinetin further increases bud number and also the length of gametophores. This effect appears to be specific, since it is not observed when kinetin is added along with 2', 3'-cAMP or 5'-AMP (Kapur & Chopra, 1983). In *Microdus brasiliensis* cAMP reduces the number of buds, and gametophores remain stunted. Co-addition of kinetin and cAMP increases the number of buds and improves the morphology of gametophores (Chopra & Sharma, 1984). The combination of cAMP and kinetin induces buds in *Dicranella coarctata* which otherwise remains bud-free on kinetin or cAMP-supplemented medium. Cyclic AMP also helps in the development of buds into normal gametophores in *Barbula gregaria* and *Bryum coronatum* (Kumra, 1981).

Adenine and Amino Acids

Adenine increases the number of buds and reduces the time taken for bud induction in *Tortella caespitosa* (Gorton & Eakin, 1957), *Leptobryum pyriforme* (Rawat, 1976), *Bryum coronatum*, *B. klinggraeffii*, and *Barbula gregaria* (Chopra & Kumra, 1978; Kumra, 1981). In *Physcomitrium coorgense* (Lal, 1961) and *Timmiella anomala* (Kapur & Chopra, 1983) adenine increases bud number, but does not affect the time of their initiation. It also induces buds in *Anoectangium thomsonii* (Rashid, 1970). *Dicranella coarctata* (Kumra, 1981), *Hyophila involuta* (Rahbar, 1983), and *Bryum atrovirens* (Vashistha, 1985) all of which fail to produce buds on basal medium. In *Funaria hygrometrica* (Overlach, 1976) and *Timmiella anomala* (Kapur & Chopra, 1983) adenosine stimulates bud formation, whereas in *Hyophila involuta* (Rahbar, 1983), *Anoectangium stracheyanum* and *Bryum atrovirens* (Vashistha, 1985) it induces buds.

Among amino acids, lysopine and octopine are stimulatory for bud induction in *Pylaisiella selwynii* (Spiess et al., 1981). Arginine favours vigorous development of leafy shoots from buds in some mosses (Burkholder, 1959; Chopra & Kumra, 1978).

Minerals and Chelates

For bryophytes the macronutrients: K, Ca, Mg, N, P, and S are essential for growth. It has been demonstrated that morphogenesis in *Funaria hygrometrica* is controlled indirectly by an appropriate concentration of calcium ions (Iwasa, 1965). The protonema of *Sphagnum* produces a thallus only when the medium is supplied with phosphate. Thallus size diminishes as phosphate concentration is decreased (Boatman & Lark, 1971). In *Pylaisiella selwynii* removal of cobaltous ions from the culture medium prevents the appearance of buds (Spiess et al., 1973).

Chelating agents increase the number of buds in some mosses which normally produce buds in cultures: *Barbula gregaria, Bryum coronatum* (Kumra, 1981), *Bartramidula bartramioides* (Rahbar & Chopra, 1983), and *Timmiella anomala* (Kapur, 1983). In addition, buds are induced by Fe-EDDHA (Chopra & Rashid, 1969b), SA or its acetyl derivative aspirin (Saxena & Rashid, 1980) in *Anoectangium thomsonii,* and by EDTA (Chopra & Sarla, 1985) in *Bryum pallescens* under non-inductive conditions. In *Pogonatum aloides* Fe-EDDHA fails to induce buds, but when added together with kinetin and IAA buds are initiated and their number becomes two-fold (Sood, 1972). The chelates also promote vegetative growth, and buds develop into normal shoots. It is considered that the effect of chelates is possibly because of an increased availability of iron. Increased concentration of ferric citrate and cupric sulphate also stimulates bud induction. The accumulation of Fe^{3+} and Cu^{2+} is facilitated by chelators. The endogenous iron content is maximum at $10^{-7}M$ EDDHA or EDTA, which is also the concentration optimal for bud induction in *Bartramidula bartramioides* (Rahbar & Chopra, 1983).

Vitamins

Effect of vitamins has also been studied. In *Pylaisiella selwynii* bud initiation is enhanced 3 to 20-fold by vitamin B_{12} at $10^{-5}M$ or by B_{12} coenzyme at $10^{-4}M$, and the time of appearance of these buds is reduced by six to 12 days as compared to control plants. At the above concentrations of B_{12} compounds, the buds formed normal gametophores (Spiess et al., 1973). Similarly in *Barbula gregaria* and *Bryum coronatum* vitamin B_{12} and B_1 reduce the time taken for bud induction and increase their number. In addition to this, both of these vitamins also induce buds in *Dicranella coarctata* which does not form buds on basal medium (Kumra, 1981). In *Timmiella anomala* vitamins B_1, B_2, and B_6 increase bud number as well as advance their initiation (Kapur, 1983). In *Bartramidula bartramioides* vitamin B_{12} enhances bud number but does not reduce the time required for their initiation (Rahbar, 1981). In *Bryum atrovirens* vitamin B_{12} induces buds (Vashistha, 1985).

Abscisic Acid

Valadon and Mummery (1971) reported that in *Funaria hygrometrica* ABA specifically lowers the number of buds produced in response to kinetin treatment. An inverse quantitative relationship was observed between the concentration of ABA and the number of kinetin-induced buds. Chopra and Rahbar (1982) observed that in *Bartramidula bartramioides* ABA inhibits bud induction and the inhibition is directly related to its concentration.

Reports to the contrary are also on record. In *Barbula gregaria* and *Bryum coronatum* ABA reduces the time taken for bud formation and increases their number. The optimum concentration for the two mosses being 10^{-7} and 10^{-6}M, respectively. Abscisic acid induces buds in *Dicranella coarctata* which remains bud-free on basal medium (Kumra, 1981).

Sugars

It has been demonstrated that moss protonemata when maintained in dark or in feeble light, continue to grow at the expense of sugar in the medium (Pringsheim & Pringsheim, 1935; Robbins, 1918; Servettaz, 1913; Ubisch, 1913). Sucrose and glucose, at moderate concentrations (0.5 to 2%) hasten bud induction and increase their number in *Pohlia nutans* (Mitra & Allsopp, 1959a). In *Funaria hygrometrica* sugars help in attaining the critical size of protonema (Bopp & Brandes, 1964). Glucose also induces buds on a mutant protonema of *Funaria* (Hatanaka-Ernst, 1966). In some mosses like *Bryum klinggraeffii, Leptobryum pyriforme* (Rawat, 1976), *Barbula gregaria* and *Bryum coronatum* (Kumra, 1981), absence of sucrose in the medium delays bud formation, whereas its addition enhances protonemal growth and increases the number of buds.

pH

It has been observed that pH of the medium also influences protonemal growth and bud initiation. At very low pH values, protonemal growth is inhibited in *Tetraplodon mnioides, Funaria hygrometrica* (Armentano &

Caponetti, 1972), and *Entodon myurus* (Sood, 1972). However, with increasing pH protonemal growth is enhanced and it is maximum at 7.8 in *Tetraplodon* and *Funaria*, and at 7.5 in *Entodon*. On the other hand in *Bryum klinggraeffii* (Rawat, 1976), *Barbula gregaria*, *Bryum coronatum* (Kumra, 1981), and *Anisothecium molliculum* (Kumra & Chopra, 1985) very low and very high pH values inhibit protonemal growth as well as bud formation. When different populations of *Funaria hygrometrica* and *Weissia controversa* are grown on varying pH values in buffered media, *Funaria* shows a clear preference for pH 8.0 to 8.5, whereas *Weissia* grows best at values between pH 7.5 to 8.0. However, the pH optimum for protonemal growth is different from that for gametophore formation. In acidic conditions the gametophores are numerous and more robust, whereas on the alkaline media they are progressively smaller and fewer (Dietert, 1979).

Influence of other Organisms

Fungi occurring as contaminants in moss cultures promote growth of protonemata and also stimulate bud formation. In some systems their effect is similar to that of GA_3. In *Pylaisiella selwynii* a virulent strain of *Agrobacterium tumefaciens* (B_6) brings about an early induction of buds on the protonema, and these buds develop into normal gametophores. The promotion of gametophore formation is directly related to the number of viable bacteria added. The exudates of *Agrobacterium* are reported to contain 2iP (Upper et al., 1970). Agrobacteria also synthesize auxin (Nester & Kosuge, 1981) as do the rhizobia (Triplett et al., 1981). *Rhizobium* spp., which are known to produce zeatin or ribosyl-zeatin (Nester & Kosuge, 1981), also induce normal gametophores in *Pylaisiella selwynii* (Spiess et al., 1971, 1975). *Rhodotorula* and *Aerobacter* enhance protonemal growth, hasten bud initiation, and increase the number of buds in *Barbula gregaria* and *Timmiella anomala*. In these respects the effects are comparable with those of cytokinins (Chopra et al., 1978). In *Philonotis glaucesens* growth is more prolific when mycorrhizal fungi (*Glomus mosseae*, *Gigaspora gigantea*, & *G. margarita*) are incorporated into the medium (Pokorny & Hendrix, cited in Parke & Linderman, 1980).

REFERENCES

ALLSOPP, A. and G.C. MITRA, *Nature*, **178**, 1063 (1956).
ALLSOPP, A. and G.C. MITRA, *Ann. Bot.*, **22**, 95 (1958).
ARMENTANO, T.V. and J.D. CAPONETTI, *Bryologist*, **75**, 147 (1972).
ASHTON, N.W. and D.J. COVE, *Mol. Gen. Genet.*, **154**, 89 (1977).
ASHTON, N.W., N.H. GRIMSLEY, and D.J. COVE, *Planta*, **144**, 427 (1979a).
ASHTON, N.W., D.J. COVE, and D.R. FEATHERSTONE, *Planta*, **144**, 437 (1979b).
BELKENGREN, R.O., *Am. J. Bot.*, **49**, 567 (1962).
BEUTELMANN, P. and L. BAUER, *Planta*, **133**, 215 (1977).
BHATLA, S.C., *Beitr. Biol. Pfl.*, **57**, 157 (1982).

BHATLA, S.C. and M. BOPP, *J. Plant Physiol.*, **120**, 233 (1985).
BOATMAN, D.J. and P.M. LARK, *New Phytol.*, **70**, 1053 (1971).
BOPP, M., *Ber. dt. bot. Ges.*, **67**, 176 (1954).
BOPP, M., *J. Linn. Soc. (Bot.)*, **58**, 305 (1963).
BOPP, M., *Ber. dt. bot. Ges.*, **78**, 44 (1965).
BOPP, M., *Ann. Rev. Pl. Physiol.*, **19**, 361 (1968).
BOPP, M., *J. Hattori bot. Lab.* No. 41, 167 (1976).
BOPP, M., *Ber. dt. bot. Ges.*, **92**, 323 (1979).
BOPP, M., "The hormonal regulation of morphogenesis in mosses." In F. Skoog, Ed., *Plant Growth Substances*, Springer-Verlag, Berlin, Heidelberg, 1980. Pp. 351-361.
BOPP, M., *J. Hattori bot. Lab.* No. 43, 159 (1982).
BOPP, M., *Z. PflPhysiol.*, **113**, 435 (1984).
BOPP, M. and H.-L. BÖHRS, *Planta*, **67**, 357 (1965).
BOPP, M. and H. BRANDES, *Planta*, **62**, 116 (1964).
BOPP, M. and J. FELL, *Z. PflPhysiol.*, **79**, 81 (1976).
BOPP, M. and B. KLEIN, *Port. Acta biol. Ser. A.*, **7**, 95 (1963).
BOPP, M., U. ERICHSEN, M. NESSEL, and B. KNOOP, *Physiol. Plant.*, **42**, 73 (1978).
BRANDES, H. and M. BOPP, *Naturwissenschaften*, **52**, 521 (1965).
BRANDES, H. and H. KENDE, *Pl. Physiol.*, **43**, 827 (1968).
BURKHOLDER, P.R., *Bryologist*, **62**, 6 (1959).
CHOPRA, R.N., "Some aspects of morphogenesis in bryophytes". In D.C. Bharadwaj, Ed., *Recent Advances in Cryptogamic Botany*, the Palaeobotanical Society, Lucknow, India, 1981. Pp. 190-201.
CHOPRA, R.N. and SADHANA DHINGRA-BABBAR, *New Phytol.*, **97**, 613 (1984).
CHOPRA, R.N. and URMILLA GUPTA, *Bryologist*, **70**, 102 (1967).
CHOPRA, R.N. and P.K. KUMRA, *Phytomorphology*, **28**, 298 (1978).
CHOPRA, R.N. and KAVITA RAHBAR, *Beitr. Biol. Pfl.*, **57**, 181 (1982).
CHOPRA, R.N. and A. RASHID, *Z. PflPhysiol.*, **61**, 192 (1969a).
CHOPRA, R.N. and A. RASHID, *Z. PflPhysiol.*, **61**, 199 (1969b).
CHOPRA, R.N. and ANITA REKHI, *Phytomorphology*, **29**, 179 (1979).
CHOPRA, R.N. and SARLA, *J. Bryol.*, **13**, 423 (1985).
CHOPRA, R.N. and POONAM SHARMA, *J. Plant Physiol.*, **117**, 293 (1984).
CHOPRA, R.N., P.K. KUMRA, and ANITA REKHI, *Curr. Sci.*, **47**, 735 (1978).
COVE, D.J., A. SCHILD, N.W. ASHTON, and E. HARTMANN, *Photochem. Photobiol.*, **27**, 249 (1978).
DIETERT, M.F., *Bryologist*, **82**, 417 (1979).
DUA, SUMAN, "Morphogenetic and physiological studies on some bryophytes," Ph.D. Thesis, Univ. Delhi, India, 1983.
ELTZ, H.V., *Nativer Morphoregulator (Faktor H) aus Moosprotonemen*, Staatesxamensarbeit, Heidelberg, 1975.
ERICHSEN, U., B. KNOOP, and M. BOPP, *Pl. Cell Physiol.*, **19**, 839 (1978).
FRIES, N., *Bot. Notiser*, **98**, 417 (1945).
GARDNER, G., M.R. SUSSMAN, and H. KENDE, *Planta*, **143**, 67 (1978).
GOEBEL, K., *S.B. bayer. Akad. Wiss.*, **26**, 447 (1896).
GORTON, B.S. and R.E. EAKIN, *Bot. Gaz.*, **119**, 31 (1957).
HAHN, M. and M. BOPP, *Z. PflPhysiol.*, **83**, 115 (1968).
HANDA, A.K. and M.M. JOHRI, *Nature*, **259**, 480 (1976).
HANDA, A.K. and M.M. JOHRI, *Pl. Physiol.*, **59**, 490 (1977).
HANDA, A.K. and M.M. JOHRI, *Planta*, **144**, 317 (1979).
HATANAKA-ERNST, M., *Z. PflPhysiol.*, **55**, 259 (1966).
HUREL-PY, G., *C.r. hebd. Séanc. Acad. Sci., Paris*, **227**, 1256 (1948).
IWASA, K., *Pl. Cell Physiol.*, **6**, 421 (1965).
JAHN, H., *Flora*, **155**, 10 (1964).
JOHRI, M.M. and S. DESAI, *Nat. New Biol.*, **245**, 223 (1973).
KAPUR, ANITA, "In vitro studies on some bryophytes", Ph.D. Thesis, Univ. Delhi, India, 1983.

KAPUR (née REKHI), ANITA and R.N. CHOPRA, *New Phytol.*, 94, 393 (1983).
KAUL, RATNUM and R.C. SACHAR, *Biochem. Biophys. Res. Comm.*, 104, 126 (1982).
KEIL, M., *Experientia*, 5, 206 (1949).
KLEBS, G., *Biol. Zbl.*, 13, 641 (1893).
KLEIN, B., *Planta*, 73, 12 (1967).
KUMRA, P.K., "Morphogenetic and physiological studies on some mosses," Ph.D. Thesis, Univ. Delhi, India, 1981.
KUMRA, SUMAN, *J. Hattori bot. Lab.* No. 59, 279 (1985).
KUMRA, SUMAN and R.N. CHOPRA, *Phytomorphology*, 35, 223 (1985).
KOFLER, LUCIE, *C.r. hebd. Séanc. Acad. Sci., Paris*, 242, 1755 (1956).
KOFLER, LUCIE, *Rev. Bryol. Lichénol.*, 28, 1 (1959).
LAL, M., "Experimental investigations on the moss *Physcomitrium*," Ph.D. Thesis, Univ. Delhi, India, 1961.
LEHNERT, B. and M. BOPP, *Z. PflPhysiol.*, 110, 379 (1985).
VON MALTZAHN, K.E., *Nature*, 183, 60 (1959).
VON MALTZAHN, K.E. and I.G. MACQUARRIE, *Nature*, 181, 1139 (1958).
MENON, M.K.C., "Morphogenetic studies on apogamy in the moss *Physcomitrium*", Ph.D. Thesis, Univ. Delhi, India, 1974.
MITRA, G.C. and A. ALLSOPP, *Phytomorphology*, 9, 55 (1959a).
MITRA, G.C. and A. ALLSOPP, *Nature*, 183, 794 (1959b).
MITRA, G.C. and A. ALLSOPP, *Phytomorphology*, 9, 64 (1959c).
MITRA, G.C., A. ALLSOPP, and P.F. WAREING, *Phytomorphology*, 9, 47 (1959).
MITRA, G.C., L.P. MISRA, and PRABHA CHANDRA, *Planta*, 65, 42 (1965).
MUROMTSEV, G.S., V.N. AGNISTIKOVA, L.M. LUPOVA, L.P. DUBOVAYA, and T.A. LEKAREVA, *Invest. Akad. nauk. SSSR. Ser. Biol. Moscow* 5, 727 (1964).
NAEF, J. and P. SIMON, "Photoregulation of the development of *Funaria hygrometrica* protonema combined with a cytokinin treatment," In J. De Greef, Ed., *Photoreceptors and Plant Development*, Antwerpen University Press, 1980. Pp. 399-404.
NAIR, H. and V. RAGHAVAN, *Bryologist*, 79, 495 (1976).
NEBEL, B.J. and A.W. NAYLOR, *Am. J. Bot.*, 55, 38 (1968).
NEHIRA, K., *J. Hattori. bot. Lab.* No. 41, 157 (1976).
NESTER, E.W. and T. KOSUGE, *Ann. Rev. Microbiol.*, 35, 531 (1981).
OEHLKERS, F., *Das Lenen der Gewachse*, Springer-Verlag, Berlin, Gottingen-Heidelberg, 1956.
OVERLACH, U., Untersuchungen über die Wirkung, den Einbau und die Metabolisierung von Cytokininen im Moosprotonema. Dissertation, Heidelberg, 1976.
PARKE, J.L. and R.G. LINDERMAN, *Can. J. Bot.*, 58, 1898 (1980).
PRINGSHEIM, E.G. and O. PRINGSHEIM, *Jb. wiss. bot.*, 82, 311 (1935).
PRUSINSKA, U., J. SCHNEIDER, and A. SZWEYKOWSKA, *Bull. Soc. Amis. Sci. Lett. Poznan Sér. D.*, 9, 3 (1969).
RAHBAR, KAVITA, "Morphogenetical and physiological studies on *Bartramidula bartramioides* Schimp. and *Hyophila involuta* (Hook.) Jaeg.," Ph.D. Thesis, Univ. Delhi, India, 1981.
RAHBAR, KAVITA, *Cryptog. Bryol. Lichénol.*, 4, 227 (1983).
RAHBAR, KAVITA and R.N. CHOPRA, *New Phytol.*, 91, 501 (1982).
RAHBAR, KAVITA and R.N. CHOPRA, *Physiol. Plant.*, 59, 148 (1983).
RASHID, A., "Morphogenetic studies on some Indian bryophytes," Ph.D. Thesis, Univ. Delhi, India, 1968.
RASHID, A., *Phytomorphology*, 20, 49 (1970).
RAWAT, M.S., "Morphogenic studies on some Bryaceae," Ph.D. Thesis, Univ. Delhi, India, 1976.
RAWAT, M.S. and R.N. CHOPRA, *Z. PflPhysiol.*, 78, 372 (1976).
RESKI, R. and W.O. ABEL, *Planta*, 165, 354 (1985).
ROBBINS, W.J., *Bot. Gaz.*, 65, 542 (1918).
SARLA and R.N. CHOPRA, *J. Bryol.*, 13, 429 (1985).

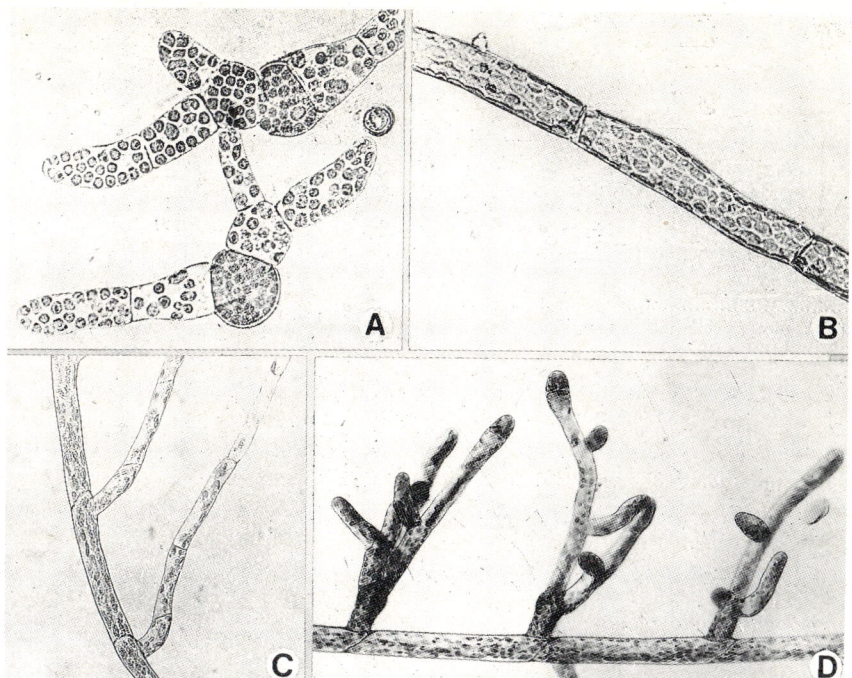

Plate 2.1. A-D. Spore germination and protonemal differentiation in mosses. **A.** Chloronemal filaments produced from germinating spores. The cells have rounded chloroplasts, and transverse septae. **B.** Portion of a chloronemal filament at a later stage. **C.** Young caulonemal filaments. **D.** Typical caulonemal filaments showing heterotrichous habit.
(**A.** After Bhatla, 1982; **B.** After Menon, 1974; **C.** After Chopra & Rekhi, 1979; **D.** From a preparation).

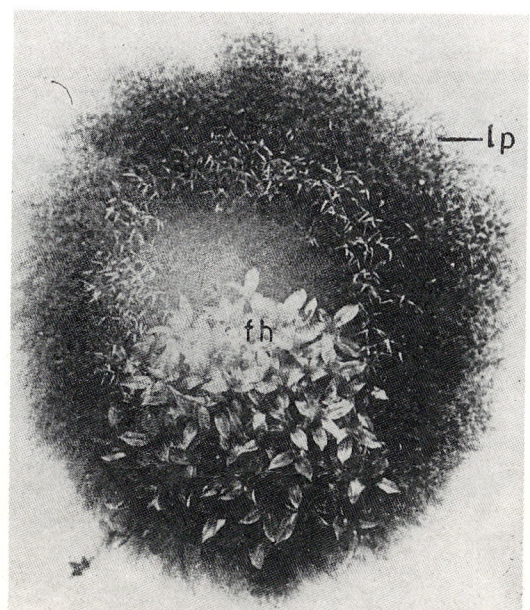

Plate 2.2. Protonema of *Leptobryum pyriforme* (lp) grown together with that of *Funaria hygrometrica* (fh). *Leptobryum* produces a large amount of Factor H which stimulates bud formation in *Funaria*.
(After Bopp, 1976).

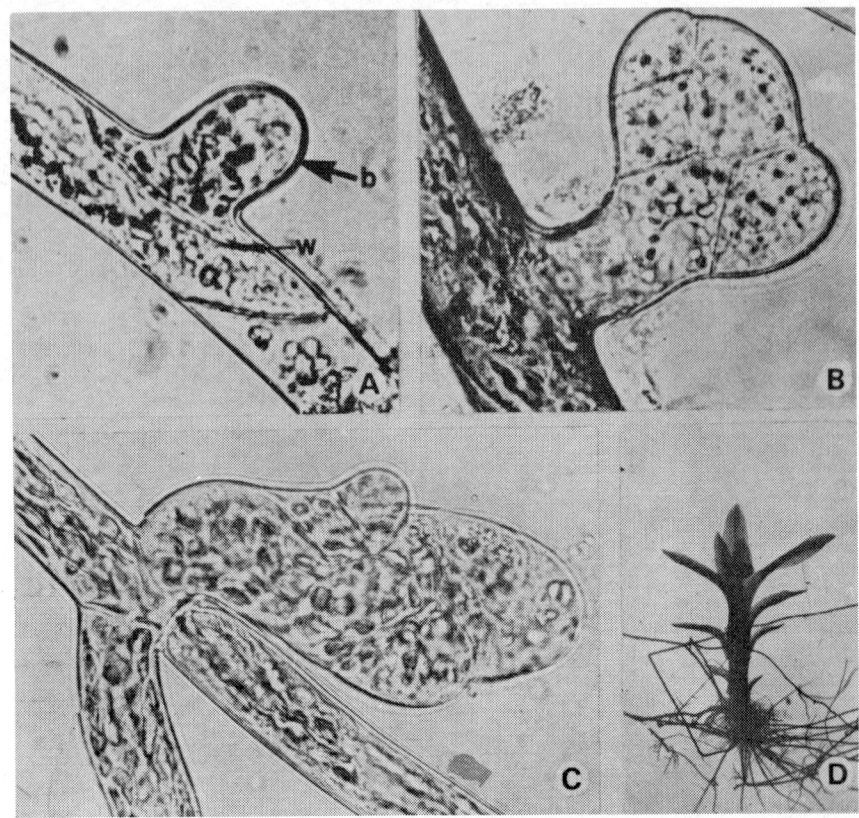

Plate 2.3 A-D. Bud initiation and gametophyte development in mosses.
A. A bud initial (b) on a caulonemal filament. A wall (w) separates the initial from the caulonema cell. **B, C.** Gametophytic buds at different stages of development. **D.** A young gametophore.
(**A, B.** After Saunders & Hepler, 1981; **C.** After Dua, 1983; **D.** After Chopra & Rekhi, 1979).

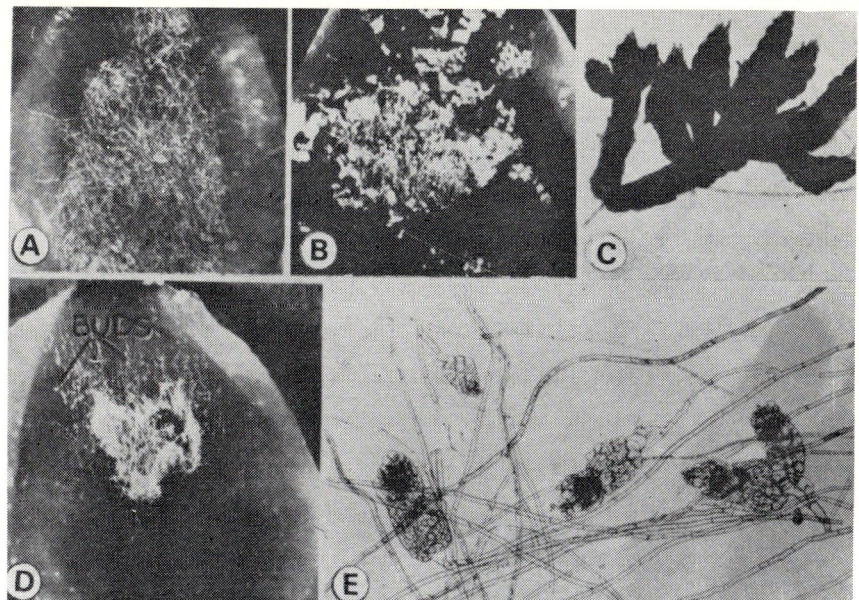

Plate 2.4 A-E. Dark-induction of buds in *Funaria hygrometrica*. **A.** Culture raised from spores in dark on medium containing sucrose (2%); buds failed to initiate. **B.** Culture on medium containing sucrose (2%) and kinetin (1 ppm); abundant gametophores (cottony white pinheads). **C.** Culture on medium containing sucrose (2%) and kinetin (0.5 ppm); branched gametophore. **D.** Culture on medium containing only kinetin (1 ppm); protonema with tiny buds. **E.** Protonema bearing buds (from culture in **D**). All photographs from five-month-old cultures.
(After Chopra & Gupta, 1967).

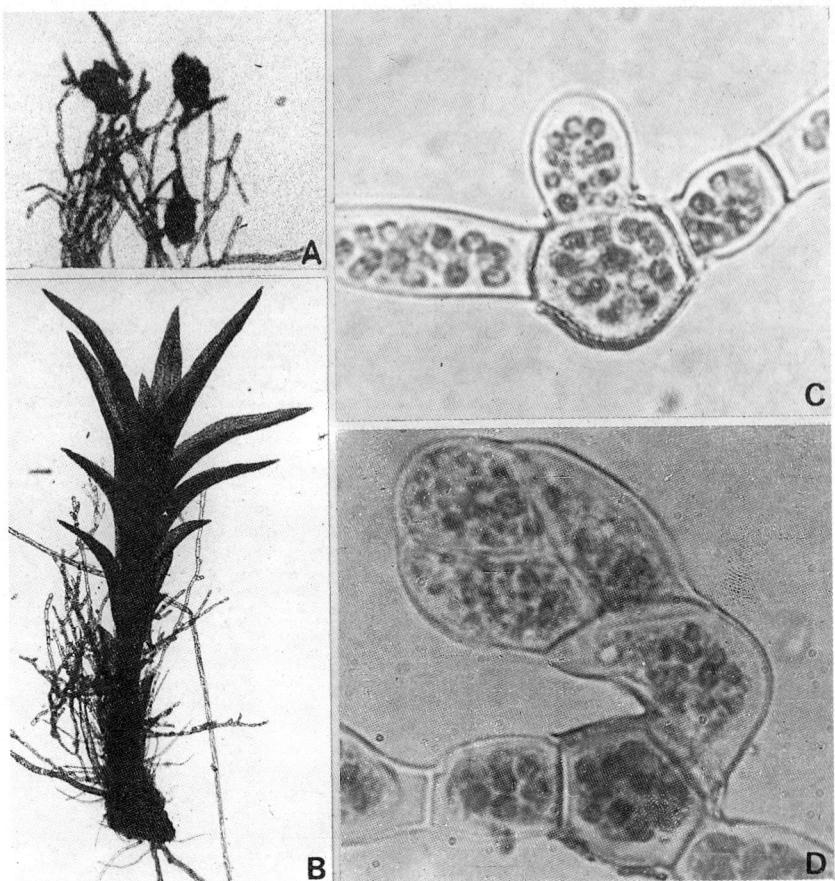

Plate 2.5 A-D. Bud formation in mosses under the influence of growth substances in light. **A, B.** Interaction between kinetin and IAA for bud formation in *Anoectangium thomsonii*. **A.** Stunted shoots developed on medium containing kinetin (2.0 ppm). **B.** Shoots developed on medium containing IAA (0.05 ppm) together with kinetin (0.5 ppm). Note the improvement in the morphology of shoots. **C, D.** Initiation of buds in *Entodon myurus*. **C.** A bipolar sporeling from a 15-day-old culture on medium containing kinetin (5 ppm). A bud initial appeared directly from the spore. **D.** A 15-day-old sporeling from kinetin (5 ppm) medium. Note the direct origin of bud from the spore.
(**A, B.** After Chopra & Rashid, 1969a; **C, D.** After Sood & Chopra, 1973).

SAUNDERS, M.J. and P.K. HEPLER, Planta, 152, 272 (1981)
SAXENA, P.K. and A. RASHID, Z. PflPhysiol., 99, 187 (1980).
SCHNEIDER, J. and A. SZWEYKOWSKA, Z. PflPhysiol., 72, 95 (1974).
SCHNEIDER, M.J., J.C.J. LIN, and F. SKOOG, Pl. Physiol., 44, 1207 (1969).
SCHNEIDER, J., A. SZWEYKOWSKA, and M. SPYCHALA, Acta Soc. Bot. Pol., 64, 607 (1975).
SERVETTAZ, C., Annls. Sci. nat. (Bot.), 17, 111 (1913).
SHARMA, POONAM and R.N. CHOPRA, Cryptog. Bryol. Lichénol., 6, 349 (1985).
SHARMA, S., R.K. JAYASWAL, and M.M. JOHRI, Pl. Physiol., 64, 154 (1979).
SIRONVAL, C., Bull. Soc. r. Bot. Belg., 79, 48 (1947).
SOOD, SNEH, "Experimental studies on some Indian mosses," Ph.D. Thesis, Univ. Delhi, India. 1972.
SOOD, SNEH, Beitr. Biol. Pfl., 51, 99 (1975).
SOOD, SNEH and R.N. CHOPRA, Z. PflPhysiol., 69, 390 (1973).
SOOD, SNEH and D. HACKENBERG, Z. PflPhysiol., 91, 385 (1979).
SOOD, SNEH, K. BRENNER, and M. BOPP, Planta, 138, 299 (1978).
SPIESS, L.D., Pl. Physiol., 55, 583 (1975).
SPIESS, L.D., Bryologist, 82, 47 (1979).
SPIESS, L.D., B.B. LIPPINCOTT, and J.A. LIPPINCOTT, Am. J. Bot., 58, 726 (1971).
SPIESS, L.D., B.B. LIPPINCOTT, and J.A. LIPPINCOTT, Am. J. Bot., 59, 233 (1972).
SPIESS, L.D., B.B. LIPPINCOTT, and J.A. LIPPINCOTT, Am. J. Bot., 60, 708 (1973).
SPIESS, L.D., B.B. LIPPINCOTT, and J.A. LIPPINCOTT, Abstr. 15 Botanical Society of America, 1975.
SPIESS, L.D., B.B. LIPPINCOTT, and J.A. LIPPINCOTT, Physiol. Plant., 51, 99 (1981).
SPYCHALA, M., I. KORCZ-ZAJCHERT, and A. SZWEYKOWSKA, Acta Soc. Bot. Pol., 55, 327 (1976).
SZWEYKOWSKA, A., Acta Soc. Bot. Pol., 31, 553 (1962).
SZWEYKOWSKA, A., J. exp. Bot., 14, 137 (1963).
SZWEYKOWSKA, A. and A. HANDSZU, Acta Soc. Bot. Pol., 34, 73 (1965).
SZWEYKOWSKA, A., T. MAĆKOWIAK, and U. PRUSIŃSKA, Zes. nauk. Uniw. Mikolaja Kopernica Torun., 23, 289 (1970).
SZWEYKOWSKA, A., E. DORNOWSKA, A. CYBULSKA, and G. WASIEK, Biochem. Physiol. Pflanzen., 162, 514 (1971).
SZWEYKOWSKA, A., I. KORCZ, B. JASKIEWICZ-MROCZKOWSKA, and M. METELSKA, Acta Soc. Bot. Pol., 41, 401 (1972).
TRIPLETT, E.W., J.J. HEITHOLT, K.B. EVENSEN, and D.G. BLEVINS, Pl. Physiol., 67, 1 (1981).
UBISCH, G. VON, Ber. dt. bot. Ges., 31, 543 (1913).
UPPER, C.D., J. HELGESON, J. KEMP, and C. SCHMIDT, Pl. Physiol., 45, 543 (1970).
VALADON, L.G.R. and R.S. MUMMERY, Physiol. Plant., 24, 232 (1971).
VAN ANDEL, Trans. Br. bryol. Soc., 2, 74 (1952).
VASHISTHA, B.D., "In vitro investigations on some Indian bryophytes," Ph.D. Thesis, Univ. Delhi, India, 1985.
WANG, T.L., D.J. COVE, P. BEUTELMANN, and E. HARTMANN, Phytochemistry, 19, 1103 (1980).
WANG, T.L., P. BEUTELMANN, and D.J. COVE, Pl. Physiol., 68, 739 (1981a).
WANG, T.L., R. HORGAN, and D.J. COVE, Pl. Physiol., 68, 735 (1981b).
WHITAKER, B.D. and H. KENDE, Planta, 121, 93 (1974).
YOSHIDA, K. and K. YAMAMOTO, Pl. Cell Physiol., 23, 737 (1982).
YOSHIDA, K. and K. YAMAMOTO, Pl. Cell Physiol., 26, 1549 (1985).

3 Regeneration

REGENERATION may be defined as restoration by an organism of tissues or organs which have been removed or damaged. It can also mean the capacity of an organ, tissue, or cell to give rise to the entire organism. Normally, regeneration of differentiated cells follows injury or isolation from the meristem (Stange, 1957; Wilmot-Dear, 1980). Isolation may be physical or physiological. Both injury and isolation disturb polarity in the plant and bring about loss of normal apical dominance.

Studies on regeneration play an important role in understanding the problems of morphogenesis. The initiation of experiments on regeneration of bryophytes dates back to the 18th century (see Giles, 1971). In the early days the interest in using mosses and liverworts for such investigations was due to the fact that these plants readily regenerate under ordinary conditions, and secondly because regenerants can be obtained from haploid as well as diploid generations. The gametophyte may produce new gametophytes or give rise to apogamous sporophytes. The sporophytes may bud out new sporophytes or may produce gametophytes aposporously (Fig. 5.6 A, B; Bopp, 1983; Kumra & Chopra, 1980). Because of these two features, experiments could be designed for understanding the processes involved in normal plant growth and development. Since regeneration prevents the normal course of events in a mature cell, i.e., ageing, information can be had about senescence. When single cells are involved, totipotency can be demonstrated. Finally, studies on experimental induction of apogamy and apospory help us to obtain polyploid plants, and also to understand the mechanisms underlying the completely different courses of differentiation in the two alternating generations (see Stange, 1964).

Regeneration should be distinguished from reactivation. In the former, mature cells de-differentiate and this is followed by re-differentiation of the products of their multiplication resulting in new plants. In the latter, well organized axillary or adventitious buds grow out when the influence of apical dominance is removed. Bünning(1955) used the term 'restitution' for the reestablishment of original constitution of the plant after interference. MacQuarrie and von Maltzahn (1959) distinguished three types of restitution in *Splachnum*:

    (1) *Regeneration:* it comprises events involving de-differentiation and re-differentiation;

    (2) *Reactivation:* it is the process in which the arrested development of embryonic cells is released by removal of correlative inhibition;

(3) *Reparation:* it involves remolding and growth which directly results in complete or partial restoration of the original form.

Stange (1964) has used the term 'embryonization' for regeneration in *Riella*, and has defined it as the process in which a cell gives up its special functional position and again enters the cycle of embryonic functions.

With the availability of new tools and techniques emphasis has shifted from observations on morphology and anatomy, to studies on metabolic and biochemical changes occurring in the regenerating cells. Such studies have given us a clearer picture of morphogenesis in some systems. The literature on regeneration is very vast, and an attempt has been made to bring the main findings and concepts together. This chapter deals with potentialities of various organs for regeneration, morphology of regenerants, factors which influence this process, and changes occurring in regenerating cells.

POTENTIALITIES OF VARIOUS ORGANS FOR REGENERATION

Species which normally produce special reproductive bodies like tubers and gemmae are also capable of regenerating from unspecialized organs like leaf and stem. Regeneration results in vegetative propagation which ensures survival of the species in the absence of sexual reproduction. Great ability of some sterile mosses for such reproduction is responsible for their wide geographic range.

Schuster (1980) opines that the high regeneration capacity of bryophytes can be regarded as a primitive character. The formation of specialized types of gemmae, cladia, and other such devices for reproduction increased as evolution proceeded. As example has been cited the family Lejeuneaceae in which the frequency of asexual propagula has increased in the advanced members. In nature regeneration is more common in mosses than in liverworts, but in both the groups many taxa remain sterile and also lack specialized propagula, and therefore in such taxa regeneration is the only method of reproduction. Practically all parts of the gametophyte and sporophyte are capable of regeneration under favourable conditions (Table 3.1).

TABLE 3.1.

Some specific examples of regeneration from organs other than thallus, protonema, stem, leaf, and seta.

Organ of the plant	Taxa	Investigator/s
Antheridial bracts and involucral bracts	*Fossombronia* and *Sewardiella*	Mehra (1976)
Antheridial stalk and jacket	*Bryum cellulare*	Narayanaswami and Lal (1957)
Archegonial neck	*Mnium*	Wettstein (1924a, cited in Ono & Tanaka, 1976).
Archegonial stalk	*Rhodobryum*	Narayanaswami and Lal (1957)

(Table contd.)

Organ of the plant	Taxa	Investigator/s
Archegonial venter	*Funaria*	Correns (1899)
Calyptra	*Barbula* and *Physcomitrium cyathicarpum*	Narayanaswami and Lal (1957), Fig. 3.1 A
	Sewardiella	Mehra (1976)
Capsule jacket	*Physcomitrium cyathicarpum*	Narayanaswami and Lal (1957)
	Funaria	Kumra and Chopra (1980)
Paraphyses	*Bryum capillare*, *Aulacomnium palustre*, and *Funaria*	Reese (1955)
Perianth	*Fossombronia* and *Petalophyllum*	Mehra (1976) Mehra and Vashisht (1950)
Perichaetial and perigonial leaves	*Octoblepharum*	Egunyomi et al. (1980)
Rhizoids	*Physcomitrium cyathicarpum* and *P. coorgense*	Narayanaswami and Lal (1957)
Vaginula	*Barbula, Physcomitrium cyathicarpum,* and *P. coorgense*	Narayanaswami and Lal (1957), Fig. 3.1 B

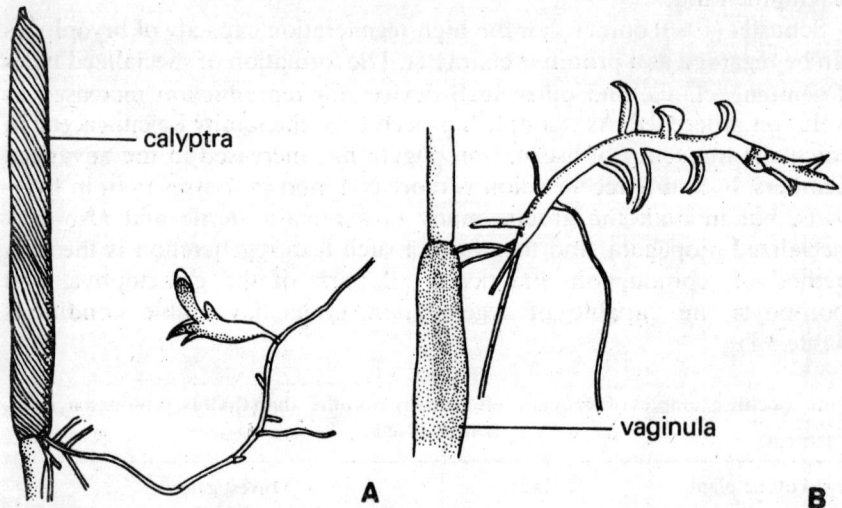

Fig. 3.1 A,B. Regeneration in *Barbula indica* from calyptra and vaginula, respectively. (After Narayanaswami & Lal, 1957).

REGENERATION FROM LEAVES

In detached leaves of mosses secondary protonemata have been observed to arise from different regions. Examples of regeneration occurring mostly or entirely from base are *Brachythecium, Tortella* (Heald, 1898), *Dicranum viride* (LaRue, 1930), *Physcomitrium pyriforme* (Kachroo, 1954), *Tortula* (Bopp, 1955), *Merceya ligulata* (Noguchi & Furuta, 1956), *Anoectangium* (Fig. 3.2A), *Gymnostomum, Herpetineuron, Molendoa, Oncophorus, Pinnatella* (Noguchi & Miyata, 1957), *Climacium, Hedwigia, Hyophila, Hypopterygium, Myuroclada, Pohlia, Rhacomitrium, Rhizogonium, Sphagnum* (Noguchi & Muraoka, 1959), *Dawsonia polytrichoides*, and *D. superba* (Selkirk, 1981).

In the following species regeneration occurred all over the leaf: *Physcomitrium turbinatum* (Meyer, 1942), *Merceya gedeana* (Noguchi & Furuta, 1956), *Campylopus* (Fig. 3.2B), *Leucobryum* (Noguchi & Miyata, 1957), *Oligotrichum, Pogonatum* (Chopra & Sharma, 1958), *Atrichum undulatum, Aulacopilum, Hookeria, Tetraphis,* and *Venturiella* (Noguchi & Muraoka,

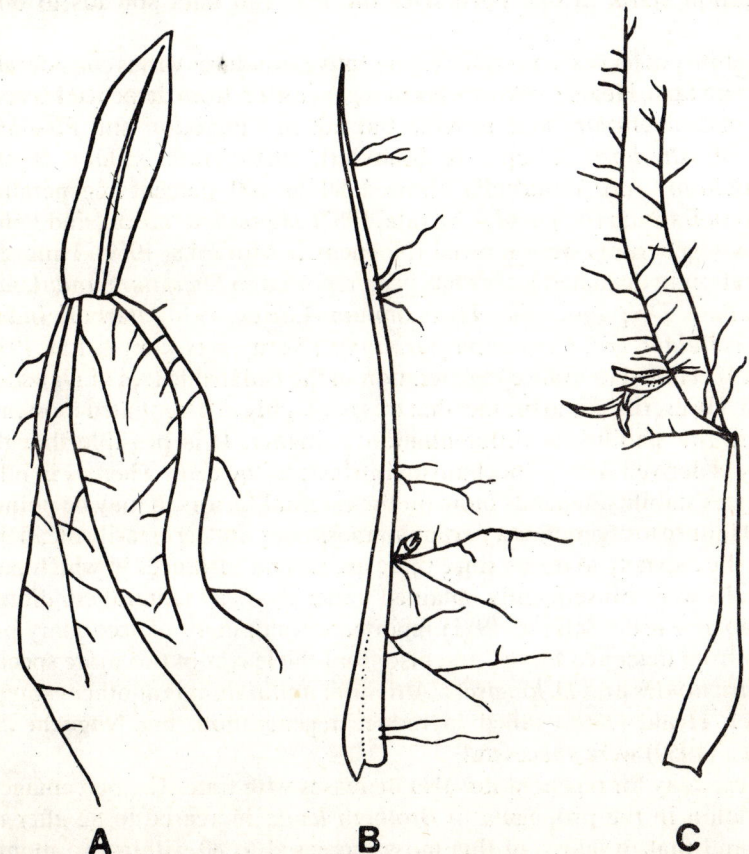

Fig. 3.2 A-C. Regeneration of leaves in *Anoectangium sublaeteviren, Campylopus atrovirens,* and *Leucobryum brevicaule.* A. Protonemal filaments produced from the base of detached leaf in *A. sublaeteviren.* B. Leaf of *C. atrovirens* with filaments all over the surface. At one place they are bearing a bud. C. In *L. brevicaule* filaments are produced from the tip of detached leaf. One of these is bearing a bud.
(After Noguchi & Miyata, 1957).

1959). In *Syrrhopodon* (Noguchi & Miyata, 1957), and *Splachnum* (von Maltzahn & MacNutt, 1958), protonemata arose mostly or only in the middle region. In *Dicranum undulatum* (LaRue, 1930) and *Ptychomitrium* (Noguchi & Miyata, 1957) regeneration was observed at the apex and base. In *Grimmia* (Noguchi & Miyata, 1957) regenerants were noticed in the middle region and base. In *Calliergon* (LaRue, 1930), *Atrichum undulatum* (Gemmell, 1953), *Funaria* (Bopp, 1955), *Leucobryum* (Fig. 3.2C; Noguchi & Miyata, 1957), and *Octoblepharum* (Egunyomi et al., 1980) protonemata arose only or mostly at the apex.

As is evident from the above data, there are relatively more taxa in which regeneration occurs at the base of detached leaves. Different species of a genus may show different responses, whereas rarely even the same species is reported to respond differently by different investigators. In some taxa regeneration starts in one portion of the leaf and later spreads to other regions.

The ability of leaves to produce regenerative structures varies considerably in different taxa. Heald (1898) observed regeneration from detached leaves in nature in *Leucobryum* and *Tortella*, but not in *Ceratodon* and *Fissidens*. Leaves of *Atrichum* (Chopra & Bhandari, 1960), *Aulacopilum*, *Bryum*, *Orthotrichum*, and *Venturiella* showed 80 to 100 percent regeneration, whereas in *Barbula* (Noguchi & Miyata, 1957), *Homaliodendron*, and *Pohlia* very few regenerants were noticed (Noguchi & Muraoka, 1959). Failure of regeneration in detached leaves has been reported in *Dicranum undulatum*, *D. flagellare*, *Drepanocladus*, *Hylocomium* (LaRue, 1930), *Barbula indica*, *Bryum cellulare*, and *Ceratodon purpureus* (Narayanaswami & Lal, 1957). Giles (1971) failed to induce regeneration in the isolated leaves of *Dawsonia superba*. He ascribed it to the fact that this is a highly differentiated moss, and therefore the stability of differentiation is higher. It is possible that this stability is derived from a mechanism intrinsic to the cells, whereas in other bryophytes stability depends more on the external factors. It may be pointed out that failure to regenerate a particular moss does not necessarily mean that it lacks the capacity to do so, since there are several instances in which positive results were subsequently obtained under changed cultural conditions. In *D. superba* itself Selkirk (1981) reported development of secondary protonema from detached leaves, and also from the leaves of two more species: *D. polytrichoides* and *D. longiseta*. *Atrichum undulatum* is another example in which Heald (1898) failed to induce regeneration, but Noguchi and Muraoka (1959) were successful.

The capacity for regeneration also increases with time. The percentage of regeneration in the propagula of *Brothera leana* increased to 82 after ten weeks, and that in leaves of this moss increased to 80 within two months (Noguchi & Furuta, 1958).

In general, there is no correlation between spore production and capacity for regeneration. However, Noguchi and Miyata (1957) reported that detached leaves are easily regenerated, especially in mosses which usually remain sterile e.g. *Brothera*, *Campylopus*, *Leucobryum*, and *Syrrhopodon*.

Heald (1898), Meyer (1940, 1942), and Noguchi and Furuta (1956) stated that leaves attached to the stem do not regenerate. MacNutt and von Maltzahn (1960) also mentioned that in leaves regeneration is generally observed when they are detached or mutilated. However, there are some reports in which regeneration is observed even in leaves attached to the stem: *Dicranum scoparium* (LaRue, 1930), *Hookeria lucens* (Gemmell, 1953), *Physcomitrium cyathicarpum*, *P. coorgense* (Narayanaswami & Lal, 1957), *Oligotrichum semilamellatum*, *Pogonatum microstomum* (Chopra & Sharma, 1958), *Polytrichum* (Ward, 1960), and *Atrichum* (Chopra & Bhandari, 1960). In *Octoblepharum* both attached and detached leaves produce protonemata in cultures (Egunyomi et al., 1980).

In the leaves of Polytrichaceae regenerants may originate from the lowermost cell of a lamella as in *Atrichum undulatum* (Gemmell, 1953), *Atrichum pallidum* (Fig. 3.3 A, B; Chopra & Bhandari, 1960), *Pogonatum perichaetiale*, *P. microstomum*, and *Oligotrichum semilamellatum* (Chopra & Sharma, 1956, 1958). Sometimes the cell of the lamina bearing the lamella gives rise to the regenerants, as in *Polytrichum commune* (Heald, 1898). In *Atrichum pallidum* a regenerant arose from one of the adaxial surface cells of leaf which did not bear any lamella (Fig. 3.3C; Chopra & Bhandari, 1960). Selkirk (1981) observed that in *Dawsonia polytrichoides* (Dawsoniaceae) protonemal filaments originated from basal cells of a lamella or upper cells of leaf surface.

REGENERATION FROM SETAE

The early studies on regeneration from setae were carried out by Campbell (1905, *cited in* Ono & Tanaka, 1976), and Marshal and Marshal (1907, *cited in* LaRue, 1930). Wettstein (1923, 1924a,b, 1925, 1928, *cited in* Ono & Tanaka, 1976) obtained many kinds of polyploid series by means of regeneration of setae in *Funaria hygrometrica* and *Physcomitrium pyriforme*, and some other species.

LaRue (1930) was able to regenerate setae in 14 moss species including those of *Amblystegium, Catharinea, Fissidens, Funaria, Mnium, Physcomitrium,* and *Polytrichum*. He, however, did not succeed in regenerating the setae of four species of *Dicranum*, although gametophytes of most species of this genus showed a very strong power of regeneration. The reverse was true of *Ceratodon* and *Polytrichum*. Earlier, Wettstein (1924b, *cited in* Ono & Tanaka, 1976) had failed to regenerate setae of these mosses, but LaRue (1930) obtained regenerants in both of them. Regeneration from setae has been reported in *Physcomitrium turbinatum* (Meyer, 1940; Redfearn & Meyer, 1949) and *P. pyriforme* (Kachroo, 1954). In a few instances jacket and setae of immature capsules in *Physcomitrium cyathicarpum* produced filamentous structures (Narayanaswami & Lal, 1957). Chopra and Sharma (1958) did not observe regeneration from setae of two members of Polytrichaceae, *Pogonatum* and *Oligotrichum*. Hughes (1969) obtained diploid protonemata

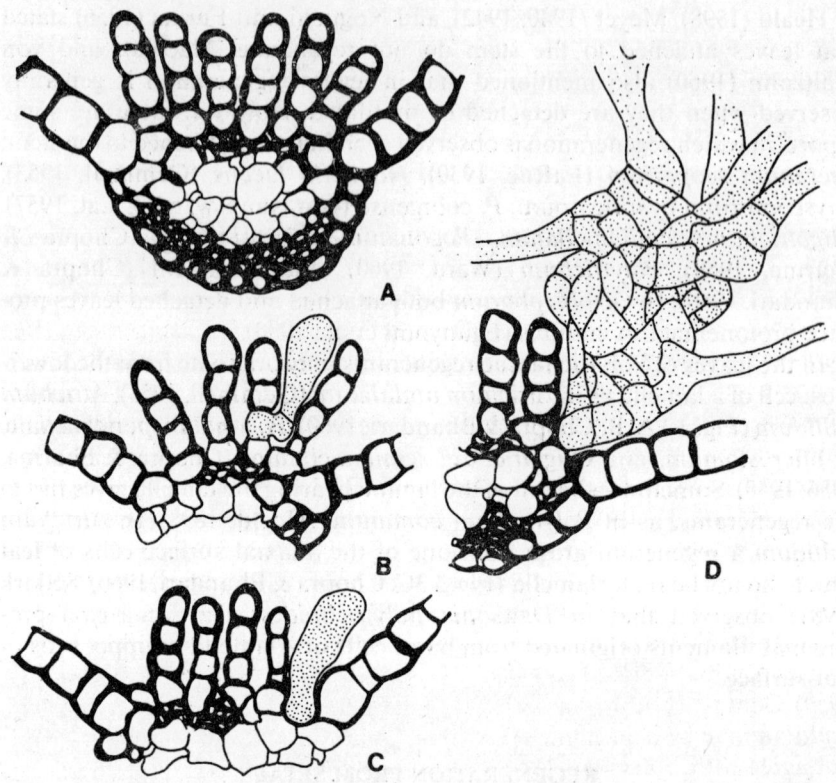

Fig. 3.3 A-D. Regeneration of leaf in *Atrichum pallidum*. A. Vertical section of mature leaf. B. Portion of vertical section of leaf with protonema developing from the basal cell of a lamella. C. Same, to show the origin of a regenerant from a cell on the adaxial surface of midrib. D. Same, showing a parenchymatous mass of cells formed by irregular divisions in a protuberance on the adaxial surface. This mass subsequently produces protonemal filaments.
(After Chopra & Bhandari, 1960).

from the regenerating seta cuttings of *Grimmia pulvinata*. Mehra and Pental (1976) also succeeded in obtaining diploid gametophytes from setae of *Athalamia pusilla*. Ono and Tanaka (1976) made a detailed study of regeneration in the setae of *Haplocladium angustifolium* taking into account the relationship between regeneration and the polarity of seta segments, the tissue from which the protonemata develop, and the cytological features of the development of protonema. Entire, relatively young setae measuring about 10 to 15 mm long were cut and used for experimentation. After 10 days of culture the protonemata began to develop from the upper ends of seta segments, and no regeneration was observed from the basal ends. After 20 days, in some such segments protonemata developed from the basal ends as well (Fig. 5.10A). Of the 133 segments, 113 (85%) showed regeneration. This response is remarkably high as compared with other mosses. Of the 113 segments, 42 (31.6%) regenerated only from the upper ends, and the remaining 71 (53.4%) from both ends. There was no segment regenerating only from the basal end.

MORPHOLOGY OF REGENERANTS

In the vast majority of mosses regeneration results in the production of secondary protonema (de-differentiation), which later bears buds (re-differentiation) (MacNutt & von Maltzahn, 1960). As examples may be cited *Physcomitrium* (Kachroo, 1954), *Merceya* (Noguchi & Furuta, 1956), *Barbula, Bryum, Rhodobryum* (Narayanaswami & Lal, 1957), and *Atrichum* (Chopra & Bhandari, 1960). Even when caulonema (the second stage of protonemal development) regenerates, it undergoes de-differentiation into chloronema (Knoop, 1973). However, variation in the nature of regenerants is on record. In some taxa direct origin of shoots has also been observed from leaves: *Fissidens, Mnium* (Heald, 1898), *Rhodobryum roseum* (LaRue, 1930), *Campylopus, Herpetineuron* (Fig. 3.4A; Noguchi & Miyata, 1957), *Rhodobryum giganteum* (Narayanaswami & Lal, 1957), *Pogonatum, Oligotrichum* (Fig. 3.5 A-C; Chopra & Sharma, 1956, 1958), *Atrichum* (Fig. 3.4 B), *Aulacopilum, Bryum* (Fig. 3.4C), *Homaliodendron, Macromitrium, Macrosporiella, Polytrichum, Rhizogonium,* and *Venturiella* (Noguchi & Muraoka, 1959). Gemmell (1953) considers the bud producing primordium in *Atrichum undulatum* to be a 'protonemal filament', but Chopra and Sharma (1958) do not agree with him, since this primordium directly passes into a bud.

Direct origin of shoot buds from stem is also known in a few mosses: *Physcomitrium coorgense, P. cyathicarpum* (Narayanaswami & Lal, 1957), *Oligotrichum* (Chopra & Sharma, 1958), *Tetraphis* (Noguchi & Muraoka, 1959), and *Atrichum* (Chopra & Bhandari, 1960).

In *Funaria* it was observed that the differentiation of regenerants depends upon the nature of the isolated part (Bopp, 1952). Isolation of protonema, very young leaves, and mature leaves resulted in different patterns of regeneration, and this was observed to be based on nutritional effects. The differentiation of regenerants depended upon the size of the photosynthesizing area of the isolated part.

In view of the above reports it is clear that protonemal filaments are not always necessary for the formation of leafy shoots. According to Noguchi and Muraoka (1959) well branched protonemata are assimilatory organs and supply nutrients to the leafy plants which are borne on them. When there are some other cells which provide nutrients to the new leafy shoots the protonema may be unnecessary in regeneration.

In the mosses studied by Noguchi and Muraoka (1959) the manner of regeneration of leaves was similar to that of the germination of spores in the taxon. For example, in the members of the Eubryales chlorophyllose filamentous protonemata were formed which later bore leafy shoots; whereas in mosses like *Sphagnum, Tetraphis,* and *Diphyscium* thalloid bodies were

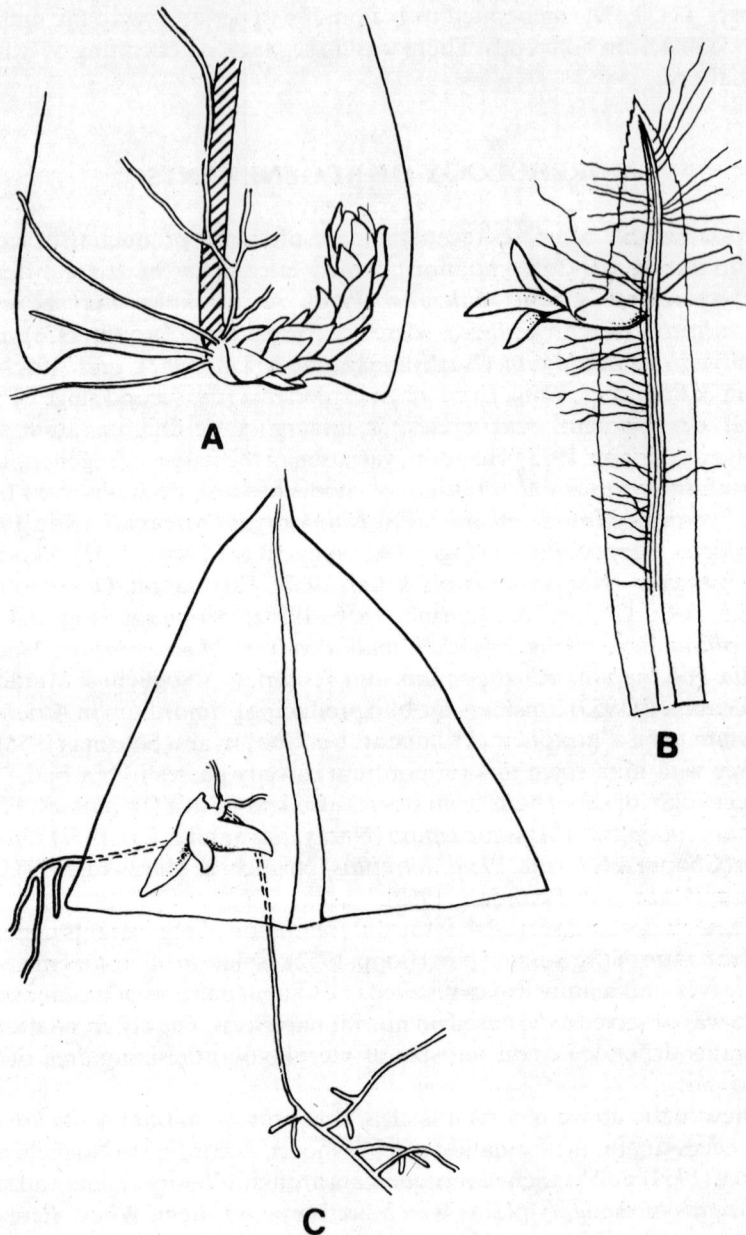

Fig. 3.4 A-C. Regeneration of leaves in *Herpetineuron toccoae*, *Atrichum undulatum*, and *Bryum capillare*. **A.** A leafy plant developed directly from the base of midrib in *H. toccoae*. Rhizoids are arising from the base of this shoot. **B.** A leafy shoot and protonemal filaments arising from the midrib of a detached leaf of *A. undulatum*. **C.** Distal part of leaf of *B. capillare* showing some protonemal filaments and a leafy shoot developed directly from a cell of lamina.
(**A.** After Noguchi & Miyata, 1957; **B, C.** After Noguchi & Muraoka, 1959).

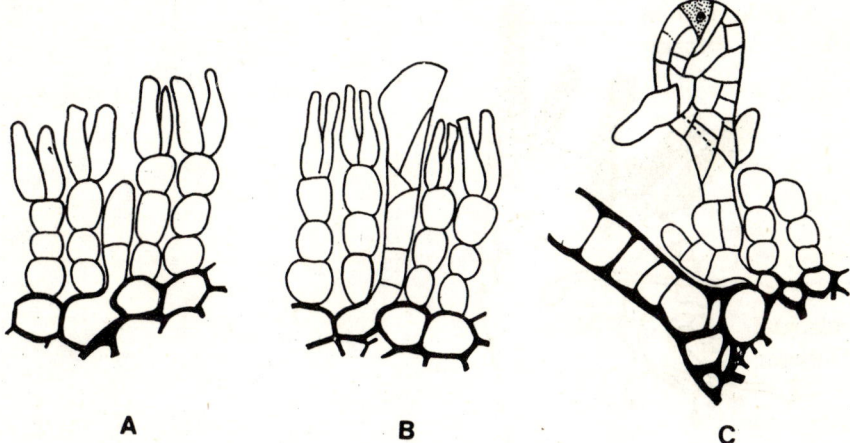

Fig. 3.5 A-C. Regeneration of leaves in *Pogonatum microstomum* and *Oligotrichum semilamellatum*. A, B. Portions of leaf in vertical section to show stages in the development of young bud directly from cells on the adaxial surface of midrib in *P. microstomum*. C. Same, to show a slightly advanced stage in the development of bud in *O. semilamellatum*.
(After Chopra & Sharma, 1958).

established on the filamentous protonemata, and these in turn produced leafy plants.

In *Atrichum pallidum*, Chopra and Bhandari (1960) observed that at times a protuberance on the adaxial surface develops into a parenchymatous mass from which protonemal branches emerge in all directions (Fig. 3.3D). Direct origin of sporogonia has also been reported from gametophytes and isolated sporophytic tissues (Chopra & Rashid, 1967; Menon & Lal, 1977).

In *Bryum cellulare* and some other mosses investigated by Narayanaswami and Lal (1957) the filamentous outgrowths were intermediate between typical rhizoids and typical protonemata. These have been designated as rhizonemata' by Narayanaswami (1957). In the leaves of *Brachythecium albicans, Funaria hygrometrica* (Heald, 1898), and *Aulacopilum japonicum* (Fig. 3.6; Noguchi & Muraoka, 1959) both rhizoids and secondary protonemata arose.

Berthier et al. (1976) have stated that in mosses there are three modalities of vegetative propagation by cuttings:

(1) System with obligatory chloronema formation e.g. lenticular gemmae of *Tetraphis* (Fig. 3.7 A,B). In this taxon the initial on the gemma produces either a chloronema (chl), or a chloronema ramified into thalloid laminae (tl), or directly a thalloid lamina. The foliar bud (b) appears secondarily at the base of the thalloid laminae.

(2) System with obligatory caulonema formation e.g. foliar cuttings of *Polytrichum* (Fig. 3.7 C,D). The preformed initials on the leaf generally produce a caulonema (c) on which foliar bud (b) appears, but under certain conditions (presence of 6-benzylaminopurine at

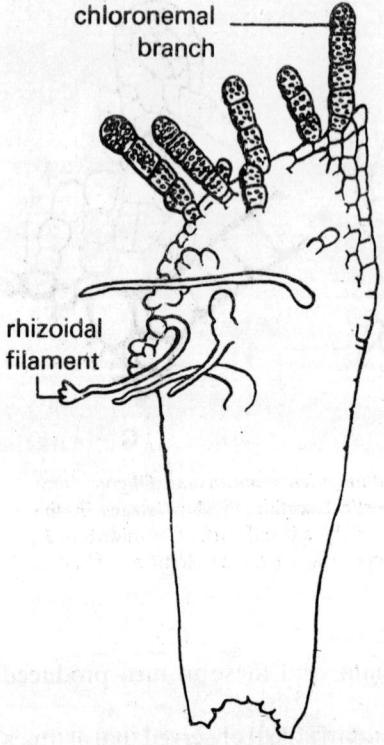

Fig. 3.6. Leaf of *Aulacopilum japonicum* with two types of regenerants arising from different cells.
(After Noguchi & Muraoka, 1959).

10^{-6} g/l) they directly develop into an adventitious bud (ab).

(3) System with preformed foliar bud e.g. bulbil of *Pohlia* (Fig. 3.7 E,F). In *Pohlia* the large propagules are with scales and have an active (normal) apical cell. These germinate into foliar stems (fs) or even into adventitious buds (ab). The smaller ones have microscopic leaves (m) and an aborted apical cells (aa). Such propagules germinate into caulonema (c), which in turn develops a bud (b).

Variations in the morphology of regenerants in liverworts and hornworts are also on record. In *Metzgeria furcata* pieces of thallus regenerated by forming adventitious shoots (Schostakowitsch, 1894). However, Goebel (1908) used weakened plants grown in dim light, and obtained reversion to a filamentous juvenile stage. Kreh (1909) considered that in liverworts the process of regeneration was generally similar to development from the spore. He, however, added that isolation of cells was necessary for protonema formation, otherwise the germ tube stage was bypassed.

According to Fulford (1956) the stages in the development of a regenerant may be a repetition of the stages of development of the sporeling and gemmaling, or one or more of these stages may be missing or the pattern of development may be entirely different. While single isolated cells behave like spores and produce a germ tube, cells in groups may produce protonemata or cell masses which directly bear buds.

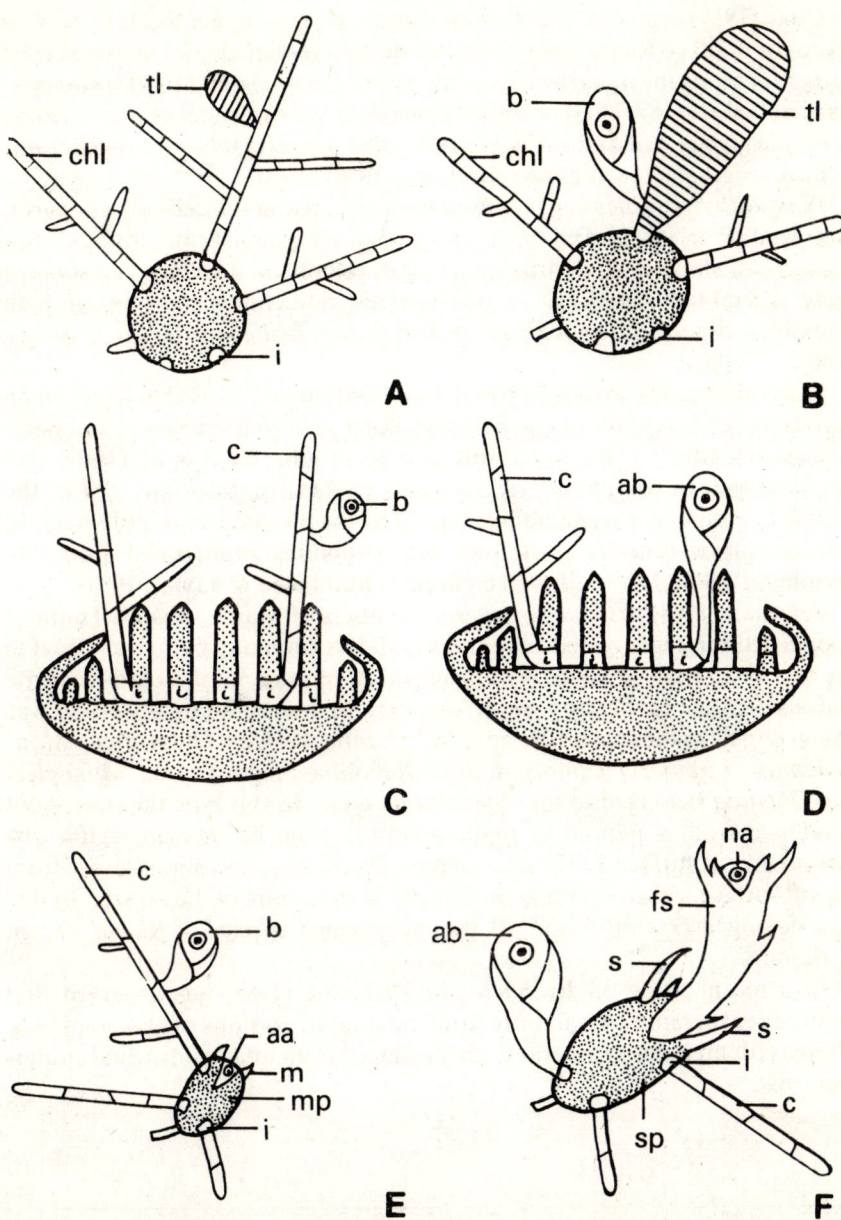

Fig. 3.7 A-F. Modalities of vegetative propagation by cuttings in mosses. **A, B.** System with obligatory chloronema: thallospores (lenticular gemmae) of *Tetraphis pellucida*. **A.** Photomorphogenesis in artificial light of 2400 ergs cm^{-2}s^{-1}. Initials (i) give rise to chloronemal filaments (chl) which branch further. After 16 days thalloid laminae (tl) appear in place of some chloronemal branches. **B.** Photomorphogenesis in sun light. The initials produce both chloronemata and thalloïd laminae. The latter bear buds (b) after some days. **C, D.** System with obligatory caulonema: foliar cuttings in *Polytrichum*. **C.** *P. juniperinum* lamellae regenerate caulonema (c) which bear buds (b) in

Udar (1957) reported that during regeneration in intact thalli as well as pieces of thalli of *Riccia crystallina* the de-differentiated cells on the ventral surface or along the margin showed the developmental patterns of sporelings. Earlier, Fellner (1875) had observed somewhat similar stages of regeneration in *Riccia glauca*, but had wrongly interpreted the germ tube-like regenerants to have originated by the transformation of rhizoids.

In *Notothylas indica* regeneration took place from the cells on the dorsal and ventral surfaces and also along the margins of the thallus. The regenerants showed wide variation in the developmental pattern, but some of these resembled sporelings of this species. Sometimes the pattern even resembled that of the sporelings of *Anthoceros fusiformis* (Udar & Singh, 1958).

Experiments on *Carrpos* by Proskauer (1961) demonstrated that any green vegetative cell is capable of regeneration and that in some instances regeneration started with a cell comparable to a germ tube. Kaul et al. (1962) also observed that in *Marchantia* regenerants could arise from any cell of the thallus or callus, and regeneration started by the production of a filament. In *Sphaerocarpos texanus* germ tube-like structures proliferated from the peripheral cells of the callus-like clumps (Montague & Taylor, 1971).

Pearman (1964) worked on several members of Marchantiales. Thalli in poor condition regenerated adventitious shoots from any part as a sequel to the loss of normal polarity, with a resulting physiological isolation of the various tissues. In healthy thalli the type of regeneration observed was characteristic of the species. Besides *Marchantia*, *Preissia*, *Dumortiera*, *Monoselenium*, *Lunularia*, *Conocephalum*, *Reboulia*, *Plagiochasma*, *Monoclea*, and *Corsinia* belonged to the 'Marchantia type'. In this type the new shoot developed from a mound of tissue ventrally from the midrib region just behind the cut surface. In the second type 'Riccia type' new shoots arose from the cut surface, either ventrally or dorsally in the centre of the midrib. In this type belonged *Oxymitra* and all the investigated species of *Riccia*, except *R. fluitans*.

In a recent study on *Riccia frostii* Vashistha (1985) has observed that thalloid regenerants arise mainly from the injured portions of gametophytes, whereas filamentous regenerants are produced from intact individual epidermal cells.

---

continuous light of 5400 ergs cm^{-2}s^{-1}. D. *P. strictum* with the addition of 6-benzylaminopurine at 10^{-6} g/l to the medium, caulonema (c) is accompanied by adventitious buds (ab) coming directly from the initial (i), in 16 h light + 8 h dark. E, F. System with preformed foliar bud: bulbils of *Pohlia muyldermansii*. Light 600 ergs cm^{-2}s^{-1}, with 16 h light + 8 h dark. E. In the small spherical propagules (mp) with microscopic leaves (m) and aborted apical cell (aa) the initials (i) produce caulonema (c), which bears buds (b). F. In the big-sized propagules (sp) with scales (s) and normal apical cell (na) the initials germinate into foliar stems (fs) or even into adventitious buds (ab).
(After Berthier et al., 1976).

FACTORS AFFECTING REGENERATION

There are several factors which influence regeneration. Among the external factors may be mentioned light, radiation, pH, season, humidity, wounding, and temperature. The chief internal factors are size of the fragment, reserve food material, location in the plant, age, correlative inhibition, polarity and apical dominance.

LIGHT

Heald (1898) observed that development of protonemata and buds takes place most readily in light. Redfearn and Meyer (1949) noticed that light is essential for the production of protonema from setae of *Physcomitrium turbinatum*. Kachroo (1954) also reported that light is essential for regeneration of detached leaves and seta segments of *P. pyriforme*. Karsten (1963) noticed that in *Riella* light in some way influences the withdrawal of the capacity of cells to divide. Demkiv (1971) observed that in the protonema of *Funaria hygrometrica* regeneration of cells was stimulated with light, and maxima were in blue and red regions. Stange (1982) reported that in *Riella helicophylla* light level influences the number of meristems, speed of the cell cycle, and the gradient of cell generation time. At 5600 lux two symmetrical active lateral meristems develop from an intercalary meristem in the gemmaling, whereas at 1400 lux only one lateral meristem is formed. At 2800 lux gemmalings with two, often asymmetrical, or with one lateral meristem are observed. The lability of meristem differentiation is represented by the unequal distribution of metaphases in the gemmaling. This in turn is brought about by the small differences in the cycle speed of adjacent cells which evokes a unidirectional transport of substances needed for growth from the slower to the faster cycling cell, thus creating local intercellular sink-source relationships. Eventually polarization emerges from the centre of highest rate of DNA synthesis.

According to Berthier et al. (1976) the photomorphogenetic effects of light during vegetative propagation can be classified into the following four categories:

(1) Photosynthetic Effects

Bopp (1952) reported that when isolated mature leaves are kept in dark only a few protonema filaments grow out from them, and these filaments do not bear buds. The number of buds is related to the time for which leaves are exposed to light after isolation and before transfer to dark. This effect of light was shown to be at least partly due to the products of photosynthesis. Relationship of regeneration to photosynthesis was also pointed out by Stange (1960).

Berthier et al. (1976) observed that in *Pohlia* the type and number of propagules which appear on the leafy axis can be modified with light level. At

4500 ergs $cm^{-2}s^{-1}$ no propagules come to maturity. At 9000 the microphyllous propagules (mp) dominate. At 18,000 more of propagules with scale-like leaves (sp) are produced, and the number of mp decreases. Finally at 36,000 a new type of green propagules (gp) with normal leaves and long axis are produced (Fig. 3.8). Thus, deficiency of light leads to increased apical dominance.

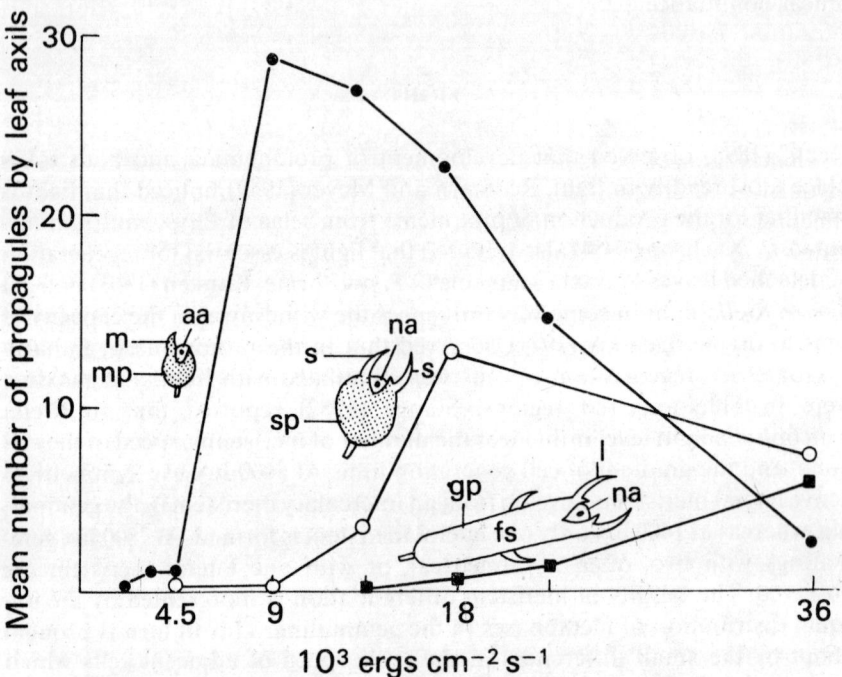

Fig. 3.8. Effect of light energy on the type of propagules produced at 14 C, by leaf axils in *Pohlia muyldermansii*. Cultures maintained in white fluorescent light, 16 h light + 8 h dark period in 24 h. gp (dark squares): green propagules with foliar stem (fs) and developed leaves (l). sp (white circles): propagules with scale leaves (s) and normal apical cell (na). mp (dark circles): propagules with microscopic leaves (m) and aborted apical cell (aa).
(After Berthier et al., 1976).

(2) Inductive Effects

In the leaves of *Polytrichum juniperinum* light plays an inductive role in the regeneration of initials at the base of leaf lamellae. For this response even a single day of light at 5400 ergs $cm^{-2}s^{-1}$ is enough (Gay, 1971, 1974).

(3) Phytochrome Action

Hahn and Miller (1966) demonstrated that chloroplast replication is a phytochrome mediated process. Giles and von Maltzahn (1967, 1968) observed that red light induced regeneration in the leaves of *Mnium affine*, whereas far-red light inhibited it. However, the phytochrome system becomes operative in regeneration only after isolation of the part from the plant, since de-differentiation of cells is not brought about by red light treatment in intact plants (Giles, 1971). The role of phytochrome has also been demonstrated in germination of initials in the propagules of *Aulacomnium* and *Tetraphis* (Larpent-Gourgaud et al., 1972, 1974). For protonemal ramification in *Funaria* phytochrome action is possible only after the photosynthetic requirement has been met with (Larpent-Gourgaud & Jacques, 1971). In the gemmae of *Tetraphis* as well, phytochrome induces lamina ramification only after these have photosynthesized for eight days (Larpent-Gourgaud et al., 1972).

(4) Blue Light (420 to 480 nm) Photoreceptor

There is some evidence that a photoreceptor of blue radiation exists in bryophytes. In *Tetraphis* this 'blue' photosystem neutralizes the phytochrome effect related to red (660 nm) light, which permits germination of initials in gemmae (Larpent-Gourgaud et al., 1974).

The latest studies of Gay (1984) on *Polytrichum juniperinum* have demonstrated that in dark only 42 ± 5 percent of the planted leaves regenerated giving rise to one to five filaments per leaf. With increasing continuous light energies, an obvious stimulation occurred. The optimum in the regeneration curve was reached with 3.6 Wm^{-2}. The action spectrum of light on regeneration revealed that blue light (480 nm) and red light (670 nm) were the most effective radiations, whereas green light (550 nm) was less effective, and far-red light (730 nm) had no effect. The further development of protonema under the inductive monochromatic lights was not as good as under white light, suggesting that induction of regeneration and stimulation of growth are not strictly correlated.

As regards photoperiod, Gay (1984) observed that with 6 h per day almost all leaves regenerated on mineral medium. If glucose was added to the substrate, a photoperiod as short as 1 h per day increased regeneration over the dark control. These observations suggest that the effect of white continuous light is primarily related to the nutritional (trophic or photosynthetic) effect.

Gay (1984) educed evidence that in *Polytrichum juniperinum* light also acts as an inductive stimulus independent of its trophic effect, and this may be related to phytochrome. He observed that the inductive light had its optimum around 24th h of culture. It can be presumed that light acts as an inductive stimulus on the metabolic products released after the isolation of leaves. Earlier, Stange (1960) also reported that light was effective for 12 h after the isolation of fragments of the thallus of *Riella*. Such an inductive effect of light through metabolic changes has also been described by Larpent-Gourgaud and Aumaître (1977). Gay (1984) concluded that light has two distinct effects on regeneration:

(1) A trophic one, directly connected with the light energy and duration; and
(2) An inductive one, related to low energy and brief sequences of light.

RADIATION

Herzfeld (1923) reported that exposure to 600 to 5400 roentgens of X-rays stimulated regeneration in gametophores of *Leptobryum pyriforme*. Gunckel and Sparrow (1961) proved that gamma radiation can either stimulate or inhibit hormone action in plants, and also that some effects of radiation are instantaneous while others are delayed. Sax (1963) also proposed that growth stimulation in irradiated plants is due to the effect of radiation on hormone balance.

Grossman and Hillson (1976) studied regeneration in *Polytrichum commune* after treatment with gamma radiation and desiccation. Cuttings of leaves and leaf-bearing stem segments were made before the start of regeneration experiments. In many of the plants that received 1600, 2200, and 2800 roentgens of gamma radiation, regeneration and growth in height appeared to be rapid at first, but became comparatively slow after several weeks. In the plants that received the lowest dose (1000 r), the amount of regeneration and growth in height of regenerants (chloronemal filaments & shoots) was more constant throughout the period of observation. On leaf cuttings proliferations arose mainly along the adaxial side of the costa. Also the upper leaves of most of these stem/leaf cuttings had more proliferation than the lower leaves. The irradiated groups at first produced more proliferation than controls. Eventually, however, controls surpassed irradiated groups in size in desiccated material, and in size and number in the non-desiccated material.

pH

Noguchi and Miyata (1957) studied the effect of pH ranging from three to eight on regeneration of leaves in several mosses. With change in pH considerable variation in the regeneration response was observed. The data for five mosses at the optimum values are given in Table 3.2.

TABLE 3.2
Effect of pH on regeneration of some mosses.

Taxa	Optimum pH	Percentage response
Anoectangium	4	100
Grimmia	3	95
Gymnostomum	4	20
Molendoa	6	61
Syrrhopodon	3	82

Thus, among the investigated mosses regeneration was best in acidic environment, and it was negligible (1 to 2%) at pH 8. *Riccia frostii* also showed profuse regeneration in acidic range (Vashistha & Chopra, 1986).

SEASON

LaRue (1930) reported that spring and early summer are best for regeneration, whereas late autumn and early winter are least favourable.

HUMIDITY

Increased humidity favours regeneration, as has been reported by Udar (1957) in *Riccia crystallina*, by Vashistha (1985) in *R. frostii*, and by Ward (1960) in *Polytrichum*.

WOUNDING

LaRue (1930) observed that stem segments of *Calliergon* and *Drepanocladus* developed rhizoids all along the sides but not from the cut ends. In *Neckera* protonema arose from epidermal cells. Meyer (1942) reported that in *Physcomitrium pyriforme* regeneration would take place from any cell of the leaf. In *Haplocladium* regeneration occurred only from parts of setae wounded by cutting (Ono & Tanaka, 1976). In *Physcomitrium turbinatum* protonemata normally developed from the internal parenchymatous cells of the seta, and even at times from the epidermis (Redfearn & Meyer, 1949). Gemmell (1953) opines that some of this regeneration is a direct result of wounding, as confirmed by Kachroo (1954). Wounding increased the amount of regeneration in *Atrichum undulatum* (Gemmell, 1953) and *Funaria hygrometrica* (Bopp, 1955). Chopra and Kumar (1961) also reported that in *Atrichum* the defoliated stems showed a higher percentage of regeneration, and this may be due to wounding. However, in *Pogonatum* and *Oligotrichum* wounding did not appear to have much significance in regeneration (Chopra & Sharma, 1958). Protonema originated from the stem surface of intact gametophores in *Barbula gregaria* (Narayanaswami & Lal, 1957). In the leaves of *Brothera leana* protonemal filaments usually grew from cells on the cut surface or adjacent portions. When entire stems with leaves attached were cultured they readily produced filaments at the cut ends, but scarcely on the entire surface of the stems (Noguchi & Furuta, 1958).

MacQuarrie and von Maltzahn (1959) demonstrated that in *Splachnum ampullaceum* isolation of leaf from stem was more important for regeneration than the wounding caused by its removal. Leaves attached to stem were cut in half at right angles to the longitudinal axis. Following this the remaining basal half-leaves were either excised or left attached to the plant. Even though the wound surface of the attached and isolated half-leaves was the same, the isolated half-leaves produced profuse protonemata, whereas the attached half-leaves did not at all regenerate. The total suppression of regeneration in attached half-leaves is interpreted to be due to the presence of correlative factors between the stem and the leaves. It was, therefore, concluded that destruction of these correlations by isolation, and not wounding, is the most important factor in initiating leaf regeneration.

The importance of isolation in regeneration is also brought out by plasmolysis experiments. Nagai (1919) obtained a large number of regenerants from each gemma of *Marchantia* and *Lunularia* by plasmolysing the gemmae in hypertonic solutions such as of ten percent potassium nitrate. Plasmolysis brings about physiological isolation of individual cells or small groups of cells (Bopp, 1953; Stange, 1957).

Stange's (1957) experiments on *Riella* also clearly demonstrated that if wounding was done without removing the connection between the apical meristem and the mature cells no regeneration occurred. MacNutt and von Maltzahn (1960) noticed that while in *Splachnum*, isolation and not wounding initiated cellular differentiation, wounding did have an effect on the position of chloroplast clusters. In *Polytrichum*, Ward (1960) left the active apical region intact and removed the leaves from rest of the stem. He did not observe any propagative growth on the denuded stem, indicating that wounding does not result in regeneration. Raudzens and Matzke (1968) considered that tissue degeneration is a pre-requisite for regeneration from setae in *Blasia pusilla*. Egunyomi et al. (1980) reported that in *Octoblepharum* injury reduced the regenerative capacity of leaves.

Thus, it may be concluded that while in some systems wounding favours regeneration, it is not a causal factor for this process.

TEMPERATURE

Rhizoidal tubers of *Ceratodon purpureus* when detached and subjected to low temperature treatment (0 or 12 C) germinated to produce protonema (Narayanaswami & Lal, 1957). Kanda (1979) concluded that the Antarctic plants of *Ceratodon purpureus* showed a specialized adaptation to low temperature with a positive growth even at the freezing point. Regeneration in these plants was very active even though its specimens had been deposited at -20 C for 2.5 years before the experiment. The developmental features of secondary protonemata, rhizoids, and new shoots produced from cultivated plants varied with conditions of culture medium, temperature, and light. The Antarctic plants of *Ceratodon* had stronger potential for regeneration in stem cells than the temperate plants.

SIZE OF THE FRAGMENT

Experiments on leaf segments of *Brothera leana* revealed that generally filaments appeared earlier from longer segments than from shorter ones. The shortest segment which produced filaments was about 0.13 mm long (Noguchi & Furuta, 1958).

MacQuarrie and von Maltzahn (1959) observed that in short decapitated stem segments of *Splachnum* (up to about 1.2 mm in length) a linear relationship existed between the length of stem segment and the number of protonemata. In segments longer than this there was no correlation of this type. It was concluded that systems of correlation, which inhibit regeneration, are present in the stem, and these become more effective as length increases. These systems are not destroyed by decapitation, and are not important in stem segments less than 1.2 mm in length. However, in leaves of *Splachnum* (MacQuarrie & von Maltzahn, 1959; von Maltzahn & MacNutt, 1958) no correlation was observed between leaf size and regeneration.

In *Riella* rectangular segments of thallus extending over the whole width were cut to three sizes (Stange, 1964): (*i*) complete width, (*ii*) divided into two halves, and (*iii*) divided into quarters. It was observed that the larger the area of the fragment the more the average number of regenerants. However, the number of regenerants was not proportional to the increase in fragment area. The number of regenerants arising from the various subdivisions of a fragment was always more than those arising from the entire fragment. By further subdivision it should be possible to get as many regenerants as the number of cells in a tissue. In *Riccardia pinguis, Moerckia flotowiana, Makinoa crispata* (Ilahi & Allsopp, 1970), and *Noteroclada confluens* (Allsopp & Ilahi, 1970) size of the fragment has also been observed to influence regeneration. Fewer regenerants were produced from the 0.5 mm pieces, but there was no appreciable difference in the number formed from the 1 and 2 mm segments.

RESERVE FOOD MATERIAL

The importance of reserve food material as a factor in regeneration on *Mnium* was suggested by LaRue (1930). In the species of *Atrichum* studied by Chopra and Bhandari (1960) prolongation of drought period shortened the time taken for regeneration and increased its percentage, as also observed by Noguchi and Furuta (1956) in *Merceya*. This may be due to increased storage of starch in plant tissues under drought conditions. In *Atrichum* the older parts showed higher percentage of regeneration and took lesser time to do so. It has been suggested that comparatively lower percentage of regeneration in the younger organs is perhaps due to scanty food reserve; food being actively utilized by these organs for growth (Chopra & Kumar, 1961).

LOCATION IN THE PLANT

On the basis of his studies on several Marchantiales Douin (1923) concluded that certain cells have greater regenerative capacity and retain their viability longer than do others. He educed evidence that these cells are distributed along four longitudinal lines which converge at the apex. Of these four lines only the ventral superficial line normally produces shoots in the active thallus.

A restriction of regeneration to specific cells of the leaf was reported in several mosses by Heitz (1925), and later by Bopp (1961) and Gay (1971) in

Polytrichum. In *Octoblepharum* protonemata develop only from the chlorophyllose cells (Egunyomi et al., 1980).

AGE

Kreh (1909) observed that liverworts show polarity during regeneration and the polar character decreases with increased age of the tissue.

From his studies on *Marchantia* gemmae Dickson (1932) concluded that the younger cells of any part of a developing thallus always dominate the older so far as the production of regenerant shoots is concerned. In *Sphagnum* (Zepf, 1952) as well the changes which accompany embryonization occurred much faster in younger cells as compared to the older ones. In *Riella* the time of appearance of cauloids and rhizoids showed a clear dependence on the age of cells (Stange, 1957). In transversely cut pieces of thalli regenerants almost always appeared at the proximal end. Thus indicating that the younger cells have a greater capacity for regeneration.

Gemmell (1953) working on *Atrichum undulatum* and Bopp (1955) on *Funaria hygrometrica* observed that although initially regeneration of leaves decreased from apex downward, there was no strict correlation between the age of leaf and its regenerative capacity. However, observations of Comeau and LeBlanc (1971) indicate that in *F. hygrometrica* leaves from upper part of stem have a higher percentage of regeneration than those from lower part. In *Splachnum ampullaceum*, von Maltzahn and MacNutt (1958) observed that leaves from the base toward the apex showed a steadily increasing number of regenerants. The age factor was even evident in young and older gametophores. Leaves from younger gametophores showed consistently a much larger number of regenerants. Subsequently, MacNutt and von Maltzahn (1960) reported that chloroplast number in the cells of the apical leaf increases more quickly and to a greater extent than in the cells of older basal leaf. In *Octoblepharum* as well regeneration decreases with increase in the age of leaf (Egunyomi et al., 1980). In *Atrichum*, on the other hand, older parts of gametophytes exhibited a higher percentage of regeneration (Chopra & Kumar, 1961).

The famous experiments of Bauer (1963) on regeneration of the hybrid sporophyte (*Physcomitrium pyriforme* × *Funaria hygrometrica*) have clearly indicated that age of the tissue not only influences the degree and time of regeneration, but also the nature of regenerants produced. In an older sporogonium all parts gave rise to only protonema filaments on regeneration. However, in very young sporogonia the regenerants were in the form of callus, seta tips, or protonema, depending upon the stage of differentiation of the zone from which the tissue was taken. The protonema in turn produced either leafy shoots or apogamous sporogonia, depending upon the exact place of its origin (Fig. 5.8).

LaRue (1930) obtained best results in mosses with setae which were nearly fully grown while the capsules were yet green. Wettstein (1923, 1924a, b, 1925, cited in Ono & Tanaka, 1976) stated that regeneration of setae was observed only at the young upper ends of segments. Redfearn and Meyer (1949) reported that in *Physcomitrium turbinatum* rarely the basal ends of setae also produced protonemata. In *Blasia pusilla* Raudzens and Matzke (1968) observed proliferations (green cell masses) from the setae of sporophytes which had reached at least the four-lobed-spore mother cell stage; setae of younger sporophytes failed to regenerate. Ono and Tanaka (1976) also suggested that in *Haplocladium angustifolium* the ability of regeneration of seta is related to the age of the tissue (see page 118).

CORRELATIVE INHIBITION

Bünning (1953) held the view that in plants embryonic activity at one place does not allow another centre of activity in its vicinity. Different regenerative behaviour of cells may be due to the presence in some cells of the prerequisites for division to a greater degree than in others, and they therefore inhibit division in other cells. In *Atrichum* (Gemmell, 1953), *Funaria* (Bopp, 1955), and *Splachnum* (von Maltzahn & MacNutt, 1958) the separated distal, middle, and proximal regions of leaf showed enhanced regenerative capacity than they had in the whole leaf. This indicates that the cells in one portion of the leaf have an inhibitory effect on the regeneration of cells in other portions. Stange (1957) also refers to correlative inhibition during regeneration in *Riella*. In this liverwort regeneration is always polar. When the thallus is cut into an apical and a basal half, cauloids appear from the axis only near the apical cut surface of the basal half. The apical half regenerates only rhizoids at its base (Fig. 3.9 A,B). Once the new thalli take a lead, regeneration is suppressed elsewhere. When the axis (rib) is separated from the wing the cells of the wing also regenerate, having been freed from the inhibitory influence of the axis. Cauloid regenerants are formed near the cut surface originally attached to the rib (Fig. 3.9 C,D). When the wing is further cut longitudinally even the cells of the marginal part of the wing acquire the capacity to produce cauloids (Fig. 3.9 E).

As a result of further work on *Splachnum ampullaceum* MacQuarrie and von Maltzahn (1959) recognized the following correlative systems: (*i*) between apex and stem, which inhibits basal regeneration; (*ii*) within leaves; (*iii*) within stem; (*iv*) between stem and leaf, presence of leaves increases regeneration from stem, but stem inhibits regeneration from leaves; and (*v*) between apex and axillary buds, which inhibits reactivation.

In *Fossombronia* the axis was segmented transversely into four pieces, each bearing a pair of leaves. None of these pieces developed regenerating buds on the leaves, even though the three posterior pieces lacked the growing point. This indicates that the axis has an inhibitory effect on the regeneration of leaves even when they are not under the influence of a growing point (Mehra, 1976).

As stated earlier, correlative inhibition is also demonstrated by plas-

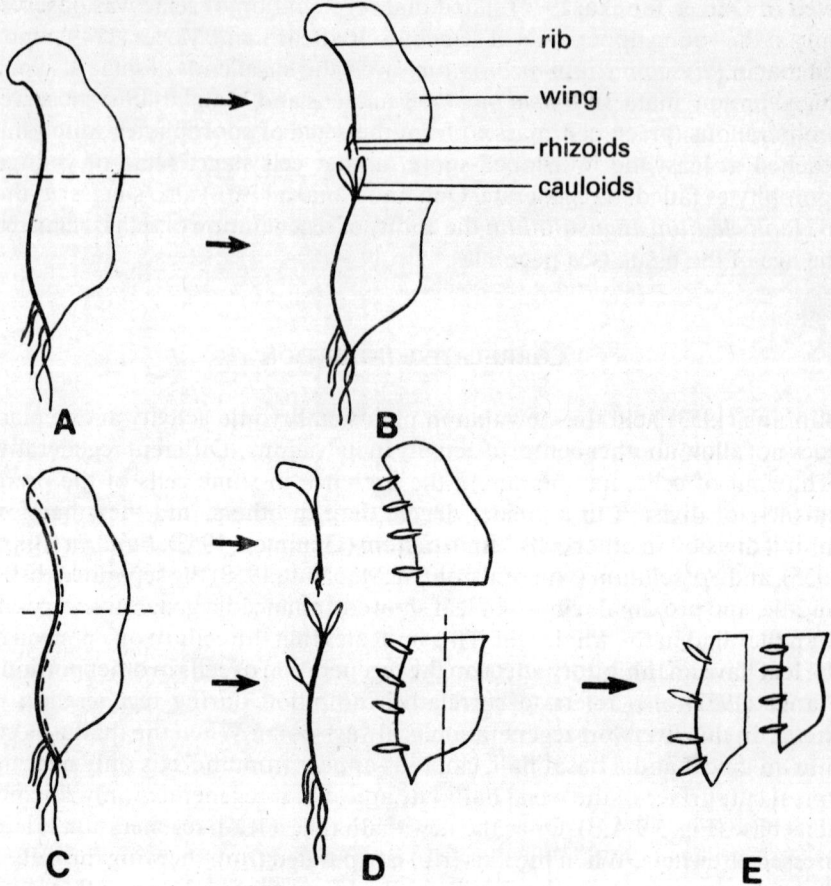

Fig. 3.9 A-E. Location and nature of regenerants after progressive subdivision of the thallus of *Riella helicophylla*.
(After Stange, 1957).

molytic experiments in which the connection between cells is effectively cut off and cells are set free from the influence of the adjoining cells, thus making regeneration possible. In *Blasia* almost all the physiologically isolated cells (in specimens treated with 100 mg/1 MH) behaved as isolated individuals and grew out as germ tubes which recapitulated the stages observed during normal sporeling development (Fig. 3.10; Allsopp & Ilahi, 1970). Earlier, Kreh (1909) had also obtained filamentous stages from the stems of various leafy liverworts following plasmolysis with 20 percent sucrose solution.

Davidonis and Munroe (1972) observed that in *Marchantia polymorpha* the unequal growth of lobes is due to the inhibitory effect of one over the other. The smaller lobe can be released from this inhibition if the two lobes are separated by a vertical cut at an early stage of growth, and then both the lobes grow to an equal size.

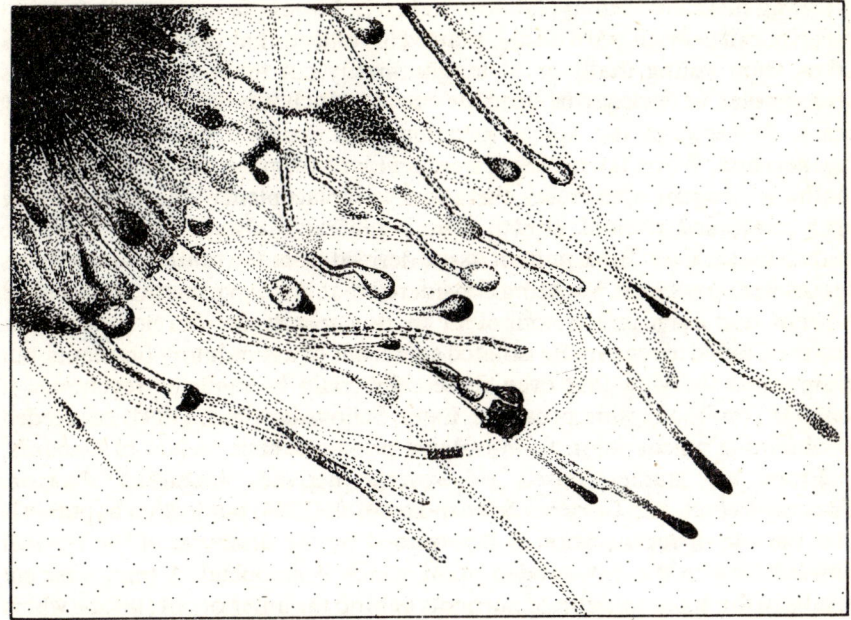

Fig. 3.10. Formation of germ tubes on shoots of *Blasia pusilla* exposed for six weeks to 100 mg/l maleic hydrazide.
(After Allsopp & Ilahi, 1970).

POLARITY AND APICAL DOMINANCE

It is now a well established observation that in an intact plant the presence of meristematic cells prevents the permanent cells from regeneration. As in higher plants, the loss of apical dominance in bryophytes can result in activation of dormant primordia (reactivation), and it also makes some mature cells to revert to the embryonic state (embryonization) and produce new plants (Stange, 1964). In plants with incomplete apical dominance regeneration occurs even with intact apical meristem. The degree of apical dominance varies with different species, and it is also influenced by the age of the plant.

Polarity and apical dominance are exhibited by the sporophytic generation as well. In mosses the seta regenerates from the cut apical end, and this capacity decreases from apical to basal poles (Wettstein, 1924a, *cited in* Ono & Tanaka, 1976). In *Physcomitrium pyriforme* the lowest percentage of regeneration (4%) was exhibited by the segments of setae with attached capsules, and the highest (65%) by whole seta segments. Regeneration invariably occurred from the apical cut region of the segment (Kachroo, 1954). In *Fossombronia* the developing sporogonia did not allow regeneration of the attached perianth (Mehra, 1976).

Most of the investigations on regeneration and polarity in hepatics have been confined to Marchantiales, *Riella* and Metzgeriales. As early as 1885

Vöchting demonstrated apical dominance in *Lunularia* and concluded that polarity exists in all parts of the plant. He noticed that in transverse pieces taken from young thalli or in longitudinally cut thalli regeneration was always polar, i.e., it occurred from the cut end which was nearer to the growing point. Removal of the apical meristem was essential for the process of regeneration. When lateral lobes were removed from the anterior part of the thallus leaving the apex intact, there was no replacement of the lobes, but the apex continued normal growth. The regenerants always arose from the midrib tissue. Even from pulp of tissue derived from the midrib adventitious shoots were produced. Vöchting considered that this experiment approached a proof that each cell is totipotent in its capacity for development. He, however, failed to regenerate isolated epidermis or assimilatory filaments, but maintained the view that even these cells could be made to regenerate if suitable conditions were provided. Later, Schostakowitsch (1894) succeeded in obtaining regeneration from isolated photosynthetic tissues of *Corsinia*.

From his studies based on several hepatics including *Preissia*, *Conocephalum*, and *Corsinia* Schostakowitsch (1894) put forth a hypothesis that the site of regeneration is determined by the direction of the normal nutrient flow in the unwounded plant, which is acropetal. A transverse cut results in the accumulation of nutrients behind the anterior cut surface where the regenerants arise. In transverse segments of *Blyttia* (syn *Pallavicinia*) as well regenerants were observed from the anterior end (Goebel, 1908). According to Goebel (1908) the meristems act as 'attraction centres' for building materials. When the permanent cells lose contact with the meristematic cells, such substances are available for cell divisions in the permanent cells themselves. From the results of his experiments on *Splachnum*, von Maltzahn (1961) assumed that accumulation of substances necessary for renewed synthetic activities leads to regeneration. In an intact plant these substances are transported to sites of incorporation at the apical meristem. Removal of this centre of growth results in the establishment of secondary sites of incorporation. Kaul et al. (1962) also consider that nutritional disruption plays a role in the regeneration from cut surface of liverwort thalli. In *Fossombronia* the isolated leaves or their parts produced buds mostly near the cut (Fig. 3.11). Mehra (1976) suggests that this is because of nutritional factors. The food produced in the leaves travels towards the axis and in detached leaves it gets blocked near the cut end which was originally in organic connection with the axis.

Westerdijk (1907) reported that moss plants produce protonemata only after removal of the apex. Evans (1910) considered that an active thallus apex inhibits the formation of adventitious branches in species of *Metzgeria*. In *Rhytidiadelphus*, *Hylocomium*, and *Dicranum* stem segments developed regenerants at the base. However, in *Fissidens* protonema was produced at the apices of segments (LaRue, 1930). Dickson (1932) observed that when a zone of cells was killed with UV light in the thallus of *Marchantia polymorpha*, regeneration occurred behind the dead tissue. He concluded that basal regeneration occurs in response to the destruction of polarity.

Apical dominance is expressed by an auxin-mediated flow of nutrients, and the midrib plays an important role in the transport of auxins.

Narayanaswami and LaRue (1955) and LaRue and Narayanaswami (1955, 1957) demonstrated that regeneration of adventitious thalli and germination (reactivation) of gemmae within the cups are inhibited by the apex in intact thalli of *Lunularia*. Application of IAA to the cut apical region substituted the effects of apical meristem. They proposed that apical dominance was expressed through the production of auxin in the apex. An explanation of the hormone mechanism of apical dominance demands a hypothesis that concentrations of auxin which promote the elongation of the main shoot also prevent growth of lateral buds (Stange, 1964). Allsopp (1967) opines that the polarized flow of auxin and other metabolites is the result and not the cause of polarity.

MacQuarrie and von Maltzahn (1959) observed that apical dominance in *Splachnum* inhibited basal stem regeneration, as also reactivation and growth of lateral branch primordia (Table 3.3). Both these influences gradually decrease with age, and later considerable amount of basal regeneration and lateral branch growth occurs in intact plants. This is possibly due to depletion of the apex during gametangial development. In this moss as well, application of IAA to the tips of decapitated plants replaced apical dominance (Table 3.4). Similar concentrations of IAA inhibited regeneration from both isolated stems and leaves; and at lower concentration some promotion was observed with IAA. Leaves isolated from intact and decapitated plants showed differences in regeneration behaviour when treated with IAA. This indicates that IAA could be involved in the

TABLE 3.3
The effect of decapitation on lateral branch reactivation and growth in *Splachnum*.

Treatment	% Bud reactivation	Branch length (mm)
Intact	29	0.47
Decapitated	97	0.63

Source. MacQuarrie and von Maltzahn, 1959.

TABLE 3.4.
The effect of decapitation and IAA application on gametophore bud reactivation in *Splachnum*.

Treatment	% Bud reactivation
Intact	12
Decapitated	57
Decapitated + IAA (10^{-5} mg/ml)	42
Decapitated + IAA (0.1 mg/ml)	12
Decapitated + IAA (1.0 mg/ml)	0

Source. MacQuarrie and von Maltzahn, 1959.

80 *Biology of Bryophytes*

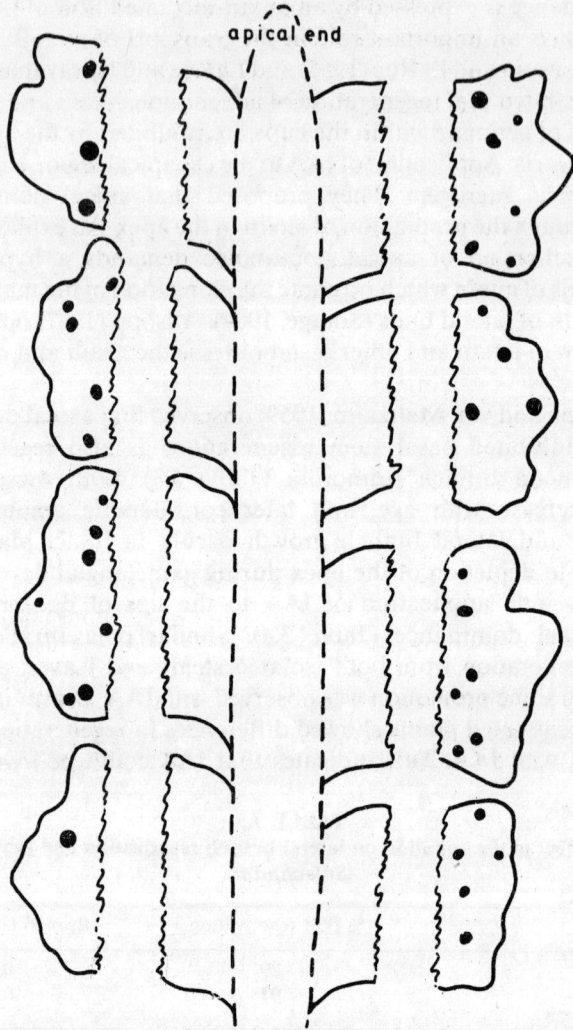

Fig. 3.11. Diagram of a shoot of *Fossombronia himalayensis* with leaves on either side cut longitudinally into halves. The adventitious buds appear only on the excised half-leaves, mostly near the cut.
(After Mehra, 1976).

inhibition or regeneration from attached leaves. However, the authors conclude that IAA cannot be the primary factor in the stem-leaf inhibition, since leaves attached to decapitated stems failed to regenerate in *Splachnum*, as also earlier reported in *Physcomitrium* (Meyer, 1942) and *Atrichum* (Gemmell, 1953).

In *Splachnum* kinetin promoted lateral bud reactivation, but it did not influence the reduction of basal regeneration caused by IAA. From these experiments MacQuarrie and von Maltzahn (1959) concluded that the mechanisms of correlation mediating apical dominance in terms of inhibition of lateral bud reactivation and the depression of basal protonemal regeneration are not identical.

Ward (1960) demonstrated that in *Polytrichum commune* regeneration results after the removal or reduction of the inhibitory forces exerted through the stem by the growing leaves and shoot apex. Further, early appearance of vigorous adventitious buds after decapitation prevents regeneration of leaves below. In *Plagiomnium cuspidatum* also (Nyman & Cutter, 1981) many of the dormant buds initially become activated on decapitation, but buds in the physiological top position of the axis reimpose inhibition on the activated buds below.

When detached leaves of *Polytrichum* were planted with base in agar medium containing 10 mg/l NAA the percentage of regeneration decreased from 100 (controls) to 57, and size of the buds was also reduced. In order to test polarity of growth substance transfer, some stem segments were inverted. The controls planted upright in the same medium (with 20 mg/l NAA) showed about three times the propagative activity in the same period (Ward, 1960). It has been demonstrated in two foliose Metzgeriales, *Noteroclada confluens* and *Blasia pusilla* (Allsopp & Ilahi, 1970) and a thalloid member, *Moerckia flotowiana* (Ilahi & Allsopp, 1970) that apical dominance can be impaired by pre-treatment of intact gametophytes with supra-optimal concentration of 2,4-D. In *Noteroclada* the dormant intercalary branch primordia were activated, whereas in *Blasia* adventitious shoots developed in response to hormone treatment. This behaviour was similar to that of decapitated shoots. Prior and Brown (1970) observed that auxin suppressed apical cell activity and cell elongation in *Marchantia*. Davidonis and Munroe (1972) demonstrated that in *Marchantia* the inhibitory effect of one lobe over the other could be replaced by the application of IAA to the smaller lobe after giving a vertical cut at an early stage of growth. They have also educed evidence that the two lobes are differentially sensitive to auxin inhibition. The larger, faster growing lobe is less sensitive to the same auxin concentration that inhibits growth of the smaller lobe. Nutrient supply also seems to affect sensitivity to auxins. Optimum nutrient supply weakens apical dominance.

Investigations of Allsopp and Ilahi (1969, 1970) on hepatics suggest that variations in exogenously supplied carbohydrates also markedly influence morphogenesis, and this effect may be related to expression of auxin control. In *Noteroclada* Allsopp and Ilahi (1970) demonstrated that apical dominance of intact shoots was impaired by treatment with supra-optimal concentration of glucose. On media with 3 or 4 percent glucose a large number of segments were formed from all parts of the shoots, in addition to the development of intercalary branches because of the reactivation of dormant branch primordia. Sucrose in the medium brought about in situ germination of gemmae in gemma cups of *Marchantia nepalensis*, and this effect increased with increase in light level. This observation indicates that sugar in some way influences auxin gradient in the thallus, which in turn annuls apical dominance (Chopra & Sood, 1970).

Work on regeneration of Marchantiales has not only revealed totipotency of the vegetative cells of thalli, but has also demonstrated that polarity is of two types: the longitudinal (apico-basal) polarity, and the dorsiventral polarity. The former is demonstrated during the regeneration process, whereas the latter is observed during the development of dorsiventral thalli of *Marchantia* from the isolateral gemmae (see Halbsguth, 1965).

Rota and Maravolo (1975) studied sucrose transport in *Marchantia polymorpha*. They observed that higher ^{14}C-sucrose levels were mobilized to the apical region during regeneration for enhanced metabolism. When the apical meristem was left intact, the level and rate of apical transport and accumulation of isotopic sucrose was substantially reduced. When IAA was supplied to the midribs of thalli without apices, sucrose transport and thallus regeneration were significantly inhibited. Thus, a correlation between auxin inhibition and nutrient availability is demonstrated, which may play a role in apical dominance. In higher plants as well, basipetal transport of auxin is known to affect nutrient transport along the entire length of the plant. Rota and Maravolo (1975) also observed that sucrose transport occurred in the midrib and predominantly acropetally. Regeneration always took place subjacent to the apical cut from the ventral surface. Further, regeneration occurred sooner in young tissues, and young plants were more efficient in movement of sucrose. These authors conclude that in *Marchantia* regeneration occurs through a polar mobilization of nutrients to the excised apical region. Auxin in the apex is responsible for apical dominance, and it inhibits regeneration by suppressing acropetal transport of sucrose.

Gourgaud (1965) noticed that young protonema of *Funaria hygrometrica* cultured in light of 1500 lux shows the phenomenon of apical dominance comparable to that in higher plants. Subsequently, Gourgaud (1966) reported that in the protonema of *Ceratodon purpureus* apical dominance becomes weak under higher light level. Knoop (1976) demonstrated that maintenance of cell differentiation in the caulonema of *Funaria hygrometrica* is also a phenomenon of apical dominance. If a specific minimum growth rate is not achieved intercalary cell divisions occur, and this represents an early step of regeneration. He was able to suppress apical dominance by application of inhibitors of protein synthesis to the protonema of this moss.

Stange (1977) experimented on the unistratose gemmae of *Riella helicophylla* and observed that the duration of the cell reproduction cycle increases from the margin to the median part of the gemmae. This polarization within the meristem disappears after addition of the antiauxin PCIB to the culture medium. This antiauxin leads to a retardation or blockage of the cell cycle during the light period of the culture. Under the influence of PCIB the amount of starch in the chloroplasts is strikingly increased, probably because of a reduction of starch degradation. Addition of sugars compensates the effect of PCIB on the cell cycle. The effects of antiauxin on cell cycle are counteracted by auxin. The results suggest that auxin plays a role in directing the transport of substances needed for the continuation of cell reproduction cycle between adjacent cells of the meristem.

von Maltzahn (1959) suggested that in the gametophore of *Splachnum* reactivation of lateral buds is controlled by an interaction between auxin and kinetin. Based on a closer SEM study of reactivation of lateral buds in *Plagiomnium cuspidatum* Nyman and Cutter (1981) reported that IAA in the presence of a cytokinin can maintain inhibition in a morphological and anatomical state which is similar to that of intact plants. Wherever growth of inhibited buds occurs, endogenous auxins and cytokinins appear to be at sub-optimal concentrations. Treatment of intact gametophores with the antiauxin TIBA, increased bud reactivation response below the treatment ring.

Experiments of Gaal et al. (1982) on *Marchantia polymorpha* using exogenous ^{14}C-IAA revealed that its transport is chiefly localized in the midrib cells of thalli. Transport occurs both acropetally and basipetally at approximately the same velocity (14 mm/h; a velocity which approximates the basipetal velocity recorded in tracheophytes). However, the basipetal transport is significantly greater in intensity. The presence of naturally occurring auxin in gametophytic tissue of *M. polymorpha* has been demonstrated both by chemical methods and bioassays (see Gaal et al., 1982). The correlative activity of endogenous growth regulators is dependent upon absolute concentrations, gradients, differential rates of synthesis, inactivation, and the direction and intensity of transport. The polar distribution of auxin in *Marchantia* has important role in apical dominance, gemma-cup formation, and rhizoid initiation through its influence on the balance between cell division and elongation (Gaal et al., 1982).

The physiological basis for the phenomenon of apical dominance/ reactivation/regeneration remains unclear. The three possibilities suggested are:
(1) basipetal flow of inhibitory substances;
(2) basipetal flow of auxin; and
(3) acropetal flow of nutrients.

The final explanation may lie in the interaction between auxin with other endogenous factors like kinins and/or carbohydrates.

CHANGES OCCURRING IN REGENERATING CELLS

In an intact plant there is great stability in the functions of mature cells, and this is responsible for normal development. De-differentiation includes processes which bring back the embryonic state in a mature cell.

MacNutt and von Maltzahn (1960) observed that in the isolated leaves of *Splachnum ampullaceum* the first visible sign of de-differentiation is an increase in chloroplast number accompanied by a decrease in their size. The size of nucleolus and nucleus also increases considerably. The chloroplasts are initially in peristrophe (arranged along the periphery). Within a few hours after isolation changes in the position of cellular components occur by the formation of an intravacuolar cytoplasmic layer. The nucleus moves to a more central position in the cell. The chloroplasts also migrate to a central position and form a tight cluster around the nucleus, leading to systrophe (Fig. 3.12 A-E). Systrophe is an important event in de-differentiation, and random distribution of such clusters denotes loss of cellular polarity. A new axis of polarity is established during re-differentiation and this is followed by cell division and formation of a regenerative protonema. The protonemal initial has denser cytoplasm and a much larger nucleus and nucleolus (Fig. 3.12 F,G). Systrophe was also reported in the leaves of *Lophocolea* by Heitz (1925) and by Mehra (1976) in regenerating cells of *Sewardiella*. Increase in nuclear and nucleolar size in regenerating cells has also been reported in *Riella helicophylla* (Stange, 1957) and *Mnium affine* (Giles, 1971). In *Funaria hygrometrica* and *Lophocolea bidentata* Heitz (1925) observed increase in nuclear volume in regenerating as well as non-regenerating cells of isolated leaves, but nucleolar volume increased only in regenerating cells. Increased nucleolar size indicates active protein synthesis.

von Maltzahn and Mühlethaler (1962a) did electron microscopic studies on *Splachnum* and noticed simple fission and budding in the chloroplasts of regenerating cells. The mitochondria also bud out small spherical electron dense bodies (von Maltzahn & Mühlethaler, 1962b). These bodies represent mitochondrial initials and are double membraned, but both the membranes are smooth. This process is named as 'budding fission' and is similar to that described for chloroplasts. These initials eventually develop into typical mitochondria (Fig. 3.13 A-D).

Before the cells in isolated fragments can reinitiate DNA synthesis, their differentiation status must change, a process that is recognizable through pronounced cytological and metabolic alterations (Stange, 1970). The activation of metabolic processes in the regenerating cells simulates that in the meristematic cells in an intact plant. It is assumed that in regenerating lower plants activation starts just after the cells are separated from the meristem (Stange, 1970; Viell, 1977). Giles (1971) concluded that the factors influencing regeneration must be of a biochemical nature, probably involving RNA metabolism. Giles and Taylor (1971) observed that during de-differentiation initiated by removal of leaves from the stem of *Funaria hygrometrica*, replication of nuclear DNA is followed by chloroplast DNA replication and formation of dumb-bell-shaped chloroplasts.

Stange (1957) finds *Riella helicophylla* to be very suitable for studies on biochemical changes accompanying regeneration. Investigations using ^{32}P revealed a steep increase in the uptake of phosphate between 12 and 36 h, and also synthesis of RNA and DNA after about 10 h of isolation of cells. Later, Stange (1970) reported that in this liverwort RNA increases exponentially from the 6th up to the 36th h, which was the last point measured. Protein synthesis also got accelerated during regeneration from the 5th up to the 35th h (Troue, 1974, *cited in* Viell, 1977).

Viell (1977) used microanalytic techniques and demonstrated that regenerating cells in small (micro) fragments of *Riella* synthesis of protein

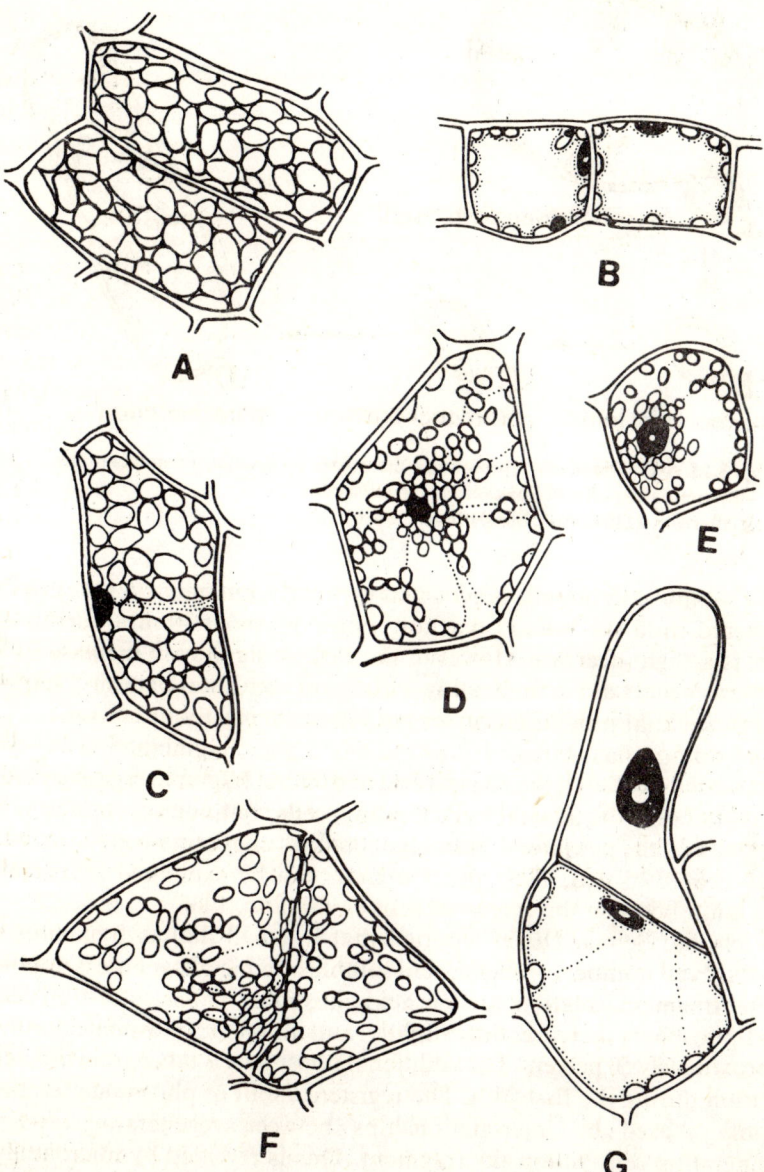

Fig. 3.12 A-G. Changes in regenerating cells of leaf in *Splachnum ampullaceum*. A. Surface view of cells immediately after leaf isolation. B. Cross-sectional view of the same. C. Surface view of a cell 6 h after isolation, showing primary cytoplasmic layer formation. D. Surface view of a cell 48 h after leaf isolation, showing systrophe. E. Cross-sectional view of the same. F. Surface view of a cell 48-60 h after leaf isolation, showing completion of new cell wall formation. G. Cross-sectional view of protonematal initial and the remaining cell. Contents other than nucleus have been omitted from the former. (After MacNutt & von Maltzahn, 1960).

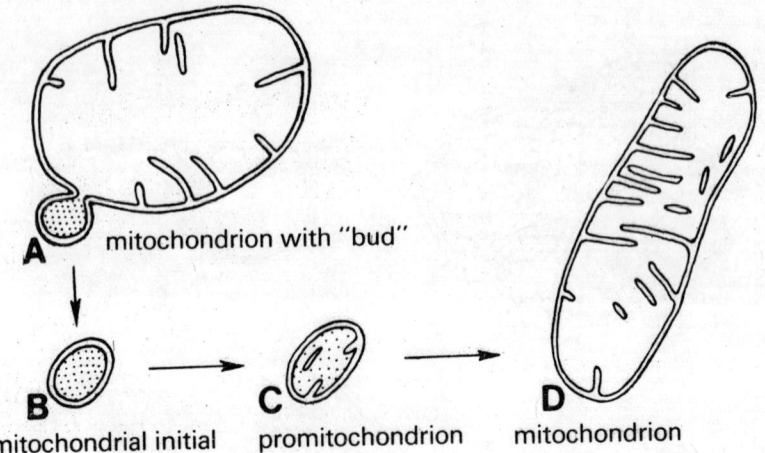

Fig. 3.13 A-D. Diagrams representing the stages in division of mitochondria in de-differentiating cells of *Splachnum ampullaceum*.
(After von Maltzahn & Mühlethaler, 1962b).

and ∝-amino compounds is activated very early, but only after the cells have recovered from the 'wound reaction' phase in about 30 min during which there is a slight decrease. However, the earliest significant increase in RNA synthesis occurs at 8 h, indicating a different activation pattern. Changes in the intracellular protein precursor pool indicate metabolic conversions long before mitosis has started. Up to the 8th h after fragmentation the free ∝-amino content of both the adaxial and peripheral fragment regions increases. Thereafter only the adaxial (regenerating) cells continue accumulating these compounds; the peripheral (non-regenerating) ones remain at the same level (Figs 3.14, 3.15 A,B). This phenomenon can be explained by correlative inhibition between the fragment cells (Lehmann, 1966).

Viell and Schaar (1979) reported that in small thallus fragments (with constant cell number) of *Riella*, acid soluble organic phosphate is doubled 48 h after fragment isolation; nucleic acid phosphate increases 20 to 30 percent; lipid-phosphate increases only slightly; and inorganic phosphate diminishes approximately 50 percent. The additional phosphate is taken from the nutrient solution during the first 24 h. The registered shift of phosphate fractions is mainly caused by interrelationships between regenerating and non-regenerating cells within the fragment. This is revealed by microanalytical analysis of the adaxial and peripheral fragment regions. The results show that in 48 h after fragment isolation inorganic phosphate is enhanced in the regenerating cells and is diminished in the non-regenerating ones, which apparently serve as phosphate donors. ^{32}P is transferred quickly between the two cell types in the isolated fragments in the same manner as in the intact plant.

In 1980 Viell demonstrated, by using the same technique, that in 72 h the regenerating cells in the small fragments of *Riella* elevate the phenolic

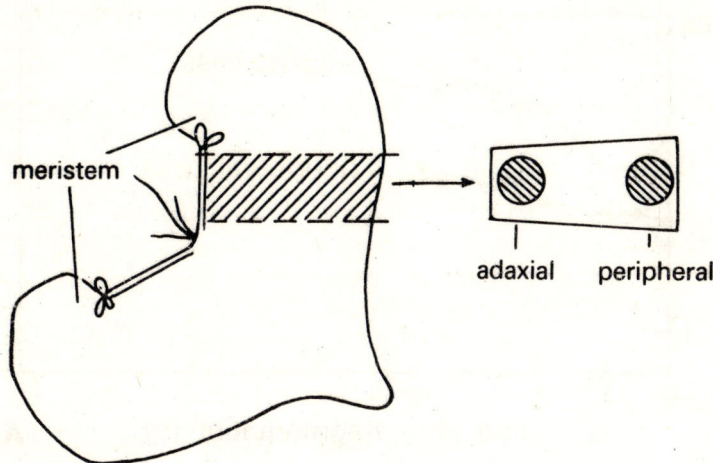

Fig. 3.14. Schematic representation of a three-week-old plant of *Riella helicophylla*. A large fragment was cut from the middle region of the thallus and later micro-fragments were analyzed from its adaxial and peripheral regions. (After Viell, 1977).

content to about 300 percent. The non-regenerating cells accumulate less phenolics. It is estimated that regenerating cells reach a phenolic level similar to that in meristematic cells of this liverwort just at a time when they reinitiate DNA synthesis.

By studying the uptake of labelled precursors (^{3}H-uridine; ^{3}H-leucine; ^{3}H-thymidine) by de-differentiating cells Gay (1982) concluded that in isolated leaves of *Polytrichum juniperinum* onset of regeneration is marked by the following metabolic changes:

(1) Protein synthesis is mainly reactivated during the first hours (between 0 to 16 h), and then goes on at a slower rate;
(2) RNA synthesis is observed in a few cells 6 h after isolation of the leaf, but the main synthesis is after 36 h;
(3) Synthesis of DNA in the nucleus occurs on the 4th day of culture, and it is a very important step since its inhibition with 5-Fd Urd at 5.10^{-4}M suppresses regeneration;
(4) Following this synthesis, mitosis occurs after five days of culture;
(5) Further incorporation of labelled precursors in nucleic acids (on the 7th and 8th day) is related to the development of the regenerated protonema.

This sequence of events is similar to that observed during regeneration in the thallus of *Riella*. In his further investigations on *Polytrichum juniperinum* Gay (1984) used specific inhibitors and noticed that RNA and then protein synthesis are required during the light stimulus mediated by phytochrome, but DNA synthesis is not needed. Earlier, RNA and protein synthesis had been shown to be essential for the effect of light on the mitotic

88 Biology of Bryophytes

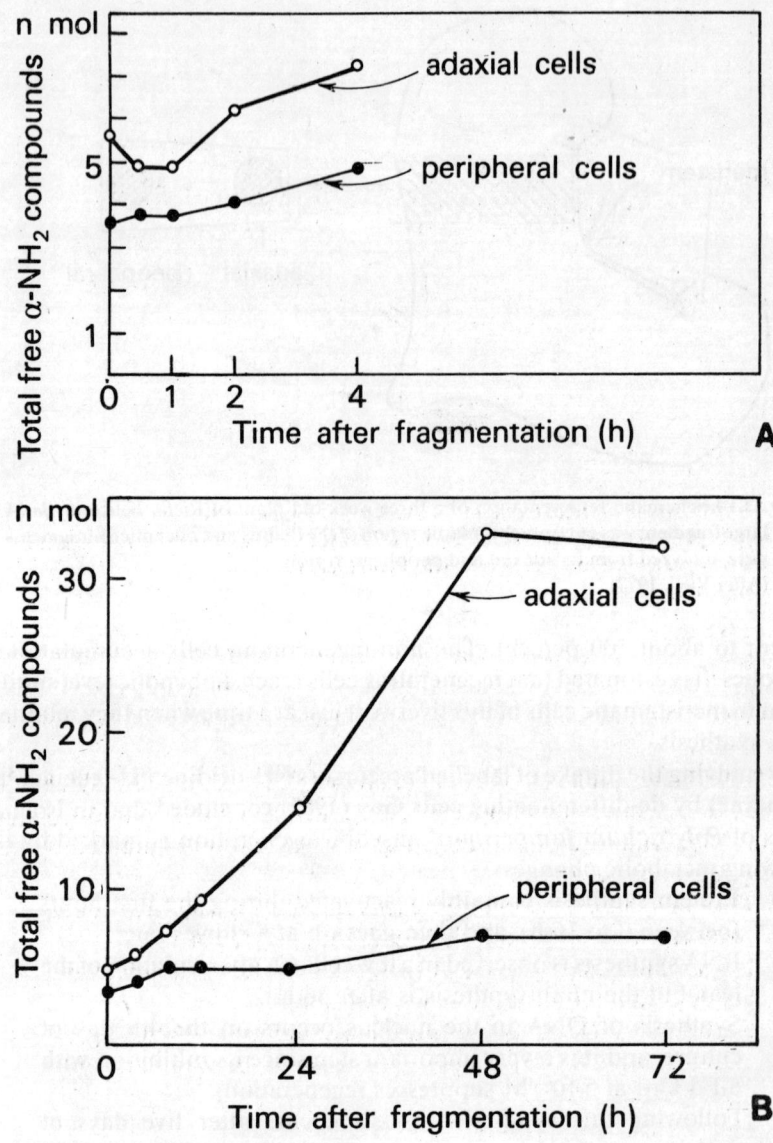

Fig. 3.15 A,B. Changes in the overall concentration of free α-amino compounds in the adaxial and peripheral regions of a regenerating larger fragment during the first (A) and later (B) hours after isolation in *Riella helicophylla*. Each value, expressed in n mol/ 5000 cells, represents the mean of nine separate determinations.
(After Viell, 1977).

factor in moss protonema (Larpent-Gourgaud & Aumaître, 1971, 1976, *cited in* Gay, 1984). For the duplication of chloroplasts in the spores of *Polytrichum* light is ineffective if protein synthesis is inhibited (Kass & Paolillo, 1974).

REFERENCES

ALLSOPP, A., *Phytomorphology*, 17, 364 (1967).
ALLSOPP, A. and I. ILAHI, *Phytomorphology*, 19, 242 (1969).
ALLSOPP, A. and I. ILAHI, *Phytomorphology*, 20, 173 (1970).
BAUER, L., *J. Linn. Soc. (Bot.)*, 58, 343 (1963).
BERTHIER, J., J.P. LARPENT, and M. LARPENT-GOURGAUD, *J. Hattori bot. Lab.* No. 41, 193 (1976).
BOPP, M., *Z. Bot.*, 40, 119 (1952).
BOPP, M., *Z. Bot.*, 41, 1 (1953).
BOPP, M., *Rev. Bryol. Lichénol.*, 24, 49 (1955).
BOPP, M., *Rev. Bryol. Lichénol.*, 30, 253 (1961).
BOPP, M.,"Developmental physiology of bryophytes". In R.M. Schuster, Ed., *New Manual of Bryology*, The Hattori Botanical Laboratory, Nichinan, Miyazaki, Japan, 1983. Pp. 276-324.
BÜNNING, E., Entwicklungs- und Bewegungs-physiologie der Pflanze, Springer-Verlag, Berlin, 1953.
BÜNNING, E.,"Regeneration bei Pflanzen". In F. Buchner, Ed., *Handbuch der allgemeine Pathologie*, Springer-Verlag, Berlin, 1955. Pp. 384-404.
CHOPRA, R.N. and A. RASHID, *Bryologist*, 70, 206 (1967).
CHOPRA, R.N. and S. SOOD, *Bryologist*, 73, 592 (1970).
CHOPRA, R.S. and N.N. BHANDARI, *Research Bulletin (N.S.) Panjab University*, 11, 87 (1960).
CHOPRA, R.S. and S.S. KUMAR, *Bull. bot. Soc. Univ. Saugar*, 13, 65 (1961).
CHOPRA, R.S. and P.D. SHARMA, *J. Indian bot. Soc.*, 35, 117 (1956).
CHOPRA, R.S. and P.D. SHARMA, *J. Indian bot. Soc.*, 37, 353 (1958).
COMEAU, G. and F. Le Blanc, *Natur. Can.*, 98, 347 (1971).
CORRENS, C., *Untersuchungen über die Vermehrung der Laubmoose durch Brutorgane und Stecklinge*, Jena, 1899.
DAVIDONIS, G.H. and M.H. MUNROE, *Bot. Gaz.*, 133, 177 (1972).
DEMKIV, O.T., *Ukr. Bot. J.*, 28, 624 (1971).
DICKSON, H., *Ann. Bot.*, 46, 683 (1932).
DOUIN, C., *Rev. gen. Bot.*, 35, 213 (1923).
EGUNYOMI, A., A.J. HARRINGTON, and S.O. OLARINMOYE, *Cryptog. Bryol. Lichénol.*, 1, 73 (1980).
EVANS, A.W., *Ann. Bot.*, 24, 271 (1910).
FELLNER, F., *Jber. akad. naturw. Ver., Graz.*, 1, 41 (1875).
FULFORD, M., *Phytomorphology*, 6, 199 (1956).
GAAL, D.J., S.J. DUFRESNE, and N.C. MARAVOLO, *Bryologist*, 85, 410 (1982).
GAY, L., *Z. PflPhysiol.*, 66, 1 (1971).
GAY, L., *Sciences*, 4, 171 (1974).
GAY, L., *Z. PflPhysiol.*, 106, 337 (1982).
GAY, L., *Physiol. Plant.*, 61, 95 (1984).
GEMMELL, A.R., *Trans. Br. bryol. Soc.*, 2, 203 (1953).
GILES, K.L., *New Zealand J. Bot.*, 9, 689 (1971).
GILES, K.L. and K.E. von MALTZAHN, *Bryologist*, 70, 312 (1967).
GILES, K.L. and K.E. von MALTZAHN, *Can. J. Bot.*, 46, 305 (1968).
GILES, K.L. and A.O. TAYLOR, *Pl. Cell Physiol.*, 12, 437 (1971).
GOEBEL, K., *Einleitung in die experimentelle Morphologie der Pflanzen*, Teubner, Leipzig, 1908.
GOURGAUD, M., *C.R. Acad. Sci., Paris*, 260, 1721 (1965).
GOURGAUD, M., *C.R. Acad. Sci., Paris*, 263, 1220 (1966).
GROSSMAN, H.H. and C.J. HILLSON, *Bryologist*, 79, 488 (1976).
GUNCKEL, J.E. and A.H. SPARROW, "Ionizing radiations: biochemical, physiological, and morphological aspects of their effects on plants". In W. Ruhland, Ed. *Handbuch der Pflan-*

zenphysiologie, Gottingen, 1961. Pp. 555-611.
HAHN, H. and J.H. MILLER, Physiol. Plant., 19, 134 (1966).
HALBSGUTH, W., "Induktion von Dorsiventralität bei Pflanzen". In W. Ruhland, Ed., Encyclopaedia of Plant Physiology, 15, Berlin, 1965. Pp. 331-382.
HEALD, F.D.F., Bot. Gaz., 26, 169 (1898).
HEITZ, E., Z. Zellforsch., 2, 69 (1925).
HERZFELD, S., Oester. Bot. Zeit., 72, 788 (1923).
HUGHES, J.G., New Phytol., 68, 883 (1969).
ILAHI, I. and A. ALLSOPP, Phytomorphology, 20, 126 (1970).
KACHROO, P., J. Indian bot. Soc., 33, 263 (1954).
KANDA, H., Memoires National Inst. Polar Res., Special issue No. 11, 58 (1979).
KARSTEN, I., Naturwissenschaften, 50, 50 (1963).
KASS, L.B. and D.J. PAOLILLO JR., Pl. Sci. Lett., 3, 81 (1974).
KAUL, K.N., G.C. MITRA, and B.K. TRIPATHI, Ann. Bot., 26, 447 (1962).
KNOOP, B., Z. PflPhysiol., 70, 22 (1973).
KNOOP, B., Z. PflPhysiol., 77, 350 (1976).
KREH, W., Nova Acta Acad. Caesar Leop. Carol., 90, 213 (1909).
KUMRA, P.K. and R.N. CHOPRA, Cryptog. Bryol. Lichénol., 1, 197 (1980).
LARPENT-GOURGAUD, M. and M.P. AUMAÎTRE, Z. PflPhysiol., 83, 467 (1977).
LARPENT-GOURGAUD, M. and R. JACQUES, C.R. Acad. Sci., Paris, 273, 162 (1971).
LARPENT-GOURGAUD, M., J.P. LARPENT, and R. JACQUES, Physiol. Veg., 10, 553 (1972).
LARPENT-GOURGAUD, M., J.P. LARPENT, and R. JACQUES, C.R. Acad. Sci., Paris, 279, 57 (1974).
LaRue, C.D., Pap. Mich. Acad. Sci., Arts and Lett., 11, 225 (1930).
LaRue, C.D. and S. NARAYANASWAMI, Bull. Torrey bot. Club, 82, 198 (1955).
LaRue, C.D. and S. NARAYANASWAMI, New Phytol., 56, 61 (1957).
LEHMANN, H., Planta, 71, 240 (1966).
MacNutt, M.M. and K.E. von MALTZAHN, Can. J. Bot., 38, 895 (1960).
MacQuarrie, I.G. and K.E. von MALTZAHN, Can. J. Bot., 37, 121 (1959).
von MALTZAHN, K.E., Nature, 183, 60 (1959).
von MALTZAHN, K.E., Nature, 192, 55 (1961).
von MALTZAHN, K.E. and M.M. MacNutt, Can. J. Bot., 36, 33 (1958).
von MALTZAHN, K.E. and K. MÜHLETHALER, Naturwissenschaften, 49, 308 (1962a).
von MALTZAHN, K.E. and K. MÜHLETHALER, Experientia, 18, 315 (1962b).
MEHRA, P.N., J. Hattori bot. Lab. No. 40, 521 (1976).
MEHRA, P.N. and D. PENTAL, J. Hattori bot. Lab. No. 40, 151 (1976).
MEHRA, P.N. and B.R. VASHISHT, Bryologist, 53, 89 (1950).
MENON, M.K.C. and M. LAL, Ann. Bot., 41, 1179 (1977).
MEYER, S.L., Am. J. Bot., 27, 221 (1940).
MEYER, S.L., Bot. Gaz., 104, 128 (1942).
MONTAGUE, M.J. and J. TAYLOR, Bryologist, 74, 18 (1971).
NAGAI, J., Bot. Mag. Tokyo, 33, 99 (1919).
NARAYANASWAMI, S., Phytomorphology, 6, 323 (1957).
NARAYANASWAMI, S. and M. LAL, Phytomorphology, 7, 244 (1957).
NARAYANASWAMI, S. and C.D. LaRue, Phytomorphology, 5, 356 (1955).
NOGUCHI, A. and H. FURUTA, J. Hattori bot. Lab. No. 17, 32 (1956).
NOGUCHI, A. and H. FURUTA, Bryologist, 61, 361 (1958).
NOGUCHI, A. and I. MIYATA, Kumamoto J. Sci. Ser. B, Sec. 2, 3, 1 (1957).
NOGUCHI, A. and S. MURAOKA, Kumamoto J. Sci. Ser. B, Sec. 2, 4, 118 (1959).
NYMAN, L.P. and E.G. CUTTER, Can. J. Bot., 59, 750 (1981).
ONO, K. and R. TANAKA, J. Jap. Bot., 51, 45 (1976).
PEARMAN, C., "Morphogenetic studies in Hepaticae", M.Sc. Thesis, Univ. Manchester, 1964.
PRIOR, P.V. and P.R. BROWN, Bryologist, 73, 687 (1970).

PROSKAUER, J., *Phytomorphology*, **11**, 359 (1961).
RAUDZENS, L. and E.B. MATZKE, *Am. J. Bot.*, **55**, 1190 (1968).
REDFEARN, P.L. and S.L. MEYER, *Bryologist*, **52**, 197 (1949).
REESE, W.D., *Bryologist*, **58**, 239 (1955).
ROTA, J.A. and N.C. MARAVOLO, *Bot. Gaz.*, **136**, 184 (1975).
SAX, K., *Radiat. Bot.*, **3**, 179 (1963).
SCHOSTAKOWITSCH, W., *Flora*, **79**, 350 (1894).
SCHUSTER, R.M., *The Hepaticae and Anthocerotae of North America*, Vol. IV, Columbia University Press, New York, 1980.
SELKIRK, P.M., *Bryologist*, **83**, 542 (1981).
STANGE, L., *Z. Bot.*, **45**, 197 (1957).
STANGE, L., *Z. Bot.*, **48**, 143 (1960).
STANGE, L., *Adv. Morphogen.*, **4**, 111 (1964).
STANGE, L., *Z. PflPhysiol.*, **63**, 84 (1970).
STANGE, L., *Planta*, **135**, 289 (1977).
STANGE, L., *J. Hattori bot. Lab.* No. 53, 249 (1982).
UDAR, R., *J. Indian bot. Soc.*, **36**, 580 (1957).
UDAR, R. and V.B. SINGH, *Curr. Sci.*, **27**, 23 (1958).
VASHISTHA, B.D., "In vitro investigations on some Indian bryophytes", Ph.D. Thesis, Univ. Delhi, 1985.
VASHISTHA, B.D. and R.N. CHOPRA, *Phytomorphology*, **36**, 299 (1986).
VIELL, B., *Planta*, **137**, 13 (1977).
VIELL, B., *Z. PflPhysiol.*, **99**, 55 (1980).
VIELL, B. and I. SCHAAR, *Biochem. Physiol. Pflanzen.*, **174**, 780 (1979).
VÖCHTING, H., *Jb. wiss. Bot.*, **16**, 367 (1885).
WARD, M., *Phytomorphology*, **10**, 325 (1960).
WESTERDIJK, J., *Botan. Zentr.*, **104**, 658 (1907).
WILMOT-DEAR, C.M., *J. Bryol.*, **11**, 145 (1980).
ZEPF, E., *Z. Bot.*, **40**, 87 (1952).

4 Reproductive Biology

PERIODICITY of vegetative growth and sexual reproduction in bryophytes is controlled by factors like light, temperature, humidity, hydration, carbohydrates, nitrogenous substances, growth regulators, chelating agents, and pH. These factors also interact with one another. Under cultural conditions induction of gametangia has been somewhat difficult. Long-day, short-day, and day-neutral plants are known, and in some taxa temperature is the main factor for the onset of reproductive phase. The physical factors seem to act either through the synthesis of growth hormones in the system, or by the removal of growth inhibitors (Chopra & Bhatla, 1983).

FACTORS AFFECTING GAMETANGIAL INDUCTION

LIGHT DURATION

After the historic discovery of photoperiodic control of flowering in 1920 by Garner and Allard, work on similar lines was initiated on bryophytes by Wann (1925) and has been followed up by others (Table 4.1). Wann grouped these plants into two categories, 'long-day' and 'short-day' plants, depending upon the length of daily illumination required for the initiation of gametangia. It was demonstrated that *Marchantia polymorpha* is a long-day plant since the plants become fertile when grown under long-photoperiods (Anthony, 1962; Benson-Evans, 1961, 1964; Miller & Colaiace, 1969; Voth & Hamner, 1940; Wann, 1925). Gametophytes grown under short-photoperiods (9 h of light) produce only gemma cups (Voth & Hamner, 1940). Some other long-day plants on record (requiring 16 to 18 h of light diurnally) are *Conocephalum conicum, Diplophyllum albicans, Lophocolea cuspidata, L. heterophylla, Lunularia cruciata, Pellia epiphylla, Preissia quadrata, Riccardia multifida,* and *R. pinguis*. On the other hand, *Riccia glauca* requires short days (6 to 8 h of light diurnally) to become fertile (Benson-Evans, 1961, 1964). Similarly, *Anthoceros laevis* exhibits gametangial formation in short-day conditions (Benson-Evans, 1964; Hughes, 1955; Ridgway, 1967). Antheridia are also initiated in *Anthoceros husnoti, A. punctatus, Phaeoceros bulbiculosus, P. laevis,* and *Notothylas orbicularis* in short days (Ridgway, 1967). In some leafy liverworts as well the onset of reproductive

TABLE 4.1.
The effect of light duration on gametangial formation in bryophytes.

Taxa	Photoperiod required for gametangial induction (h)	Category	Investigator/s
HEPATICOPSIDA (HEPATICAE)			
Cephalozia media	18	Long-day	Lockwood (1975)
Conocephalum conicum	16-18	—do—	Benson-Evans (1961, 1964)
Cryptothallus mirabilis	6-18	Day-neutral	—do—
Diplophyllum albicans	16-18	Long-day	Benson-Evans (1964)
Lophocolea cuspidata	—do—	—do—	—do—
Lophocolea heterophylla	—do—	—do—	—do—
Lunularia cruciata	8-18*	—do—	Benson-Evans (1961, 1964)
Marchantia polymorpha	>10	—do—	Wann (1925)
	18	—do—	Voth and Hamner (1940)
	8-18*	—do—	Benson-Evans (1961, 1964)
	>12	—do—	Anthony (1962)
	24	—do—	Miller and Colaiace (1969)
Pellia epiphylla	8-18*	—do—	Benson-Evans (1961, 1964)
Preissia quadrata	16-18	—do—	Benson-Evans (1961)
Riccardia multifida	16	—do—	Benson-Evans (1961, 1964)
Riccardia pinguis	8-18*	—do—	—do—
Riccia crystallina	6-24	Day-neutral	Chopra and Sood (1973a)
Riccia gangetica	—do—	—do—	Dua (1983)
Riccia frostii	8-24	—do—	Vashistha (1985)
Riccia glauca	6-8	Short-day	Benson-Evans (1964)
ANTHOCEROTOPSIDA (ANTHOCEROTALES)			
Anthoceros husnoti	4-12	Short-day	Ridgway (1967)
Anthoceros laevis	<12	—do—	Hughes (1955)
	6-8	—do—	Benson-Evans (1964)
	4-12	—do—	Ridgway (1967)
Anthoceros punctatus	—do—	—do—	—do—
Phaeoceros bulbiculosus	—do—	—do—	—do—
Phaeoceros laevis	—do—	—do—	—do—
Notothylas orbicularis	—do—	—do—	—do—
BRYOPSIDA (MUSCI)			
Barbula gregaria	8-24	Day-neutral	Kumra and Chopra (1983)

(Table contd.)

Bartramidula bartramioides	—do—	—do—	Rahbar (1981)
Bryum argenteum	—do—	—do—	Chopra and Bhatla (1981a)
Bryum coronatum	—do—	—do—	Kumra and Chopra (1983)
Funaria hygrometrica	>4	—do—	Monroe (1965)
	12-24	—do—	Nakosteen and Hughes (1978)
	—do—	—do—	Dietert (1980)
Leptobryum pyriforme	10-24	—do—	Chopra and Rawat (1977)
Physcomitrella patens	12-24	—do—	Nakosteen and Hughes (1978)
Physcomitrium pyriforme	—do—	—do—	—do—
Polytrichum aloides	8-18	—do—	Benson-Evans (1964)
	—do—	—do—	Hughes (1955)
Sphagnum plumulosum	6-8	Short-day	Benson-Evans (1964)

* There is no 'critical' light requirement for gametangial induction, and the degree of response increases with an increase in the duration of light. These plants have been classified as long-day plants by the investigator, but they should be regarded as day-neutral, since the range of photoperiod in which they become fertile is from 8 to 18 h.

phase is primarily controlled by photoperiod (Morris, 1961, *cited in* Monroe, 1965). *Cryptothallus mirabilis, Riccia crystallina, R. gangetica,* and *R. frostii* are day-neutral (Benson-Evans, 1964; Chopra & Sood, 1973a; Dua, 1983; Vashistha, 1985).

Among mosses day length is not usually the governing factor in the production of gametangia. As examples may be mentioned *Polytrichum aloides* (Benson-Evans, 1964; Hughes, 1955); *Funaria hygrometrica* (Dietert, 1980; Monroe, 1965; Nakosteen & Hughes, 1978); *Physcomitrella patens, Physcomitrium pyriforme* (Nakosteen & Hughes, 1978); *Leptobryum pyriforme* (Chopra & Rawat, 1977); *Bryum argenteum* (Chopra & Bhatla, 1981a); *Bryum coronatum, Barbula gregaria* (Kumra & Chopra, 1983), and *Bartramidula bartramioides* (Rahbar, 1981), all of which are day-neutral. However, in *Leptobryum, Bryum* spp., *Barbula,* and *Bartramidula* the degree of response increases with increasing photoperiod. In *Sphagnum plumulosum,* on the other hand, gametangia appear under short-photoperiods (Benson-Evans, 1964).

From the above account it is apparent that liverworts are relatively more sensitive to photoperiod than are mosses. Most of the investigated liverworts behave as long-day plants. Members of Anthocerotales require short days for gametangial induction. Among mosses, all but *Sphagnum plumulosum* appear to be independent of photoperiod for the onset of reproductive phase.

LIGHT LEVEL

Liverworts are more shade tolerant than mosses. In *Marchantia* and *Lunularia* gametangiophores are produced throughout the range of 25 to 1000 ft.c. in long days at 20 C. However, high light level enhances gametangial induction in these liverworts. Similarly, in *Conocephalum conicum*, *Pellia epiphylla*, and *Riccardia pinguis* the rate of gametangial production increases with increasing light level (Benson-Evans, 1964). An increase in light level also increases gametangial production in *Riccia crystallina*; in dark and in diffuse light no gametangia are observed (Chopra & Sood, 1973b). In *Riccia gangetica* and *R. frostii* also gametangial induction increases with an increase in level of continuous light (Dua et al., 1982; Vashistha, 1985). In *Leptobryum pyriforme* maximum gametangial formation is observed by increasing light level with a simultaneous lowering of temperature. Gametangia are formed earlier, and their number is highest at 4500 lux (Chopra & Rawat, 1977). Increase in light level up to 3500 and 4500 lux increases gametangial induction in *Barbula gregaria* and *Bryum coronatum*, respectively (Kumra & Chopra, 1983). In *Bartramidula bartramioides* no gametangia are observed at 80 to 100 lux, but gametangial production increases with increasing light level. The response is directly proportional to light level (Rahbar, 1981). In contrast to these systems, gametangial formation in *Bryum argenteum* is observed in as low levels as 100 lux, but maximum response occurs at 3500 to 4000 lux (Chopra & Bhatla, 1981a).

LIGHT QUALITY

Under fluorescent tubes, thalli of *Marchantia polymorpha* remain vegetative even if they are provided with long-day conditions. However, if incandescent light of 300 W is provided, primordia of both male and female gametangiophores are initiated within a month (Courtoy, 1964, 1966).

TEMPERATURE

Temperature acts as a critical factor in the induction of gametangia in day-neutral plants, and sometimes has marked effect even in those requiring specific day length.

In *Lunularia cruciata* a low temperature requirement (4 to 10 C) must be fulfilled before the stimulus of rise in temperature (by 10 C) can be effective in bringing about the onset of reproductive phase. This can be compared with the phenomenon of vernalization in higher plants (Benson-Evans, 1961, 1964; Benson-Evans & Hughes, 1955). The gametophytes of *Cryptothallus mirabilis* also become fertile with a rise in temperature (21 C), after a low temperature requirement has been met with (Benson-Evans, 1961, 1964).

Mosses have variable temperature requirements, irrespective of day length (Table 4.2). In *Polytrichum aloides* a specific temperature (21 C) is responsible for the initiation of gametangia (Benson-Evans, 1964). Temperature is also the critical factor for induction and maturation of gametangia in *Funaria hygrometrica* (Monroe, 1965; Nakosteen & Hughes, 1978), *Physcomitrella patens* (Engel, 1968; Nakosteen & Hughes, 1978), and *Physcomitrium pyriforme* (Nakosteen & Hughes, 1978). A similar response is shown by *Philonotis turneriana* (Plate 4.1 A,B) which requires a critical temperature of 18 C for gametangial induction (Kumra & Chopra, 1983).

In contrast to these observations some mosses do not require a specific temperature, but exhibit response over a wide range of temperature. *Leptobryum pyriforme* remains vegetative at 25 C, and only a few plants turn fertile at 20 C. Lowering of temperature to 10 C results in early gametangial induction in a large number of gametophytes. Increased light level or longer photoperiods also enhance fertility. Antheridia appear first and archegonia develop 15 to 20 days later in the same head. Archegonia continue to form even at 25 C provided that antheridia had been initiated under the inductive temperatures (Chopra & Rawat, 1977). Similarly, when uninduced gametophytes of *Bryum argenteum* (Chopra & Bhatla, 1981a), *Barbula gregaria*,

TABLE 4.2.
The effect of temperature, irrespective of day length, on gametangial induction in bryophytes.

Taxa		Temperature requirement (C)	Investigator/s
I	Species requiring a low temperature pretreatment		
	Cryptothallus mirabilis	≥ 21	Benson-Evans (1961, 1964)
	Lunularia cruciata	≥ 14	Benson-Evans and Hughes (1955) Benson-Evans (1961, 1964)
II	Species requiring a specific inductive temperature		
	Funaria hygrometrica	10	Monroe (1965)
		7	Nakosteen and Hughes (1978)
	Philonotis turneriana	18	Kumra and Chopra (1983)
	Physcomitrella patens	17	Nakosteen and Hughes (1978)
	Physcomitrium pyriforme	7	—do—
	Polytrichum aloides	21	Benson-Evans (1964)
III	Species turning fertile in a broad temperature range		
	Barbula gregaria	12-30 (24)*	Kumra and Chopra (1983)
	Bartramidula bartramioides	18-30 (25)*	Rahbar (1981)
	Bryum argenteum	10-25 (25)*	Chopra and Bhatla (1981a)
	Bryum coronatum	12-30 (24)*	Kumra and Chopra (1983)
	Leptobryum pyriforme	10-20 (10)*	Chopra and Rawat (1977)
	Riccia crystallina	8-15 (8)*	Chopra and Sood (1973a)
	Riccia gangetica	18-25 (25)*	Dua et al. (1982)

( )* Optimal temperature for gametangial production.

Bryum coronatum (Kumra & Chopra, 1983), and *Bartramidula bartramioides* (Rahbar, 1981) are subjected to different temperatures, gametangia are produced in almost all the regimes (Plate 4.1 C-E).

In some taxa the differential response to temperature can be correlated to their natural habitats. For example, *Philonotis turneriana* occurs at higher altitudes and requires a lower inductive temperature in cultures, whereas *Barbula gregaria* and *Bryum coronatum* which occur at lower as well as higher altitudes develop gametangia over a wide temperature range. This type of behaviour is also noticed in species of *Riccia* which occur in Delhi. *R. crystallina* grows during winter, and *R. gangetica* grows in summer. In cultures gametangial induction is maximum at 8 to 15 C in the former, whereas the response is optimum at 25 ± 2 C in the latter (Chopra & Sood, 1973b; Dua et al., 1982).

TEMPERATURE-PHOTOPERIOD INTERACTION

Bryophytes also respond to a temperature-photoperiod interaction (Table 4.3). *Conocephalum conicum* and *Marchantia polymorpha* produce gametangia only at 21 C in long-day conditions. They remain vegetative at 10 C in long days or in short days at either temperature. Similarly in *Riccia glauca* gametangia are produced only under short days and at 10 C; a temperature of 21 C suppresses gametangial production even in short days (Benson-Evans, 1964). In *Asterella tenella* archegoniophores are induced

TABLE 4.3.
The effect of temperature-photoperiod interaction on gametangial formation in bryophytes.

	Taxa	Day length requirement	Temperature requirement (C)	Investigator/s
I	Species requiring a specific inductive temperature			
	Conocephalum conicum	Long days	21	Benson-Evans (1964)
	Marchantia polymorpha	—do—	21	—do—
	Riccia glauca	Short days	10	—do—
II	Species turning fertile in a broad temperature range			
	Anthoceros husnoti	Short days	5-27 (15)*	Ridgway (1967)
	Anthoceros laevis	—do—	10-21 (21)*	Benson-Evans (1964)
		—do—	5-27 (15)*	Ridgway (1967)
	Anthoceros punctatus	—do—	—do—	—do—
	Pellia epiphylla	Long days	10-21 (21)*	Benson-Evans (1964)
	Phaeoceros bulbiculosus	Short days	5-27 (15)*	Ridgway (1967)
	Phaeoceros laevis	—do—	—do—	—do—
	Preissia quadrata	Long days	10-21 (21)*	Benson-Evans (1964)
	Notothylas orbicularis	Short days	5-27 (15)*	Ridgway (1967)

( )* Optimal temperature for gametangial production.

under short days (10 h light) and low temperatures (15 C); the temperature during the dark period (14 h) being 10 C (Plate 4.2 A,B; Bostic, 1981). Thus, these four liverworts require specific inductive temperature and photoperiod.

However, there are taxa which do not require a specific inductive temperature during the inductive photoperiod. *Preissia quadrata* and *Pellia epiphylla* develop gametangia under long days in a broad temperature range (Benson-Evans, 1964). Similarly, some members of Anthocerotales turn fertile under short days over a wide temperature range. In these species photoperiod is the critical factor in the initiation of gametangia, whereas temperature has no appreciable effect (Ridgway, 1967).

In *Riccia crystallina*, a day-neutral plant, both light level and temperature are important (Chopra & Sood, 1973a). There is no critical day length, and gametangia are produced so long as a certain minimum period of illumination is provided. Lower temperatures increase gametangial production, irrespective of the light period. However, in *Riccia gangetica* gametangial production is maximum at 25 ± 2 C, and it decreases at 18 ± 2 C and 12 ± 2 C (Dua et al., 1982). Fertile thalli of both the species of *Riccia* produce sporophytes with viable spores, under ordinary cultural conditions (Plate 4.3 A-F).

In *Sphagnum plumulosum* higher temperature combined with higher light level induces gametangial heads two months earlier than in the field under short days (Benson-Evans, 1964).

HUMIDITY

In *Marchantia polymorpha* relatively high humidity hastens the sexual response and relatively low humidity tends to retard the production of archegoniophores (Klebs, 1903; Voth & Hamner, 1940; Wann, 1925).

HYDRATION

Hydration of nutrient medium, as altered by variation in agar concentration, also affects gametangial formation. *Riccia crystallina* raised on media with different concentrations of agar (0.2 to 4%) shows normal growth up to one percent. Higher concentrations retard vegetative growth and consequently inhibit gametangial formation (Chopra & Sood, 1973b). In *Bartramidula bartramioides*, presence of agar not only delays initiation of gametangia, but also reduces the response per culture (Rahbar, 1981). Similarly, in male clones of *Barbula gregaria* and *Bryum coronatum* the time required for antheridial induction is minimum, and the response per culture is maximum in liquid medium (Kumra, 1981). This may be due to greater availability of nutrients. Vashistha (1985) observed that female thalli of *Riccia frostii* remain sterile in liquid medium, but produce archegonia on agar cultures. Archegonial production is maximum in cultures with 0.8 percent agar.

CARBOHYDRATES

Excess of carbohydrates promotes the onset of sexuality in plants. In higher plants flowering depends on the carbohydrate-nitrogen ratio (Klebs, 1903; Kraus & Kraybill, 1918). In *Marchantia polymorpha* a relatively high carbohydrate-nitrogen ratio in the culture medium results in the formation of sexual branches (Wann, 1925). It has also been noticed that there is an increase in the carbohydrate-nitrogen ratio during the transition from vegetative to reproductive phase in female plants of the investigated liverworts (Rao & Das, 1968). Nutritional factor may also be involved in the production of male and female gametangia at different times on the same thallus of *Reboulia hemisphaerica*. Cultures growing on medium containing one percent glucose form male receptacles, whereas those on media without glucose form isolated female gametangiophores (Allsopp, 1964).

Carbohydrate supply also affects gametangial production in *Riccia crystallina*. Sucrose supports optimum growth and gametangial production, whereas fructose and glucose are less effective. No gametangia are produced in the absence of sucrose, or in the presence of eight percent sucrose (Chopra & Sood, 1973b). Similarly, in *Fossombronia himalayensis* gametangial formation is inhibited in the presence of excess sucrose (4, 6, & 8%). Both types of gametangia are formed on basal medium supplemented with one percent sucrose (Mehra & Pahwa, 1976). In *Riccia gangetica* gametangial induction is inhibited at two and four percent glucose and mannose. With sucrose and fructose inhibition is observed only at four percent. This appears to be due to greater inhibition of vegetative growth by glucose and mannose at two percent. In general, maximum gametangial production is observed at concentrations of sugars optimum for vegetative growth (Dua et al., 1982). Of the six sugars tested (fructose, galactose, glucose, mannose, sucrose, and xylose) only sucrose and glucose favour archegonial formation in *Riccia frostii* (Vashistha, 1985). It has also been observed that in *Bryum argenteum* (Bhatla & Chopra, 1979), *Barbula gregaria*, *Bryum coronatum* (Kumra, 1981), and *Bartramidula bartramioides* (Rahbar, 1981) gametangial induction is specifically inhibited with concentrations of sucrose higher than one percent.

NITROGENOUS SUBSTANCES

Nitrogen availability influences the rate of flower development in higher plants. Keeping this in view some investigators studied the effect of inorganic (e.g. in the form of nitrates) as well as organic nitrogen sources (urea, amino acids, casein hydrolysate) on gametangial induction in bryophytes. In *Pohlia nutans* archegonia do not appear on plants without urea, and their appearance varies with urea concentration (Mitra, 1967). However, no antheridial production is observed under these conditions. Similarly, in *Riccia crystallina* archegonial production increases with increasing concentration of urea, but antheridial formation is not significantly affected.

Archegonial production is enhanced by peptone, yeast extract, and casein hydrolysate, but antheridial production is not affected (Sood, 1974). In the female thalli of *Riccia frostii* yeast extract and casein hydrolysate stimulate formation of archegonia at 0.5 and 50 ppm, respectively, but prove inhibitory at higher levels. In this species peptone inhibits archegonial production (Vashistha, 1985).

Amino acids influence gametangial production in *Riccia* spp. (Table 4.4).

TABLE 4.4.
The effect of some amino acids on gametangial formation in three species of *Riccia*.

Amino acid	Increase in number of Antheridia/Archegonia		
	R. crystallina (Sood, 1974)	*R. gangetica* (Chopra & Kumra, 1984)	*R. frostii** (Vashistha, 1985)
Alanine	Archegonia	—	—
Asparagine	Archegonia	Archegonia	—
Aspartic Acid	Antheridia	Antheridia	Archegonia
Glutamic Acid	Archegonia	Antheridia	—
Glycine	Antheridia	Antheridia	—
Hydroxyproline	Archegonia	—	—
Leucine	Archegonia	—	—
Serine	Archegonia	Archegonia	—
Threonine	Archegonia	—	Archegonia
Tryptophan	Antheridia	Archegonia	Archegonia
Valine	Antheridia	—	—

* effect is studied only on female thalli.
— indicates amino acid not tried.

At low concentration, arginine, cystine, and tryptophan together with kinetin overcome the photoperiodic control of the reproductive response in *Cephalozia media* (Lockwood, 1975). Thus, in general, it has been observed that organic form of nitrogen favours gametangial production (especially archegonial) in liverworts.

Effect of $Ca(NO_3)_2$, KNO_3, and NH_4NO_3 has also been studied. In *Cephalozia media* gametangial formation is stimulated only when the plants are grown on medium devoid of inorganic nitrogen (Lockwood, 1975). Similarly, in four strains of *Riccia fluitans* and seven other Ricciaceous members gametangial induction occurs earlier when the concentration of $NaNO_3$ is lowered to half of its full strength (Woodfin, 1976). Plants of *Riccia duplex* remain vegetative on medium containing full strength KNO_3, but turn fertile in its absence (Selkirk, 1979). Thus, it is evident that depletion of inorganic forms of nitrogen significantly favours gametangial formation in liverworts. However, in *Riccia crystallina* archegonial production as well as the fresh weight of thalli increases with increasing concentration of KNO_3. Nitrates significantly enhance archegonial production, but the number of antheridia is not appreciably affected (Sood, 1974). Ammonium ions in the form of NH_4Cl cause inhibition of gametangial induction in *Bryum argenteum* (Bhatla, 1980).

GROWTH REGULATORS

Growth regulators play an important role in the control of flowering in higher plants. Some reports are also available regarding their effect on gametangial induction in bryophytes. Application of auxins (2,4-D & NAA) to the vegetative plants of *Marchantia polymorpha* induces structures morphologically comparable to the receptacles. In *Conocephalum conicum* as well 2,4-D induces receptacles but these do not bear gametangia (Benson-Evans, 1961). In *Fossombronia himalayensis* auxins (IAA, IBA, IPA, 2,4-D, & NAA) induce gametangia at 0.05 ppm, but fail to do so at higher levels (Mehra & Pahwà, 1976). In *Riccia crystallina* all auxins tried (IAA, NAA, 2,4,5-T, & 0-CPA) promote archegonial production. Antheridial formation is not significantly influenced by auxins. The two antiauxins (MH & TIBA) are inhibitory for gametangial induction (Chopra & Sood, 1973b). Auxins also enhance femaleness in *Riccia gangetica* (Kumra & Chopra, 1984) and *R. frostii* (Vashistha, 1985). Of the auxins tested IAA elicits maximum response in *R. crystallina* and *R. gangetica*, whereas in *R. frostii* NAA has proved best. In *R. gangetica* auxins not only increase the number of archegonia, but also reduce the time required for their initiation. It has been noticed that in female gametophytes of *Asterella angusta*, *Exormotheca tuberifera*, *Plagiochasma articulatum*, *Reboulia hemisphaerica*, and *Pallavicinia canaras* appearance of archegonia is accompanied with an increase in endogenous IAA level. In male plants a reverse change is observed (Rao & Das, 1968). On the other hand, among the mosses IAA is either ineffective, as in *Leptobryum pyriforme* (Chopra & Rawat, 1977), or enhances the production of antheridia as in male clones of *Bryum argenteum*, *Barbula gregaria*, and *Bryum coronatum* (Bhatla & Chopra, 1981; Chopra & Kumra, 1983). In contrast to these observations, IAA inhibits gametangial formation in the monoecious moss *Bartramidula bartramioides* (Rahbar, 1981). Thus, auxins have variable effects on bryophytes and no generalizations can be made at present.

Gibberellins promote onset of reproductive phase in many higher as well as lower plants. Their effect on ferns is well documented (Brandes, 1973). Antheridium formation in these plants is controlled by certain endogenous substances (antheridiogens) which are gibberellin-like (Schraudolf, 1962). Gibberellins (GA_3, GA_{13}, & $GA_{4/7}$) enhance antheridial production in *Riccia crystallina*. GA_3 is more effective than GA_{13} and $GA_{4/7}$; it also causes a slight increase in archegonial number (Chopra & Sood, 1973b). GA_3 also markedly favours antheridial formation in *R. gangetica* (Chopra & Kumra, 1986) but it does not affect archegonial production in *R. gangetica* and *R. frostii* (Vashistha, 1985). Among the mosses, GA_3 increases antheridial number in *Bryum argenteum* but it does not affect gametangial production in *Leptob-*

ryum pyriforme (Bhatla & Chopra, 1981; Chopra & Rawat, 1977). In *Barbula gregaria* and *Bryum coronatum* antheridial formation is stimulated by GA_3. Gibberellic acid also induces antheridial formation in *Philonotis* which remains vegetative on basal medium (Chopra & Kumra, 1983). Stimulation of antheridial production in *Bryum argenteum* and *Barbula gregaria* is more with GA_3 than with IAA, whereas the reverse is true for *Bryum coronatum* (Bhatla & Chopra, 1981; Chopra & Kumra, 1983). It is clear from the above findings that as in most higher plants and ferns, gibberellins are involved in antheridial induction in bryophytes. They not only stimulate antheridial production but in some systems they also induce antheridia under non-inductive conditions.

Cytokinins also influence sexual reproduction in plants, but they are not as effective as auxins or gibberellins. In *Riccia crystallina* kinetin enhances archegonial production, and to a slight extent even antheridial formation (Chopra & Sood, 1973b). Similar observations have been recorded in *R. gangetica* (Kumra & Chopra, 1984). In the female thalli of *R. frostii* all the tested cytokinins (BAP, 2iP, & kinetin) enhance archegonial production (Vashistha, 1985). In *Bryum argenteum* kinetin and 2iP stimulate archegonial production, but inhibit antheridial formation (Bhatla & Chopra, 1981). In the male clones of *Barbula gregaria* and *Bryum coronatum* antheridial production is inhibited at higher concentrations of kinetin (Chopra & Kumra, 1983). IAA and kinetin show a synergistic effect on gametangial induction in *Bryum argenteum*. Their combination also nullifies the inhibition of gametangial formation caused individually by kinetin and IAA, in male or female clone, respectively (Bhatla & Chopra, 1981).

CHELATING AGENTS

Chelating agents like EDDHA, EDTA, their iron salts, and salicylic acid affect flowering in higher plants. There are also some reports of their effect on gametangial production in bryophytes. In *Riccia crystallina* iron salts of EDDHA and EDTA appreciably increase archegonial production, but there is a slight increase in antheridial number (Sood, 1974). In *Riccia gangetica*, on the other hand, EDDHA, EDTA and their iron salts significantly enhance antheridial formation, and to a lesser extent archegonial production; EDDHA being most effective for production of both types of gametangia (Chopra & Kumra, 1985). Plants of *Sphaerocarpos texanus* growing on liquid organic basal medium (containing ferric chloride) exhibit male involucres, but on replacing the iron source with ferric potassium EDTA, growth of thalli is poor and male involucres become rare (Diller et al., 1955). On the other hand, EDDHA, EDTA, and their iron salts significantly enhance the percentage of fertile gametophytes in *Bryum argenteum*, and the response is more in male clone than in female clone (Bhatla & Chopra, 1983). In *Barbula gregaria*, *Bryum coronatum*, and *Microdus brasiliensis* EDDHA, EDTA, and their iron salts significantly increase antheridial production. The response elicited by the iron salts of EDDHA and EDTA is more than by the chelates themselves (Kumra, 1982; Sharma & Chopra, 1985). In *Bartramidula bartramioides* EDDHA and EDTA stimulate gametangial production (Rahbar, 1981). Salicylic acid is a phenol and induces flowering in higher plants under non-inductive conditions. It enhances gametangial formation in *B. bartramioides*. In *Riccia gangetica* SA increases antheridial formation, but archegonial production is only slightly stimulated at 10^{-7} and 10^{-6}M (Chopra & Kumra, 1985). It enhances antheridial production in *Microdus brasiliensis* (Sharma & Chopra, 1985). However, in *Bryum argenteum, Barbula gregaria*, and *Bryum coronatum* SA lowers the percentage of fertile gametophytes (Bhatla & Chopra, 1983; Kumra, 1982). Ferric citrate also affects gametangial induction in some mosses. The percentage response per culture increases with increasing concentration of ferric citrate (Bhatla & Chopra, 1983; Kumra, 1982; Sharma & Chopra, 1985). However, in the liverwort *Riccia crystallina* ferric citrate has no marked effect on sexuality (Sood, 1974). Chopra and Kumra (1985) reported that in *Riccia gangetica* gametangial number increased significantly at higher levels of ferric citrate (20 & 30 mg/l); 20 mg/l being optimal.

In *Barbula gregaria* endogenous levels of iron and copper were determined in male clone subjected to varying concentrations of ferric citrate. EDDHA, or Fe-EDDHA. With an increase in gametangial induction the iron content increases, and that of copper decreases, but no definite correlation can be observed between their endogenous levels and the gametangial induction response. However, it is likely that in this moss iron plays an important role in antheridial induction, and the chelates might be influencing its bioavailability. In this respect, Fe-EDDHA is most effective (Kumra, 1981). In *Bryum argenteum* as well EDDHA and EDTA bring about an increase in iron content of the plants during gametangial induction (Bhatla & Chopra, 1983). In *Microdus brasiliensis* EDTA and Fe-EDTA-stimulated antheridial formation is also associated with a corresponding increase in endogenous iron. Copper content increases only at higher levels of EDTA and Fe-EDTA, and there is no correlation with the antheridial induction response (Sharma & Chopra, 1985). Thus, in general, chelates markedly enhance gametangial formation in bryophytes.

pH AND OTHER FACTORS

Growth and development of plants is affected by pH of the nutrient medium. Among bryophytes, growth of male and female clones of *Sphaerocarpos donnellii* is accompanied with definite changes in pH of the medium. Male plants grow only when the medium is alkaline or near neutral, whereas female plants can tolerate lower pH values. During their growth, the two types of thalli bring about a differential alteration in pH of the medium, the males shifting it rapidly to acid values and the female tending to hold it near neutral (Machlis, 1962). Maximal antheridial and archegonial production in *Riccia crystallina* is exhibited in cultures maintained at pH 4.5 and 6.5, respectively (Chopra & Sood, 1973b). In *Riccia gangetica* as well maximum antheridia were noticed at pH 4.5, whereas pH 7.5 elicited optimal response for archegonial production (Dua, 1983). In *Bryum argenteum* a sharp decrease in pH of the medium is observed during the transition from vegetative to reproductive phase. However, since exogenously modified pH does not affect gametangial induction it indicates that change in pH of the medium is only an effect of this morphogenic change and is not its cause (Bhatla, 1981).

Concentration of mineral salts in the medium also affects gametangial formation in *Riccia crystallina*. Antheridial production is more at lower concentrations (1/4 and 1/8th of the original concentration), whereas higher concentrations (1/2, full, and twice the full strength) favour archegonial formation (Sood, 1974). In *Riccia duplex* gametangial induction is also affected by the strength of the nutrient solution (NS). Sporophytes appear after eight weeks only in the 10^{-2}NS (Selkirk, 1979). Varying concentrations of nutrient solution also affect gametangial formation in *Bartramidula bartramioides*. The response is optimum at full strength; whereas lower (up to 1/4 dilution) and higher (up to four times the full) concentrations are inhibitory (Rahbar, 1981).

Effect of some cyclic nucleotides and other purine derivatives (adenine, adenosine, 3'-AMP, ADP, & ATP) has also been studied on gametangial induction in *Bryum argenteum*. Cyclic AMP stimulates gametangial formation. The effect of other purine derivatives is not so significant. Cyclic AMP also overcomes the inhibitory effect of high concentrations of sucrose on gametangial induction in this moss (Chopra & Bhatla, 1981b). In *Barbula gregaria* and *Bryum coronatum* the effect of cAMP is not well marked (Kumra, 1981). However, co-addition of cAMP and kinetin brings about considerable enhancement in antheridial production in *Microdus brasiliensis* (Chopra & Sharma, 1985).

Effect of animal sex hormones (testosterone & progesterone) has also been studied on gametangial induction in two mosses: *Barbula gregaria* and *Bryum coronatum*. Testosterone enhances antheridial production at lower concentrations, whereas it inhibits the response at higher levels. Progesterone delays antheridial induction in the male clones of these mosses (Chopra & Kumra, 1983). In *Riccia gangetica* antheridial production was enhanced by both the hormones, and testosterone was more effective (Dua, 1983).

Sex expression in mosses is influenced both by the genotype and environment. From the point of view of developmental physiology, the latter aspect is more interesting. The dioecious moss *Splachnum rubrum* has genotypically determined female plants, but when senescent female plants (when formation of gametangia is inhibited for a period) are subsequently subcultured on fresh medium, male as well as normal female plants arise. These male plants are completely different from the normal males arising from spores. They are considerably smaller (Plate 4.4 A-C), and are similar to 'dwarf-males' of other mosses. Further, these dwarf-males become sexually mature much earlier, but have smaller antheridia which only occasionally emit motile spermatozoids. All the dwarf-males are alike and are mostly sexually stable, even after several subcultures. It is important to note that after being transferred to fresh medium the dwarf-males occasionally revert to the normal-sized female plants (Bauer, 1963).

Studies on reproductive biology of bryophytes are comparatively very recent and so far very few taxa have been critically investigated. The variable response of different taxa to identical treatments can possibly be explained after estimating the nature and level of endogenous growth regulators and other metabolites in uninduced and induced plants.

REFERENCES

ALLSOPP, A., *Phytomorphology*, 14, 1 (1964).
ANTHONY, R.E., *Turtox News*, 40, 2 (1962).
BAUER, L., *J. Linn. Soc. (Bot.)*, 58, 337 (1963).
BENSON-EVANS, KATHRYN, *Nature*, 191, 255 (1961).
BENSON-EVANS, KATHRYN, *Bryologist*, 67, 431 (1964).
BENSON-EVANS, KATHRYN and J.G. HUGHES, *Trans. Br. bryol. Soc.*, 2, 513 (1955).
BHATLA, S.C., "Physiology of sexual reproduction in *Bryum argenteum* Hedw.", Ph.D. Thesis, Univ. Delhi, India, 1980.
BHATLA, S.C., *Curr. Sci.*, 50, 960 (1981).
BHATLA, S.C. and R.N. CHOPRA, *Z. PflPhysiol.*, 92, 375 (1979).
BHATLA, S.C. and R.N. CHOPRA, *J. exp. Bot.*, 32, 1243 (1981).
BHATLA, S.C. and R.N. CHOPRA, *Ann. Bot.*, 52, 755 (1983).
BOSTIC, S.R., *Bryologist*, 84, 89 (1981).
BRANDES, H., *Ann. Rev. Pl. Physiol.*, 24, 115 (1973).
CHOPRA, R.N. and S.C. BHATLA, *New Phytol.*, 89, 439 (1981a).
CHOPRA, R.N. and S.C. BHATLA, *Z. PflPhysiol.*, 103, 394 (1981b).
CHOPRA, R.N. and S.C. BHATLA, *Bot. Rev.*, 49, 29 (1983).
CHOPRA, R.N. and P.K. KUMRA, *J. Bryol.*, 12, 491 (1983).
CHOPRA, R.N. and POONAM SHARMA, *J. Plant Physiol.*, 117, 293 (1985).
CHOPRA, R.N. and SUMAN KUMRA, *J. exp. Bot.*, 35, 1537 (1984).
CHOPRA, R.N. and SUMAN KUMRA, *Beitr. Biol. Pfl.*, 60, 367 (1985).
CHOPRA, R.N. and SUMAN KUMRA, *Beitr. Biol. Pfl.*, 61, 99 (1986).
CHOPRA, R.N. and M.S. RAWAT, *Beitr. Biol. Pfl.*, 53, 353 (1977).
CHOPRA, R.N. and SNEH SOOD, *Bryologist*, 76, 278 (1973a).
CHOPRA, R.N. and SNEH SOOD, *Phytomorphology*, 23, 230 (1973b).
COURTOY, R., *Bull. Acad. Roy. Belg. Cl. Sci.*, 50, 1341 (1964).
COURTOY, R., *Photochem. Photobiol.*, 5, 441 (1966).
DIETERT, M.F., *Am. J. Bot.*, 67, 369 (1980).
DILLER, V.M., M. FULFORD, and H. KERSTEN, *Bryologist*, 58, 173 (1955).
DUA, SUMAN, "Morphogenetic and physiological studies on some bryophytes", Ph.D. Thesis, Univ. Delhi, India, 1983.
DUA, SUMAN, NEETA SINGAL, and R.N. CHOPRA, *Cryptog. Bryol. Lichénol.*, 3, 189 (1982).
ENGEL, P P., *Am. J. Bot.*, 55, 438 (1968).
GARNER, W.W. and H.A. ALLARD, *J. agric. Res.*, 18, 553 (1920).
HUGHES, J.G., "Physiology of sexual reproduction in the bryophytes", Ph.D. Thesis, Univ. Wales, Cardiff, U.K., 1955.
KLEBS, G., *Willkürlicke Entwicklungsänderungen bei Pflanzen*, Fischer, Jena, 1903.
KRAUS, E.J. and H.R. KRAYBILL, *Oreg. Agric. Exp. Stat. Bull.*, 149, 1 (1918).
KUMRA, P.K., "Morphogenetic and physiological studies on some mosses", Ph.D. Thesis, Univ. Delhi, India, 1981.

KUMRA, P.K., *Ann. Bot.*, **50**, 771 (1982).
KUMRA, P.K. and R.N. CHOPRA, *Bot. Gaz.*, **144**, 533 (1983).
KUMRA, SUMAN and R.N. CHOPRA, *Ann. Bot.*, **54**, 605 (1984)
LOCKWOOD, L.G., *Am. J. Bot.*, **62**, 893 (1975).
MACHLIS, L., *Physiol. Plant.*, **15**, 354 (1962).
MEHRA, P.N. and M.S. PAHWA, *J. Hattori bot. Lab.* No. **40**, 371 (1976).
MITRA, G.C., *Curr. Sci.*, **36**, 134 (1967).
MILLER, M.W. and J. COLAIACE, *Bryologist*, **72**, 45 (1969).
MONROE, J.H., *Bryologist*, **68**, 337 (1965).
NAKOSTEEN, P.C. and K.W. HUGHES, *Bryologist*, **81**, 307 (1978).
RAHBAR, KAVITA, "Morphogenetical and physiological studies on *Bartramidula bartramioides* Schimp. and *Hyophila involuta* (Hook.) Jaeg.", Ph.D. Thesis, Univ. Delhi, India, 1981.
RAO, M.P. and V.S.R. DAS, *Z. PflPhysiol.*, **59**, 87 (1968)
RIDGWAY, J.E., *Bryologist*, **70**, 203 (1967).
SCHRAUDOLF, H., *Biol. Zbl.*, **81**, 731 (1962).
SELKIRK, P.M., *Bryologist*, **82**, 37 (1979).
SHARMA, POONAM and R.N. CHOPRA, *J. exp. Bot.*, **36**, 494 (1985).
SOOD, SNEH, *Phytomorphology*, **24**, 186 (1974).
VASHISTHA, B.D., "In vitro investigations on some Indian bryophytes", Ph.D. Thesis, Univ. Delhi, India, 1985.
VOTH, P.D. and K.C. HAMNER, *Bot. Gaz.*, **102**, 169 (1940).
WANN, F., *Am. J. Bot.*, **12**, 307 (1925).
WOODFIN, C.M., *J. Hattori bot. Lab.* No. **41**, 179 (1976).

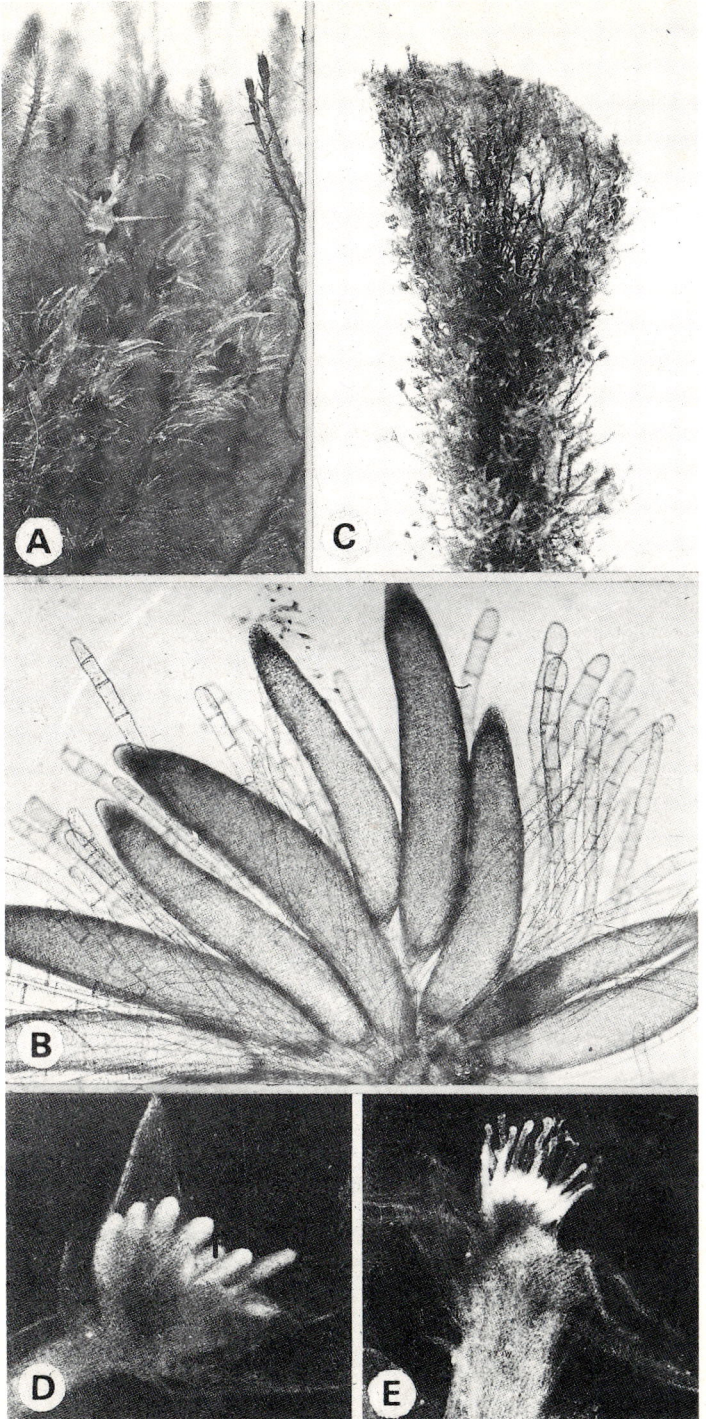

Plate 4.1 A-E. Gametangial induction in mosses. **A.** Plants of *Philonotis turneriana* with antheridial heads induced at 18 C. **B.** A dissected antheridial head from **A. C.** A male clone of *Bryum coronatum* with fertile gametophytes induced under ordinary cultural conditions (3500 to 4500 lux light & 25 ± 2 C temperature). **D.** Whole mount of antheridial head of *Bryum argenteum*. **E.** Archegonial head of the same moss. Both types of gametangia were induced under ordinary cultural conditions.
(**A-C.** After Kumra & Chopra, 1983; **D, E.** After Chopra & Bhatla, 1981a).

Plate 4.2 A, B. *Asterella tennella*. **A.** Fertile plants under field conditions. **B.** Laboratory culture with young archegoniophores produced under short days and low temperature.
(After Bostic, 1981).

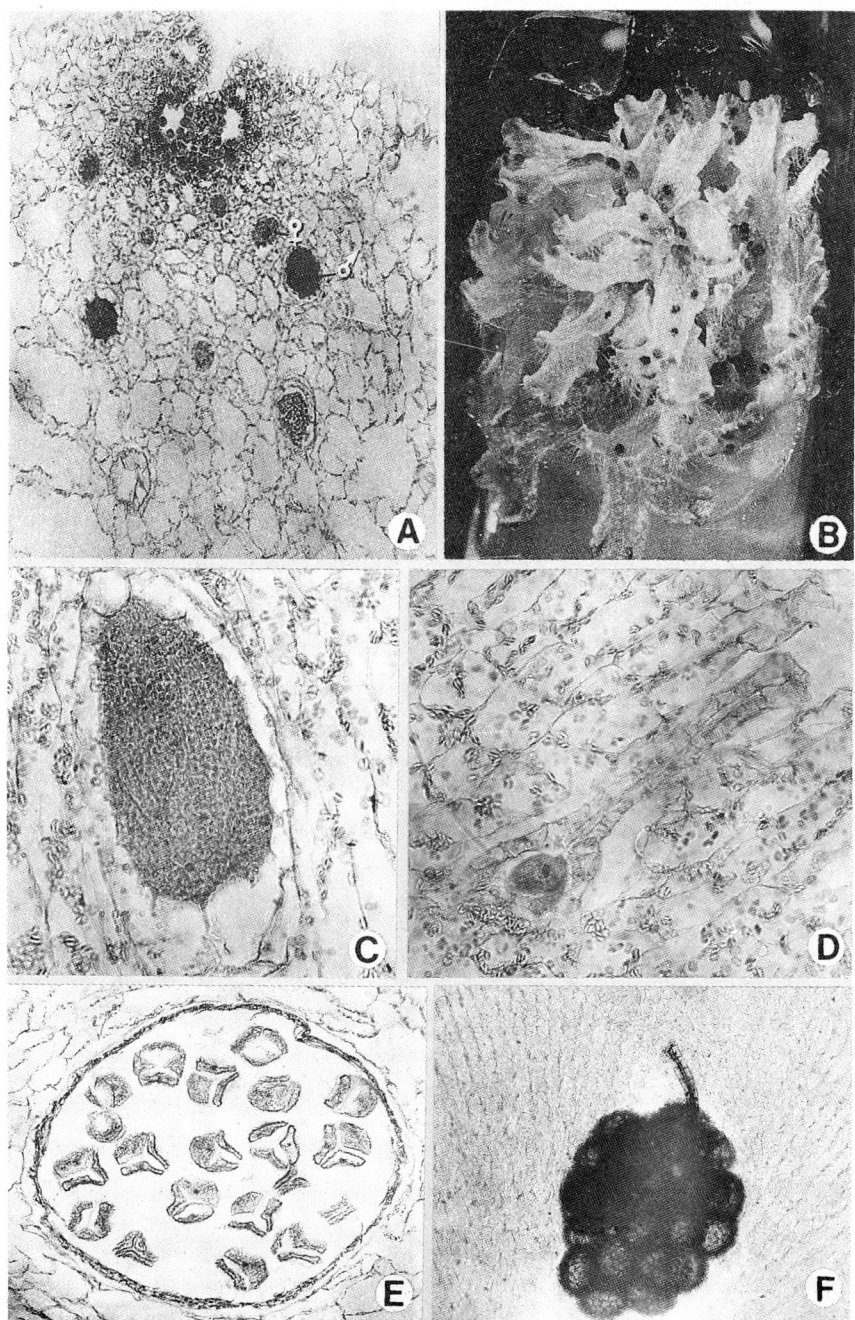

Plate 4.3 A-F. Gametangial induction and sporophyte production in *Riccia* spp. **A, B.** *R. crystallina*. **A.** Paradermal section of a 30-day-old thallus with antheridia and archegonia cut transversely. **B.** 60-day-old culture showing mature sporophytes (visible as black dots). **C, D.** *R. gangetica*. Vertical sections of thalli showing antheridium and archegonium. **E.** Paradermal section of a 60-day-old thallus of *R. crystallina* with 'sporophyte' containing normal spore tetrads. **F.** Cleared whole mount of the thallus of *R. gangetica* showing 'sporophyte' with mature spores.
(A, B, E. After Chopra & Sood, 1973; C, D, F. After Chopra & Kumra, 1986).

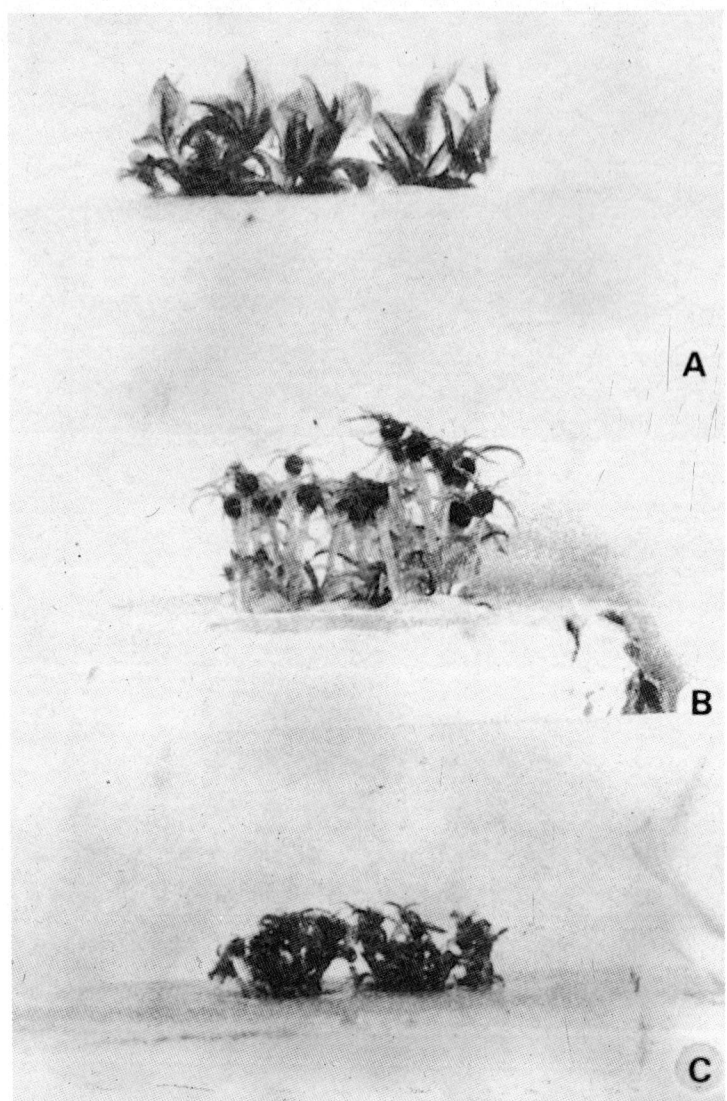

Plate 4.4 A-C. Gametangial induction in the haploid plants of *Splachnum rubrum*. **A.** Females. **B.** Primary males. **C.** Secondary males (dwarf males).
(After Bauer, 1963).

5 Alternative Pathways in Life Cycle

BRYOPHYTES display a well-defined alternation of heteromorphic generations. The gametophytic generation is predominant, and the attached sporophytic generation is dependent on it to varying degrees. In some instances alternative pathways to the normal life cycle are exhibited under cultural conditions: (1) apogamy—formation of sporophyte directly from the gametophytic cells without intervention of gametes, and (2) apospory—development of gametophyte from the vegetative cells of sporophyte without intervention of spores. The occurrence of these phenomena in nature is very rare, and so their significance in the life cycle is meagre. However, such studies are of considerable interest because of the following reasons:

(1) Differentiational processes at the cellular level can be observed from their inception.
(2) Tissues with haploid and diploid complements can be obtained and employed for comparative experimental studies.
(3) Direct derivation of one generation from the other is likely to help in a better understanding of the basis of the phenomenon of alternation of generations.
(4) Apogamy and apospory are usually preceded by the formation of callus. Controlled differentiation of callus into gametophytes or sporophytes is possible by modifying the cultural conditions, and the genetic component imposes no restriction. Thus, the role of external factors regulating differentiation can be studied with precision.
(5) Maintenance of reasonably homogenous cell clumps under controlled environmental conditions is very suitable for studies on production of secondary metabolites.

Many liverworts of the order Jungermanniales contain considerable amounts of sesquiterpenoids. Steam distillation of the cultured cells of *Calypogeia granulata* yields a considerable amount of an essential oil, the main component of which is 1,4-dimethyl azulene. Besides this, a number of other sesquiterpenoids have been isolated and identified (Takeda & Katoh, 1981). Thus, research on the biosynthetic problems of these compounds employing bryophytes as experimental material is significant, since their formation is quite limited in cultured cells of vascular plants.

APOGAMY

OCCURRENCE OF APOGAMY IN DIPLOPHASE AND HAPLOPHASE

Formation of apogamous sporophytes in mosses was first reported by Springer (1935) on the leaf tips of naturally occurring diploid gametophytes of *Phascum cuspidatum*. Till today it appears to be the only instance of apogamy in vivo. The swelling on the leaf tips of this moss showed two alternative paths of differentiation. In the presence of abundant moisture protonema was produced, but under drier conditions or with increased salt concentration in the nutritive medium these swellings developed into apogamous sporogonia. Further, 'dryness' or 'reduced hydration' of the medium (by using 3% agar) increased the number of apogamous sporophytes per unit area of protonemal patch in *Phascum* (von Wettstein, 1942). Similarly, in *Georgia pellucida* diploid protonema derived aposporously by regeneration of seta, directly produced numerous sporogon primordia bypassing the gametophytic stage. This study also reveals that differentiation of apogamous sporophytes is to a great extent favoured by relative dryness of the nutritive medium (Bauer, 1956). In *Physcomitrium pyriforme* young sporophytes regenerated to form new sporophytes instead of protonema (Bauer, 1957). Bauer (1959a) obtained apogamous sporophytes exclusively or in addition to normal vegetative buds, from the diploid protonema regenerated from intergeneric hybrid sporophyte *Physcomitrium pyriforme* × *Funaria hygrometrica*. The behaviour of diploid protonema is quite interesting. It produces apogamous sporophytes so long as it is in organic contact with the parent sporophyte. If the connection is removed, only normal-looking gametophytes are formed. This led Bauer to postulate that a factor elaborated by the diploid parent sporophyte is translocated to the aposporous protonemal filaments, and fresh daughter apogamous sporophytes are induced. Normally, regeneration of sporogonia leads to protonemal differentiation. However, in one experiment on *Physcomitrium pyriforme* Bauer (1959b) modified the conditions in such a way that a cluster of daughter sporogonia was produced directly on the sporophyte. The prerequisites are as follows:

(1) The sporogon should be young.
(2) Agar concentration in the medium should be high.
(3) The apical portion of the sporogon with the meristem and its immediate derivatives has to be excised (a kind of apical dominance).

Up to 80 percent of the sporophytes were observed to regenerate in this manner, and the morphological nature of regenerants depended upon the age of the regenerating zone (Bauer, 1959b).

Reports on apogamy in diplophase and haplophase in mosses are summarized in Table 5.1. Apogamous sporogonia in the diplophase of *Desmatodon ucrainicus* and *D. randii* have been shown in Fig. 5.1 A,B, and those

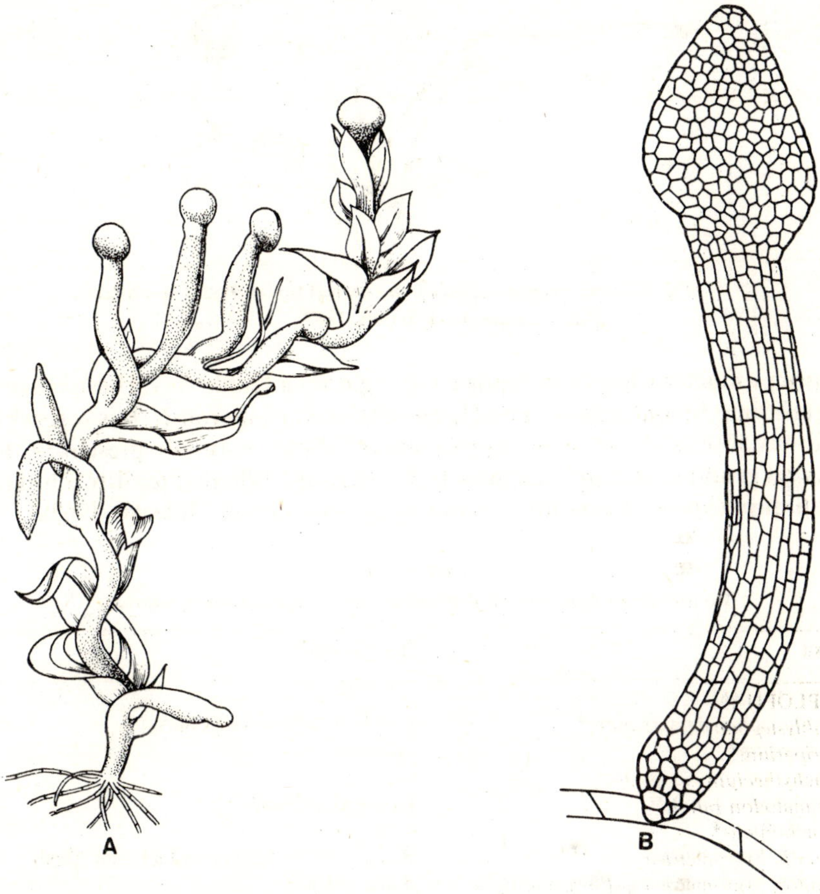

Fig. 5.1 A,B. Apogamy in diplophase in *Desmatodon*. A. Gametophyte of *D. ucrainicus* bearing several apogamous sporophytes. B. An enlarged apogamous sporophyte of *D. randii*.
(After Lazarenko, 1960).

of *Pottia lanceolata* in Fig. 5.2. Apogamous sporophytes in the haplophase of *Physcomitrium pyriforme* and *Funaria hygrometrica* have been shown in Plate 5.1 A,B.

SPORE PRODUCTION IN APOGAMOUS SPOROPHYTES

Formation of viable spores in haploid apogamous sporophytes is rare. *Physcomitrium coorgense* is possibly the only instance on record in which 50 percent apogamous sporogonia were reported to produce viable spores (Plate 5.2 A,B; Lal, 1963)[+] In apogamous sporophytes developing on diploid tissues

---

[+] The haploid nature of these apogamous sporophytes has not been confirmed cytologically. Absence of meiosis was taken as an indication for their being haploid.

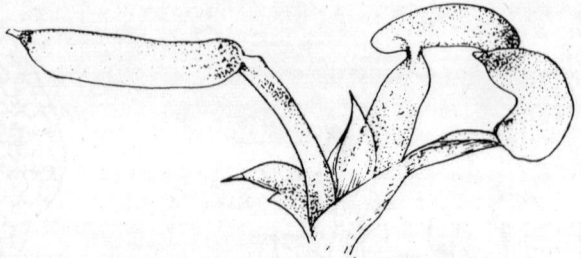

Fig. 5.2. Apogamous sporophytes on leaf tips of *Pottia lanceolata*. (After Lazarenko & Svitlyk, 1969).

spore production has been reported in eight taxa (Table 5.1). These mosses were either diploid strains or had been obtained aposporously. It is suggested that because of the lack of complementary alleles normally present in the second genome, the haploids show lack of fertility. Whether fertility is linked with duplication of genome remains to be ascertained (Tulecke, 1965).

TABLE 5.1.
Incidence of apogamy in diplophase and haplophase in mosses.

Taxa	Investigator/s
DIPLOPHASE	
*Amblystegium juratzkanum**	Lazarenko (1961, 1963, 1965a)
*A. riparium**	—do—
*Brachythecium campestre**	—do—
*Desmatodon randii**	Lazarenko (1960)
*D. ucrainicus**	—do—
Funaria hygrometrica	Bauer (1959a); Kumra and Chopra (1980)
Funaria hygrometrica × *Physcomitrium pyriforme* (Hybrid)	Bauer (1959a)
Grimmia pulvinata	Hughes (1969)
Georgia pellucida	Bauer (1956)
Phascum cuspidatum	Springer (1935); von Wettstein (1942)
*Physcomitrium pyriforme**	Bauer (1957, 1959a,b)
*Pottia intermedia**	Lazarenko (1961, 1963, 1965a)
P. lanceolata	Lazarenko and Svitlyk (1969)
*Splachnum ovatum**	Lazarenko (1961, 1963, 1965a)
S. pedunculatum	—do—
HAPLOPHASE	
Bryum sp.	cited in Bauer (1967)
Desmatodon randii	Lazarenko et al. (1961); Lazarenko (1965b)
Funaria hygrometrica	Chopra and Rashid (1967)
*Physcomitrium coorgense**	Lal (1961a)
P. pyriforme	Menon and Lal (1972)
Pottia intermedia	Lazarenko (1963)
Splachnum sp.	cited in Bauer (1967)
Tetraphis pellucida (=*Georgia pellucida*)	Hughes (1969)

* Apogamous sporophytes produce viable spores.

DIFFERENTIATION OF APOGAMOUS SPOROPHYTES FROM CALLUS

Besides producing protonema, the sporophytes of *Physcomitrium pyriforme*, *Funaria hygrometrica*, and their hybrid sporophyte give rise to calli which later differentiate into daughter sporophytes (Bauer, 1957, 1961a). In *Physcomitrium coorgense* the callus obtained from haploid gametophytic axes and archegonia also differentiated into apogamous sporophytes (Lal, 1961 a,b, 1963). Similarly, callus formed from the haploid protonema of *F. hygrometrica* produced apogamous sporophytes and/or gametophytes depending upon the environmental conditions (Kumra, 1981).

FACTORS CONTROLLING DIFFERENTIATION OF APOGAMOUS SPOROPHYTES

Exogenous Factors

These include light, hydration, sugars, chloral hydrate, growth hormones, inorganic nutrients, and other growth adjuvants.

Light. The gametophytic callus of *Physcomitrium coorgense* produces sporophytes and gametophytes in diffuse light, whereas in dark only apogamous sporophytes differentiate. It has been suggested that light plays a formative role in determining the behaviour of the apical cell (Lal, 1963). The frequency of apogamy in the diploid protonema of *Phascum cuspidatum* (obtained from seta cuttings) is low in day light, while in yellow filtered fluorescent light it is greatly increased (Hughes, 1969). In *Physcomitrium pyriforme* apogamy is suppressed at high light level irrespective of the wavelength (Menon, 1974). Whereas in low light level of complete spectrum apogamous sporophytes as well as gametophytes differentiate from protonemal filaments (Plate 5.3).

Hydration. Reduced hydration of the medium favours apogamy in *Phascum cuspidatum* (Springer, 1935; von Wettstein, 1942), *Georgia pellucida* (Bauer, 1956), *Physcomitrium pyriforme* (Bauer, 1957; Menon, 1974), *Desmatodon ucrainicus* (Lazarenko, 1960), and *Splachnum ovatum* (Lazarenko, 1961). When the plants of *Splachnum sphaericum* are transferred to very dry culture media, the young male and female gametangia transform into sporogonium-like structures (Bauer, 1963). Drying of the medium also induces apogamy in *Funaria hygrometrica* (Chopra & Rashid, 1967).

Sugars. Sucrose, and to some extent glucose, influences the induction of apogamous sporophytes in some mosses. In *Physcomitrium coorgense* the cultures raised on mineral agar medium do not produce apogamous sporogonia, but on a sucrose-medium apogamy is regularly observed (Lal, 1961a). In *Funaria hygrometrica* an increase in concentration of sucrose (1 to 4%) in the medium promotes induction of apogamous sporophytes from stem axes (Rashid & Chopra, 1969). Young, intact capsules of *F. hygrometrica* produce only protonema on media without sucrose or with low concentration of sucrose (0.5 or 1%), whereas two percent sucrose favours formation of apogamous sporophytes × in addition to protonema (Plate 5.4A,B; Kumra & Chopra, 1980). Apogamous sporophytes also differentiate from the callus obtained from protonema of *F. hygrometrica* on media containing higher concentrations of sucrose (Kumra, 1981). Apogamous sporophytes are initiated on the callus of *Physcomitrium pyriforme* only in the presence of sucrose (0.5 to 2%). Protonemal cells cut off some small intercalary cells with zygote-like potentialities. Such cells divide anticlinally to develop two cutting faces, and eventually form apogamous sporophytes either directly or after producing callus (Menon, 1974). It seems that the sporophytic generation requires more sugar, on account of higher respiratory rate and greater osmotic concentration of its cells. The enhanced energy level is a prerequisite for apogamy in some systems, and it tilts the balance of differentiation to the sporophytic development. On the basis of their work on *P. pyriforme* Menon and Lal (1972, 1977) postulated that sucrose may be exercising some sort of 'hormone-like' control over the production of a factor for apogamy, possibly by interacting with the endogenous growth substances.

Chloral hydrate. *Splachnum luteum* produces apogamous sporophytes spontaneously in addition to normal buds, but their percentage is very low. This capability increases up to 80 percent with the addition of chloral hydrate to the medium (Bauer, 1959c). The protonema regenerated from the hybrid sporophyte *P. pyriforme* × *F. hygrometrica* produces apogamous sporophytes only when the isolated protonema is cultured on a medium containing chloral hydrate (Bauer, 1966). Since this effect is race specific, it is attributed to the genetic imbalance brought about by the intergeneric cross.

Growth hormones. In *Georgia pellucida* the number of sporogons per unit area of protonema increases with addition of low concentrations of IAA (Bauer, 1956). In *Funaria hygrometrica* the capacity of gametophytes to produce sporophytes is influenced to varying degrees by growth substances. At low concentrations, GA_3, IAA, kinetin, and a combination of kinetin and IAA appreciably increases the number of sporophytes per culture. Higher concentrations of kinetin are inhibitory for gametophyte as well as sporophyte differentiation, but GA_3 and IAA inhibit sporophyte formation without affecting gametophyte growth (Rashid & Chopra, 1969). Callus of *F. hygrometrica* differentiates apogamous sporophytes and gametophytes in response to BAP (Kumra, 1981). In *Physcomitrium pyriforme* IAA and NAA act somewhat synergistically with sucrose in inducing apogamy/callus on protonema. In the presence of sucrose, kinetin (a well established gametophytic bud-inducer) does not suppress the formation of apogamous sporophytes. Abscisic acid is inhibitory for protonemal growth as well as for differentiation of apogamous sporophytes (Menon, 1974).

Inorganic nutrients. Inorganic salts have variable influence on the induction of apogamous sporophytes. Nitrogen promotes apogamy in *Splachnum luteum* (Bauer, 1959c). In *Physcomitrium coorgense* higher concentrations of salts inhibit the formation of apogamous sporophytes (Lal, 1961a). On the other hand, in *F. hygrometrica* increase in the phosphate and nitrate content, and even of all the mineral nutrients (up to four fold), has no effect on sporophytic or gametophytic growth (Rashid, 1968). Similarly, in *Physcomitrium pyriforme* change in nitrogen source or doubling the nitrate and chloride concentration does not alter the incidence of apogamy (Menon, 1974).

Other growth adjuvants. Coconut milk stimulates differentiation of apogamous sporophytes from the callus of *Physcomitrium coorgense* (Lal, 1961b). In *Funaria hygrometrica* casein hydrolysate enhances apogamy at lower concentrations, but at higher levels sporophyte production is inhibited without affecting growth of gametophytes (Rashid & Chopra, 1969). On the other hand, yeast extract suppresses apogamy in this system (Rashid, 1968).

Endogenous Factors

Sporogon factor. The concept of sporogon factor was formulated by Bauer (1959b). The protonema obtained from the hybrid sporogonium (*P. pyriforme* × *F. hygrometrica*) produces apogamous sporophytes only when it is in organic union with the parent sporophyte. Bauer suggested that a sporogon factor emanating from the diploid sporophyte is translocated into the aposporous protonema. This factor has the following properties:
(1) Its production depends upon the genotype (haploid or diploid condition).
(2) It multiplies in the tissue.
(3) It reveals a quantitative gradient.
(4) It is inherited during vegetative propagation.
(5) It is labile (destructible) and therefore a constant source is essential.

High concentration of sugar in the medium, and dry conditions favour the production of this factor. Lazarenko (1960) also postulated that there is a factor in the diploid protonemata of mosses which on attaining a threshold concentration results in the differentiation of apogamous sporophytes.

It has been suggested that in *Physcomitrium coorgense* differentiation of gametophytes and apogamous sporophytes is controlled by two distinct substances. The one responsible for gametophytic induction is formed only in light, whereas the 'sporogon factor' (responsible for the production of sporophytes) is produced in light as well as in dark. That is why the gametophytic callus of *P. coorgense* differentiates both gametophytes and apogamous sporophytes in light, but in dark it does not form gametophytes (Lal, 1963). There is also evidence for a factor for apogamy in *F. hygrometrica*, and its production seems to be favoured by high concentration of sucrose. After the initiation of apogamous sporogonia in dry cultures, the gametophores continue to produce new sporophytes after subculture on fresh medium (Rashid & Chopra, 1969). In *Physcomitrium pyriforme* the sporogon factor accumulates in leaves and initiates differentiation of sporophytes. High light level and absence of sucrose in the medium inhibit the production of this factor (Menon & Lal, 1977, 1981).

The sporogon factor seems to be of hormonal nature, and may be a mixture of hormones. Bauer (1966) extracted 'bryokinin' (an adenine-type cytokinin) from the callus of the hybrid sporophyte *P. pyriforme* × *F. hygrometrica*. The application of bryokinin enhances apogamy in *Splachnum ovatum*, which produces apogamous sporophytes spontaneously as well (Bauer, 1961b). Hartmann (1970) observed that red light increases synthesis of bryokinin in the hybrid callus, and also promotes apogamy.

Products of moss metabolism. In *Georgia pellucida* sporogons are not formed on protonema while its growth is rapid, but only when it passes to the second phase of growth. i.e., caulonema. Decline in growth seems to be in response to accumulation of certain uncharacterized products of metabolism. Accordingly, sporogon formation is suppressed indefinitely by frequent transfers to fresh media (Bauer, 1956).

Age of the tissue. Difference in regenerative behaviour also seems to depend on the age of the tissue. Bauer (1963) observed that when sporogonia attain ageing they regenerate only protonema, irrespective of the zone. An embryonic sporophyte, on the other hand, yields variable results. The tissues regenerated from the extreme tip (zone A) retain embryonic features and proliferate into a callus mass. A portion lower than this (zone B_1) forms seta tips and some protonema. The zone below this (B_2) shows a transitional behaviour. It produces protonema which develops apogamous sporophytes instead of gametophytes. The three lowest zones (C,D,E) form only protonema, which gives rise to gametophytes (Fig. 5.3).

Genetic constitution. Chromosome number seems to play an important

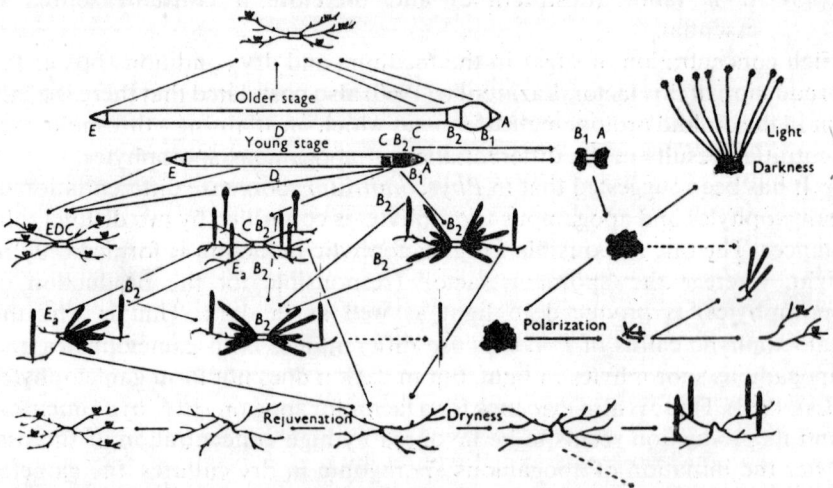

Fig. 5.3. Regeneration products of different zones of the hybrid sporophyte *Physcomitrium pyriforme* × *Funaria hygrometrica*. In the young stage A,—gives rise to apical cell and adjacent cells; B_1,—forms the capsule proper; B_2,—contributes to the apophysis; C,—extension zone; D,—gives rise to fully differentiated part of the seta; E,—matures into foot of a normally developed sporogonium. For behaviour on regeneration *see* text. (After Bauer, 1963 with modification).

role in apogamy. As stated earlier, apogamy is known in vivo only in one diploid moss-*Phascum cuspidatum*, and in cultures it is induced with comparative ease in diploid systems, or in mosses with high chromosome numbers. In *Physcomitrium pyriforme* Menon (1974) observed as many as 72 chromosomes in some callus cells derived from haploid secondary protonema. Earlier, three different chromosome numbers had been reported for *P. pyriforme* i.e., $n=9$; $n=18$; and $n=36$ (Bryan, 1957; Crum & Anderson, 1955; Khanna, 1960; Pande & Chopra, 1957). Bauer (1959b) proposed that apogamy is preceded by diploidization of chromosomes, since in his work with *Funaria hygrometrica* only spontaneous diploids formed apogamous sporophytes on regeneration. Later, he (Bauer, 1963; 1967) put forward the hypothesis that the pathway of differentiation (sporophyte or gametophyte) is controlled by the nature of nucleocytoplasmic interactions. Lazarenko (1960) and Lazarenko et al. (1961) obtained apogamous sporophytes in both haplophase and diplophase in an allopolyploid species of *Desmatodon randii*. There are no reports of apogamy in species in which the chromosome number is monoploid (true haploid). Therefore, it seems that polyploidization is an important factor in apogamy. This hypothesis is further supported by the fact that in liverworts, which have lower chromosome numbers, there are no reports of apogamy. On the other hand, this phenomenon is much more common in ferns which in general have much higher chromosome numbers.

Ripetsky and Matasov (1973) hold a different view. According to them increase in ploidy as such is not an important factor in apogamy. They observed that in *Pottia intermedia* the aposporously derived diploid gametophytes produce apogamous sporophytes. The spores obtained from these apogamous sporophytes produce protonema and gametophytes, but do not bear any apogamous sporophytes. According to them, aposporous strains completely lose the ability to produce apogamous structures. They postulated that some factors added during fertilization and sporophyte development are responsible for this phenomenon.

Ripetsky (1979a,b) later reported stable maintenance of the ability of *Pottia intermedia* in the diplophase (aposporic) to form apogamous sporophytes. This is regarded to be a consequence of changes in the gene activity (epigenetic changes) which bring about sporophytic morphogenesis. It is interesting to note that aposporic gametophytes of this moss regenerated from the tissue of a young spore sac wall, in contrast to clones from other parts of sporogonium, lack the capacity for apogamy (Ripetsky, 1980). It has, therefore, been concluded that ontogenic variations associated with determination of sporophyte growth are not a result of simple chromosome duplication, but are due to heritable variations (Ripetsky, 1983). Ripetsky (1985) further adds that even though the capacity to produce apogamous sporophytes is maintained at the cellular level, the actual formation of apogamic structures takes place only in suitable environmental conditions.

DIFFERENTIATION OF SPOROPHYTE AND GAMETOPHYTE

Generally the differentiation of gametophores in mosses is preceded by establishment of an apical cell with three-cutting faces, whereas sporophytes have an apical cell with two-cutting faces (Parihar, 1972; Smith, 1938; Watson, 1971). Apical cells are very labile and can even revert to the protonemal phase with one-cutting face (Bopp, 1953). It is postulated that an apical cell with three-cutting faces passes through a stage with two-cutting faces. If factors are conducive to apogamy, the apical cell with two-cutting faces becomes stabilized. If, on the other hand, factors for development of apical cell with three-cutting faces are available, a gametophyte results. In other words, development of a sporophyte requires elimination of the gametophytic factors present in the cells (Menon & Lal, 1974). However, both apogamous sporophytes and gametophytes have been observed to arise from the same protonema in *Physcomitrium pyriforme* (Plate 5.5; Menon & Lal, 1972).

ROLE OF CALYPTRA IN SPOROGON DEVELOPMENT

A comparative study of normal sporophytes and diploid apogamous sporophytes of *Phascum cuspidatum* reveals that the gametophyte exerts a controlling influence on the developing sporophyte (Hughes, 1969). This is mainly through calyptra which is usually a membranous, cap-like structure formed after fertilization by the reinforced venter wall, and it covers the capsule until the latter is nearly mature.

The various effects of calyptra on the developing sporogon are as follows:

(1) Protection from mechanical injury.
(2) Prevention of desiccation by reducing transpiration.
(3) Retention and circulation of water in the space between calyptra and capsule wall.
(4) Phototropic and geotropic responses of sporophyte are enhanced after removal of calyptra.
(5) Calyptra may also regulate the symmetry of moss capsule. In *Polytrichum juniperinum* splitting of the inner sheathing layer of calyptra leads to asymmetrical expansion and subsequent development of the normal bilaterally symmetrical capsule. When unexpanded sporophytes are cultured without the sheathing layer, expansion of capsule becomes radially symmetrical. However, in *Funaria hygrometrica* no such correlation is observed. In this moss symmetry of capsule is not determined by the splitting of calyptra, since the capsule attains a strong asymmetrical profile before calyptra begins to rupture (Paolillo, 1968). Bopp (1957b) had also reported earlier that bilateral symmetry of the capsule of *Funaria* is not because of the effect of calyptra.
(6) French and Paolillo (1975a) reported that calyptra also influences the

plane of division in guard cell mother cells in the capsule of *Funaria* and *Physcomitrium*. In the developing capsules enclosed by calyptra (before the division of guard cell mother cells), the axes of the stomata are oriented parallel to the axis of the capsule. Removal of calyptra from an elongating sporophyte leads to an essentially random orientation of stomata, and a variety of other stomatal abnormalities.

(7) Detailed work on the effect of calyptra on sporogon has been done mainly in *Funaria hygrometrica*, but the overall regulating role of calyptra on the normal growth and development of seta and capsule has been demonstrated in several mosses (Bopp, 1954a, 1956; Herzfelder, 1921, 1923; Zielinski, 1910). The nature and degree of the effect depends upon the taxon and the age of sporogon at which calyptra is removed. Bopp (1954a) observed that if calyptra of *Funaria* is removed before its sporophyte has attained the required 'physiological age', the sporophyte fails to form a capsule. In conformity with these findings, French and Paolillo (1976) also noticed that when calyptra is removed from a very young sporophyte of *Funaria* (5 to 10 mm long) the seta does not elongate but instead shows dramatic thickening, and the capsule fails to develop. The thickening in the seta is because of changes in the plane of cell division as well as directions of cell expansion in the intercalary (subapical) meristem. Removal of calyptra in a very young sporophyte also results in indeterminate activity in the subapical meristem. In nature the presence of calyptra leads to regular cell division and cell expansion in this meristem, and so the elongation of seta is normal. Once the sporophyte is beyond its early development, removal of calyptra has only minor effects on the timing and quality of capsule expansion. Late in elongation (25 to 35 mm long sporophyte), absence of calyptra even seems to promote earlier and more vigorous capsule expansion. The results were, however, different with *Polytrichum* in which an early determination of the capsule takes place, and removal of calyptra does not upset the developmental balance of the parts in the sporophyte (Paolillo, 1968).

In his earlier experiments on *Funaria hygrometrica* Bopp (1954b) observed that the thickening of seta which results on removal of calyptra can be checked by application of fresh extract of calyptrae to naked sporogons. This thickening could also be prevented by application of the antiauxin maleic hydrazide, indicating that there may be a hormonal antagonism between calyptra and sporogon. However, in a subsequent publication he (Bopp, 1957a) demonstrated that the effect of calyptra in preventing the thickening of seta is not of a hormonal nature, but is due to physical restraints imposed by the calyptra on the developing capsule. This report of Bopp was later confirmed by French and Paolillo (1975b), who also noticed that even when the original calyptrae of *Funaria* were replaced by dead (chemically extracted) calyptrae the development of capsules proceeded normally, proving thereby that calyptra does not control capsule development via chemical regulation.

In more recent experiments of French and Paolillo (1975c) benzyladenine or zeatin was applied to the growing sporophytes of *Funaria* with intact calyptrae. This led to indeterminate growth of setae without the formation of capsules, indicating that in the normal course of events the influence of calyptra may have a chemical basis. These authors opine that there is a hormonal interaction between the apical region of the capsule and the intercalary meristem, and even if calyptra has no direct chemical role, its presence is essential for changing the hormonal status of the plant.

Significance of the role of calyptra is also proved by studies on apogamous sporophytes, which develop outside the confines of calyptra. Such sporophytes are different in morphology from the sexually produced sporophytes and are also usually sterile (*see* Chopra & Rashid, 1967; Rashid & Chopra, 1969).

APOSPORY

Since the discovery of apospory in *Hypnum serpens. H. cupressiforme, Bryum caespitosum* (Pringsheim, 1876), and *Ceratodon purpureus* (Stahl, 1876), development of diploid gametophytes from the vegetative tissues of sporophyte (i.e., capsule wall, seta) has been demonstrated in some other mosses as well (LaRue, 1930; Marchal & Marchal, 1907, 1909). Marchal and Marchal (1907, 1909) raised polyploid races in *Amblystegium* (octoploid) and *Bryum* (tetraploid) by this method. The technique of seta regeneration has been employed to obtain diploid gametophytes for genetic studies in *Funaria hygrometrica, Physcomitrium pyriforme*; the intergeneric hybrid of these two taxa, and some other mosses (von Wettstein, 1923, 1924a,b, 1925). In *Physcomitrium turbinatum* the highest percentage of regeneration was obtained from entire setae and their top-halves cut into segments. The lowest percentage of regeneration was noticed in the region just above the foot (Redfearn & Meyer, 1949). Similar observations were made by Kachroo (1954) on *Physcomitrium pyriforme*. Hughes (1969) reported formation of leafy shoots on the diploid protonema obtained from seta cuttings of *Grimmia pulvinata*. Regeneration from the capsule wall has also been reported in *Physcomitrium cyathicarpum* (Narayanaswami & Lal, 1957) and *Funaria hygrometrica* (Kumra & Chopra, 1980). In *Haplocladium angustifolium* wounding is necessary for seta regeneration (Plate 5.6A), and protonemata always develop from the internal parenchymatous cells (Ono & Tanaka, 1976). The protonema regenerated from the seta in *Funaria hygrometrica* bears aposporous gametophytes (Plate 5.6B). These gametophytes grow very slowly and are much smaller as compared to haploid ones. Gametangia fail to form on them under conditions in which haploid gametophytes reproduce readily (Kumra & Chopra, 1980).

There are not many reports of apospory in liverworts. Aposporous outgrowths were observed in *Anthoceros* (Borenhagen, 1926; Lang, 1901; Rink, 1935), but the majority failed to develop into gametophytes (Schwarzenbach, 1926). *Marchantia polymorpha* and the anacrogynous liverwort *Blasia pusilla*, also developed diploid thalli aposporously (Burgeff, 1943; Matzke & Raudzens, 1968). Degeneration of tissue is a prerequisite for apospory in *B. pusilla*. It is suggested that when sufficient number of cells die and integration within the organ fails, the surviving cells act as individuals and not as a part of the whole. Similar behaviour had been reported earlier in some other liverworts: *Pellia neesiana, Aneura palmata*, and *A. latifrons* grown in cultures (Schwarzenbach, 1926). In *B. pusilla* the initiation of apospory has been observed on a glucose-free medium (reverse of apogamy), and these proliferations grow into characteristic *Blasia* gametophytes with flask-shaped gemma receptacles on transfer to glucose-supplemented medium. The growth rate of diploid plants is the same as that of the haploids. The aposporous plants are all female and they become fertile. The induction of apospory is fairly high (59%), and in every instance the aposporous outgrowths develop into mature diploid gametophytes. Apospory has also been reported in *Pellia epiphylla* and *Pallavicinia lyellii* (Matzke & Raudzens, 1968, 1969; Raudzens & Matzke, 1968), and some other members of the Jungermanniaceae: *Lophocolea heterophylla, Radula complanata, Scapania nemorosa, Nowellia curvifolia, Ptilidium pulcherrimum, Jungermannia lanceolata*, and *Porella pinnata* (Simone, 1966). Mehra and Pental (1976) obtained aposporous, diploid gametophytes from the setae of *Athalamia pusilla*, and some of these thalli produced archegoniophores.

Since the diploid gametophytes obtained from the cells of sporophytes retain the growth pattern of haploid gametophytes, these findings support the homologous rather than the antithetic concept of alternation of generations. It is suggested that increase in ploidy could have come about, at least occasionally in mosses, as a result of apospory (Matzke & Raudzens, 1969). Since the formation of diploid gametophytes is rare in hepatics, it explains the almost complete absence of polyploidy in this group (Berrie, 1960).

CALLUS FORMATION AND ITS DIFFERENTIATION

FORMATION OF CALLUS

Polarity is essential for differentiation. Its suppression by neutralization of internal gradients of nutrients and growth substances results in the formation of callus. In bryophytes callus has been induced from different parts of the gametophyte and sporophyte. The available reports are summarized in Table 5.2.

Callus can be induced by a variety of factors. Systems in which mere increase in sugar concentration is sufficient are: *Polytrichum commune*,

TABLE 5.2.
Callus formation from different parts of gametophyte and sporophyte in bryophytes.

Taxa	Portion from which callus raised	Investigator/s
LIVERWORTS		
Asterella angusta = A. wallichiana	Thallus	Rashid (1971); Kumra (1984)
Athalamia pusilla	—do—	Mehra and Pental (1976)
Blasia pusilla	—do—	Allsopp and Ilahi (1970)
Calypogeia granulata	—do—	Takeda and Katoh (1981)
Cephaloziella sp.	Spore	Meyer (1953)
Fossombronia himalayensis	Thallus	Mehra and Pahwa (1976)
F. pusilla	Spore	Allsopp (1957)
Jungermannia subulata	Thallus	Ohta et al. (1981)
Lophocolea heterophylla	—do—	Ohta and Hirose (1982)
Makinoa crispata	—do—	Ilahi and Allsopp (1970)
Mannia dichotoma	—do—	Kapur (1983)
Marchantia nepalensis	—do—	Kaul et al. (1961, 1962); Tripathi (1978)
M. polymorpha	Gemmae	Ono (1973); Ohta et al. (1977); Katoh et al. (1980)
Moerckia flotowiana	Thallus	Ilahi and Allsopp (1970)
Noteroclada confluens	—do—	Allsopp and Ilahi (1970)
Petalophyllum ralfsii	—do—	Ilahi (1972)
Reboulia hemisphaerica	Spore	Allsopp (1957)
Riccardia multifida	Thallus	Ilahi (1972)
R. pinguis	—do—	Ilahi and Allsopp (1970)
Riccia crystallina	—do—	Chopra and Sood (1973)
R. frostii	—do—	Vashistha (1985)
Scapania nemorosa	—do—	Basile (1965)
S. parvidens	—do—	Ohta and Hirose (1982)
Sphaerocarpos donnellii	—do—	Miller and Machlis (1968); Grusak et al. (1980)
S. texanus	—do—	Montague and Taylor (1971)
MOSSES		
Gametophytic Callus		
Atrichum undulatum	Protonema	Ward (1960); Spiess (1976)
Anoectangium thomsonii	—do—	Saxena and Rashid (1982)
Bryum coronatum	—do—	Kumra and Chopra (1982)
	Gametophytic bud	—do—
Climacium americanum	Protonema	Spiess (1976)
Entodon seductrix	—do—	—do—
Funaria hygrometrica	Spore	von Wettstein (1953); Bünning and von Wettstein (1953)
	Protonema	Spiess (1975); Kumra (1981)
	Gametophytic bud	Kumra (1981)
Heterophyllum haldaneanum	Protonema	Spiess (1976)
Hyophila involuta	—do—	Rahbar and Chopra (1980)
Physcomitrium coorgense	Stem	Lal (1961a)
	Archegonial venter	—do—

(Table contd.)

Taxa	Portion from which callus raised	Investigator/s
P. pyriforme	Protonema	Menon (1974)
	Leaf	—do—
Polytrichum commune	Protonema	Ward (1960); Spiess (1976)
	Stem	Ward and Frederick (1967)
	Leaf	—do—
Polytrichum juniperinum	Protoplasts isolated from leaves	Gay (1976)
Pottia lanceolata	Leaf	Lazarenko and Svitlyk (1969)
Pylaisiella selwynii	Protonema	Spiess (1976)
	Gametophytic bud	Spiess et al. (1973)
Thuidium delicatulum Sporophytic Callus	Protonema	Spiess (1976)
Dicranella schreberi	Aposporous protonemata	Ulychna (1971)
Funaria hygrometrica	Apogamous sporophyte	Rashid and Chopra (1969); Kumra (1981)
Georgia pellucida (=Tetraphis pellucida)	Aposporous protonemata	Bauer (1956); Demkiv et al. (1981)
Physcomitrium pyriforme	Sporogonial tip	Bauer (1957, 1961a,b)
Hybrid of Funaria hygrometrica × Physcomitrium pyriforme	Sporophyte	Beutelmann (1973)
Pottia intermedia	Aposporous protonemata	Ripetsky (1979a)

Atrichum undulatum (Ward, 1960); *Physcomitrium coorgense* (Lal, 1961a); *P. pyriforme* (Menon, 1974); *Fossombronia pusilla, Reboulia hemisphaerica* (Allsopp, 1957); *Riccardia multifida, R. pinguis, Noteroclada confluens* (Allsopp & Ilahi, 1969, 1970; Ilahi & Allsopp, 1970); *Asterella angusta* (Rashid, 1971); *Riccia crystallina* (Chopra & Sood, 1973); *Calypogeia granulata* (Takeda & Katoh, 1981); *Jungermannia subulata, Lophocolea heterophylla*, and *Scapania parvidens* (Ohta & Hirose, 1981). A schematic procedure for callus induction from members of Jungermanniales has been suggested by Ohta and Hirose (1982):

Mature capsules
 ↓ 0.1%—Benzalkonium chloride solution, 5 to 10 min
   2%—Sodium hypochlorite solution, 3 to 5 min
   sterilized distilled water, three times
Aseptic capsules
 ↓ Broken on Knop's agar medium
   Cultured at 25 C in light
Gametophytes
 ↓ Transferred to a medium containing 4% glucose
Callus

Such calli or cells in suspension show growth characteristics similar to those reported for cells of Marchantiales (Ohta & Hirose, 1982):
(1) Callus is composed mainly of haploid cells.
(2) In general, the cultured cells grow actively in light and have many, well developed chloroplasts (Katoh et al., 1979; Ohta et al., 1977).
(3) De-differentiation and re-differentiation is controlled by the amount of sugar in the medium (Allsopp, 1957; Ono, 1976; Takeda & Katoh, 1981).

When cultured cells of *Jungermannia subulata*, and *Calypogeia granulata* are transferred to glucose-free medium, gametophytes are obtained (Takeda & Katoh, 1981). The cultured cells of *J. subulata* and *Lophocolea heterophylla* show an unusual behaviour in the utilization of nitrogen sources. Ammonium is taken up preferentially, particularly at the earliest stage of growth, whereas only a negligible amount of nitrate is utilized as long as ammonium is present in the medium. This unbalanced utilization of the two nitrogen sources causes an abrupt drop in pH of the medium (Ohta et al., 1981).

Darkness generally favours callus formation in bryophytes. Callus has also been raised from the X-ray irradiated spores of *Polytrichum commune* and *Atrichum undulatum* (Ward, 1960). In *Bryum coronatum* and *Funaria hygrometrica* callus develops after irradiating the protonema with UV-light. In *B. coronatum* and *F. hygrometrica* kinetin, IAA, and vitamins are essential for callus induction (Plate 5.7 A-E). Ultraviolet radiation substitutes the requirement for growth regulators (Kumra & Chopra, 1982). It also reduces the time required for callus initiation and enhances callus growth in *Asterella wallichiana* (Chopra & Kumra, 1986). Callus has also been induced with growth regulators, individually or in combination. Auxins induce callus in *Funaria* (von Wettstein, 1953), *Marchantia nepalensis* (Kaul et al., 1962; Tripathi, 1978), *Sphaerocarpos texanus* (Montague & Taylor, 1971), and *Riccia frostii* (Vashistha, 1985), whereas cytokinins do so in *Pylaisiella selwynii* (Spiess, 1979) and *Tetraphis pellucida* (Demkiv et al., 1981). Some of the hepatics form callus only by an interaction of growth regulators: auxins (IAA, NAA, & 2,4-D) with kinetin in *Athalamia pusilla* (Mehra & Pental, 1976); and IAA with GA_3 in *Marchantia polymorpha* (Ono, 1973).

Addition of coconut milk to the callus-inducing media enhances callus growth in *Funaria hygrometrica* (Rashid & Chopra, 1969), *Athalamia pusilla* (Mehra & Pental, 1976), and *Asterella wallichiana* (Dua, 1983). Callus can also be obtained with the synergism of high concentration of sucrose and coconut milk, as in *Fossombronia himalayensis* (Mehra & Pahwa, 1976). Vitamin B_1 at high concentrations induces callus in *Funaria hygrometrica* (Bünning & von Wettstein, 1953; von Wettstein, 1953). However, vitamins together with ammonium nitrate, chelated iron, and sucrose are required for callusing in *Hyophila involuta* (Plate 5.8 A-D; Rahbar & Chopra, 1980). In *Mannia dichotoma* callus is induced only in the presence of ammonium nitrate, growth regulators (kinetin & IAA), and vitamins (Kapur, 1983).

Macroelements of MS medium, sucrose (2%), and growth regulators (kinetin & IAA) are essential for callus induction in *Asterella wallichiana* (Kumra, 1984). Peptone, yeast extract, and casein hydrolysate retard normal differentiation of thalli in *Riccia crystallina*, and a deep green callus is formed (Sood, 1974). In *Asterella wallichiana* (Dua, 1983) and *Riccia frostii* (Vashistha, 1985) casein hydrolysate also favours callus formation.

DIFFERENTIATION IN CALLUS

Differentiation in callus usually results after the causal factor has been removed. In *Cephaloziella* (Meyer, 1953), *Riccia* (Sood, 1974), and *Funaria* (Kumra, 1981) calli differentiate on the same medium in which they are induced. The pattern of differentiation is influenced by light level, light quality, temperature, and chemical constitution of the medium.

Cultures grown in light, show differentiation of both sporophytes and gametophytes from the callus of *Physcomitrium coorgense*, whereas only sporophytes develop in dark. In *P. pyriforme* and *Funaria hygrometrica* low level of light favours differentiation of sporophytes. In *F. hygrometrica* low temperature also enhances sporophytic differentiation (Kumra, 1981). It is opined by some workers that endogenous factor/s for sporophytic differentiation accumulate in callus cells during the inductive treatments (Kumra, 1981; Lal, 1961b; Menon, 1974), and that these factors are of hormonal nature (Gautheret, 1971; Pierik, 1967).

Higher concentration of sugars prevents differentiation in calli of *Fossombronia pusilla*, *Reboulia hemisphaerica* (Allsopp, 1957); *Riccardia multifida*, *R. pinguis* (Ilahi & Allsopp, 1970); and *Physcomitrium pyriforme* (Menon, 1974). On media containing higher concentrations of sucrose, the callus in *Bryum coronatum* remains undifferentiated, but in *Funaria hygrometrica* apogamous sporophytes are formed (Plate 5.9 A-C; Kumra, 1981). Addition of auxins (NAA, IAA, or 2,4-D) to the medium results in the production of aberrant gametophytes from calli of *Polytrichum* and *Atrichum* (Ward, 1964). On coconut milk-supplemented medium only sporophytes differentiate in light as well as in dark from the callus of *Physcomitrium coorgense* (Lal, 1961b). Similarly, whereas in *P. pyriforme* callus growing in low light level on medium with two percent sucrose differentiates into sporophytes as well as gametophytes, addition of coconut milk results in a predominantly sporophytic differentiation (Plate 5.10 A; Menon & Lal, 1977). In liquid medium containing coconut milk the calli of *P. coorgense* and *P. pyriforme* (Plate 5.10 B,C) dissociate into single cells and cell aggregates, which initially organize into short filaments and later give rise to sporophytes (Lal, 1961b; Menon, 1974). In *Asterella angusta* growing on basal medium, free cells regenerate directly into thalli without forming cell aggregates (Plate 5.11 A,B; Rashid, 1971). These studies indicate that free cells can behave as spores and are totipotent.

CONTROLS IN DIFFERENTIATION

Morphogenesis in vivo or in vitro is regulated by three controls: (1) exogenous, (2) cytoplasmic, and (3) nuclear. These are interdependent and work in intimate coordination.

The exogenous control comprises factors like light, temperature, and mineral nutrients. In the in vitro experiments with callus, it is the composition of the medium that is of prime importance and this must include carbohydrates in assimilable form, minerals such as potassium, nitrates, phosphates, trace elements, and in addition certain complex substances like vitamins and hormones. Calli have an innate capacity to synthesize their own growth substances, provided the requisite nutrients are available. Thus, if a tissue is endowed with the capacity to manufacture auxins, it may need only an appropriate supply of cytokinins for differentiation, and vice versa. Certain calli may be able to manufacture both the substances in sufficient amounts and, therefore, are independent of the exogenous supply of growth substances for differentiation.

The cytoplasmic control mediates between the exogenous factors and the nuclear control, acting as a sort of buffer.

The third and the most important is the nuclear control, exerted by the genes of the organism. All the genes are not active at the same time in a given tissue or organ. Some are in a state of activity, while others are quiescent or repressed. Nevertheless they have the capacity to be triggered to activity when exposed to appropriate biochemical environment. It has been shown that organ-specific proteins and enzymes are formed by activated genes.

According to Mehra's 'gene block' hypothesis there is no fundamental difference in the sporophytic or gametophytic phase of the life cycle (Mehra, 1972). Both are merely expressions of the different facets of a labile genome which can be switched one way or the other, depending upon the nutritive and biochemical status of the cells. Partial activation of both the facets of the genome can occur in the intermediate environments inducing organization of synthetic structures with part sporophytic and part gametophytic characters. The number of genomes by themselves do not determine the sporophytic or gametophytic phase. A gametophyte may be X, 2X, 3X, 4X, and so on, and likewise a sporophyte may have corresponding and identical constitution. Yet they are so different in their morphologies. In these two generations, it is the different gene systems of the genome that are in an active state.

ALTERNATION OF GENERATIONS

In bryophytes, the spore results in the formation of the gametophyte (or the gamete bearing generation) and the fertilized egg gives rise to the sporophyte (or the spore bearing generation). A remarkable feature of this group of plants is the conspicuous, independent, green. nutritionally self-sufficient gametophyte to which the sporophyte is permanently attached. The two generations are variously organized, consistent with their mode of nutrition and function. The alternation of generations in the life cycle was elucidated by Hofmeister (1851) and the associated reduction and duplication in chromosome number was demonstrated by Strassburger (1894). There is, therefore, an alternation of morphologies as well as level of ploidy.

Views of various workers concerning the origin of the bryophytic sporophyte are largely influenced by how they adhere to the two conflicting theories concerning the nature of the sporophyte. According to the homologous theory (also referred to as modification or transformation theory) (Pringsheim, 1878) the sporophyte is a direct modification of the gametophyte when developed under different environmental conditions. Supporters of this concept hold that the sporophyte is to be interpreted as a neutral generation, and one whose primary function is to produce spores. Evidences in support of the modification theory have been drawn from algae, bryophytes, and pteridophytes. These include the isomorphic alternation of generations in certain algae; photosynthesis in sporophytes of pteridophytes and bryophytes; apogamy; apospory; and formation of structures intermediate between two generations. Furthermore, these studies suggest the ability of tissues to retain the potential to form both the gametophytic and sporophytic generations, the manifestation of which is determined by the environment. Lazarenko (1965b) reported the co-existence of apogamous sporophytes with normal sporophytes on the haploid plants of *Desmatodon randii* (Fig. 5.4). He postulated that mosses are amphomorphic i.e., they are able to manifest both sporophytic and gametophytic potential. Thus, it is clear that for alternation of generations a change in the levels of ploidy is not obligatory, but there has to be an alternation of morphologies. Ripetsky (1985) also regards the alternation of generations in the moss life cycle to be due to a change in the dominance of gametophytic and sporophytic patterns of morphogenesis.

According to the antithetic theory (also referred to as intercalation or interpolation theory), suggested by Celakovsky (1874) and developed by Bower (1889), the sporophyte is to be interpreted as an entirely new structure intercalated between two successive gametophytic generations in connection with the migration of plants from water to land. It is thought that the bryophytic sporophyte did not make its appearance until after a gamete-bearing plant had evolved an archegonium (*see* Chopra, 1981). The zygote retained inside an archegonium did not divide meiotically, but divided mitotically to form a number of diploid cells, each of which later divided meiotically to form four spores. The result was a rudimentary type of sporophyte in which all cells were sporogenous and it was morphologically different from the gametophyte. From this was evolved a sporophyte in which the superficial cells were sterile instead of sporogenous. Further evolution of the sporophyte came about as a result of additional sterilization of the potential sporogenous tissue. In what are considered more advanced bryophytic sporophytes the sterile tissues resulted in a differentiation of foot, seta, and

126   Biology of Bryophytes

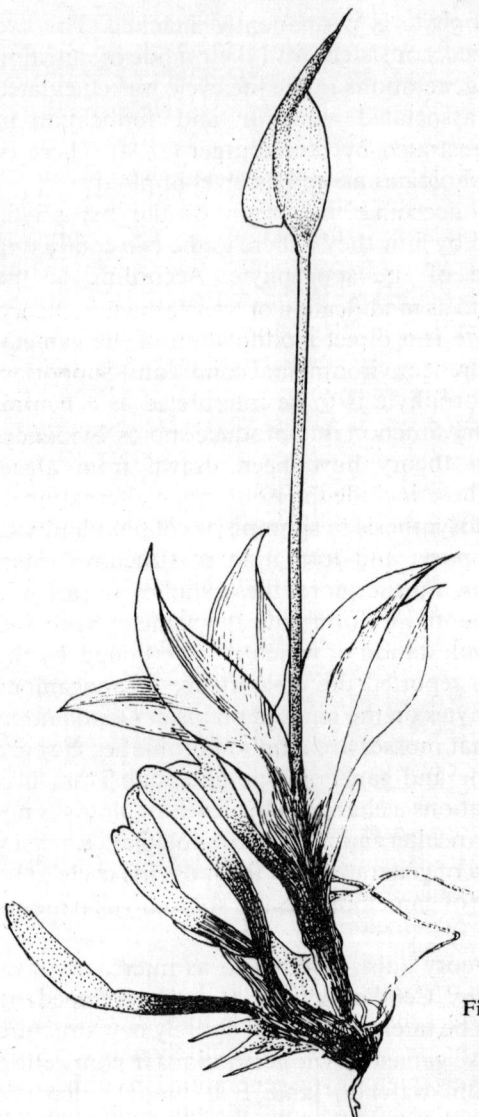

Fig. 5.4. Haploid gametophyte of *Desmatodon randii* bearing a cluster of apogamous sporophytes at the base, and the normal terminal sporophyte. (After Lazarenko, 1965b).

the parts of capsule connected with photosynthesis, dehiscence, and regulated dispersal of spores. Those supporting the intercalation theory believe that there has been a progressive evolution in the sporophytic generation of bryophytes from a simple to a more complex one.

REFERENCES

ALLSOPP, A., *Nature*, **179**, 681 (1957).
ALLSOPP, A. and I. ILAHI, *Phytomorphology*, **19**, 242 (1969).

ALLSOPP, A. and I. ILAHI, *Phytomorphology.* 20, 9 (1970).
BASILE, D.V., *Am. J. Bot.*, 52, 443 (1965).
BAUER, L., *Planta,* 46, 604 (1956).
BAUER, L., *Ber. dt. bot. Ges.*, 70, 424 (1957).
BAUER, L., *Naturwissenschaften,* 46, 154 (1959a).
BAUER, L., *Naturwissenschaften,* 46, 154 (1959b).
BAUER, L., *Planta,* 53, 628 (1959c).
BAUER, L., *Naturwissenschaften,* 48, 507 (1961a).
BAUER, L., *Biol. Zbl.,* 80, 353 (1961b).
BAUER, L., *J. Linn. Soc. (Bot.),* 58, 343 (1963).
BAUER, L., *Beitr. Biol. Pfl.,* 42, 113 (1966).
BAUER, L., "Determination von Gametophyt und Sporophyt." In W. Ruhland, Ed., *Encyclopedia of Plant Physiology,* Vol. XVIII, Springer-Verlag, Berlin, Heidelberg, New York 1967. Pp. 235-256.
BERRIE, G.K., *Trans. Br. bryol. Soc.,* 3, 688 (1960).
BEUTELMANN, P., *Planta,* 112, 181 (1973).
BOPP, M., *Z. Bot.,* 41, 1 (1953).
BOPP, M., *Z. Bot.,* 42, 331 (1954a).
BOPP, M., *Naturwissenschaften,* 41, 234 (1954b).
BOPP, M., *Ber. dt. bot. Ges.,* 69, 455 (1956).
BOPP, M., *Z. Indukt. Abstamm. -u. VererbLehre,* 88, 600 (1957a).
BOPP, M., *Z. Indukt. Abstamm. -u. VererbLehre,* 88, 608 (1957b).
BORENHAGEN, H., *Biol. Zbl.,* 46, 578 (1926).
BOWER, F.O., *Ann. Bot.,* 3, 305 (1889).
BRYAN, V.S., *Bryologist,* 60, 103 (1957).
BÜNNING, E. and D. VON WETTSTEIN, *Naturwissenschaften,* 40, 147 (1953).
BÜRGEFF, H., *Genetische Studien an Marchantia,* Gustav-Fisher, Jena, 1943.
ČELAKOVSKÝ, L., *Über die Verschiedenen Formen und die Bedeutung des Generations Wechschel der Pflanzen,* Prague, 1874.
CHOPRA, R.N. and SUMAN KUMRA, *Cryptog. Bryol. Lichénol.* 7, 249 (1986).
CHOPRA, R.N. and A. RASHID, *Bryologist,* 70, 206 (1967).
CHOPRA, R.N. and SNEH SOOD, *Phytomorphology,* 23, 230 (1973).
CHOPRA, R.S., *Misc. Bryol. Lichenol.,* 9, 1 (1981).
CRUM, H. and L.E. ANDERSON, *Bryologist,* 58, 1 (1955).
DEMKIV, O.T., R.T. RIPETSKY, D. FEDYK, JR., and K.O. ULYCHNA, "Control of the morphogenetic patterns and the stability of the epigenetic state in bryophytes (Abstr.)." In *The Proceedings of XIII International Botanical Congress,* Sydney, Australia, 1981. Pp. 145.
DUA, SUMAN, "Morphogenetic and physiological studies on some bryophytes," Ph.D. Thesis, Univ. Delhi, India, 1983.
FRENCH, J.C. and D.J., JR. PAOLILLO, *Ann. Bot.,* 39, 233 (1975a).
FRENCH, J.C. and D.J., JR. PAOLILLO, *Bryologist,* 78, 438 (1975b).
FRENCH, J.C. and D.J., JR. PAOLILLO, *Bryologist,* 78, 433 (1975c).
FRENCH, J.C. and D.J., JR. PAOLILLO, *Am. J. Bot.,* 63, 492 (1976).
GAUTHERET, R.J., "Action de variations de temperature sur la rhizogenése des tissues de Topinambour cultivés in vitro." In *Les Culture des Tissues de Plantes.* Colloq. Int. Centre natn. Rech. Sci., Paris, 1971. Pp. 187-199.
GAY, L., *Z. PflPhysiol.,* 79, 33 (1976).
GRUSAK, M.A., R.J. THOMAS, and B.H. MARSH, *J. Cell Sci.,* 43, 167 (1980).
HARTMANN, E., "Über den Stoffwechsel und das Differenzierungsverhalten des Laubmooskallus in Abhängigkeit von Licht," Ph.D. Thesis, Johannes Guttenberg, Univ. Mainz, Germany, 1970.
HERZFELDER, H., *Flora,* 114, 383 (1921).
HERZFELDER, H., *Flora,* 116, 476 (1923).
HOFMEISTER, Q.E., *Vergleichende Untersuchungen,* Leipzig, 1851.

HUGHES, J.G., *New Phytol.*, **68**, 883 (1969).
ILAHI, I., *Biologia*, **18**, 127 (1972).
ILAHI, I. and A. ALLSOPP, *Phytomorphology*, **20**, 126 (1970).
KACHROO, P., *J. Indian bot. Soc.*, **33**, 262 (1954).
KAPUR, ANITA, "In vitro studies on some bryophytes", Ph.D. Thesis, Univ. Delhi, India, 1983.
KATOH, K., Y. OHTA, Y. HIROSE, and T. IWAMURA, *Planta*, **144**, 509 (1979).
KATOH, K., M. ISHIKAWA, K. MIYAKE, Y. OHTA, Y. HIROSE, and T. IWAMURA, *Physiol. Plant.*, **49**, 241 (1980).
KAUL, K.N., G.C. MITRA, and B.K. TRIPATHI, *Curr. Sci.*, **30**, 131 (1961).
KAUL, K.N., G.C. MITRA, and B.K. TRIPATHI, *Ann. Bot.*, **26**, 447 (1962).
KHANNA, K.R., *Caryologia*, **13**, 559 (1960).
KUMRA, P.K., "Morphogenetic and physiological studies on some mosses", Ph.D. Thesis, Univ. Delhi, India, 1981.
KUMRA, P.K. and R.N. CHOPRA, *Cryptog. Bryol. Lichénol.*, **1**, 197 (1980).
KUMRA, P.K. and R.N. CHOPRA, *Z. PflPhysiol.*, **108**, 143 (1982).
KUMRA, SUMAN, *J. Plant Physiol.*, **115**, 263 (1984).
LAL, M., *Phytomorphology*, **11**, 263 (1961a).
LAL, M., "Experimental investigations on the moss *Physcomitrium*", Ph.D. Thesis, Univ. Delhi, India, 1961b.
LAL, M., "Experimental induction of apogamy and the control of differentiation in gametophytic callus of moss *Physcomitrium coorgense* Broth." In P. Maheshwari and N.S. Rangaswamy, Eds, *Plant Tissue and Organ Culture—A Symposium*, Int. Soc. Pl. Morphologists, Delhi, 1963. Pp. 363-381.
LANG, W.H., *Ann. Bot.*, **15**, 503 (1901).
LaRUE, C.D., *Pap. Mich. Acad. Sci.*, **11**, 225 (1930).
LAZARENKO, A.S., *Dokl. Akad. nauk. SSSR*, **134**, 1240 (1960).
LAZARENKO, A.S., *Dopov. Akad. nauk. Ukr. RSR*, **1**, 99 (1961).
LAZARENKO, A.S., *Dopov. Akad. nauk. Ukr. RSR*, **11**, 1524 (1963).
LAZARENKO, A.S., *Cytologia and Genetica*, **1965**, 158 (1965a).
LAZARENKO, A.S., *Dokl. Akad. nauk. SSSR*, **162**, 962 (1965b).
LAZARENKO, A.S. and E.E. SVITLYK, *C. R. Dopov. Acad. Sci. Ukr. RSR Sér. B.*, **5**, 459 (1969).
LAZARENKO, A.S., C.T. PASHUK, and E.N. LESHYAK, *C. R. Dopov. Akad. nauk. Ukr. RSR*, **10**, 1381 (1961).
MARCHAL, ÉL. and ÉM. MARCHAL, *Bull. Acad. Roy. Belg. Cl. Sci.*, **7**, 765 (1907).
MARCHAL, ÉL. and ÉM. MARCHAL, *Bull. Acad. Roy. Belg. Cl. Sci.*, **12**, 1249, (1909).
MATZKE, E.B. and L. RAUDZENS, *Proc. natn. Acad. Sci. (U.S.A.)*, **59**, 752 (1968).
MATZKE, E.B. and L. RAUDZENS, "Apospory in the Hepaticae". In J.E. Gunckel, Ed., *Current Topics in Plant Science*, Academic Press, New York, 1969. Pp. 117-119.
MEHRA, P.N., *Res. Bull. Panjab Univ.*, **23**, 221 (1972).
MEHRA, P.N. and M.S. PAHWA, *J. Hattori bot. Lab.* No. **40**, 371 (1976).
MEHRA, P.N. and D. PENTAL, *J. Hattori bot. Lab.* No. **40**, 151 (1976).
MENON, M.K.C., "Morphogenetic studies on apogamy in the moss *Physcomitrium*", Ph.D. Thesis, Univ. Delhi, India, 1974.
MENON, M.K.C. and M. LAL, *Naturwissenschaften*, **59**, 514 (1972).
MENON, M.K.C. and M. LAL, *Planta*, **115**, 319 (1974).
MENON, M.K.C. and M. LAL, *Ann. Bot.*, **41**, 1179 (1977).
MENON, M.K.C. and M. LAL, *Proc. Indian natn. Sci. Acad. B*, **47**, 115 (1981).
MEYER, D.E., *Naturwissenschaften*, **40**, 297 (1953).
MILLER, D.H. and L. MACHLIS, *Pl. Physiol.*, **43**, 714 (1968).
MONTAGUE, M.J. and JANE TAYLOR, *Bryologist*, **74**, 18 (1971).
NARAYANASWAMI, S. and M. LAL, *Phytomorphology*, **7**, 244 (1957).
OHTA, Y. and Y. HIROSE, "Induction and characteristics of cultured cells from some liverworts

Plate 5.1 A, B. Apogamy in haplophase of mosses. **A.** *Physcomitrium pyriforme.* **B.** *Funaria hygrometrica.*
(**A.** After Menon, 1974; **B.** After Chopra & Rashid, 1967).

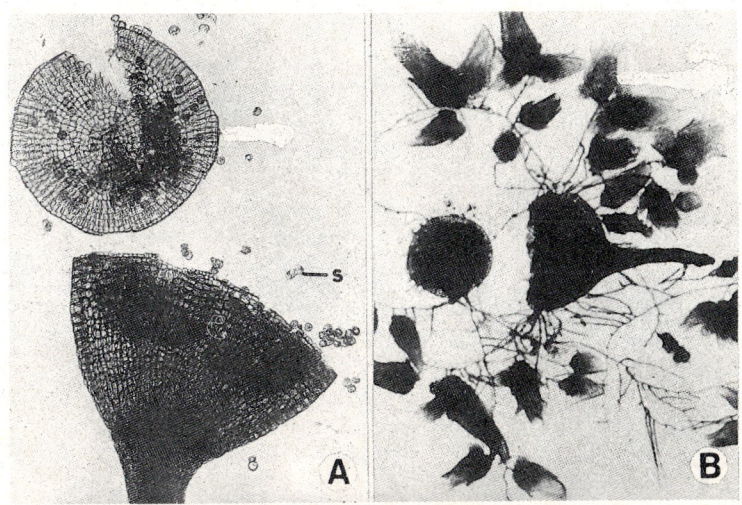

Plate 5.2 A, B. Production of viable spores in apogamous sporophytes of *Physcomitrium coorgense.* **A.** Upper portion of a mature, apogamous sporophyte containing viable spores(s). **B.** Gametophytes formed on protonema obtained from such spores.
(After Lal, 1961b).

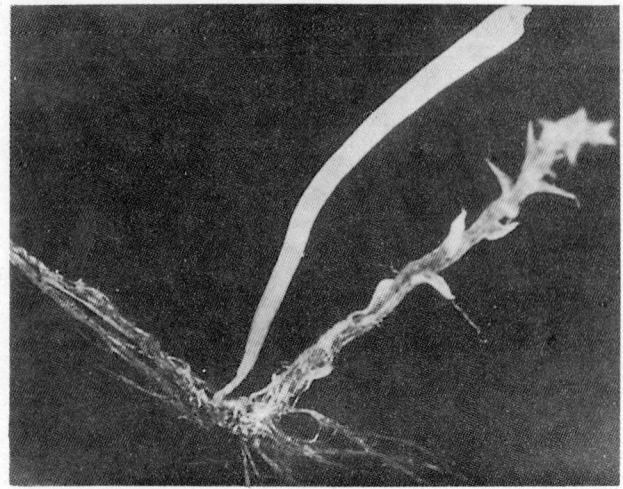

Plate 5.3. Differentiation of gametophyte and apogamous sporophyte on secondary prótonema of *Physcomitrium pyriforme* in low light level.
(After Menon, 1974).

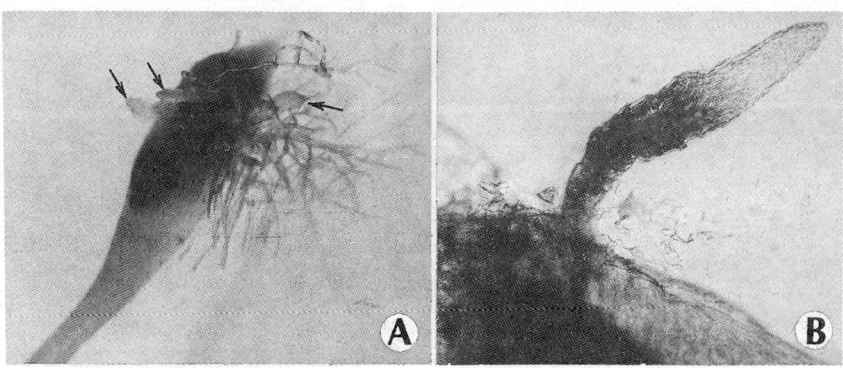

Plate 5.4 A, B. Apogamy in *Funaria hygrometrica*. **A.** Three apogamous sporophytes (arrow-marked) arising from capsule wall, along with aposporous protonema. **B.** Magnified view of one apogamous sporophyte from another culture.
(After Kumra & Chopra, 1980).

Plate 5.5. Formation of apogamous sporophytes and gametophytes on the same protonemal filament of *Physcomitrium pyriforme*.
(After Menon & Lal, 1972).

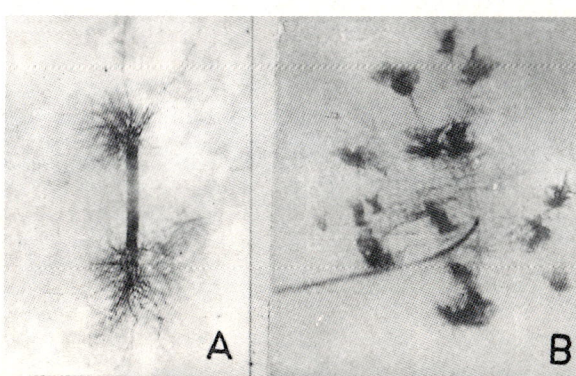

Plate 5.6 A, B. Apospory in mosses. **A.** Regeneration from seta in *Haplocladium angustifolium*. **B.** The aposporous protonema from seta of *Funaria hygrometrica* bearing young gametophytes.
(**A.** After Ono & Tanaka, 1976; **B.** After Kumra, 1981).

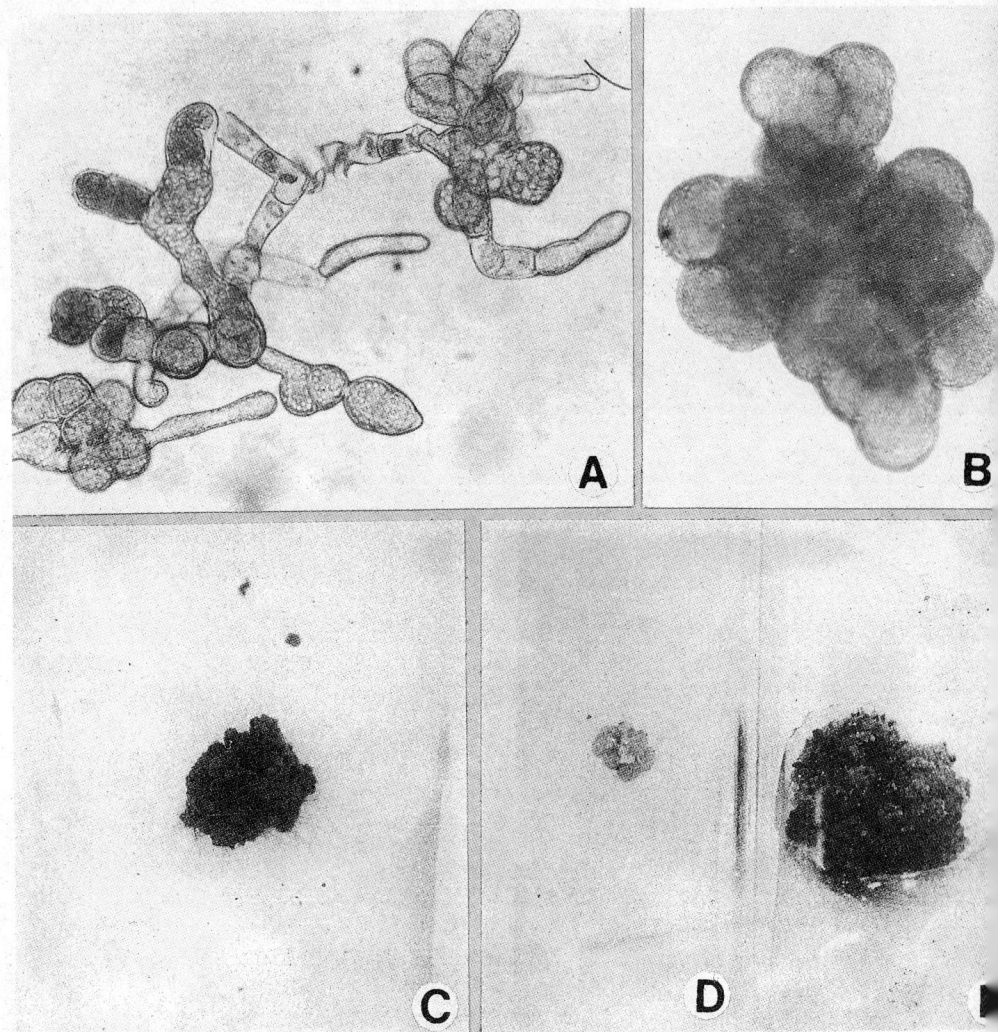

Plate 5.7 A-E. Callus formation in mosses. **A-C.** *Funaria hygrometrica*. **A.** Initiation of callus from protonema. **B.** Young isolated callus mass. **C.** 40-day-old callus. **D, E.** 40 and 120-day-old calli of *Bryum coronatum*.
(**A-C.** After Kumra, 1981; **D, E.** After Kumra & Chopra, 1982).

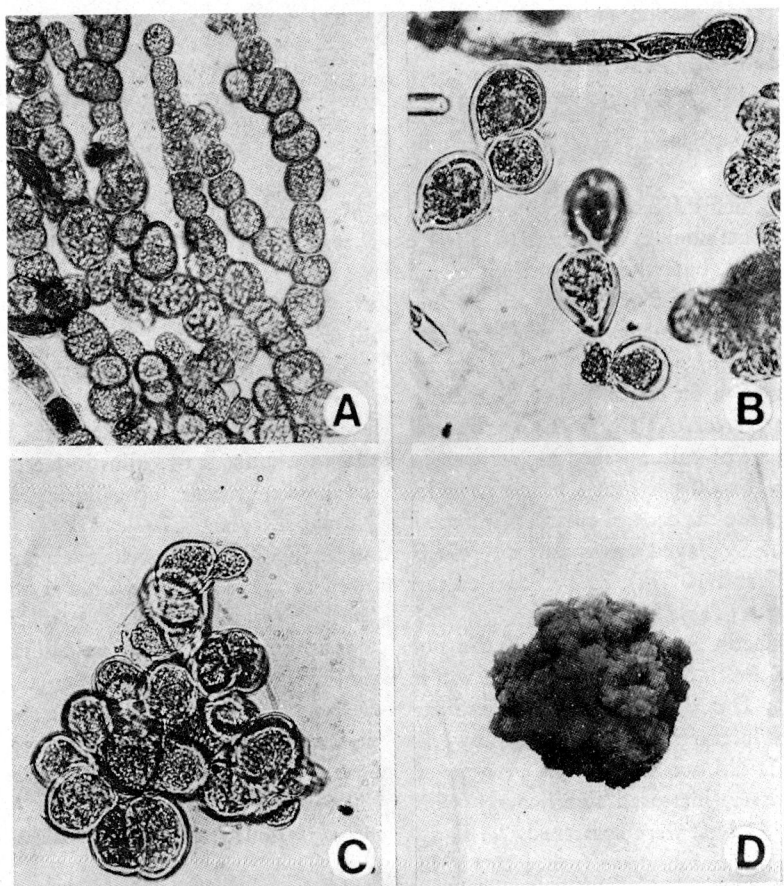

Plate 5.8 A-D. Development of callus from protonema in *Hyophila involuta*. **A.** Chloronemal filaments with rounded cells. **B.** Isolated cells from the same. **C.** Initial stage in the formation of callus. **D.** 10-week-old callus mass.
(After Rahbar & Chopra, 1980).

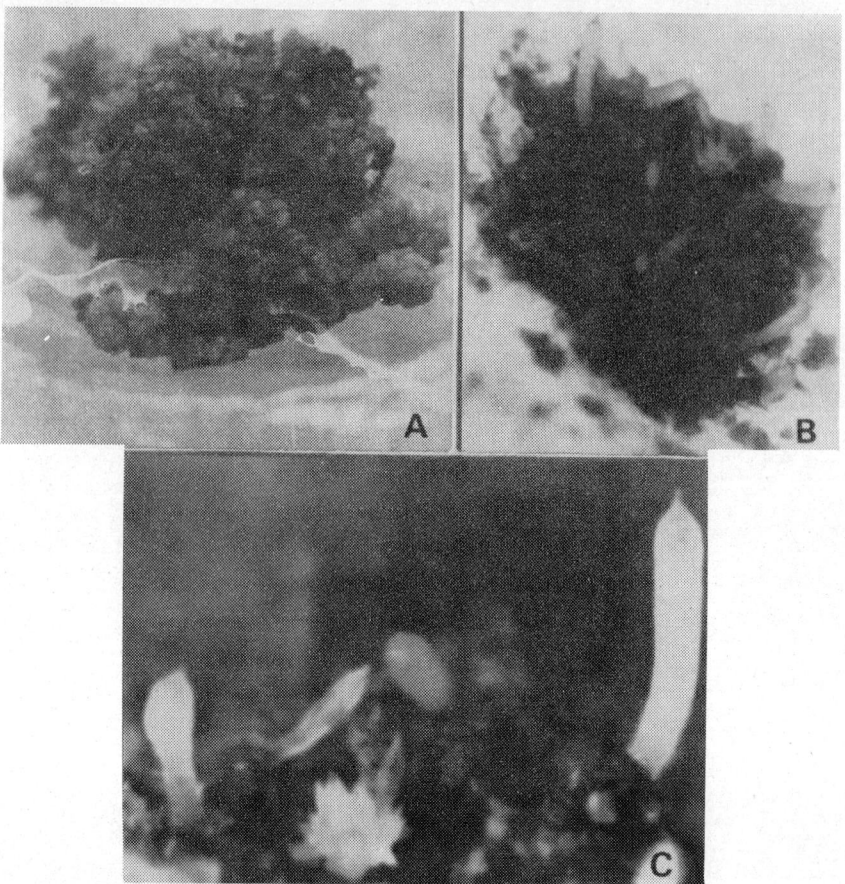

Plate 5.9 A-C. Effect of sucrose on growth and differentiation of callus in *Funaria hygrometrica*. **A.** Callus grown on medium containing 2% sucrose. **B.** Differentiation of apogamous sporophytes from callus on medium containing 4% sucrose. **C.** Magnified view of part of **B**.
(After Kumra, 1981).

Plate 5.10 A-C. Differentiation of callus in *Physcomitrium pyriforme*. **A.** Localized formation of gametophytes and apogamous sporophytes from callus grown on coconut milk-supplemented medium. **B.** Apogamous sporophytes with long and short setae from a culture with liquid basal medium without sucrose. **C.** Stunted sporophytes formed in medium with sucrose and coconut milk.
(After Menon, 1974).

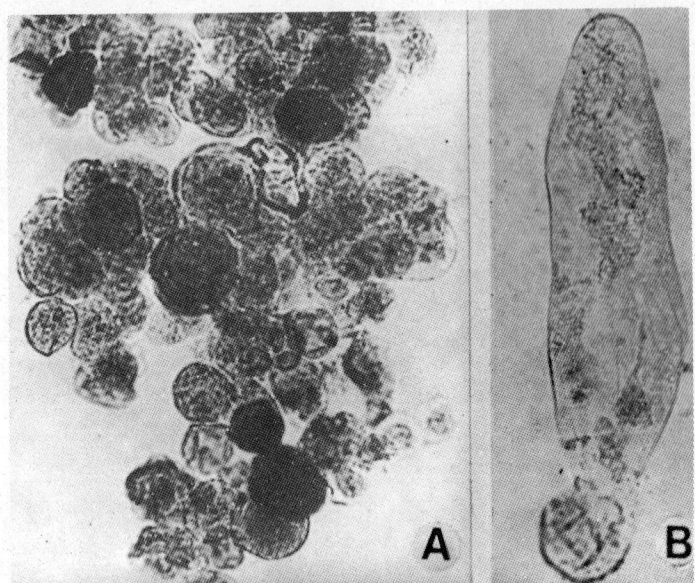

Plate 5.11 A, B. Callusing and regeneration in *Asterella angusta*. **A.** Free-cells and small cell-groups in suspension. **B.** Formation of germ-filament from a free-cell.
(After Rashid, 1971).

of Jungermanniales (Abstr.)". In The Proceedings of XIII International Botanical Congress, Sydney, Australia, 1981. Pp. 144.
OHTA, Y. and Y. HIROSE, J. Hattori bot. Lab. No. 53, 239 (1982).
OHTA, Y., K. KATOH, and K. MIYAKE, Planta, 136, 229 (1977).
OHTA, Y., M. ISHIKAWA, S. ABE, K. KATOH, and Y. HIROSE, Pl. Cell Physiol., 22, 1533 (1981).
ONO, K., Jap. J. Genet., 48, 69 (1973).
ONO, K., Jap. J. Genet., 51, 11 (1976).
ONO, K. and R. TANAKA, J. Jap. Bot., 51, 45 (1976).
PANDE, S.K. and N. CHOPRA, J. Indian bot. Soc., 36, 241 (1957).
PAOLILLO, D.J. JR., Bryologist, 71, 327 (1968).
PARIHAR, N.S., An Introduction to Embryophyta, Vol. I, Bryophyta, Central Book Depot, Allahabad, India, 1972.
PIERIK, R.L.M., Mededel. Landbouwhogeschoole Wageningen, 67, 1 (1967).
PRINGSHEIM, N., Monatsber. Königl. Akad. Wiss., Berlin, 1876, 425 (1876).
PRINGSHEIM, N., Jb. wiss. bot., 11, 1 (1878).
RAHBAR, KAVITA and R.N. CHOPRA, Z. PflPhysiol., 99, 199 (1980).
RASHID, A., "Morphogenetic studies on some Indian bryophytes", Ph.D. Thesis, Univ. Delhi, India, 1968.
RASHID, A., Experientia, 27, 1998 (1971).
RASHID, A. and R.N. CHOPRA, Phytomorphology, 19, 170 (1969).
RAUDZENS, L. and E.B. MATZKE, Am. J. Bot., 55, 1190 (1968).
REDFEARN, P.L. JR. and S.L. MEYER, Bryologist, 52, 197 (1949).
RINK, W., Flora, 130, 87 (1935).
RIPETSKY, R.T., Ukr. bot. Zh., 36, 1 (1979a).
RIPETSKY, R.T., Cytologia & Genetica, 13, 347 (1979b).
RIPETSKY, R.T., Ukr. bot. Zh., 37, 30 (1980).
RIPETSKY, R.T., Ukr. bot. Zh., 39, 43 (1983).
RIPETSKY, R.T., Ontogeny, 16, 229 (1985).
RIPETSKY, R.T. and V.I. MATASOV, Ontogeny, 4, 404 (1973).
SAXENA, P.K. and A. RASHID, Beitr. Biol. Pfl., 57, 301 (1982).
SCHWARZENBACH, M., Arch. Julius Klaus-Stift. VererbLehre-Forsch., 2, 91 (1926).
SIMONE, L.D., Am. J. Bot. (Abstr.), 53, 609 (1966).
SMITH, G.M., Cryptogamic Botany, Vol. II, McGraw-Hill, New York, 1938.
SOOD, SNEH, Phytomorphology, 24, 186 (1974).
SPIESS, L.D., Pl. Physiol., 55, 583 (1975).
SPIESS, L.D., Pl. Physiol., 58, 107 (1976).
SPIESS, L.D., Bryologist, 82, 47 (1979).
SPIESS, L.D., B.B. LIPPINCOTT, and J.A. LIPPINCOTT, Am. J. Bot., 60, 708 (1973).
SPRINGER, E., Z. Indukt. Abstamm. -u. VererbLehre, 69, 249 (1935).
STAHL, E., Bot. Ztg., 34, 689 (1876).
STRASSBURGER, E., Ann. Bot., 8, 281 (1894).
TAKEDA, R. and K. KATOH, Planta, 151, 525 (1981).
TRIPATHI, B.K., Rev. gén. Bot., 85, 91 (1978).
TULECKE, W., Symp. Soc. Dev. Biol., 24, 217 (1965).
ULYCHNA, K.O., Ontogeny, 2, 207 (1971).
VASHISTHA, B.D., "In vitro investigations on some Indian bryophytes", Ph.D. Thesis, Univ. Delhi, India, 1985.
WARD, M., Science, 132, 1401 (1960).
WARD, M., Nature, 204, 400 (1964).
WARD, M. and S.E. FREDERICK, Phytomorphology, 17, 371 (1967).
WATSON, E.V., The Structure and Life of Bryophytes, Hutchinson University Library, London, 1971.
von WETTSTEIN, D., Z. Bot., 41, 199 (1953).

von WETTSTEIN, F., *Biol. Zbl.*, **43**, 71 (1923).
von WETTSTEIN, F., *Z. Indukt. Abstamm.-u. VererbLehre*, **33**, 1 (1924a).
von WETTSTEIN, F., *Biol. Zbl.*, **44**, 145 (1924b).
von WETTSTEIN, F., *Biblphia genet.*, **1**, 1 (1925).
von WETTSTEIN, F., *Ber. dt. bot. Ges.*, **60**, 394 (1942).
ZIELINSKI, F., *Flora*, **100**, 1 (1910).

6 Photomorphogenesis

PHOTOMORPHOGENETIC responses of plants have been of wide interest in photobiological research. According to Mohr (1962, 1982) photomorphogenesis includes all the non-photosynthetic effects of light on growth, development, and differentiation of plants. As such, photoperiodic and phototropic responses come under the purview of photomorphogenesis. These responses are evoked by different regions of light spectrum, and in contrast to photosynthesis only small numbers of light quanta are required in photomorphogenesis. Within photomorphogenesis different photoreactive response systems can be distinguished either on the basis of the nature of the photoreceptors involved, or the energy levels required to elicit the responses. The red/far-red photoreversible response system which requires very low-energy of irradiation is known as the "phytochrome system", or "low-energy reaction system" since for these responses the photoreceptor is phytochrome. The other responses are known as high-energy reactions; and for these the pigment involved may be called the "blue-far-red pigment system". In the high-energy reaction system reversibility is not involved (Smith, 1975).

The presently known region of the electromagnetic spectrum involved in photoregulation of growth and development in plants under natural conditions occurs between 290 and 850 nm. This is somewhat broader than the spectral region (380 to 760 nm) generally recognized to be utilized in human vision. Quanta of this spectral range have a relatively low energy of about 90 down to 36 Kcal Einstein^{-1}, depending upon the particular wavelength. The energy required for photomorphogenetic responses is derived from metabolism of food. The light signal is merely a switching mechanism to regulate the metabolic reactions.

Plants contain photochemical systems capable of sensing changes in the direction, duration, irradiance level, and spectral quality of light signals. These systems activate processes which result in qualitative and quantitative modifications of growth and developmental pattern of the plant in response to variations of the light conditions. One of the most prominent systems involves the pigment phytochrome, which exists in two interconvertible forms with absorption in red (660 nm) and far-red (730 nm) region of the spectrum. However, a number of photomorphogenetic responses are mediated by other pigments, which are yet to be characterized. These responses have, therefore, been described on the basis of most effective spectral regions. The processes influenced by light in bryophytes are spore germination, growth, vegetative propagation, metabolism, senescence, bud induction, and polarotropism (see Tables 6.1 to 6.5)*

TABLE 6.1.
Photoreceptors involved during spore germination of bryophytes.

Taxa	Light quality	Photoreceptor/s	Investigator/s
Ceratodon purpureus	Red and Far-red	Phytochrome	Valanne (1966)
Dicranum scoparium	—do—	—do—	—do—
Octoblepharum albidum	Red, Far-red, and Blue	—do—	Egunyomi (1979)
Physcomitrella patens	Red	—do—	Cove et al. (1978)
Sphaerocarpos cristatus	Red, Far-red, Green, and Blue	—	Doyle (1963)
S. donnellii	Red, Far-red, and Blue	Phytochrome and Blue light absorbing pigment	Mohr (1963); Steiner (1964)

— indicates photoreceptor not named.

TABLE 6.2.
Photoreceptors involved during growth of bryophytes.

Taxa	Light quality	Photoreceptor/s	Investigator/s
Anoectangium thomsonii	Blue	—	Rashid (1970)
Ceratodon purpureus	Red	—	Larpent-Gourgaud and Aumaitre (1977a)
Funaria hygrometrica	Red and Blue	Phytochrome	Pringsheim and Pringsheim (1935); Demkiv and Ripetsky (1970)
Lunularia cruciata	Red, Far-red, and Blue	—do—	Schwabe and Valio (1970)
Marchantia nepalensis	—do—	Phytochrome and Blue light absorbing pigment	Kaul and Kaul (1974); Shukla and Kaul (1978)
M. polymorpha	Red and Far-red	Phytochrome	Fredericq and DeGreef (1966, 1968)
Physcomitrella patens	—do—	—do—	Cove et al. (1978)
Physcomitrium turbinatum	—do—	—do—	Nebel (1968)
Plagiochasma appendiculatum	Red and Orange	—do—	Shukla and Kaul (1978)
Pohlia nutans	Red and Blue	—do—	Mitra et al. (1959)
Polytrichum juniperinum	Red and Far-red	—do—	Krisko and Paolillo (1972)

(Table contd.)

* Tables prepared by Vashistha (1981) have been up-to-dated.

Taxa	Light quality	Photoreceptor/s	Investigator/s
Riccia discolor	Red, Blue, and Yellow		Dagar et al. (1980)
Sphaerocarpos donnellii	Red and Blue	Phytochrome and Blue light absorbing pigment	Steiner (1963); Miller and Machlis (1968a,b)

— indicates photoreceptor not named.

TABLE 6.3.
Photoreceptors involved in vegetative propagation of bryophytes.

Taxa	Light quality	Photoreceptor/s	Investigator/s
Aulacomnium palustre	Red and Blue	Phytochrome	Larpent-Gourgaud et al. (1974a)
Bryum klinggraeffii	Red, Blue, and Green	—	Kumra (1977)
Didymodon recurvus	Blue	—	Vashistha (1985)
Funaria hygrometrica	Red and Blue	Phytochrome	Demkiv (1971)
Lunularia cruciata	Red and Far-red	—do—	Valio and Schwabe (1969)
Marchantia nepalensis	Red, Orange, and Green	—	Kaul and Kaul (1974)
M. polymorpha	Red and Far-red	Phytochrome	Ninnemann and Halbsguth (1965); Otto and Halbsguth (1976)
Mnium affine	—do—	—do—	Giles and von Maltzahn (1967)
Tetraphis pellucida	Red and Blue	—do—	Larpent-Gourgaud et al. (1974a)

— indicates photoreceptor not named.

TABLE 6.4.
Photoreceptors involved in metabolism and senescence of bryophytes.

Response	Taxa	Light quality	Photoreceptor/s	Investigator/s
Metabolism	Bryum klinggraeffii	Red, Blue, and Green	—	Chopra and Kumra (1978)
	Funaria hygrometrica	Red and Far-red	Phytochrome	Demkiv (1972)
	Hybrid callus (Funaria hygrometrica × Physcomitrium pyriforme)	—do—	—do—	Hartmann (1971); Hartmann and Kilbinger (1974)
	Lunularia cruciata	—do—	—do—	Valio et al. (1969)
Senescence	Marchantia polymorpha	Red and Far-red	Phytochrome	DeGreef et al. (1971)

— indicates photoreceptor not named.

TABLE 6.5.
Photoreceptors involved in bud induction and polarotropism of bryophytes.

Response	Taxa	Light quality	Photoreceptor/s	Investigator/s
Bud induction	Anoectangium thomsonii	Red	—	Rashid (1970)
	Didymodon recurvus	—do—	—	Vashistha (1985)
	Funaria hygrometrica	—do—	—	Pringsheim and Pringsheim (1935); Jahn (1964)
		Red and Far-red	Phytochrome	Simon and Naef (1981)
	Physcomitrella patens	—do—	—do—	Cove et al. (1978)
	Physcomitrium turbinatum	—do—	—do—	Nebel and Naylor (1964)
	Pohlia nutans	Red	—	Mitra et al. (1959)
Polarotropism	Ceratodon purpureus	Red	Phytochrome	Hartmann (1984); Hartmann et al. (1983)
	Funaria hygrometrica	—do—	—do—	Jaffe and Etzold (1965)
	Physcomitrella patens	—do—	—do—	Jenkins and Cove, (1983a,b,c)
	Physcomitrium turbinatum	Red and Far-red	—do—	Nebel (1968, 1969)
	Sphaerocarpos donnellii	Blue	Blue light absorbing pigment	Steiner (1969a,b)

— indicates photoreceptor not named.

SPORE GERMINATION

It has long been known that spores of most lower archegoniates do not germinate in dark on inorganic media. An absolute requirement of light for germination has been observed by some investigators (Bittner, 1905; Borodin, 1868; Heald, 1898; Mohr, 1956; Schulz, 1902; Stephan, 1928). However, Kessler (1914), Laage (1906), Meyer (1948), Patschovsky (1926), Servettaz (1913), Treboux (1905), and von Ubisch (1913) reported that spores of some mosses also germinate in dark on inorganic media.

The significance of exposure to light of different spectral regions during germination of spores of bryophytes was recognized by Listowski (1927), Schulz (1902), and Stephan (1928), but their results were to some extent contradictory. The first two investigators observed spore germination in light of all spectral regions, but Stephan reported germination in green and red light.

LIVERWORTS

Spores of liverworts do not germinate in dark even in the presence of sugar, and in many taxa blue light plays an important role (Inoue, 1960). Germination of spores in *Sphaerocarpos donnellii* is controlled by at least two photoreactive systems, and can be induced by both blue and far-red radiations (Mohr, 1963). The action spectra indicate that germination is probably controlled by phytochrome as well as the high-energy reaction, and there is an interaction between the two systems (Steiner, 1964).

Irradiation sequences with red and far-red, and the simultaneous activation of the phytochrome system and the high-energy reaction indicate that the effect of the high-energy reaction is markedly influenced by phytochrome. It is further remarkable that germination in blue and green light can be promoted by a simultaneous irradiation, whereas germination in far-red light is inhibited under these conditions. The effectiveness of both phytochrome and high-energy reaction, as well as the kind of interaction between these two systems is influenced by the endogenous germination ability. In phytochrome-mediated spore germination photoperiodic behaviour is observed with red light, but not with far-red radiation. Therefore, it seems that the mechanism of the high-energy reaction inducing germination is at least partially different from the action of phytochrome.

After induction of photoperiodism by red light, the course of light sensitivity with short irradiations of red and far-red exhibits a 'circadian rhythmic behaviour'. Red light is more effective at the beginning of day and far-red at the end. With blue light there is no indication of periodical change of light sensitivity (Steiner, 1964).

The endogenous rhythms are induced only by levels of red light which are insufficient for induction of germination (Steiner, 1969a).

In *Sphaerocarpos cristatus*, red, far-red, green, and blue lights are equally effective in promoting spore germination, but there is no evidence to suggest the existence of the reversible red/far-red system mediated by phytochrome (Doyle, 1963).

MOSSES

The process of light regulated germination in mosses can be divided into two stages: the first leads to an increase in volume due to uptake of water, followed by rupture of exospore and intensive greening of the plastids; the second step is characterized by the initiation of rhizoids on the chloronema or spore. In some populations of *Funaria* the first stage of germination can occur in darkness.

Bauer and Mohr (1959) established that light requirement of moss spores is associated with the phytochrome system, and suggested that photosynthesis might be involved in the process at a later stage of germination. The phytochrome system has been detected in *Ceratodon purpureus*, *Dicranum scoparium* (Valanne, 1966), and *Octoblepharum albidum* (Egunyomi, 1979) but not in *Funaria hygrometrica*. Bauer and Mohr (1959) had earlier reported that in *Funaria* germination is influenced by the red/far-red reversible photoreactions.

Valanne (1966) obtained peaks in the blue (415 nm) and far-red (710 nm) regions of the germination curve of *Ceratodon* and *Funaria* spores, and suggested these to be an indication of photosynthesis and high-energy reaction. Krupa (1967) also observed effectiveness of blue and far-red light in inducing germination of *Funaria* spores, when a higher irradiance or a longer period of illumination was used. In *Octoblepharum* red light (630 nm) supported spore germination, though the protonema was morphologically distorted (Egunyomi, 1979). From the above data it is clear that mosses differ in their response to different light spectra.

In *Physcomitrella patens* red light (640 to 680 nm) is much more effective in inducing spore germination than are longer and shorter wavelengths, but red-far-red reversibility of germination has not been demonstrated (Cove et al., 1978).

The chloroplasts of germinating spores of *Polytrichum* replicate to a lesser extent in dark than in light. A stable number of plastids is attained by 48 h in dark i.e., the plastids become light-dependent for replication. The promotive effect of red light can be completely reversed by far-red light (Kass & Paolillo, 1974). Light also stimulates the incorporation of [^{3}H]-uridine and [^{3}H]-thymidine in addition to plastid replication in germinating *Polytrichum* spores. Significant amounts of both these are also incorporated in dark, but not to the same extent as in light. Plastids continue to produce nucleic acids when their capacity to multiply is suspended in the absence of light (Kass & Paolillo, 1977). According to Paolillo and Kass (1977), in *Polytrichum* phytochrome mediates nuclear control of plastid replication and synthesis of cytoplasmic proteins in plastids.

GROWTH

LIVERWORTS

In *Sphaerocarpos donnellii* red light promotes branching of thalli and leads to normal growth and morphogenesis. In blue light the thalli remain unbranched. Branching can be induced in young thalli by short irradiation with red, and to a lesser extent with far-red. It has been demonstrated that three photoreactive systems are involved: (1) photosynthesis, (2) a red light-dependent photoreactive system, and (3) a blue light-dependent photoreactive system. Photosynthesis is a pre-requisite for growth of thalli. Morphogenesis of thalli as also the size of chloroplasts is specifically dependent on the red and blue reaction systems. The quantitative effect of red and blue light is, however, strictly dependent on the rate of photosynthesis under these conditions of irradiation. Under low level of illumination the size of chloroplasts is the same in blue, red, or white light. With increasing level of illumination, however, the size of chloroplasts increases in blue, and decreases in red light. In low levels of white light the blue reaction system is predominant; with increasing levels the red reaction system becomes predominant. Therefore, chloroplast size initially increases to a maximum and then decreases.

The red light effect is mediated by phytochrome. It can be brought about by a 2-min exposure (every 12 h) to low level of red light; and far-red light reverses the effect of red light. The increased growth rate affected by red light is related to a change in the morphology of thallus. Dark-grown plants form compact balls of tissue consisting of rounded and thick lobes, which exhibit an abnormal callus type growth, with a few well defined meristematic regions. The lobes of plants grown in red light have a morphology more typical of normal *Sphaerocarpos* thalli (Miller & Machlis, 1968a,b).

Growth of *Marchantia polymorpha* thalli exhibits phytochrome-mediated response. Fredericq and DeGreef (1966, 1968) observed that in *M. polymorpha* after 8 h of light a 5-min exposure to far-red during the following 16 h dark period causes curling of lobes. If this treatment is continued, further growth is orthotropic, and narrower and more elongated lobes result as compared to controls. After one month, the effects of this daily far-red exposure are spectacular. Thalli receiving 2½-min red light after the 5-min far-red remain as flat as controls and have the same morphology. With a sequence of alternate irradiations at the end of the photoperiod, the last one completely reverses the action of the preceding exposure. Even 1-min far-red at the end of 16 h day has a clear-cut effect. Chlorophyll content of far-red treated thalli is much lower than that in controls. This depressive influence of far-red is reversed by red light. For a given far-red exposure the degree of orthotropism and narrowness is stronger with shorter photoperiods; the same holds true for the chlorophyll decrease. In higher plants also the far-red action is weaker after longer photoperiods. The presence of P_{fr} form (under white fluorescent light) of phytochrome, considered to be the physiologically active form, determines flat growth, and plays a role in the maintenance of broad lobes in *M. polymorpha*. Even in the middle of 16 h night there is probably sufficient P_{fr} left, because reversing the pigment at this moment by 5-min far-red light results in a marked orthotropism, reversible by red light.

In *Lunularia cruciata* growth and dormancy are controlled by day length. Short days promote active growth, whereas long days or light break treatments induce dormancy (Wilson & Schwabe, 1964). In this species 8 h photoperiod is optimal for growth. There is no growth in dark or in continuous light which causes rapid onset of dormancy. Short-day cycles intercalated among a series of continuous light cycles promote growth. In cycles longer than 24 h, very long dark periods are deterimental. With very short photoperiods (5-min) red light promotes growth more effectively than white light at higher level. Far-red light acts as dark. The growth effects of red and far-red light breaks (3-min) depend on the time of application. Red light proves inhibitory in the middle, but promotes growth at the beginning of 16 h dark period. Far-red light has the opposite effect. In each case red and far-red effects are reversible by the other wavelength. Blue light gives the same response as red, inducing the reversibility of far-red light effect and vice versa. Reversal of red light effect by far-red light strongly suggests that phytochrome is involved in the photoperiodic response mechanism of *Lunularia cruciata*. Surprisingly, significant effects of 5-min red, blue, and far-red irradiations are also observed in the middle of the main high level white light period. Red and blue light promote growth, far-red reduces it, and again there is ready reversibility of the effects (Schwabe & Valio, 1970).

In *Marchantia nepalensis* growth is better in far-red than in red, orange, or green light (Kaul & Kaul, 1974). Shukla and Kaul (1978) reported that growth rate is maximum in blue light, whereas in red light growth response is the poorest. It seems that in *M. nepalensis* growth rate is controlled by two pigment systems: phytochrome, and blue light absorbing system. Phytochrome also appears to be involved in the control of growth in *Plagiochasma appendiculatum*, in which growth is inhibited by red light and it is at its best in orange light (Shukla & Kaul, 1978). Dagar et al. (1980) reported that the growth of newly formed branches in *Riccia discolor* is favoured by blue light as compared to yellow or red light. Green light retards growth.

In *Riccia frostii* red light proved most effective in promoting vegetative growth in terms of dry weight yield. Further, more filamentous regenerants were produced in red light and the resulting thalli were green, long, thin, narrow, and unbranched. In blue light more thalloid regenerants were formed, and these developed into dark-green, short, thick, broader, and sparingly branched thalli. In far-red light only filamentous regenerants were produced in large numbers, and a few of these gave rise to pale, thin, and unbranched thalli (Vashistha, 1985).

MOSSES

Growth of Protonema

It has been known since the early work of Klebs (1893) that growth and development of moss protonema is greatly influenced by light level. The significance of different spectral regions of light was recognized by Listowski (1927) and Stephan (1928).

Pringsheim and Pringsheim (1935) observed that in *Funaria* protonemal growth is fairly good in blue light, but in green light the protonema has sparingly branched, pale-green filaments which closely resemble those produced in dark. In *Pohlia nutans* blue and red light have promotory effect on protonemal growth, whereas green light is inhibitory In blue light branching is more profuse than in red light, but in both treatments protonema shows heterotrichous habit. In red light prominent nutation patterns are observed both in the prostrate and erect filaments. Protonema is very attenuated and yellowish-green as a result of scanty development in green light, and there is no indication of heterotrichous habit. Far-red light has no effect on protonemal growth in *Pohlia nutans* (Mitra et al., 1959). According to Mitra et al. (1965) blue light with red light produces greater protonemal growth than red light alone.

Nebel (1968) observed that in *Physcomitrium turbinatum* the growth rate of lateral branches of chloronema is controlled by the quality of light. Growth is maximum in red light and minimum in far-red. According to him, growth and orientation in this species is dependent upon an interaction between the photosynthetic and the phytochrome systems. The physiological activity of phytochrome in photo-orientation of growth does not result from a certain amount of P_{fr} or P_{fr}/P_r ratio but rather from the simultaneous excitation and consequent cycling of P_r and P_{fr}. By using polarized red and far-red light, Nebel (1969) concluded that the active photoreceptors are dichroic and are coin- or disc-shaped. Synergistic interaction between red and far-red irradiations supports the hypothesis that the photoreceptor is phytochrome and that the growth response is proportional to the total quanta absorbed by the pigment and/or the consequent cycling of the phytochrome between P_r and P_{fr}.

Blue light favours branching and promotes growth of protonema in *Anoectangium thomsonii*. In red light the protonema is pale-green and sparse (Rashid, 1970). Red light induces the development of lateral outgrowths and additional growth primordia, thereby increasing the amount of cytoplasmic RNA in apical chloronemal cells of *Funaria hygrometrica*. The light spectrum region in which maximum apical cell branching is noticed corresponds to the action spectrum of the phytochrome reaction. Treatment with far-red immediately after red light reduced the promotory effect of the latter. This indicates that phytochrome system is also involved in the induced morphological effect (Demkiv & Ripetsky, 1970). Red light also causes an increase in the volume of nuclei and nucleoli. Effect of red light on the early stages of the moss ontogeny proved to be associated with its ability to activate RNA and probably protein synthesis as well (Demkiv et al., 1971). Red light stimulates growth and ramification of protonema of *Ceratodon purpureus* (Larpent-Gourgaud & Aumaître, 1977a). Cove et al. (1978) reported that protonemal growth in *Physcomitrella patens* is promoted by red light, and this effect is reversible by far-red, suggesting the phytochrome-mediated control of protonemal growth. In *Didymodon recurvus* red light was more effective than blue light in promoting protonemal growth (Vashistha, 1985).

Growth of Sporophytes

The expansion of sporophyte in mosses is also controlled by light. Haberlandt (1886) noticed that in *Funaria* light is required for proper capsule development. In dark the spore sacs failed to expand even if the seta and gametophyte were exposed to light. The role of light in controlling capsule expansion in *Polytrichum* appears to be morphogenetic rather than photosynthetic (Paolillo & Bazaz, 1968). Red light is more effective in promoting expansion than white, blue, or green light of equal energy (Krisko & Paolillo, 1972). Relatively low irradiances are required for capsule expansion in *Polytrichum* (Krisko & Paolillo, 1972) and *Funaria* (French & Paolillo, 1976).

Growth and Differentiation of Cultured Cells

Hartmann (1973) used the callus obtained by Bauer (1961) from the seta of the hybrid sporophyte *Physcomitrium pyriforme* × *Funaria hygrometrica*, for photomorphogenetic studies. In blue light the callus continued to grow in an undifferentiated state, but when it was transferred to red light, chains of short, broad cells developed. The terminal cell of each chain differentiated into an apical cell with two cutting faces. On the other hand, the callus grown in dark formed normal protonemal filaments. Brief exposures of red light in darkness prevented differentiation of protonemal filaments from callus. However, far-red light resulted in the formation of protonemal filaments, but these were morphologically different from those produced in darkness. Menon et al. (1980) also observed light-induced differentiation of callus in *Physcomitrium pyriforme*. At low levels of white light, callus differentiated predominantly into apogamous sporophytes, whereas at higher light levels it produced protonemal filaments and gametophores.

VEGETATIVE PROPAGATION

Light plays an important role in vegetative propagation by promoting the formation of propagules as well as controlling their regeneration (see Chapter 3).

LIVERWORTS

Vegetative propagation of liverworts through gemmae is a light-controlled response, as gemmae do not germinate in dark. The formation of rhizoids on gemmae and further growth of gemmae in *Marchantia polymorpha* is controlled by the phytochrome system. Red light is necessary for the formation of rhizoids on gemma in this species (Ninnemann & Halbsguth, 1965; Otto & Halbsguth, 1976). In *M. nepalensis* germination of gemmae occurs only in red, orange, or green light (Kaul & Kaul, 1974). In *Lunularia cruciata* rhizoid formation (the first stage of incipient growth) is also very sensitively controlled by the red/far-red system. Only a few seconds' exposure to red light is required for a high percentage of rhizoid production. Far-red light given for a very short period completely inhibits rhizoid formation, as does total darkness (Valio & Schwabe, 1969).

MOSSES

Protonemal regeneration from cells of isolated leaves of *Mnium affine* is under the control of a phytochrome-mediated system. Phytochrome (P_{fr}) induced by short periods of red light, stimulates protonemal regeneration of mature leaf cells. The system is red/far-red reversible. Action of blue light is similar to that of red light and is reversible by far-red light (Giles & von Maltzahn, 1967). Phytochrome has been isolated from two species of *Mnium: M. hornum* and *M. undulatum* (Giles & von Maltzahn, 1968). In *Funaria hygrometrica* regeneration in protonemal cells occurs in red and blue light. It is probably influenced by phytochrome (Demkiv, 1971). Phytochrome controls the germination of propagules in *Aulacomnium palustre* and *Tetraphis pellucida*. Red light induces germination, but a second photoreceptor for blue light nullifies the stimulating effect of red light and inhibits germination (Larpent-Gourgaud et al., 1974 a,b).

In *Bryum klinggraeffii* maximum gemmae are produced in red light. The next in order of effectiveness are blue light and green light (Kumra, 1977).

Jenkins and Cove (1983a) studied the light requirements for regeneration of protoplasts of *Physcomitrella patens*. The protoplasts follow a simple regenerative sequence, namely: cell wall synthesis, formation of an asymmetric cell, division of the asymmetric cell, and further extension and division to produce a new chloronemal filament. Only cell wall formation occurs independently of light. Relatively high levels of monochromatic red or blue light are required for the production of an asymmetric cell, and the process ceases upon transfer to darkness. The subsequent stages of regeneration require much lower levels, and red light is considerably more effective than blue or far-red light in permitting cell division. Thus, it is suggested that phytochrome may be involved in controlling division of these cells, although far-red reversibility has not yet been demonstrated.

METABOLISM

Synthesis of a number of endogenous morphoregulatory substances, which directly or indirectly affect growth and development in bryophytes, has been shown to be controlled by the quality of light.

LIVERWORTS

Lunularic acid was isolated from *Lunularia cruciata* This compound controls growth and drought resistance. It also arrests the growth of gemmae while in the gemma cup. The endogenous level of this inhibitor is quantitatively controlled by a phytochrome-mediated day length response (Valio et al., 1969).

Three endogenous gibberellin-like substances were detected in vegetative thalli of *Marchantia polymorpha* by Melstrom et al. (1974). Increase in photoperiod from 12 to 18 h elicited quantitative differences in the activity of these gibberellins, and resulted in increased thallus elongation and orthogeotropic growth.

MOSSES

Synthesis of acetylcholine in the moss callus, regenerated from seta of the hybrid sporophyte *Funaria hygrometrica* × *Physcomitrium pyriforme*, is regulated by phytochrome. Red light promotes synthesis of acetylcholine, whereas far-red inhibits it. In alternating red and far-red light treatments the ultimate effect is determined by the nature of last treatment; if it is red then a high concentration of acetylcholine results, if far-red then a low concentration is observed (Hartmann, 1971; Hartmann & Kilbinger, 1974).

Red light causes an increase in the concentration of cytoplasmic RNA and total proteins in the apical cells of young protonema of *Funaria hygrometrica* (Demkiv, 1972). In the same species cell permeability is under phytochrome control. Red light increases cell permeability to cation stains (acridine orange and neutral red) and an anion stain (uridine). Far-red light partly reverses the action of the red light, solely during the initial stages of irradiation (Demkiv & Fedyk, 1977). Light induces the synthesis of a cellular division factor by the protonema cells of *Ceratodon purpureus* (Larpent-Gourgaud & Aumaitre, 1977b).

In *Bryum klinggraeffii* protonema produces a morphoregulatory substance which controls gemma formation in this species. In red light production of this substance is maximum, whereas it is minimum in green light (Chopra & Kumra, 1978).

SENESCENCE

Senescence in *Marchantia polymorpha* is controlled by the quality of light. The thalli remain green when given a daily 1 h photoperiod of white light. The tissue is, however, induced to bleach when each daily 1 h photoperiod is terminated with a brief irradiation of far-red light. Bleaching occurs due to the loss of chlorophyll and is accompanied by a breakdown of cell organelles and cytoplasm. Bleaching is not observed when each irradiation of far-red light is followed by a brief irradiation with red light. Phytochrome is clearly implicated in the control of senescence by light. It has also been found that a 5-min exposure to red light given once a day is as effective as the 1 h photoperiod with white light in preventing bleaching, and that bleaching is induced when red light is followed by a 10-min irradiation with far-red light. DeGreef et al. (1971) suggested that the rate of dark decay of P_{fr} is probably a factor controlling senescence.

BUD INDUCTION IN MOSSES

Klebs (1893) discovered that in *Funaria hygrometrica* light controls the initiation of shoot-buds. Importance of light in the development of buds was further confirmed by Goebel (1896), Servettaz (1913), von Ubisch (1913), and Robbins (1918).

The quality of light also plays an important role in bud induction. Pringsheim and Pringsheim (1935) observed that in *F. hygrometrica* shoot buds are formed on the protonema in white or red light but not in blue or green light. Mitra et al. (1959) reported that in *Pohlia nutans* buds are formed readily in day light and in red light but do not appear in blue light, green light, dark, and continuous infra-red radiation even in the presence of sucrose. These results suggest that red light has a specific formative effect on the initiation of buds on protonema, but its effect is not reversible by far-red light.

Red light induces buds in *Pohlia nutans* but it does not support their development into leafy shoots. For the latter response, 11 h of red light followed by 6 h of blue light in a 24 h cycle seems to be the best. Like kinetin, red light counteracts the inhibitory effects of coconut milk on bud formation. This indicates the similarity in action of red light and kinetin in bud formation (Chopra & Gupta, 1967; Mitra et al., 1965). In *Physcomitrium turbinatum* bud initiation is controlled by the phytochrome system (Nebel & Naylor, 1964). In *Anoectangium thomsonii* buds are not produced in white light, but red light elicits this response (Rashid, 1970).

Cove et al. (1978) observed that buds can be induced in *Physcomitrella patens* by growing cultures in a 23 h dark/1 h red light cycle. The effect of this 1 h red light can be reversed by 15-min exposure to far-red light, indicating the involvement of phytochrome in bud initiation. The involvement of phytochrome in bud induction has also been reported in *Funaria hygrometrica* by Simon and Naef (1981). The role of light in bud formation has also been dealt with in Chapter 2.

In *Didymodon recurvus* red light favoured bud formation, whereas blue light proved inhibitory. On the other hand, blue light favoured formation of gemmae on protonema, and red light inhibited this response (Vashistha, 1985).

TROPIC RESPONSES

LIVERWORTS

The germ tubes of *Sphaerocarpos donnellii* grow at an angle of 90° to the plane of vibration of the electrical vector of linearly polarized light applied from above. If the plane of vibration is turned, the germ tubes rapidly and correspondingly change their direction of growth. The germ tubes of *S. donnellii* do not respond polarotropically to irradiation with wavelengths greater than 550 nm. A pre-, post-, or simultaneous irradiation with red or far-red, has no significant effect on the blue light-mediated polarotropic response. Likewise, germ tubes do not show any change in sensitivity of the system mediating the polarotropic response after a dark period (Steiner, 1969b). The shapes of action spectra for polarotropism in *Sphaerocarpos* germ tubes, in general, are similar to the pattern of action spectra reported for other blue-UV-mediated photoresponses, like phototropism of *Avena* coleoptiles (Shropshire & Withrow, 1958) and of *Dryopteris* sporelings (Steiner, 1969c). The details of action spectrum strongly suggest a flavoprotein as a photoreceptor. The photoreceptor molecules must be highly oriented in a dichroic structure, presumably close to the surface of the cell (Steiner, 1969c).

MOSSES

Pringsheim and Pringsheim (1935) were among the first to demonstrate that moss protonemata have a strong tendency to grow towards a unilateral light source, a response termed positive phototropism. Hartmann et al. (1983) demonstrated phytochrome-mediated phototropism in dark-grown protonemal filaments of *Ceratodon purpureus*. Irradiation of the apical region of vertically positioned protonemal filaments with red light induced the tip cells to swell slightly and then grow towards the light source. The effect of red light was completely reversed by far-red light.

When spores of *Funaria* are irradiated with polarized red light (10^1 to 10^3 ergs cm^{-2} sec^{-1}) the chloronemal axis develops preferentially in a direction perpendicular to the plane of polarization. Polarotropism in this intensity range is controlled by rod-shaped photoreceptors located in or near the cell surface and oriented tangentially to it (Jaffe & Etzold, 1965).

Protonemal filaments of *Physcomitrium turbinatum* respond phototropically to red and far-red irradiation. Both red and far-red light striking the growing cell surface perpendicularly, are absorbed independent of their angle of polarization. However, radiation striking the flank tangentially is absorbed if the plane of polarization (vibrational plane of the electric vector) is parallel to the cell surface. The blue light absorbing pigment system, responsible for polarotropism in virtually all other groups of plants, is inactive or absent in this species (Nebel, 1968, 1969).

Jenkins and Cove (1983b,c) examined polarotropism of primary chloronemata of wild-type as well as mutant strains of *Physcomitrella patens*, and observed that response is dependent on the irradiance as well as the wavelength of monochromatic light. The mutants and wild-type show the same qualitative tropic responses, but differ with respect to the light conditions under which they are expressed. In both, the responses are controlled by phytochrome. In monochromatic red light, at low irradiance, wild-type primary chloronema grows positively phototropically in unidirectional light or perpendicular to the electrical vector (E) in polarized light; at high irradiance, growth is lateral to the unidirectional incident light and it is parallel to E in polarized light. The mutants, however, show only the lateral phototropic responses at all irradiances of red light tested. In far-red light, the wild-type primary chloronema adopts a positive phototropic or a perpendicular polarotropic response; the mutants show the same responses but in a lower percentage of filaments. From these experiments it has been concluded that mutants of this moss can adopt the range of tropic orientations exhibited by the wild-type.

In *Ceratodon purpureus* the positive phototropism of dark-grown protonema is controlled by membrane bound phytochrome (Hartmann, 1984; Hartmann et al., 1983). Irradiation of the apical region of vertically positioned protonema with red light induces a swelling of the tip region, and this is followed by growth towards the light source. If the red light irradiation is followed by exposure to far-red light the phototropic response is not observed. Hartmann (1984) opines that components of the cytoskeleton might be in some way involved in the reorientation of the growth centre. The effects of light on development are considered to be through the following series of events: (1) perception of light stimulus, (2) transduction of this signal by means of physiological reactions in the cell, and finally (3) the observable response.

Experiments done by Hartmann (1984) indicate that phytochrome is able to control fluxes of Ca^{2+} ions, which in turn influence cytomorphogenesis. However, it is added that modulation of Ca^{2+}-fluxes may not be the primary step in signal transduction.

REFERENCES

BAUER, L., *Naturwissenschaften*, 48, 507 (1961).
BAUER, L. and H. MOHR, *Planta* 54, 68 (1959).
BITTNER, KAROLINA, *Ost. bot. Z.*, 55, 302 (1905).
BORODIN, J., *Bull. Acad. Imp. Sci. de St. Petersbourg*, 12, 432 (1868).
CHOPRA, R.N. and URMILLA GUPTA, *Bryologist*, 70, 102 (1967).
CHOPRA, R.N. and P.K. KUMRA, *Phytomorphology*, 28, 298 (1978).
COVE, D.J., A. SCHILD, N.W. ASHTON, and E. HARTMANN, *Photochem. Photobiol.*, 27, 249 (1978).
DAGAR, J.C., A.S. AHLAWAT, and V.P. SINGH, *Cryptog. Bryol. Lichénol.*, 1, 305 (1980).
DeGREEF, J., W.L. BUTLER, T.F. ROTH, and H. FREDERICQ, *Pl. Physiol.*, 48, 407 (1971).
DEMKIV, O.T., *Ukr. bot. Zh.*, 28, 624 (1971).

DEMKIV, O.T., *Ukr. bot. Zh.*, **30**, 511 (1972).
DEMKIV, O.T. and Y.D. FEDYK, *Biophysics*, **22**, 824 (1977).
DEMKIV, O.T. and R.T. RIPETSKY, *Ukr. bot. Zh.*, **27**, 758 (1970).
DEMKIV, O.T., R.T. RIPETSKY, and Y.D. FEDYK, *Ukr. bot. Zh*, **28**, 309 (1971).
DOYLE, W.T., *Bryologist*, **66**, 238 (1963).
EGUNYOMI, A., *Nova Hedwigia*, **31**, 319 (1979).
FREDERICQ, H. and J. DeGreef, *Naturwissenschaften*, **53**, 337 (1966).
FREDERICQ, H. and J. DeGreef, *Physiol. Plant.*, **21**, 346 (1968).
FRENCH, J.C. and D.J. PAOLILLO, JR., *Bryologist*, **79**, 457 (1976).
GILES, K.L. and K.E. von MALTZAHN, *Bryologist*, **70**, 312 (1967).
GILES, K.L. and K.E. von MALTZAHN, *Can. J. Bot.*, **46**, 305 (1968)
GOEBEL, K., *S.B. bayer Akad. Wiss.*, **26**, 447 (1896).
HABERLANDT, G., *Jb. wiss. Bot.*, **17**, 359 (1886).
HARTMANN, E., *Planta*, **101**, 159 (1971).
HARTMANN, E., *Beitr. Biol. Pfl.*, **49**, 1 (1973).
HARTMANN, E., *J. Hattori bot. Lab.* No. **55**, 87 (1984).
HARTMANN, E. and H. KILBINGER, *Biochem. J.*, **137**, 249 (1974).
HARTMANN, E., B. KLINGENBERG, and L. BAUER, *Photochem. Photobiol.*, **38**, 599 (1983).
HEALD, F.D.F., *Bot. Gaz.*, **26**, 25 (1898).
INOUE, H., *J. Hattori bot. Lab.* No. **23**, 148 (1960).
JAFFE, L.F. and H. ETZOLD, *Biophys. J.*, **5**, 715 (1965).
JAHN, H., *Flora*, **155**, 10 (1964).
JENKINS, G.I. and D.J. COVE, *Planta*, **157**, 39 (1983a).
JENKINS, G.I. and D.J. COVE, *Planta*, **158**, 357 (1983b).
JENKINS, G.I. and D.J. COVE, *Planta*, **159**, 432 (1983c).
KASS, L.B. and D.J. PAOLILLO, JR., *Pl. Sci. Lett.*, **3**, 81 (1974).
KASS, L.B. and D.J. PAOLILLO, JR., *J. Cell Sci.*, **28**, 61 (1977).
KAUL, A. and RITA KAUL, *J. Hattori bot. Lab.* No. **38**, 435 (1974).
KESSLER, B., *Beih. bot. Zbl.*, **31**, 358 (1914).
KLEBS, G., *Biol. Zbl.*, **13**, 641 (1893).
KRISKO, M.E.P. and D.J. PAOLILLO, JR., *Bryologist*, **75**, 509 (1972).
KRUPA, J., *Acta Soc. Bot. Pol.*, **36**, 57 (1967).
KUMRA, P.K., "Experimental studies on some Delhi mosses", M.Sc. Thesis, Univ. Delhi, India, 1977.
LAAGE, A., *Beih. bot. Zbl.*, **21**, 76 (1906).
LARPENT-GOURGAUD, M. and M.P. AUMAÎTRE, *Experientia*, **33**, 1601 (1977a).
LARPENT-GOURGAUD, M. and M.P. AUMAÎTRE, *Z. PflPhysiol.*, **83**, 467 (1977b).
LARPENT-GOURGAUD, M., J.P. LARPENT, and R. JACQUES, *C.r. Acad. Sci., Paris*, **279**, 57 (1974a).
LARPENT-GOURGAUD, M., J.P. LARPENT, and R. JACQUES, *Bull. Soc. bot. Fr., Coll. Bryologie*, **121**, 153 (1974b).
LISTOWSKI, A., *Bull. Acad. Pol. Sci. II*, **6B**, 631 (1927).
MELSTROM, C.E., N.C. MARAVOLO, and J.R. STROEMER, *Bryologist*, **77**, 33 (1974).
MENON, M.K.C., I. GRASMÜCK, and E. HARTMANN, "Effect of light on the callus cells of the moss *Physcomitrium*. In J. DeGreef, Ed., *Photoreceptors and Plant Development*. Antwerpen Univ. Press, Antwerp, Belgium, 1980, Pp. 557-569.
MEYER, S.L., *Bryologist*, **51**, 213 (1948).
MILLER, D.H. and L. MACHLIS, *Pl. Physiol.*, **43**, 714 (1968a).
MILLER, D.H. and L. MACHLIS, *Pl. Physiol.*, **43**, 723 (1968b).
MITRA, G.C., A. ALLSOPP, and P.F. WAREING, *Phytomorphology*, **9**, 47 (1959).
MITRA, G.C., L.P. MISRA, and CHANDRA PRABHA, *Planta*, **65**, 42 (1965).
MOHR, H., *Planta*, **46**, 534 (1956).
MOHR, H., *Ann. Rev. Pl. Physiol.*, **13**, 465 (1962).
MOHR, H., *J. Linn. Soc. (Bot.)*, **58**, 278 (1963).

MOHR, H., "Phytochrome and gene expression." In C. Helene, M. Charlier, Th. Montenay-Garestier, and G. Laustriat, Eds, Trends in Photobiology. Plenum Publishing Corporation, 1982. Pp. 515-530.
NEBEL, B.J. *Planta*, 81, 287 (1968).
NEBEL, B.J. *Planta*, 87, 170 (1969).
NEBEL, B.J. and A.W. NAYLOR, *Pl. Physiol.*, 39 (Suppl.), 51 (1964).
NINNEMANN, H. and W. HALBSGUTH, *Naturwissenschaften*, 52, 110 (1965).
OTTO, K.R. and W. HALBSGUTH, *Z. PflPhysiol.*, 80, 197 (1976).
PAOLILLO, D.J. JR. and F.A. BAZAZ, *Bryologist*, 71, 335 (1968).
PAOLILLO, D.J. JR. and L.B. KASS, *J. exp. Bot.*, 28, 457 (1977).
PATSCHOVSKY, N., *Zschr, Abst, Vereb.*, 46, 112 (1926).
PRINGSHEIM, E.G. and O. PRINGSHEIM, *Jb. wiss. Bot.*, 82, 311 (1935).
RASHID, A., *Phytomorphology*, 20, 49 (1970).
ROBBINS, W.J., *Bot. Gaz.*, 65, 543 (1918).
SCHULZ, N., *Beih. bot. Zbl.*, 11, 81 (1902).
SCHWABE, W.W. and I.F.M. VALIO, *J. exp. Bot.*, 21, 122 (1970).
SERVETTAZ, C., *Annls. Sci. nat. (Bot.) IX*, 17, 111 (1913).
SHROPSHIRE, W. JR. and R.B. WITHROW, *Pl. Physiol.*, 33, 360 (1958).
SHUKLA, R.M. and A. KAUL, *Rev. Bryol. Lichénol.*, 44, 133 (1978).
SIMON, P.E. and J.B. NAEF, *Physiol. Plant.*, 53, 13 (1981).
SMITH, H., *Phytochrome and Photomorphogenesis*. McGraw-Hill, London, UK, 1975.
STEINER, A.M., *Z. Bot.*, 51, 399 (1963).
STEINER, A.M., *Z. Bot.*, 52, 245 (1964).
STEINER, A.M., *Z. PflPhysiol.*, 61, 184 (1969a).
STEINER, A.M., *Planta*, 86, 334, (1969b).
STEINER, A.M., *Planta*, 86, 343 (1969c).
STEPHAN, J., *Planta*, 5, 381 (1928).
TREBOUX, O., *Ber. dt. bot. Ges.*, 23, 397 (1905).
VON UBISCH, G., *Ber. dt. bot. Ges.*, 31, 543 (1913).
VALANNE, N., *Ann. Bot. Fenn.*, 3, 1 (1966).
VALIO, I.F.M. and W.W. SCHWABE, *J. exp. Bot.*, 20, 615 (1969).
VALIO, I.F.M., R.S. BURDON, and W.W. SCHWABE, *Nature*, 223, 1176 (1969).
VASHISTHA, B.D., "Photomorphogenesis in lower archegòniatae", M. Phil. Thesis, Univ. Delhi, India, 1981.
VASHISTHA, B.D., "In vitro investigations on some Indian bryophytes", Ph.D. Thesis, Univ. Delhi, India, 1985.
WILSON, J.R. and W.W. SCHWABE, *J. exp. Bot.*, 15, 368 (1964).

7 Ultrastructural Studies

In recent years extensive investigations have been carried out on the ultrastructure of bryophytes. This chapter deals with the ultrastructure of spore, protonema, gametophore, gametangia, gametogenesis, sporogenesis, sporophyte-gametophyte interaction zone, and seta, as revealed by transmission electron microscope. Histoenzymological studies on the sporophyte-gametophyte junction have also been included.

SPORE

Four wall layers are of common occurrence in the spores of bryophytes, as reported in *Hypnum, Polytrichum, Mnium,* and *Fissidens* (Eyme & Suire, 1969; Genevès, 1968; McClymont & Larson, 1964; Mueller, 1974; Paolillo, 1969). However, the terminology used varies with the investigator.

(1) The innermost layer of sporoderm is fibrillar and electron-translucent, and is designated as intine. It shows variation in thickness at different places, and this feature has been interpreted as an indicator of internal polarity.
(2) External to the intine is a very thin electron-opaque layer which is considered to be a subunit of the exine by some workers, whereas others regard it as second layer of intine.
(3) Outside this, there is a very thin layer, the exine. This layer has tripartite strips or plates usually of unit membrane thickness.
(4) The outermost layer, the perine, appears as a homogenous, yellowish layer which is extremely opaque. It does not appear to be uniform in density in electron micrographs.

The spore wall of *Riccia* consists of a thin intine and a three-layered, lamellate exine. The inner exine lamellae are thin and closely spaced, giving this zone a fibrillar appearance. Lamellae of the middle region are thick, whereas those of the outer region are of intermediate thickness, but both have a relatively thick electron-dense coating. Pores usually occur at the junctures of the triradiate ridge arms and the equatorial rim. These do not penetrate the entire exine, and represent localized areas where the outermost group of lamellae has not been formed (Steinkamp & Doyle, 1979). In contrast to *Riccia*, the exine of other Marchantiales and of *Sphaerocarpos* (Sphaerocarpales) has fewer, thicker, more widely spaced, or variously arranged lamellae in its outer portion. Thinner lamellae, more similar to those of *Riccia*, are usually found in the inner exine region of other Marchantiales (Denizot, 1971; Steinkamp, 1973). *Riella* and *Geothallus* (Sphaerocarpales) spores have groups of compact lamellae that are structurally different from those of *Riccia* (Steinkamp, 1973). Exine in the spores of Jungermanniales is usually thin and variously made of fine 'slip' or 'sheet' lamellar and granular elements (Heckman, 1970). Spore walls of the Metzgeriales have been extensively studied. The exine of the investigated genera differs from one another as well as from *Riccia* in the arrangement of lamellar components (Horner et al., 1966; Steinkamp, 1973).

Steinkamp and Doyle (1981) examined the mature spores of *Athalamia hyalina* with transmission electron microscope. They observed that hollow dome-like projections, sometimes having small pores and a coarsely granular surface texture, stud the spore surface, usually in a pattern of concentric circles. In section, the spore wall has an intine and two-layered exine. Intine-like material separates some lamellae of the inner exine, which is joined to the outer exine around the dome bases. Cavities of domes are often filled with a loose network of granular material (Plates 7.1 A-C; 7.2A,B).

The chloroplasts in ungerminated spores of *Ceratodon purpureus* have poorly developed lamellae with a few starch grains. The most pronounced structural transformation in the interior of the spore occurs during the swelling phase, when the grana in the chloroplasts become clearly differentiated and vesicle-like particles appear in the stroma. Most of the starch grains of the chloroplasts disappear during this phase. Spore distension begins with the orientation of the organelles towards that side of the spore where the intine is thickest and where exospore generally ruptures during the swelling phase. During normal germination, the lipid particles of the plasma and the starch grains of the chloroplasts disappear at a place where the germ tube is differentiated. Generally, formation of the first cross wall follows distension. The changes in structure of the organelles seem to be independent of the formation of the first cross wall (Valanne, 1966).

In *Funaria hygrometrica* gradual changes occur in the fine structural organization during germination, and one nuclear division occurs during the first 24 h. The plastids undergo rapid changes and within 2 h after immersion of spores in water, the plastids begin to elongate. In about 4 h the amount of starch increases considerably, but the number and complexity of the chloroplast lamellae remains about the same. The chloroplasts divide by the 12th h when the lamellae reach their maximum development. The mitochondria elongate and divide about the 4th h after immersion. Very simple Golgi bodies are still noticed 2 h after spore immersion, and a marked increase in the number of component cisternae is evident by the 4th h. and it continues. The single membrane-bounded bodies do not change greatly in size or appearance throughout germination, but apparently increase in number. No significant change is observed in the amount, appearance, or position of the endoplasmic reticulum. The lipid bodies are somewhat depleted as germination proceeds. Vacuoles develop by the 6th h after immersion (Monroe, 1968). Sequence of germination phases in mosses have also been studied with electron microscope in three isosporous species viz. *Polytrichum commune*, *Ceratodon purpureus*, and *Funaria hygrometrica* and in one anisosporous species *Macromitrium sulcatum* (Olesen & Mogensen, 1978). In all these the exine is thinner outside the conspicuous thickening of the fibrillar intine, and in this area a lamellar structure is present in the intine. There are 8 to 10 lamellae in the small spores of *M. sulcatum* and only one in *C. purpureus*, and these are embedded in a granular or fibrillar matrix. In *P. commune* the lamellae are less conspicuous and appear to be thinner. The intine consists of fibrillar material and shows a conspicuous disc-shaped local thickening. The aperture region is characterized by: (1) thickening of the intine, (2) decrease in thickness of the exine, (3) lamellate structure subjacent to the thin part of the exine, and (4) accumulation of electron-dense material into a thin separating layer between the intine and exine. After water absorption through the aperture the subsequent stages are swelling, spore wall rupturing, protrusion, and spore distension. The transition from spore swelling to protrusion seems to consist of a series of substages forming a gradual sequence of processes. Contrary to this, the transition from protrusion to distension is interrupted by a 'recovery' stage during which no structural or histochemical changes could be demonstrated.

Brown et al. (1982) reported that spore wall of *Sphagnum* is unique in having two layers in the exine: an inner lamellate layer, A-layer (ea), and a thick, homogenous outer layer, B-layer (eb). The exine of other mosses consists of only the outermost homogenous layer and, at most, a thin ill-defined opaque layer. In *Sphagnum*, during development of the A-layer-exine and the intine, a cortical system of evenly spaced microtubules underlies the plasma membrane. The A-layer-exine develops evenly around the young spore immediately after cytokinesis. As the intine is deposited centripetally inside it, the homogenous B-layer-exine is deposited outside the first-formed A-layer. The B-layer is responsible for the primary sculpturing of the spore surface (Plate 7.3A-D).

PROTONEMA

In the protonema of *Funaria hygrometrica* the longitudinal walls have a microfibrillar structure with no layer differentiation. The transverse walls are considerably thinner, and numerous plasmodesmata penetrate them, indicating close connections between the protoplasts in the long protonemal filament. Various structures such as Golgi apparatus, lipid bodies, and microbodies show features similar to the corresponding organelles in higher plants. The Golgi apparatus is composed of one to five cisternae. Vesicles with higher electron density than the ground cytoplasm are observed at the distended ends and around the cisternae. Endoplasmic reticulum is sparingly represented in the form of single tubules which occur along the cell walls. A few oval, kidney-, or pea-shaped bodies of about 1 µm consisting of a single membrane surrounding a homogenous, densely granulated content are also observed (Młodzianowski, 1970).

The nucleus is irregular in shape and the nucleoplasm is homogenous and finely granular. Oval and rod-like profiles of mitochondria are frequently observed in the neighbourhood of chloroplasts. They possess tubular structure and contain homogenous, finely granular matrix.

The structure of chloroplasts in the protonema of *Funaria* is more primitive as compared to that in higher plants. The chloroplasts are surrounded by a double membrane. The grana discs are more or less laterally displaced, forming a characteristic fan-shaped configuration. The matrix is uniformly and finely granulated (Młodzianowski, 1970). Some variations occur in other mosses e.g. cell walls are thick and of uniform texture in *Buxbaumia aphylla*. The cells comprising the protonema are highly vacuolated (Mueller, 1972).

A striking feature of the apical cells of the protonemal filaments of *Dawsonia superba* is the extensive development of dictyosomes and their intense vesicular activity. Microtubules are present throughout the inner cytoplasm. Wall microfibrils are oriented parallel to the peripheral cytoplasmic microtubules (DeMaggio & Stetler, 1977). The observations on protonemal cells of *Ceratodon purpureus* reveal that the apical cells have a cytoplasm of distinctly higher electron density, and contain a considerably higher number of cell organelles, the form of which indicates higher metabolic activity. The nuclei are more or less spherical, the fibrillar and granular components are present in approximately equal amounts, and are uniformly distributed. A less active form of organelles is observed in the intercalary cells. They are highly vacuolized, and their cytoplasm shows a low grade of electron density with a smaller number of ribosomes as compared to the apical cells. The nuclei are smaller, usually ellipsoid; the nucleolar substructures show a high degree of dispersion, which indicates a low metabolic activity of the nucleus. In the apical cells, mitochondria are numerous and are distributed throughout the cell; endoplasmic reticulum is in the form of short scattered fragments. On the other hand, in the intercalary cells mitochondria occur in the neighbourhood of other cell structures (most often plastids), and endoplasmic reticulum is observed only occasionally. In both the apical and the intercalary cells microbodies containing fine granular, homogenous matrix are observed mainly near the plastids, and Golgi structures are infrequently present. In the apical cells, complexes of small strongly osmiophilic globules are arranged in rows along microfilaments, and these are rarely observed in the intercalary cells (Idzikowska &

Szweykowska, 1978; Młodzianowski, 1970).

The terminal region of the tip cell in the caulonema of *Funaria hygrometrica* consists of a clear cap with a tip body free of plastids. Cell polarity is not only expressed by the unequal distribution of cell organelles, but also by their differentiation. The apical cell divides when it reaches a maximum length of about 400 μm. It grows exclusively at the tip and divides every six to seven hours. Subapical cells divide in a rhythm of about 12 to 14 h. They have a similar, but less distinct polar cytoplasmic organization. Starch content of chloroplasts within subapical cells decreases in a basal direction, as it does in the apical cell (Schmiedel & Schnepf, 1980).

The formation and development of protonemal branches was studied with particular attention to the distribution of microtubules and the polar organization of the cytoplasm in *Funaria hygrometrica*. A side branch is initiated 40 to 50 um above the nucleus, on the cell flank opposite the nucleus. The cell contains a large central vacuole; the nucleus is oval and is surrounded by numerous organelles. The cell wall is about 0.2 μm thick and consists of an outer dense layer with a distinct fibrillar structure and a less dense inner one in which the microfibrils appear to be placed more irregularly. In the first stage of side branch formation the cell wall bulges slightly and thickens by the deposition of new wall material, and the other wall layer becomes less dense. The cytoplasm contains active dictyosomes and vesicles but does not have a distinct tip body. The mass of the cytoplasm increases and it contains more microtubules than in former stages (Schmiedel & Schnepf, 1979). The new side branch breaks through the wall of the mother cell (Plate 7.4A). The nucleus rounds up and migrates into the centre of the cell, initially without moving in apical direction. Then the surrounding cytoplasm divides the central vacuole. The apex of the developing side branch lacks microtubules. The cell outgrowth is filled with cytoplasm and contains only a few vacuoles. The chloroplasts are positioned along the cell periphery (Plate 7.4B).

STEM

Eschrich and Steiner (1967, 1968) studied the ultrastructure of the stem of *Polytrichum commune*. The central core of hydroids, the water conducting cells, is surrounded by the eight longitudinal strands of leptoids. In the young part of the stem, the leptoids and parenchyma cells are nearly identically equipped with cell organelles. Both contain a nucleus, chloroplasts, mitochondria, and dictyosomes. The endoplasmic reticulum builds characteristic cisternae in the form of hollow cylinders extending from one end wall to the other. The cisternae are interconnected with many plasmodesmata, which occur only in the end walls. Leptoids have oblique end walls with 16 to 20 plasmodesmata per um^2, whereas the parenchyma cells show cross walls perpendicular to the axis with 9 to 12 plasmodesmata per μm^2.

Early in the development the leptoids undergo a change in their protoplasmic structure. Two to three cm below the apical cell their protoplasts degenerate and show lysosome-like structures. The endoplasmic reticulum and other structures get deformed or are dissolved; the plasmodesmata are constricted by callose deposits. At the same level the parenchyma cells still retain the original structure of their protoplasts. Hébant (1970, 1979) studied the structure and development of conducting cells (hydroids & leptoids) in some of the most highly developed mosses, especially gametophytes of various Polytrichales (see page 297). The hydroids constitute an axial strand in the leafy stem, and show resemblance to tracheids (Haberlandt, 1886; Vaisey, 1888a,b, 1891, all cited in Hébant, 1970).

Scheirer (1973, 1975, 1977) studied the ultrastructure of the conducting cells of *Dendroligotrichum dendroides* (Plates 7.5A-C; 7.6A-C). He observed that the lateral walls of hydroids appear to be lamellated and can be distinguished into three regions: outer, middle, and inner. The outer and middle regions appear polylamellate. The boundary between these regions is not sharp. The lateral walls of the leptoids of *Dendroligotrichum* differ from leptoid cells of other mosses in that they display collenchyma-like thickenings.

Ligrone et al. (1980a) have reported the ultrastructural differentiation of the cauloid (stem) in *Timmiella barbuloides*. Three histological complexes are distinguished at maturity, a peripheral band of stereids, a middle parenchymatic cortex, and a central strand of hydroids. Leptoids are lacking. The hydroids as well as cortical cells are dead and empty at maturity. On the other hand, members of Polytrichales retain some living cells in their cauloids (Hébant, 1974, 1976, 1977; Scheirer, 1976). Further, in *T. barbuloides* both the transversal and the longitudinal walls of hydroids remain thin and undergo a partial hydrolysis in the last stages of differentiation. In Polytrichales only the transversal cell walls are hydrolyzed, and the longitudinal ones remain intact (Hébant, 1974, 1977; Scheirer, 1973, 1976). Ligrone et al. (1980b) have observed for the first time granular spherical inclusions in the apical cells of the cauloid in *T. barbuloides*. These bodies are always present in association with mitochondria and microbodies.

LEAF

Leaves of mosses may exhibit three special types of cells—deuter cells, stereids, and hydroids. Besides, some mosses like *Sphagnum* and *Leucobryum* have dead and empty cells. Paolillo and Reighard (1967) demonstrated that in the leaves of *Polytrichum* and *Atrichum* cells of green lamellae attached to the midrib parenchyma are typical chlorenchyma, and contain a nucleus, mitochondria, modestly developed endoplasmic reticulum, Golgi bodies, and plastids. A complex vacuolar system is interspersed among the organelles. The elongate parenchyma cells of the midrib have numerous protoplasmic connections that traverse the highly perforated cross walls. All the three parenchyma layers share this feature. However, the middle layer of parenchyma distinguishes itself from the others by the larger diameter of its cells. The middle and third layers contain cells that are more elongate than those of the adaxial (first) parenchyma layer and their plastids are less complex. Stereids intervene the first and the middle layer of parenchyma in the median portion of the midrib. Hydroids are present between the second and third parenchyma layers. These are mostly free from protoplasm and are separated from one another by exceptionally thin, non-perforate cross walls. Luttge and Krapf (1968) observed that in *Mnium* the appearance of cytoplasm and the endoplasmic reticulum membranes differs with the age of leaves. The leaves of *Leucophanes candidum* comprise three layers: two outer rows of thin-walled hyaline empty cells (leucocysts) and an inner row of thick-walled, living cells containing remnants of cytoplasm and a few chloroplasts (chlorocysts). The walls of adjacent chlorocysts are perforated by a number of plasmodesmata which appear contoured by a clear sheath, most likely callose. Plasmodesmata are absent between chlorocysts and leucocysts (Favali & Bassi, 1978). Phylloids of *Leucobryum* are also constituted by leucocysts and chlorocysts. The leucocysts are arranged in two or more layers, have pores on their cell walls and form a complex open system of cavities that extends along the whole length of the phylloids. The chlorocysts are embedded among the leucocysts and form a central green layer. They have very thick-cell walls and contain microbody-like organelles, most likely peroxisomes. The morphology and the arrangement of the leucocysts as well as the particular position of their external pores on the phylloid surface, suggest that the main function of these cells may be to retain the absorbed water (Castaldo et al., 1979a, b).

Castaldo and Martino (1968-69) carried out ultrastructural studies on the hyaline cells of the leaves of *Sphagnum recurvum* with special attention to the pores. The covering membrane of the pores represents the outermost part of the primary cell wall. The annular and spiral thickenings of the wall of the hyaline cells play a fundamental role as supporting bars of the cells. The bars also seem to have an important influence on the absorption and conduction of water. Microbody-like organelles are also reported in the leaves of mosses. These are numerous and rod-shaped in *Leucobryum* sp., but are generally spherical and less numerous in *Timmiella barbuloides* (Gambardella et al., 1979). Gambardella et al. (1980) reported catalase activity in the microbodies of the photosynthetic cells of the phylloids of *T. barbuloides*.

Recently, Scheirer (1983) observed that in the leaf of *Polytrichum commune* the deuter cells, which are large-diametered parenchymatic elements, exhibit dense cytoplasm, polysomes, papillate wall ingrowths characteristic of transfer cells, and a peripheral network of endoplasmic reticulum associated with the wall ingrowths (Plate 7.7A-D).

GAMETANGIA

ANTHERIDIUM

In the parenchymatous cells of the antheridial wall of *Sphagnum* the main components are nucleus, starch-laden plastids, mitochondria, fat bodies, tubules, Golgi bodies, vesicles, and granules (Manton, 1957). The fat bodies appear dark, and range in diameter from 0.25 to 0.5 um with a length up to twice this dimension.

Duckett (1973) studied the jacket cells of mature antheridia in *Anthoceros laevis*. These cells resemble transfer cells in having wall ingrowths and other general cytological characteristics like dense cytoplasm adjacent to the ingrowths, abundant endoplasmic reticulum and ribosomes, and numerous mitochondria and Golgi bodies with small vesicles. Production of nutrients for the developing spermatocytes is a likely function of the antheridial jacket cells. It is also suggested that the ingrowths may form an integral part of the dehiscence and spermatozoid discharge mechanism in *Anthoceros*.

Extensive studies on the development and maturation of antheridia in *Polytrichum juniperinum* have been conducted by Paolillo (1975) and Hausmann and Paolillo (1977, 1978). The mature antheridia of this moss are resistant to plasmolysis because they are protected from the environment by a cuticle. Ultrastructural investigations have demonstrated that a specialized fluorescent boundary is formed across the stalk of an antheridium as a distinctive secondary wall layer at the time when an antheridium has attained less than half its final length. Hausmann and Paolillo (1978) stated that a fluorescent boundary at the base of the antheridium is of widespread occurrence in mosses. This boundary cuts off the antheridium from the stalk and thus makes the protection of the antheridium complete. This wall comprises light and dark lamellations and possesses plasmodesmata. It is likely that the natural role of this boundary is in maintaining hydrostatic pressure in the antheridium and in regulating the movement of solutes. The cells of stalk below the boundary are rich in lipid droplets, whereas those in the base of the antheridium have markedly fewer and smaller lipid droplets but have abundant rough endoplasmic reticulum. During maturation of the antheridium a fluid collects beneath and around the sperm mass, and this fluid plays an important part in sperm release. However, the cells surrounding this fluid show no special morphological adaptations related to secretion.

GAMETOGENESIS

SPERMATOGENESIS

The spermatogenous cells become cytologically distinct from the cells of the jacket layer of the antheridium. These cells are interconnected by plasmodesmata, but are isolated from the jacket cells. Strict synchrony in development is restricted to the derivatives of each spermatogenous cell in *Marchantia* and *Riccia* (Górska-Brylass, 1969). Androgones (the multiplying cells of a growing sperm mass) are arranged in groups or blocks, surrounded by walls that are thicker than the walls around individual androgones. A callose wall is formed around each mother cell, and this disintegrates after maturation of the spermatozoids. The plastids of the spermatogenous cells usually elongate and at times get trapped across a cell plate as observed in *Pogonatum* (Bonnot, 1967).

As the spermatozoid develops, blepharoplast appears. During the morphogenesis of blepharoplast in *Marchantia* two centrioles become the basal bodies and eventually produce flagella (Kreitner, 1977a). Subjacent to the basal bodies, a multilayered structure (mls) develops which attracts and holds a mitochondrion. One plastid and one mitochondrion together form the limisphere, which becomes closely appressed to the side of the nucleus. In the spermatocytes the first notable event is the lateral displacement of the nucleus and the aggregation of the rest of the organelles (chiefly the plastids. mitochondria, & Golgi bodies) in the other half of the cell (Lal & Bell, 1975; Paolillo, 1965; Paolillo & Cukierski, 1976; Paolillo et al., 1968a, b). Some of the microtubules associated with the upper layer of the mls lengthen considerably (Carothers & Kreitner, 1968). The nucleus elongates at this stage and signals the maturation of the spermatid. The flagella also emerge. They run posteriorly, more or less in alignment with the microtubules of the microtubular ribbon (Plate 7.8) which issues from the multilayered structure. The mls in a spermatid is morphologically significant, and only its upper portion persists in the mature spermatozoid as reported in *Physcomitrium* (Lal & Bell, 1975). The lower region of the mls seems to behave as a specialized organizing centre of microtubules (Kreitner & Carothers, 1976). In *Phaeoceros* an exceptionally short mls is present which is equally wide both in front and rear (Moser, 1970; Moser et al., 1977). The shape of the mls and the disorganization of the lower strata exert controlling influences on the locomotor ability of spermatids. As mls becomes completely differentiated, chromatin condensation begins in the nucleus, and spermatozoids of *Pellia* show nuclear envelope diverticula, which probably are a truss of endoplasmic reticulum overlain by a band of microtubules (Suire, 1970). It is also proposed that in *Marchantia* coiling of the spermatid nucleus is related to a reduction of spermatid cytoplasm during maturation (Kreitner, 1977b). In the spermatozoid of *Sphagnum* there is a coiled nucleus, a vestigial plastid, putative mitochondria at the two extremities of the cell, and two flagella (Manton, 1957). Transverse and longitudinal sections of a flagellum of mature sperm of *Polytrichum* reveal a highly differentiated structure (Paolillo, 1967). The axoneme of a mature flagellum in a sperm of *Polytrichum juniperinum* contains a double helix of two 500 A° ribbons, between the central fibers and the peripheral doublets. The double helix is bonded to the peripheral doublets and appears late in the maturation of the flagellum. Each ribbon is composed of five strands of subunits, and this feature has been incorporated into the tentative model of the axoneme (Plate 7.9).

The spermatozoids of *Bryum* sp. and *Funaria hygrometrica* have two flagella connected to a blepharoplast. Each flagellum consists of 11 fibrils: nine double peripheral fibrils and two single central fibrils, as also observed in *Physcomitrium coorgense* (Plate 7.8). The blepharoplast is 2 to 2.5 um in length and 0.3 to 1 μm in width. In the centre of the spermatozoid there is a single chloroplast, about 2 μm in length, and it is surrounded by a membrane. The lamellae within the chloroplast are united in pairs. Osmiophilic granules, 0.04 μm in diameter, are also present. The crescent-shaped nucleus is eccentrically located, lying behind the chloroplast and occupying the posterior portion (Sun, 1964).

When first extruded from the antheridium, the spermatozoids are surrounded by a shell of gelatinous mucilage (Lal & Bell, 1975). It is indeed noteworthy that specialized mucilage papillae are associated with gametangia of bryophytes. The development of these structures has been studied in some Polytrichaceous mosses (Hébant & Bonnot, 1974) and in *Marchantia* (Galatis & Apostolakos, 1977). Active secretory phase of mucilage papillae is marked by a hyper secretory activity of dictyosomes producing abundant vesicles containing polysaccharides, which possibly have a regulatory activity in bryophyte morphogenesis.

OOGENESIS

Not much work has been done on oogenesis in bryophytes, but the available data are of considerable interest. The earliest electron microscopic study on oogenesis in liverworts is that of Diers (1965a). He investigated in details, the fine structural changes manifested during the development of archegonium of the liverwort *Sphaerocarpos donnellii*. The development was traced from the initiation of the archegonium to the formation of the secondary central cell (Diers, 1965a, b).

Initially the archegonial initial (or the primary cell) possesses all the usual organelles of a typical somatic cell, such as plastids, mitochondria, endoplasmic reticulum, small vesicles with single membranes, and larger bodies which eventually form vacuoles. At the archegonial initial stage a noteworthy feature is the occurrence of small projections from the cell wall into the cytoplasm. Inside the wall-projections there are small vesicles bounded by a single membrane. The primary cell divides into a cap cell and a primary central cell or mother cell. The cytoplasm of the primary central cell and of the jacket cells is regularly connected by plasmodesmata. The primary central cell undergoes a transverse division, to produce a secondary central cell and a neck canal mother cell. As the neck canal mother cell divides, the secondary central cell enlarges considerably. The plasmodesmata between the wall cells and the two-celled axial row are lost, and subsequent development of axial cells into central cell and neck canal cells occurs in an isolated environment. The central cell divides giving rise to the egg cell and the ventral canal cell, and the plasmodesmata are formed de novo in their cross wall. The egg during its maturation separates from the wall of the venter leaving a space between the egg and the associated wall cells. This space is apparently filled, in all the investigated taxa, with an amorphous substance, a hydrophilic polysaccharide of homogenous mucilage (Barbier, 1972; Lal & Bell, 1977; Vian et al., 1970).

The mature eggs of *Sphaerocarpos* (Diers, 1965a, b, 1970) and *Marchantia* (Zinsmeister & Carothers, 1974) are characterized by cytoplasm rich in ribosomes and other organelles. Nucleocytoplasmic interaction is emerging as a regular cytological feature of oogenesis in lower archegoniates (Bell, 1975), though the mechanism varies in different groups. This is attributed to the massive evaginations of the nucleus into the cytoplasm during the differentiation of the egg cell in the two taxa named above. The final fate of these evaginations is still not known, but these are considered to be the progenitors of organelles donated by the maternal nucleus to the developing sporophytes (Plate 7.10).

In *Mnium undulatum* the ultrastructural characters of the developing ventral cell and the young oosphere are quite distinct. The plastids, vesicles, mitochondria, and dictyosomes are uniformly distributed in the cytoplasm of ventral cell. The typical distribution of cellular organelles and the inclusions in the oosphere show that it acquires a distinct polarity. The nucleus is present in the lower half and is surrounded by plastids, mitochondria, dictyosomes, and endoplasmic reticulum, while vesicles (bounded by single membrane) are present in large number in the upper half of egg. Electron dense nucleolus-like bodies (2 um in diameter) are also present in the cytoplasm of the female gamete (Barbier, 1972, 1979). A pectinaceous material is deposited around the oosphere. Vesicles are secreted from the endoplasmic reticulum of the oosphere, and the smaller round vesicles contain starch grains.

In *Physcomitrium coorgense* the boundary of the nucleus becomes indistinct during maturation. The surface of the nucleus is thrown into extensive folds, and nuclear pores are prominent in all parts of the envelope (Lal & Bell, 1977). The nucleolus remains distinct throughout the maturation of the egg. The plastids undergo differentiation during oogenesis. The cytoplasm of the mature egg is characterized by the presence of closely packed ribosomes, large vacuoles, and lipid bodies. It is believed that vacuoles contribute to the production of mucilage in the venter around the egg as they move towards the periphery of cell during maturation. The moss egg is unique in being freely suspended in a matrix in the cavity of the venter. Differentiation of the egg involves its progressive isolation from the surrounding cells, and important fine structural changes in the cytoplasm (Plates 7.11A,B; 7.12 A-D).

Fertilization in *P. coorgense* is followed by the formation of a conspicuous mucilage plug in the lower part of the archegonial neck. As the zygote enlarges to occupy the ventral cavity, remnants of the matrix persist as a concave at its upper end. A new cell wall is formed around the zygote which has a irregular nucleus placed in the upper half of the cell.

Lal et al. (1982a) described oogenesis in *Physcomitrium cyathicarpum*. The egg cell is isolated from the surrounding jacket cells at a very early stage. Isolation begins with the breaking of the plasmodesmata at the ventral cell stage. At the same time there is a retraction of the ventral cell cytoplasm. The plasmodesmata between the young egg cell and the ventral canal cell are broken soon afterwards. Maturation of the egg is accompanied by an expansion of the egg chamber resulting in the formation of a mucilage-filled space around the egg and the ventral canal cell. The mucilaginous hemicellulosic material in which the egg is suspended is PAS-positive, indicating its polysaccharide nature. It is suggested that the egg itself is solely responsible for the secretion of mucilage in the venter, and the jacket cells of the archegonium do not contribute to its formation. This is clear from the following facts: (1) the maturing egg abounds in vesicles, some of which are seen to be discharging their contents into the venter cavity (reverse pinocytosis), (2) the formation of hemicellulosic mucilage coincides with the disappearance of starch from plastids in the mature egg, (3) presence of irregular tubules, 20 nm in diameter, radiating into the matrix, some of which originate in a cytoplasmic vesicle.

In *Physcomitrium cyathicarpum* secretion of mucilage around the egg is the only means of isolation of the egg from the surrounding tissues. It is considered that such a barrier is essential for maturation of the egg. It would not allow any external influences to reach the egg and also prevent the essential metabolites peculiar to the egg from diffusing into the surrounding tissues (Lal et al., 1982a). The deposition of callose around the ventral cell, the neck canal cells, and later even around the egg and the ventral canal cell has also been reported as a means of isolation in liverworts like *Marchantia polymorpha* and *Riccia glauca* (Górska-Brylass, 1969).

Unlike in the investigated liverworts, nuclear evaginations are absent in the eggs of mosses so far studied: *P. coorgense* (Lal & Bell, 1977), *P. cyathicarpum* (Lal et al., 1982a), and *Mnium undulatum* (Barbier, 1972, 1979). As in *Mnium* (Barbier, 1972), the distribution of organelles in the egg cytoplasm of *Sphaerocarpos* (Diers, 1966) and *Marchantia* (Bell, 1975) is believed to lead to a strong polarity. However, in *P. cyathicarpum* there is considerable variation in the organelle distribution (Lal et al., 1982a).

Lal et al. (1982b) observed an unusual archegonium with persistent ventral canal cell in *P. cyathicarpum*. Ultrastructural investigations revealed a close similarity between the egg and the ventral canal cell. The similarity in fine structure provides an additional support to the notion that the ephemeral ventral canal cell in the archegonia of present day bryophytes is in fact an additional egg which has become non-functional during the course of evolution.

Recently, Lal et al. (1985a,b) have studied the ultrastructure of ventral canal cell, neck cells, neck canal cells, and venter cells in *Physcomitrium cyathicarpum*. The ventral canal cell (VCC) possesses smaller plastids with fewer starch grains as compared to those in the egg. In addition, it has increased number of vacuoles, and does not have rough ER. The fine structure of VCC indicates that it may also contribute toward the formation of the mucilaginous matrix (in which the mature egg is suspended), most of which is secreted by the egg itself during its maturation. The ephemeral neck canal cells show ultracytology similar to that of meristematic cells, whereas the neck cells are characterized by irregular projections on the inner tangential walls. These investigators suggest that neck canal cells play an important role in the production of mucilage which possibly aids the process of fertilization.

SPOROGENESIS

The cells of the capsule which form spores undergo a lot of fundamental structural changes which are unique in some aspects. A comparison of this process with microsporogenesis in angiosperms reveals that some features are common to the two groups. The sporogenous tissue results as a consequence of prolific meristematic activity of specialized cells (archesporial cells), which originate as a single layer from either of the two fundamental embryonic layers—amphithecium and endothecium. In Andreaeales and also in most of the members of Bryales the archesporial cells are derived from endothecium, but in Sphagnales the archesporium originates from amphithecium.

The cells of the spore sac can be easily distinguished from the sporogenous cells by their shape, size, and ultrastructural features. Spore sac is usually a single layer of cells. However, in some taxa like *Fissidens limbatus* there is a single-layered inner spore sac and a two-layered outer spore sac, the inner layer of which stains darkly and resembles the inner spore sac. The cells in the outer layer have characteristically many spindle-shaped chloroplasts containing large quantities of starch (Mueller, 1974). In *Physcomitrium cyathicarpum* the outer spore sac is two-layered but a distinct inner spore sac is lacking, so that the sporogenous tissue directly abuts the central sterile tissue (Lal & Bhandari, 1968). The sporogenous cells in most mosses contain dense cytoplasm, a disproportionately large and centrally placed nucleus, a large number of mitochondria, dictyosomes, endoplasmic reticulum, ribosomes, a few vesicles, and most strikingly, a single plastid. This is in contrast to the situation in the cells of the adjacent spore sac and wall layers which show many plastids. The sporogenous cells are separated from each other by their homogenous walls, whereas the walls which demarcate the spore sac cells from sporogenous cells are coarse and fibrillar in composition. The premeiotic sporogenous cells are interconnected by plasmodesmata, but vegetative and sporogenous cells remain wholly unconnected. The sporogenous cells which are the product of last mitotic divisions are termed

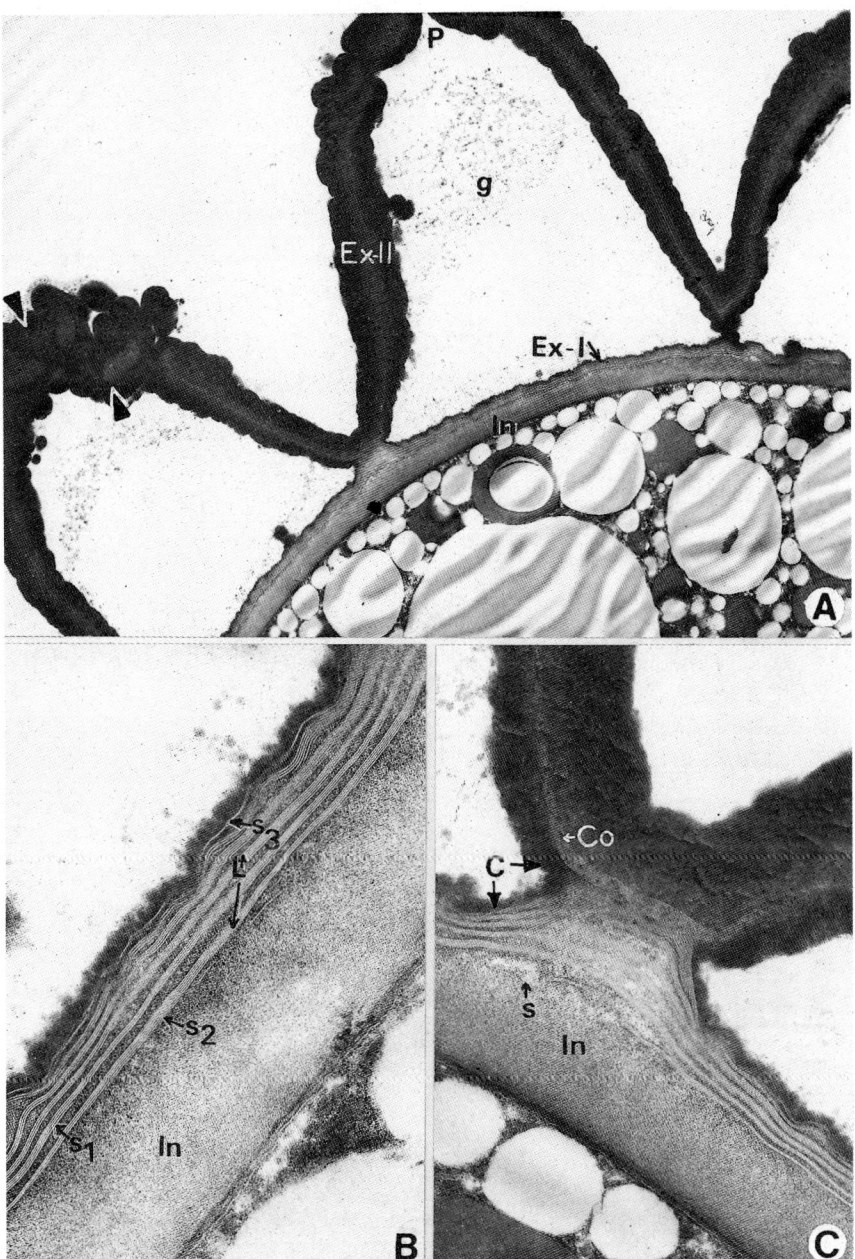

Plate 7.1 A-C. Spore wall interior in the liverwort *Athalamia hyalina*. **A.** The distal face sporoderm consists of the intine (In) and exine regions I and II (Ex-I, Ex-II). A pore (P) penetrates exine region II at apex of the centre dome. Skewed exine II lamellar segments (arrowheads) are present around the left dome apex. A loose meshwork of finely granular or flocculent material (g) is scattered throughout the cavities formed by domes. **B.** Magnified view of intine and exine I underlying the central area of a dome. The intine (In) has a moderately electron-dense finely granular to fibrillar appearance. Exine I lamellae (L) are composed of closely parallel groups of electron-lucent subunits (s), the inner of which in each lamella may terminate abruptly (s_1), or fade out (s_2). Electron-dense granular material coats the outermost lamellar subunits (s_3). **C.** Region of contact between exines I and II
(Contd.)

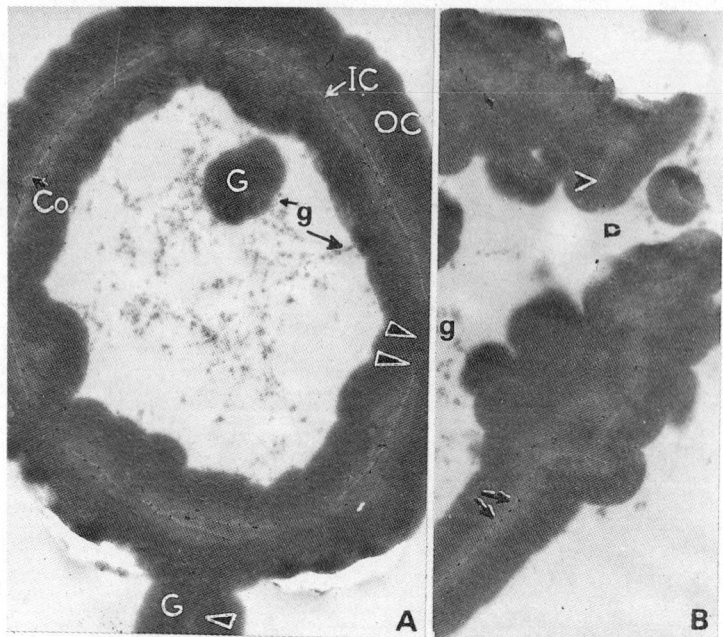

Plate 7.2 A, B. Spore wall interior in the liverwort *Athalamia hyalina*. **A.** Transverse section of a dome approximately at midheight. Lateral branching of the core (Co) and interruptions of the core and inner coating (IC) are apparent (double arrowheads). A large granule (G) containing a core segment (arrowhead) projects from the dome exterior. The finely granular or flocculent material (g) of the core interior joins outer coating material of the exine II lamella and large internal granule (small arrows). **B.** Cross-section of a dome apex and pore. Skewed lamellar segments are frequent around the dome apex and pore (P) and provide a coarsely granular surface texture to the area. The inner lamellar coating is very thin across the terminus of some core branches (arrowhead). Small electron-dense granules are scattered in the lamellar core (double arrows).
(After Steinkamp & Doyle, 1981).

Plate 7.1 A-C (Contd.)

at a dome base. Lamellar coatings (c) of the two exine regions disappear in the contact area, thereby allowing lamellar subunits to become aligned (arrowheads); coatings are, however, contiguous at the margin of the contact area. Subunits are resolved in the core of the exine II lamella (Co) and also are found in the outer region of the intine (In at s). Intine-like material separates the exine I lamellae.
(After Steinkamp & Doyle, 1981).

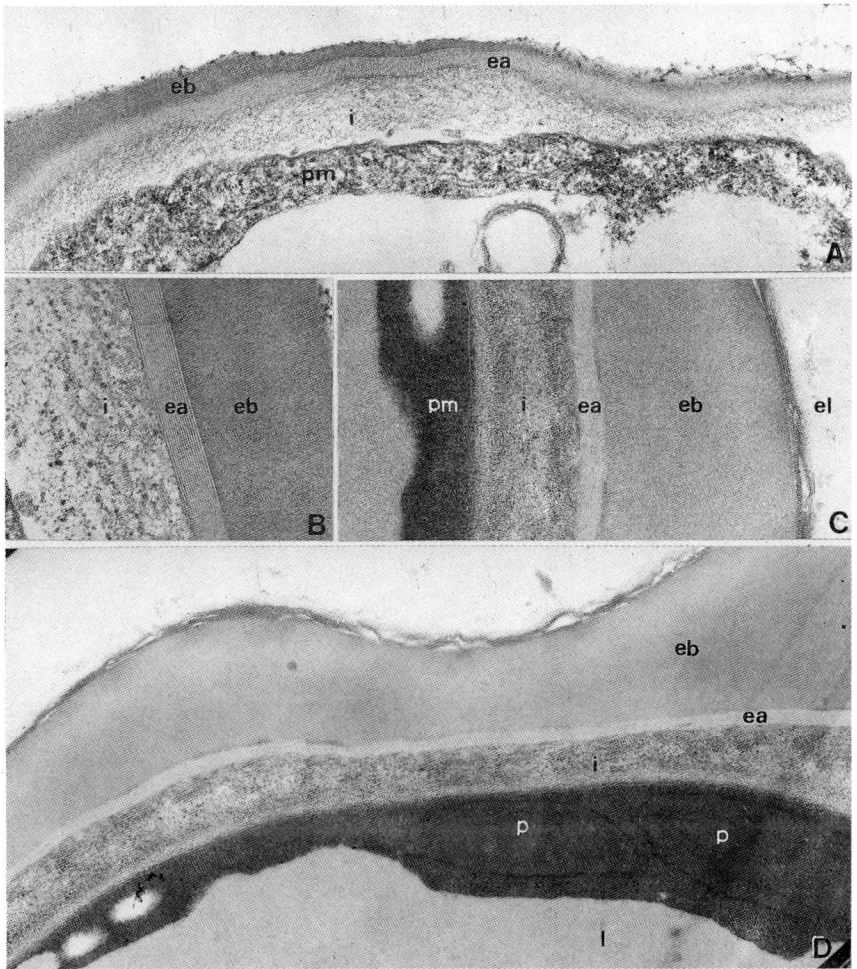

Plate 7.3 A-D. Development of exine in spores of *Sphagnum lescurii*. **A.** Early deposition of B-layer exine on the outside of the A-layer exine. **B.** Radial feature in A-layer exine. **C.** Spore wall with nearly mature B-layer. Stacks of tripartite lamellae (el) appear on outer surface. **D.** Nearly mature spore with completed B-layer, A-layer, and intine. Two highly reduced plastids are seen in the dense peripheral cytoplasm. The entire central portion of the spore is filled with storage lipids. ea—A-layer exine; eb—B-layer exine; el—tripartite lamellae; i—intine; l—lipid; p—plastid; pm—plasma membrane.
(After Brown et al., 1982).

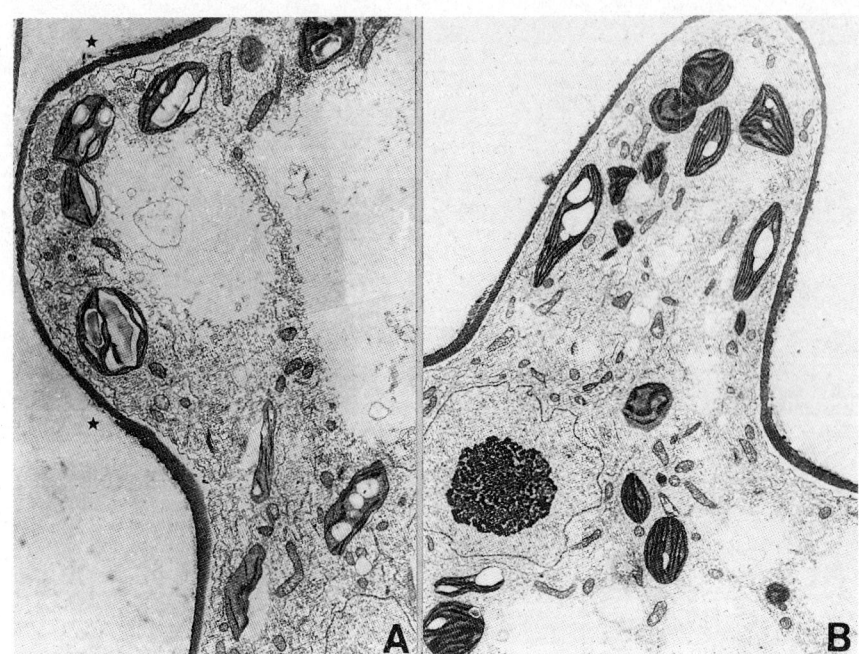

Plate 7.4 A, B. Formation of side branch in the protonema of *Funaria hygrometrica*. **A.** A side branch has ruptured the old cell wall (asterisks). **B.** A developing side branch with the nucleus (in preprophase) moved to its base. Chloroplasts begin to become chloronematic. Tip of the outgrowth is without vesicle accumulation.
(After Schmiedel & Schnepf, 1979).

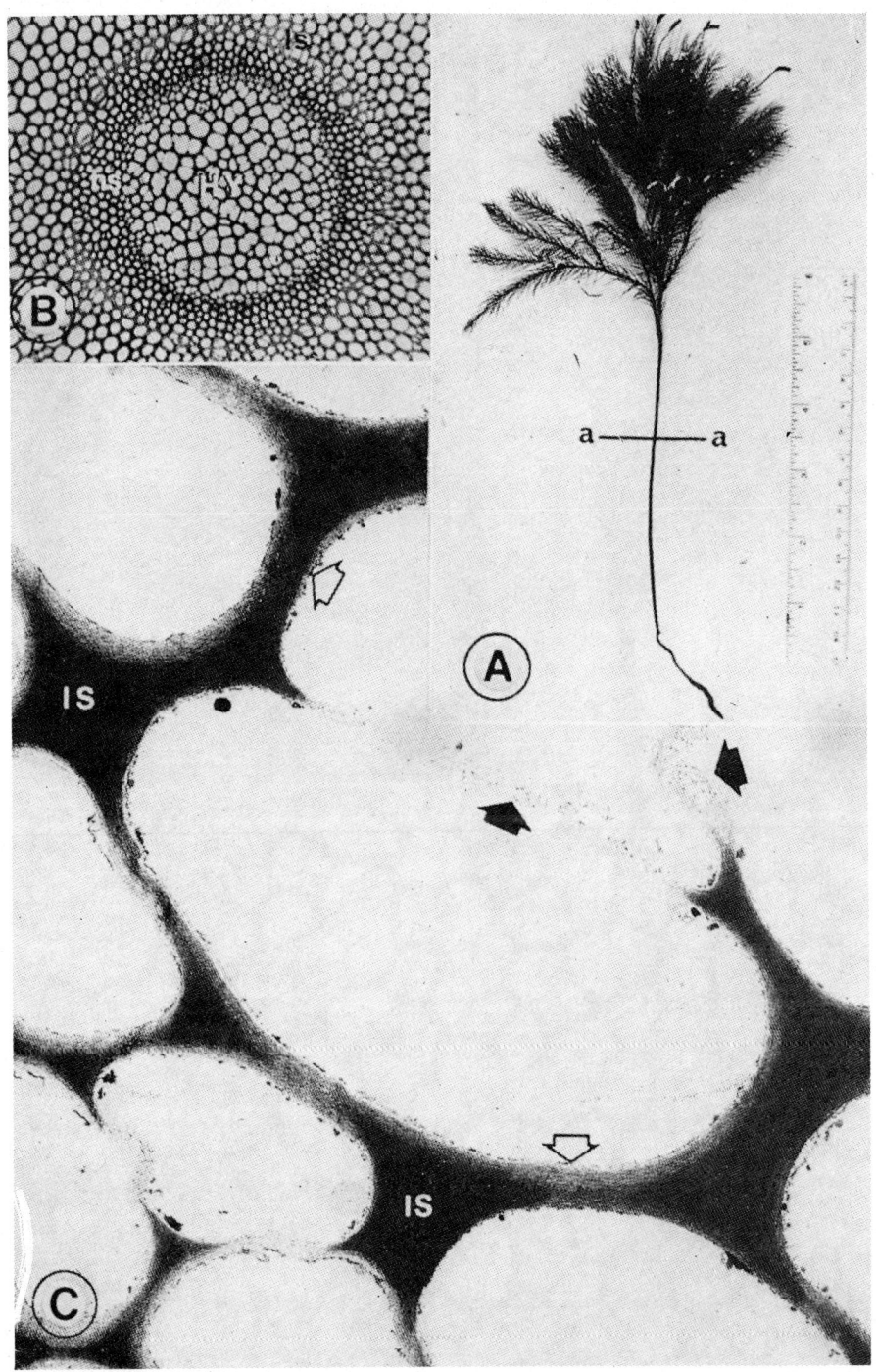

(See legend on reverse).

Plate 7.5 A-C. Anatomy of stem in *Dendroligotrichum dendroides*. **A.** Gametophore illustrating general morphology. Note lower leaf-less portion, "dendroid" upper portion, and attached sporophytes. **B.** Photomicrograph of transverse section through plane a-a of Figure **A.** Central cylinder of gametophore consists of hydroids (HY), hydrom sheath (hs) and leptom sheath (ls). **C.** Electron micrograph of transverse section of hydroids of central cylinder. Two morphologically distinct cell walls are apparent; hydrolysed end walls (closed arrows) and thickened lateral walls (open arrows) with prominent intercellular substance (IS).
(After Scheirer, 1975).

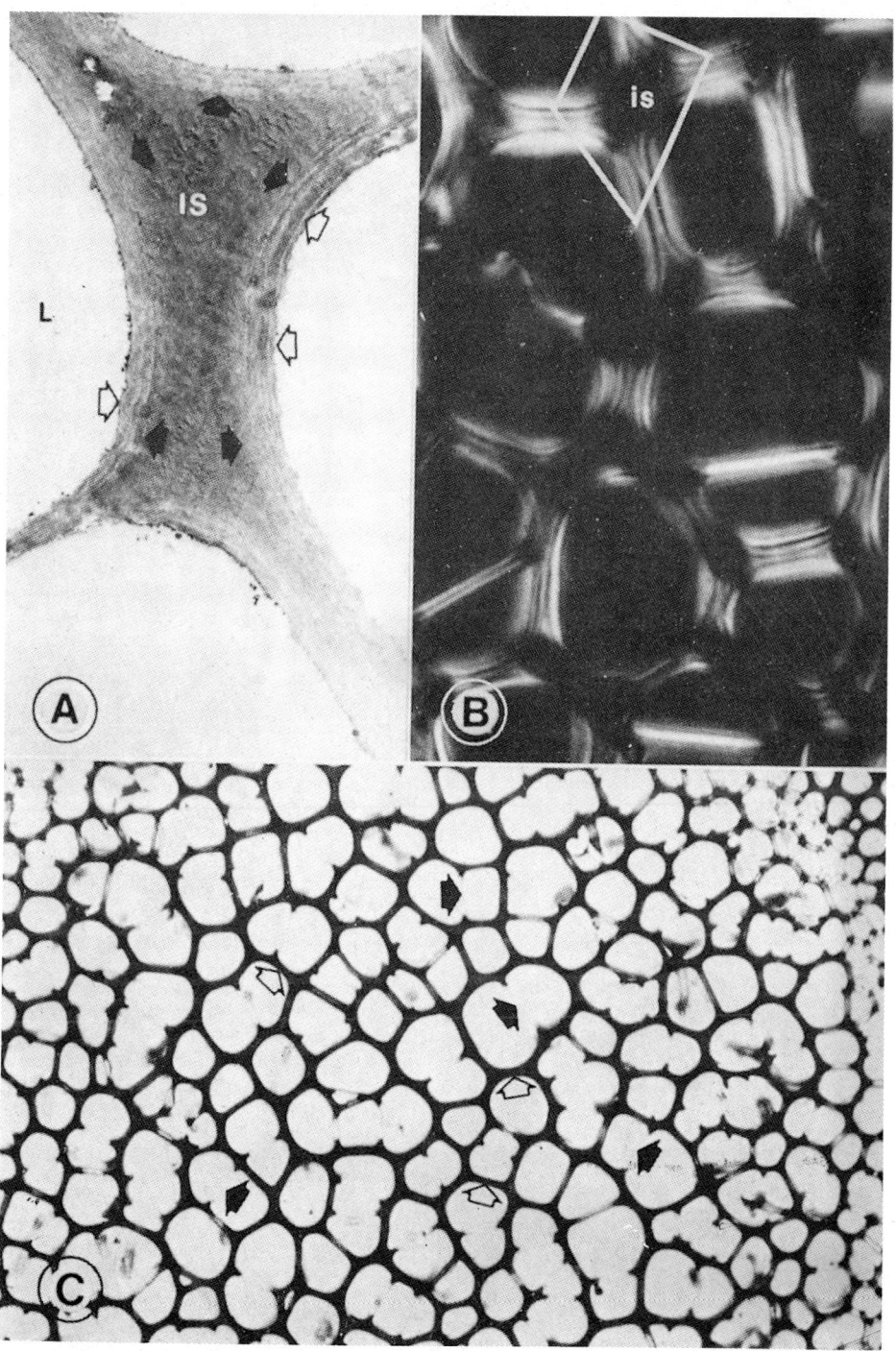

Plate 7.6 A-C. Anatomy of stem in *Dendroligotrichum dendroides*. **A.** Electron micrograph of a transverse section of hydroids illustrating juncture of four hydroids with intercellular substance (IS) prominent between hydroids. Apparent thickness of the lateral walls is enhanced by this electron-dense intercellular substance. Outer boundary of the lateral walls (closed arrows) is seen appressed to intercellular substance, whereas laminated lateral walls (open arrows) are seen as alternating electron-translucent and electron-dense layers. L—lumen (phosphate buffered gluteraldehyde and osmium fixation. No post staining). **B.** Transverse section of hydroids as photographed with polarized light. Note birefringence of the lateral walls and isotropic intercellular substance (is). Area delimited by white lines is similar to the area of hydroids depicted in Figure **A**. **C.** Transverse section of hydroids of central cylinder treated with silver-hexamine reagent. Blackend nature of lateral walls (open arrows) is due to deposition of silver grains by reducing substances in the walls. Hydrolysed end walls (closed arrows) are unreactive.
(After Scheirer, 1975).

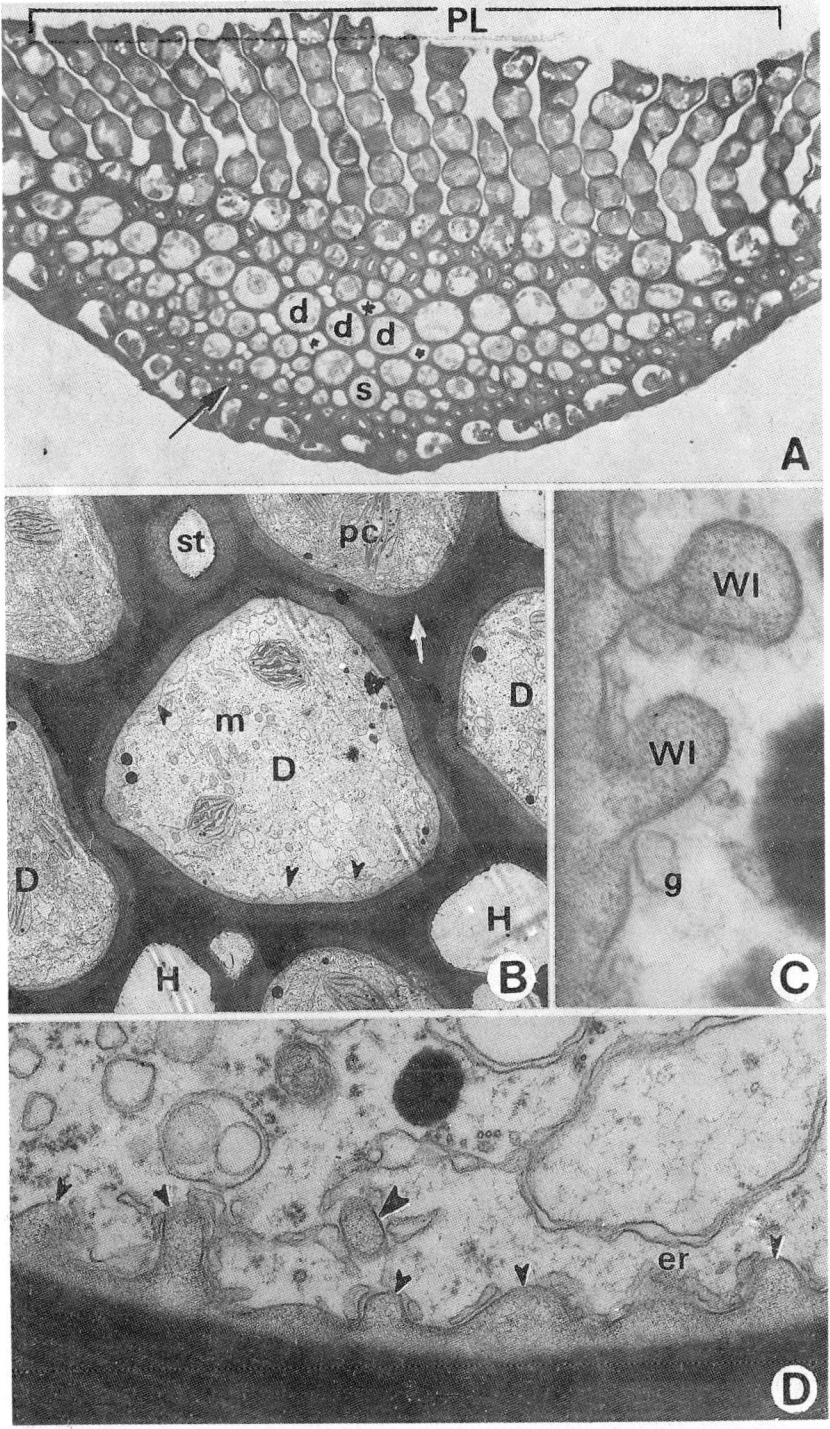

(See legend on reverse)

Plate 7.7 A-D. Anatomy of leaf in *Polytrichum commune*. **A.** Photomicrograph of transverse section of leaf illustrating histological organization. Deuters (d) are prominent in the median region and extend laterally to the adaxial stereids. Asterisks indicate position of hydroids, and the arrow marks the band of stereids under the abaxial epidermis. A similar group of stereids can be seen beneath the adaxial epidermis. (The photosynthetic lamellae (PL) arise from the adaxial epidermis. Parenchyma abaxial to the deuters are the socci (s). **B-D.** Electron micrographs. **B.** Large diametered deuters (D) from leaf bundle. Papillate wall ingrowths (arrowheads) penetrate into cytoplasm. Numerous mitochondria (m) are seen in the central region of the deuter. The arrow indicates the direction of the photosynthetic lamellae. Stereids (st) and hydroids (H) are also labelled. **C.** Magnified view of wall ingrowths (WI) from a deuter of the leaf bundle. Plasma membrane follows the contour of wall ingrowth. **D.** Small, papillate wall ingrowths (small arrowheads) of a deuter extend into the cytoplasm. Elliptical profile of a wall ingrowth (large arrowhead) appears distinct in the cytoplasm. Smooth endoplasmic reticulum (er) is seen in close proximity to the wall ingrowths.
(After Scheirer, 1983).

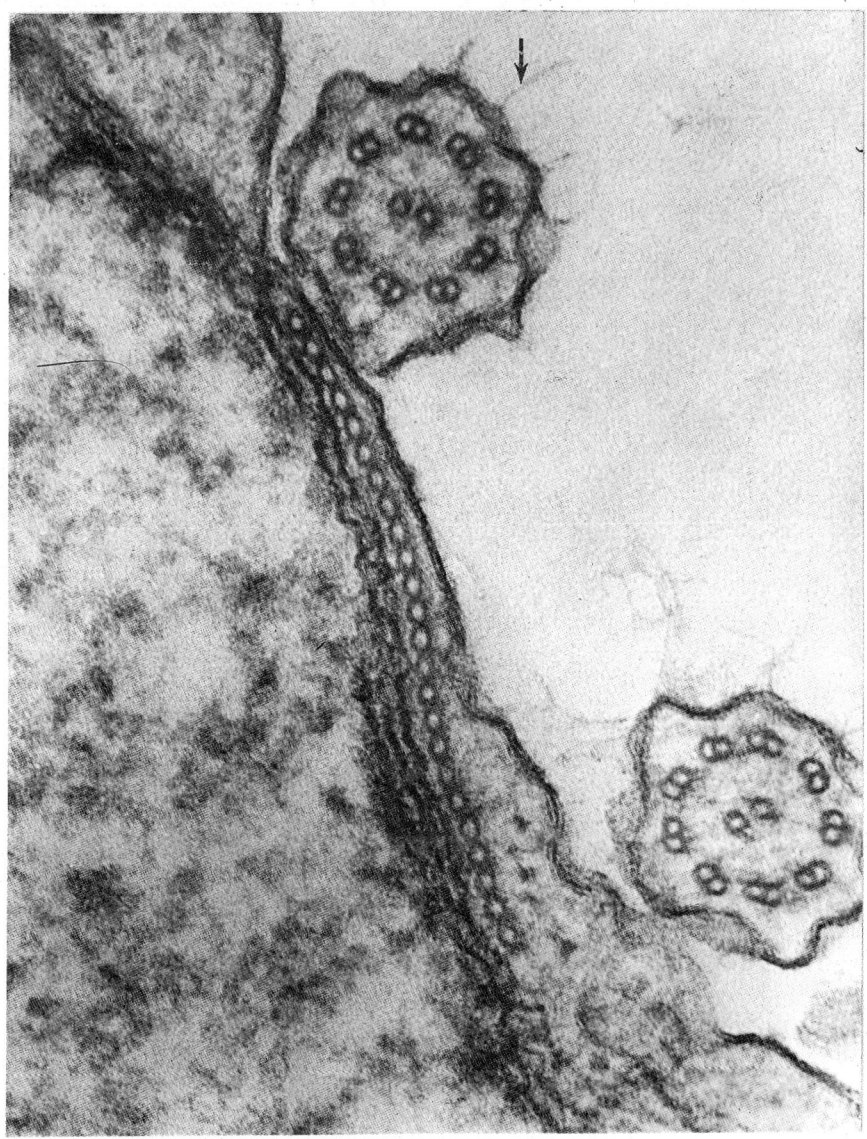

Plate 7.8. Transverse section of the microtubular ribbon adjacent to the nuclear envelope (left of ribbon) and flagellae in a spermatocyte of *Physcomitrium coorgense*. There are occasional indications of marginal hairs (arrow). (After Lal & Bell, 1975).

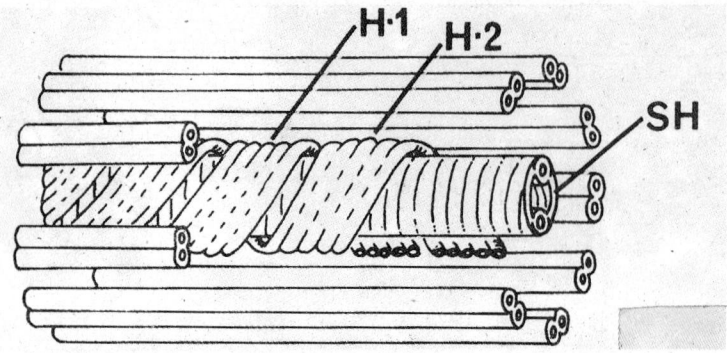

Plate 7.9. Diagram of axoneme structure in flagella of *Polytrichum juniperinum*. Two ribbons of five strands each (H-1, H-2) comprise the double helix. The sheath of the central fiber complex is also represented (SH).
(After Paolillo, 1967).

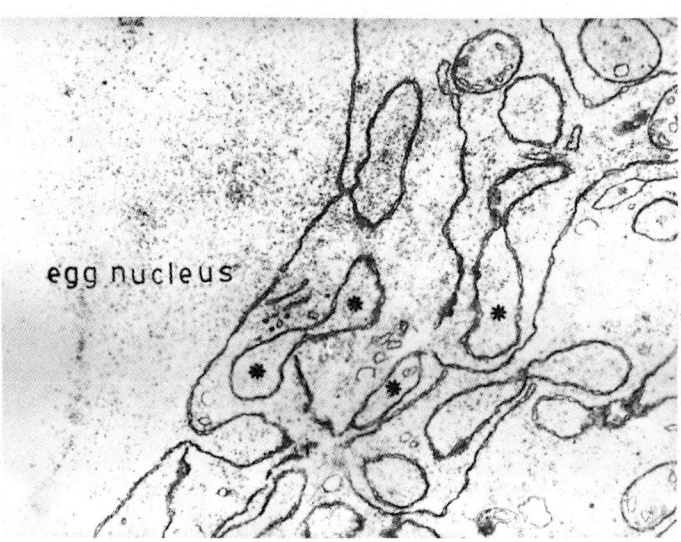

Plate 7.10. Structure of egg in *Marchantia polymorpha*. Sac-like evaginations attached to the egg nucleus. The asterisks indicate profiles which are probably of evaginations connected with the nucleus at another level.
(After Bell, 1975).

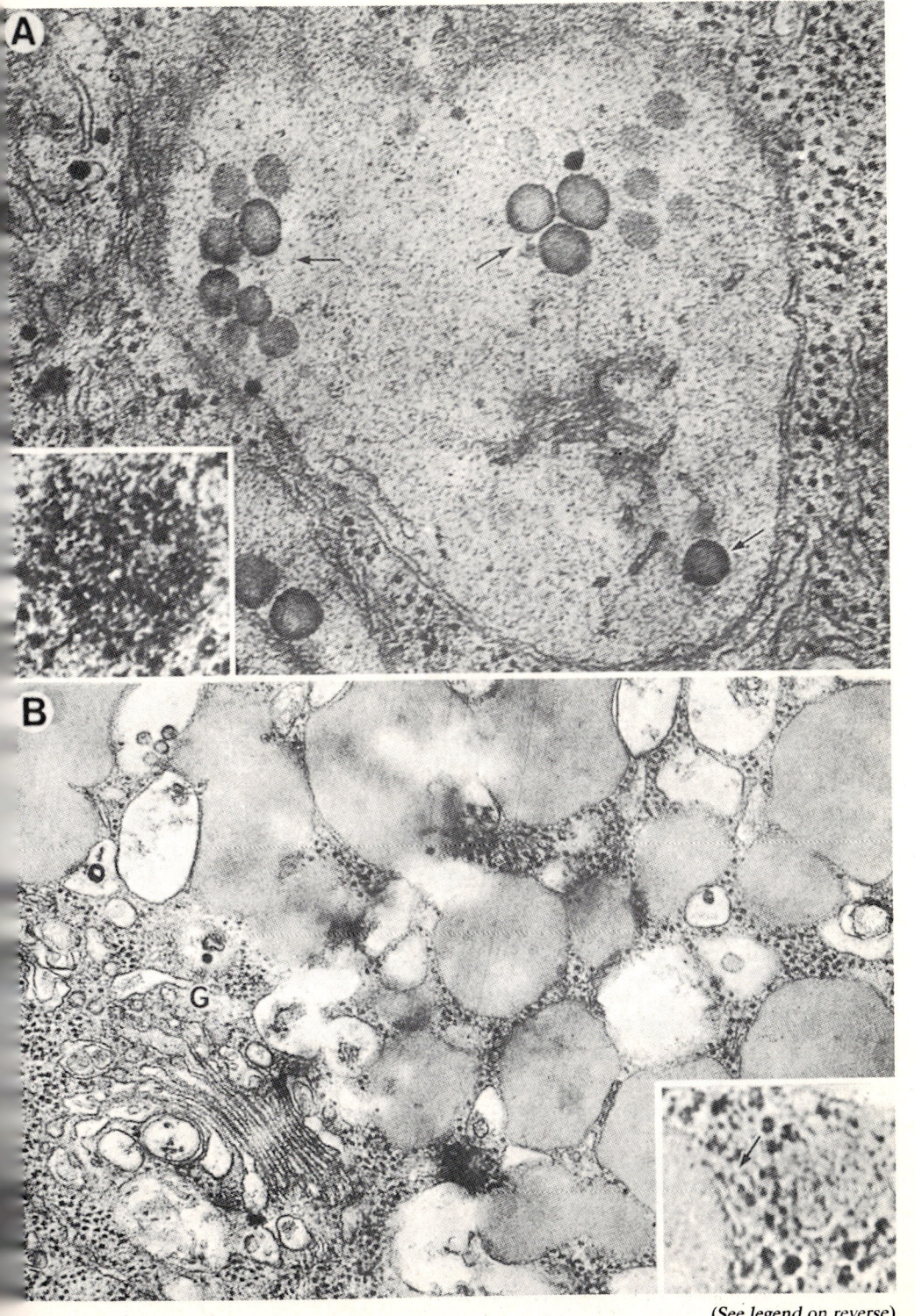

Plate 7.11 A, B. Oogenesis in *Physcomitrium coorgense*. **A.** Plastid in a maturing egg cell. Internal lamellae are few and fragmentary. Droplets (arrows) are conspicuously more frequent than earlier in oogenesis, and are more osmiophilic. Inset. Small granular body representative of those scattered in the cytoplasm of maturing egg. **B.** Part of a cluster of vesicles towards the periphery of a maturing egg. A golgi body (G) lies adjacent to the cluster. Inset. The boundary of a vesicle containing osmiophilic material showing evidence of continuity with endoplasmic reticulum (arrow).
(After Lal & Bell, 1977)

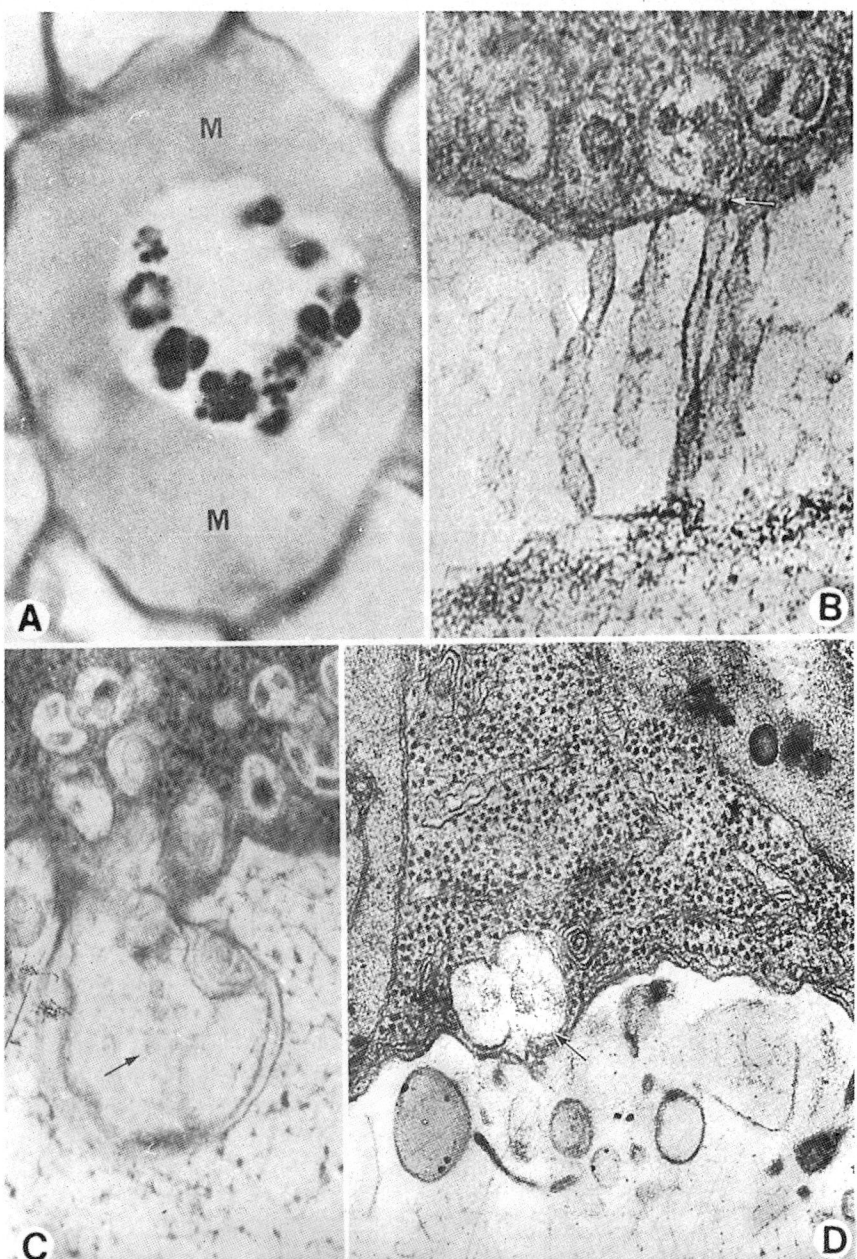

Plate 7.12 A-D. Oogenesis in *Physcomitrium coorgense*. **A.** Tangential section (striking only the cytoplasm) of a maturing egg surrounded by a large amount of mucilage (M). The 1.5 μm resin section has been subjected to the PAS reaction, and subsequently photographed in green light. Red-stained material appears dark. The deeply staining nodular bodies in the cytoplasm of the egg correspond with the clusters of vesicles. **B.** Part of the periphery of a maturing egg showing vesicles close to the plasmalemma, and tubular and fibrillar configurations in the surrounding mucilage. The arrow indicates a tubule appearing to come from a vesicle. **C.** Large vesicle containing fibrillar material (arrow) apparently being ejected from the cytoplasm. **D.** Vesicle containing fibrillar material (arrow) possibly about to discharge at the surface of the protoplast, together with apparently intact vesicles in the mucilage.
(After Lal & Bell, 1977).

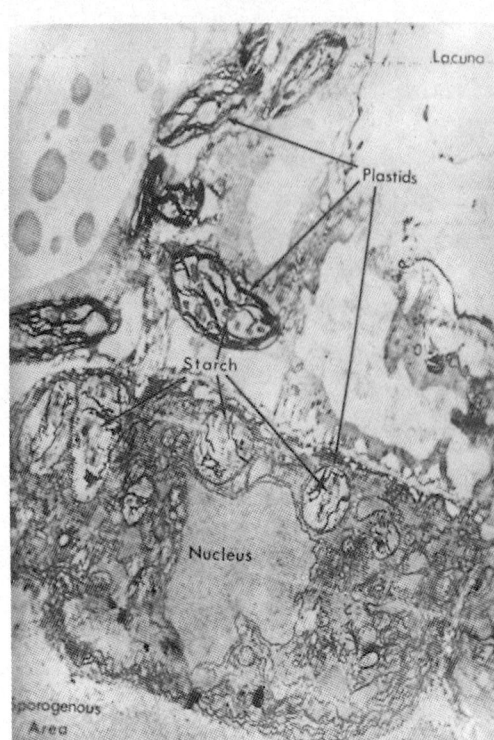

Plate 7.13. Sporogenesis in *Polytrichum commune*. Transverse section through tapetum. The upper cell in the micrograph is adjacent to the lacuna, the lower cell is adjacent to the sporogenous area. (After Paolillo, 1964).

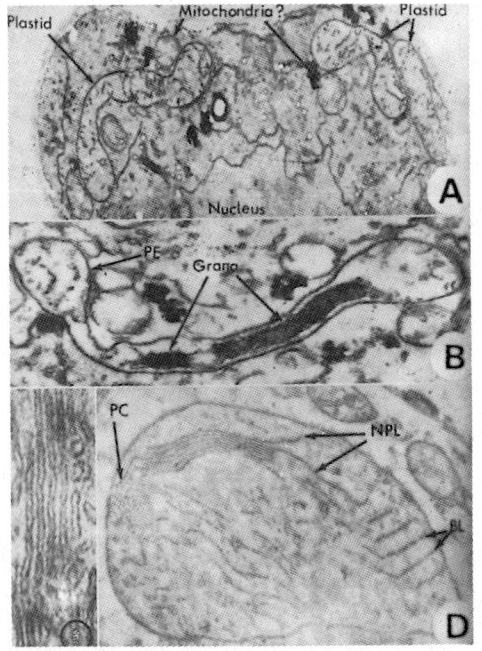

Plate 7.14 A-D. Sporogenesis in *Polytrichum commune*. **A.** Two lobes of a sporocyte. Certain of the profiles might represent mitochondria. The plastids are almost certainly represented by more than one fragment each. **B.** Profile of a sporocyte plastid, showing two stacks of discs (Grana). **C.** Portion of a sporocyte granum showing the wide interdisc spacing. **D.** Profile of a spore plastid. BL—beaded lamellae; NPL—non-parallel lamellae; PC—plastid centre; PE—plastid envelope. (After Paolillo, 1964).

sporocytes. As they approach maturity the walls between these cells lyse. In the mean time the protoplast of each cell recedes from the wall and secretes a thick polysaccharide coat around itself. The sporocyte with a fresh wall is called the spore mother cell which undergoes meiotic division to produce a tetrad of spores.

A perusal of the literature reveals that the work on sporogenesis in bryophytes has centered around five broad facets, viz. spore sac or tapetum; fluctuations in plastid number during sporogenesis; structural changes in plastids, cytoplasm, and other organelles during sporogenesis; meiosis in the sporogenous tissue; and the mode of spore wall formation.

SPORE SAC OR TAPETUM

The cell layers limiting the archesporium on outer and inner sides are characterized by the richness of cell contents, and it is clear that they provide nutrition to the archesporium and the sporocytes (Goebel, 1905), and function like the anther tapetum in angiosperms. Chevallier et al. (1977) and Jensen and Hulbary (1978) have in fact designated the spore sac as tapetum. The ultrastructural features of the cells of the spore sac justify the new nomenclature.

The spore sac cells show a large nucleus, numerous vesicles (presumably of dictyosomal origin), well developed endoplasmic reticulum scattered throughout the cytoplasm or in large isolated stacks (Jensen & Hulbary, 1978). These features strongly indicate the secretory nature of the spore sac. Jensen and Hulbary (1978) suggested that the 'moss tapetum' is involved in the production of carbohydrates that diffuse into the spore mother cells and are then converted into the starch within the plastid stroma. Paolillo (1964) observed that in *Polytrichum commune*, tapetum is typically two cell layers thick, and a distinct differentiation of these layers is noticed. The layer that adjoins the lacuna contains chloroplasts that have well developed grana and small starch grains. On the other hand, the layer that adjoins the spores or spore mother cells contains plastids that have larger starch grains, smaller grana, and more extensive stroma (Plate 7.13).

It has also been reported that the tapetal secretion has a definite role to play in the ornamentation of the developing spores (Brown & Lemmon, 1980a; Geneves, 1972; Mueller, 1974; Seabury, 1975). Mueller (1974) observed that the cells of spore sac in *Fissidens limbatus* are quite enlarged and have extremely dense cytoplasm. At the time of perine formation, the cells of the spore sac show profuse layering of endoplasmic reticulum, often six or more layers along the walls adjacent to the spore mass. Jarvis (1974) and Neidhart (1975) demonstrated that the perine material is derived from an external source.

FLUCTUATION IN PLASTID NUMBER

Fluctuation in the number of plastids during sporogenesis was observed as early as 1911 by Sapĕhin. Only one plastid remains at the spore mother cell stage and the number increases to four just before meiosis. These are subsequently distributed to the four young spores, where they multiply till the attainment of specific chloroplast number of the species (Butterfass, 1979). So far, monoplastidy in sporocytes of mosses has been reported in *Hypnum rusciforme* (Genevès, 1966); *Atrichum undulatum, Eurhynchium striatum, Mnium cuspidatum, M. undulatum* (Eymé & Suire, 1969, 1971); *Polytrichum commune* (Paolillo, 1964, 1969); *Funaria hygrometrica* (Neidhart, 1975); *Fissidens limbatus* (Mueller, 1974); *Aulacomnium heterostrichum, Bartramia pomiformis, Mnium cuspidatum, M. medium, M. rostratum, Timmia megapolitana* (Jensen & Hulbary, 1978), and *Ditrichum pallidum* (Brown & Lemmon, 1980b). Two possible mechanisms have been postulated for monoplastidy in the spore mother cells. This condition may arise due to fusion of all the plastids into a single plastid (Jensen & Hulbary, 1978), or the number of plastids per cell gets reduced through continued mitotic divisions of the sporogenous cells which are not accompanied by divisions of the plastids (Mueller, 1974).

STRUCTURAL CHANGES IN PLASTIDS, CYTOPLASM, AND OTHER ORGANELLES

The plastids of sporogenous cells are differentiated from those of vegetative cells by the lesser complexity of their lamellar systems and by a diversity of grana. The discs, or compartments, within the sporocyte plastids are wider than are the grana or compartments of the chloroplasts in other cells of the capsule. Plastid profiles in the sporocytes of *Polytrichum commune* indicate that they are flattened, lobed structures, and may be devoid of lamellated structure and starch (Plate 7.14A). Like the grana of chloroplasts, the lamellated structures within the sporocyte plastids are composed of parallel discs or compartments (Plate 7.14B,C). In the spores of a tetrad, the 50 to 60 A° interdisc spacing of the lamellated structures persists (Plate 7.14D), but the plastid profiles of the spores are different from those of the sporocytes. A plastid centre (PC) is adjacent to a granum. The lamellae of the granum are prolonged in non-parallel fashion to the right (NPL). Beaded lamellae (BL) are also present. These lamellae might be another aspect of those extending from the granum (Paolillo, 1964).

On the other hand, the sporocyte possesses a single, elongated, flattened, cup-shaped plastid in *Fissidens limbatus* (Mueller, 1974). It has a well developed grana fretwork system, many starch grains, and lipid droplets. However, these plastids become more flattened as the sporogenous cells divide and are seen placed around the interphase nucleus. Concomitant with this is the reduction in their lamellar system and starch grains. The monoplastid stage is short-lived, and immediately the plastid divides twice to produce four plastids. Each of these newly formed plastids then migrates to one of the four poles. At this stage they show a progressive simplification in their grana fretwork system which eventually becomes reduced to just a single thylakoid immediately prior to meiosis. Just after meiosis the plastids do not show any appreciable change in their pattern of organization from that seen in the proceeding sporocytes. However, in a young spore the plastids reveal distinct lobes indicating fission. In addition, there has been a concomitant formation of new membranes within the stroma. Re-establishment of grana fretwork system into functional chloroplasts with fully developed photosynthetic machinery is achieved in the mature spores. Jensen and Hulbary (1978) observed as many as 15 plastids in cross sections of spores and these still appeared to be dividing. The final number of plastids is obviously very large in each spore. The formation of proplastids and their maturation into chloroplasts during spore development is the most striking feature of mosses. The plastids become oriented around the nucleus and this configuration is very distinct in mature spores of many mosses.

During sporogenesis, the nucleoplasm is heterogenous in the sporogenous tissue and young spores, but takes a beaded form in the sporocyte before meiosis. The nucleolus is large in the sporogenous cells during sporocyte formation, it becomes less prominent in the sporocyte and in early spore development, and regains its original proportions in the mature spore. There are numerous vesicles in the sporocyte and immature spores, which become fewer as spores mature. In mature spores mitochondria are oval to slightly elongate and are generally located peripherally. Spores lack extensive vacuolation (Brown & Lemmon, 1980b). There is a dilation of the mitochondrial cristae around midprophase of meiosis (Eymé & Suire, 1971). Although these structural states have so far not been correlated with their metabolic activities, they are probably indicative of some functional change (Opik, 1974).

Sterile cells (nutritive cells & elaters) are very common in the sporogonia of liverworts. Elaters have spiral thickenings in the wall, whereas nutritive cells lack them. The first indication of the ultrastructural differentiation of intracapsular cells from undifferentiated meristematic tissue during sporogenesis is also of interest. In the spore mother cells lipids accumulate in the spherosomes, whereas vacuolation and plastids with starch are characteristic of nutritive cells in the spore sac of *Sphaerocarpos* (Siler, 1934). Within each spore mother cell, cytomictic channels interconnect the sporocytes through the infurrowing walls in *Riccardia pinguis* (Horner et al., 1966).

MEIOSIS

The reduction division in mosses is characterized by several unique features. Sufficient information has been contributed by different investigators by light as well as ultrastructural studies (Allen, 1916; Brown & Lemmon, 1980b, c, 1982; Dill, 1964; Lambert, 1974; Paolillo, 1969; Weier, 1931).

Brown and Lemmon (1980b) described six sub-stages of meiotic prophase in *Ditrichum pallidum*.

Stage 1: Prior to meiosis each sporocyte produces a thick layer of fibrillar polysaccharide, the sporocyte wall, between the protoplast and archesporial cell wall, which persists throughout spore formation. Sporocytes are released into the spore chamber by lysis of the archesporial cell walls. Ribosomes are distributed throughout the cytoplasm and are particularly abundant around the oil bodies.

Stage 2: This stage, which has many cytoplasmic features similar to the preceding stage, is distinguished by the initial appearance of synaptinemal complexes.

Stage 3: As prophase proceeds, nuclei containing well developed synaptinemal complexes migrate to sporocyte periphery and assume the acentric "Bouquet" appearance. The nucleus contains very fine, tightly packed unpaired chromosomes, which later assume a typical pachytene morphology.

Stage 4: Pachytene nuclei return to a more central position.

Stage 5: Nuclei enter a diffuse stage with chromatin relaxed; cytoplasmic furrows produce lobes in tetrahedral arrangement; microtubules appear and proliferate around the plastid in each of the four lobes and finally ensheath the nucleus. Similarly, Allen (1916) also observed that the spore mother cell of *Catharinea* becomes four-lobed prior to meiosis. This process was termed as precocious cytoplasmic lobing. Allen also reported that each lobe of the spore mother cell has one chloroplast.

Stage 6: The nuclear membrane is dissociated, chromatin recondenses into chromosomes, and kinetochore-microtubule attachment appears.

The diplotene phase starts with chiasmata formation in the bivalents.

In *Hypnum circinale* the diplonema stage is quite prolonged and may persist for a week or more (Dill, 1964). Though it is very difficult to get even one or two bivalents separated from the chromosomal mass of six bivalents, in rare spore mother cells all the six bivalents can be seen.

Working on three species of mosses viz. *Hypnum circinale, Brachythecium frizidium,* and *Calopodium crispifolium* Dill (1964) reported for the first time the 'dictyotene stage' in the meiotic prophase, similar to the one commonly observed during prophase I of oogenesis in many animals. He suggested that since this stage was present in members of three different families, it might be of wider occurrence among mosses. The dictyotene stage is characterized by an elongation of the paired chromosomes after diplotene. The chromosomes lose their discreteness and the nuclear contents look like those of the interphase stage. However, the fine threads of chromatin and chiasmata can be discerned with the help of phase contrast microscope. The chromatin body also stains intensely. It, therefore, appears that the basic diplotene structure is retained during dictyotene. Dill (1964) also reported that the dictyotene stage may last for two or three days after which meiosis proceeds normally. However, the subsequent meiotic divisional stages in all the spore mother cells are not synchronous. The bivalents get condensed rapidly. Initially the chromatin is extremely diffuse, but later becomes dense and contracted. During diakinesis the clumped chromosomes are easily distinguishable.

Events of meiosis after prophase have been reported in *Ditrichum pallidum* by Brown and Lemmon (1980c). During metaphase I the bivalents are distributed along the equator of an open spindle consisting of continuous microtubules. Microtubules are attached at kinetochores which appear similar in structure to those reported in sporocytes of *Mnium* by Lambert (1974, 1977). The distribution of both plastids and mitochondria is associated with the lobing pattern in the cytoplasm of metaphase I sporocyte. Most notable is the positioning of one of the four plastids into each of the cytoplasmic lobes. Mitochondria are concentrated in the cytoplasm along the cleavage furrows. Structurally simple spindle poles are located in the cytoplasm within opposite cleavage furrows between pairs of cytoplasmic lobes into which telophase II nuclei will be distributed. No discrete structure is observed to be associated with lobing or with spindle microtubules orientation. After a short intrameiotic interphase, meiosis II occurs in the undivided cytoplasm with spindles perpendicular to each other. Each of the poles is located near the plastid within each of the four lobes. Intersporal septae are formed by coalescence of vesicles simultaneously along cleavage planes predetermined by precocious lobing. Fused vesicle membranes contribute to the formation of plasma membrane on proximal surface of the spores, and the fibrillar contents of the vesicles, indistinguishable from the polysaccharide fibrils of the sporocyte wall, separate the young spores within the tetrad. The spores are invariably arranged tetrahedrally as reported in *Ditrichum pallidum* (Brown & Lemmon, 1980c), *Catharinea* (Allen, 1916), and *Polytrichum commune* (Weier, 1931). This arrangement is correlated with the lobing of cytoplasm occurring precociously.

Ripetsky and Matasov (1975) observed postmeiotic changes in *Physcomitrella patens*, *Desmatodon randii*, *Funaria hygrometrica*, *Physcomitrium pyriforme*, and *Pottia intermedia*, during sporogenesis with special reference to spore tetrad formation. Intensification of aniline blue reaction under UV to callose was noticed in the tetrads of young moss spores during their development. The callose polyglucosan is localized in the middle of tetrads as thin, net-like wall between the protoplasts. Local accumulation is also observed in the protoplast of certain species. Occasionally accumulation of the newly synthesized starch was also observed in the young spores before they had fully developed chloroplasts.

SPORE WALL FORMATION

With the gradual softening of walls of tetrads, the spores become somewhat rounded and start forming their own walls. At first this wall is quite thin but after complete dissolution of the old wall, the spores become free and ornamented.

The ultrastructural details of spore wall formation have so far been investigated in only a few mosses like *Hypnum rusciforme* (Geneves, 1972), *Fissidens limbatus* (Mueller, 1974), *Funaria hygrometrica* (Jarvis, 1974;

Neidhart, 1975), *Archidium, Bruchia, Ephemerum* (McClymont & Larson, 1964), and *Ditrichum pallidum* (Brown & Lemmon, 1980a).

In *Fissidens limbatus* the exine is first apparent during spore enlargement and chromatin dispersion and is made of a homogenous layer of electron-dense material. The enlarged spore is later coated by a more densely staining material, the perine, which is deposited in the form of irregular electron-dense masses outside the exine. After perine formation, the large number of vesicles present at this time indicates advent of intine formation, which is marked by a thin layer of deposits inside the exine. Initially, it is less electron-dense than the exine and consists of a coarse fibrillar material. The exine and intine are delineated in the mature spore by a darker (outer) ring of intine, which forms along the interface of the exine and more recently deposited (inner) material of the intine.

Ultrastructural investigations have revealed a well marked cytoplasmic polarity during the wall and aperture ontogeny in spores of *Ditrichum pallidum* (Plates 7.15A-C; 7.16A-D; Brown & Lemmon, 1980a). An extensive accumulation of microtubules underlies the entire distal spore surface where exine deposition occurs shortly after cytokinesis. These microtubules appear to be focussed on the plastid. Very shortly shallowly lobed distal spore surface becomes thickened by exine deposition, whereas the equally irregular proximal face remains naked and is without underlying microtubules. Exine extends outward towards the distal margins and thickens considerably before gradually extending to the proximal surface.

As the lamellate exine thickens and extends to the proximal surface, the plastids and associated nucleus migrate to the proximal surface, where an elaborate system of microtubules-involved aperture development is generated. The aperture so formed in the spore, is a complex localized modification of these layers on the proximal surface. It consists of a pore containing a fibrillar material and is surrounded by a thin annulus. The aperture plug is continuous with the intine. The separating layer is distinct between the annulus and aperture plug, and in this layer the exine lamellae are not consolidated as are those of the exine proper. The uneven thickness of the wall layers has been reported to be due to the spore polarity. This polarity is established as early as during metaphase of meiosis. The cytoskeleton, which interconnects plastids and also encloses the nucleus, contributes to the establishment of sporocyte polarity (Brown & Lemmon, 1982).

There is considerable interest in finding out the exact source of origin of the perine. Some workers have found its origin to be exinous and others have reported it to be non-exinous. McClymont and Larson (1964) noticed that the ornamentation of the spore wall was totally exinous in origin in *Archidium, Bruchia,* and *Ephemerum,* whereas non-exinous origin has been observed in *Polytrichum, Encalypta, Physcomitrium, Phascum* (McClymont & Larson, 1964), *Ceratodon* (Valanne, 1971), *Dicranum* (Valanne et al., 1976), *Bryum* (Nurit, 1974), *Cinclidium artichum, C. stygium, C. subrotundom* (Mogensen, 1978), *Ditrichum pallidum* (Brown & Lemmon, 1980a), *Eurhynchium*

(Neidhart, unpublished), *Fissidens limbatus* (Mueller, 1974), *Grimmia*, *Ulota*, and *Hylocomium* (Gullvåg, 1967). In general it would appear that most mosses have a non-exinous perine. However, in some mosses the exine only forms the extreme basal part of the sculpturing elements. The non-exinous origin is supported by cytohistochemical studies which have also revealed that the exine and perine are quite different from each other, although both are resistant to acetolysis. In non-acetolysed spores the perine, unlike the exine, reacts with silver pectinate after periodic treatment coupled with thiosemicarbazide (Thiéry, 1967). With uranyl acetate staining, the outer part of perine appears electron-opaque unlike electron transparent nature of the corresponding part of the exine.

The occurrence of dense globules throughout the spore mass at the time of formation of the sculptured perine in *Fissidens* is considered to be a good evidence for an external source of this layer (Mueller, 1974). In contrast to other spore wall layers, the perine material is formed by the sporocyte and not by the spore.

SPOROPHYTE-GAMETOPHYTE JUNCTION

Electron microscopic studies on mosses have revealed the presence of extensive wall labyrinths in the epidermal cells of the foot. The study of the placental region in bryophytes has contributed significantly not only to our understanding of the nutrition of the sporophyte, but also toward the formulation of the present concept of 'Transfer cells'. Earlier anatomical investigations on the sporophyte-gametophyte junction of bryophytes gave important clues to the recognition of the foot as an absorptive organ which serves as a haustorium. In *Diphyscium* and *Buxbaumia* the foot is provided with special tubular outgrowths which are chambered by cross-walls and may be so far branched that they look like rhizoids. With regard to the absorption of water the sporogonium in most Musci depends permanently on the gametophyte, yet rarely there are forms like *Eriopus remotifolius* which are able to take up water through the abundant hair-like outgrowths of the stalk of the sporogonium (Goebel, 1905).

In *Polytrichum formosum* labyrinths in the outer walls of peripheral cells in the foot originate from the lamellated cell wall., There is a correlation between the lysis of the cell wall and the initiation of new ingrowths. These are seen to increase with the age of the growing sporophyte, and to fuse at their extremities to form a complex labyrinth enclosing pockets of cytoplasm (Maier, 1967). Similar wall labyrinths have also been observed in the cells of the sporophyte-gametophyte junction of *Mnium cuspidatum* (Eymé & Suire, 1967). In this moss extensive cell wall projections have been observed at two sites, viz., the foot and the surrounding vaginular cells, as also in *Funaria* (Wiencke & Schulz, 1975), *Dawsonia*, *Dendroligotrichum* (Hébant, 1975), and the liverwort *Sphaerocarpos donnellii* (Kelley, 1969). In a species of

Sphagnum the transfer cells have been reported to be altogether absent (Maier, see Gunning & Pate, 1974).

These studies have highlighted two important points: (1) the characteristic wall protuberances (ingrowths) are responsible for greatly increasing the surface area of the plasma membranes of the cells possessing them. The presence of such wall projections on both sides of the sporophyte-gametophyte junction serves to further increase the surface area available, not only for absorption but also for translocation, thereby efficiently increasing the potential rate of solute movement between plant cells which may lack plasmodesmata, and (2) association of a large number of mitochondria with the wall labyrinth reflects a high energy requirement for absorption and movement of nutrients along these cells, thus suggesting the existence of active transport mechanisms across the contiguous walls of the sporophyte and the gametophyte. The term 'Transfer cell' is coined in a wider context for those cells which have wall labyrinth (Gunning, 1977; Gunning & Pate, 1969, 1974; Gunning et al., 1968; Luttge & Higinbotham, 1979; Pate & Gunning, 1972). The wall ingrowths differentiate just at the time of commencement of intensive transport, and the extent of their development is commensurate with the magnitude of conspicuous endoplasmic reticulum in the transfer cells which may regulate the exchange of solutes across the plasma membrane. Electron microscopic studies made on *Polytrichum piliferum* also highlighted the presence of wall labyrinth in the haustorium and the surrounding tissue of the gametophyte (Maier & Maier, 1972). In this moss the role of foot epidermis has been correlated with the absorptive epithelium. It has also been shown that the growth of the wall labyrinth occurs in two phases. The first phase includes the origin of smooth, finger-like protuberances emerging individually from the cell wall proper. During the second phase the smooth-surfaced walls become more and more rough.

In *Funaria hygrometrica* the basal part of the foot shows less developed wall ingrowths as compared to the extensively developed cell wall labyrinth in the middle part (Wiencke & Schulz, 1975). The structure of plastids and the degree of vacuolation are markedly different in the cells constituting these two zones. Whereas the cells in the basal zone show a large central vacuole and dumb-bell-shaped plastids with poorly developed grana fretwork system, the cells in the middle zone are characterized by very small vacuoles and plastids with well developed thylakoids and occasional starch.

Studies of the sporophyte-gametophyte junction in *Funaria hygrometrica* have revealed that the development of transfer cell occurs in three stages:

(1) Primary stage: It coincides with the time of detachment of calyptra from the ripened archegonium and involves the formation of wall ingrowths.

(2) Secondary stage: It is characterized by the deposition of amorphous inclusions in the wall labyrinth of the transfer cells.

(3) Tertiary stage: As the sporophyte ripens de-differentiation of transfer cell wall labyrinth takes place with the formation of heavily incrusted, outer cell wall (Browning & Gunning, 1979).

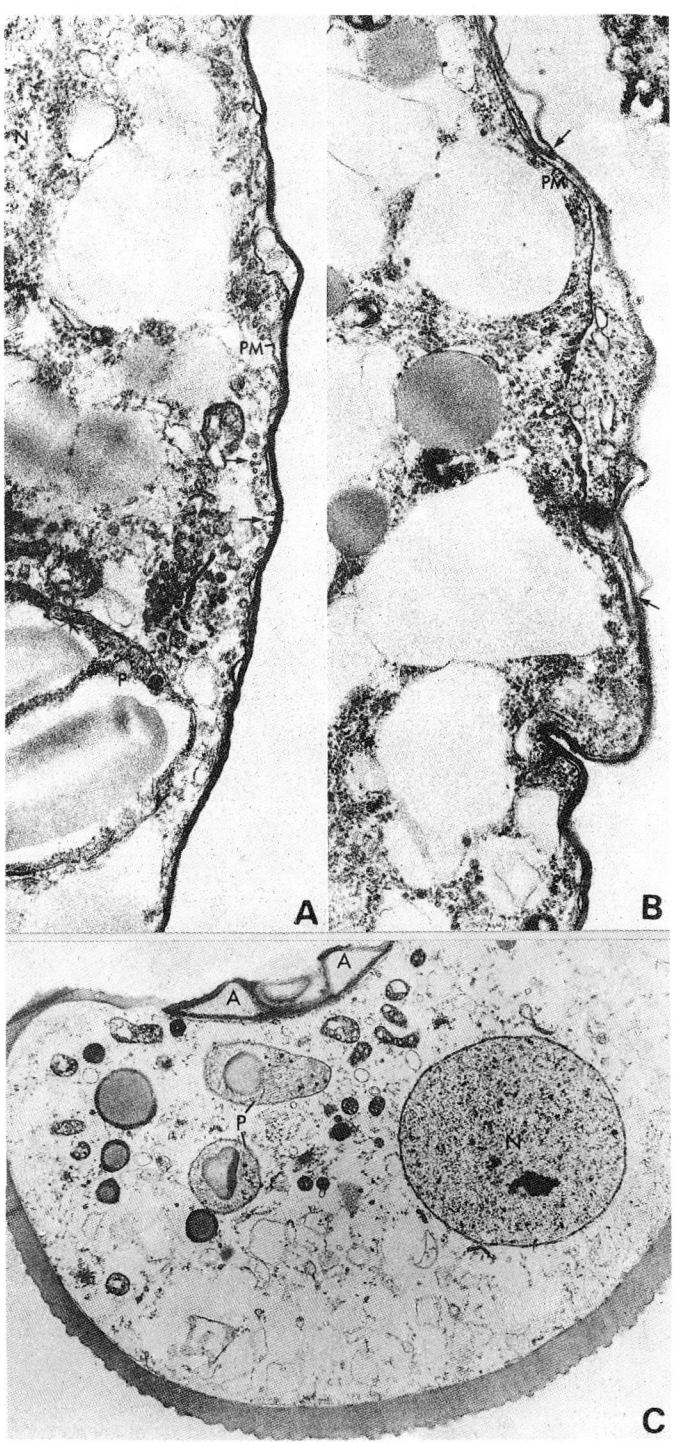

(See legend on reverse)

Plate 7.15 A-C. Sporogenesis in *Ditrichum pallidum*. **A, B.** Early aperture formation in spore. **A.** Aperture is initiated on proximal surface in the vicinity of nucleus and plastid. Note the underlying microtubules (arrows). **B.** Early expansion of aperture. The plasma membrane is displaced from the exine and the intervening space (between arrows) is filled with vesicles and a granular substance. Note the separation of the exine lamellae at the aperture margin (arrows). **C.** Spore with thickened papillate exine on distal surface and aperture on proximal side.

A—annulus; N—nucleus; P—plastid; PM—plasma membrane.
(After Brown & Lemmon, 1980a).

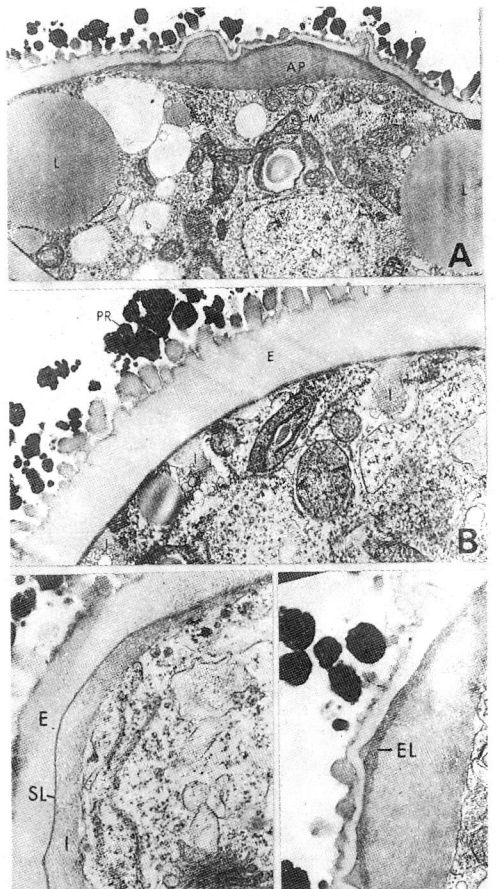

Plate 7.16 A-D. Sporogenesis in *Ditrichum pallidum*. **A, B.** Nearly mature spores. **A.** Spore viewed in linear median section through aperture. **B.** Portion of a spore with pockets of developing intine. **C, D.** Details of mature spores. **C.** Endoplasmic reticulum, dictyosomes, and vesicles in association with developing fibrillar intine. The four wall layers are visible in this section. The separating layer is distinct between intine and exine. **D.** Aperture region in near median section. Exine lamellae are distinct in the separating layer only in the region of the aperture.

AP—aperture pore; E—exine; EL—exine lamella; I—intine; L—lipid body; M—mitochondria; PR—perine; SL—separating layer.

(After Brown & Lemmon, 1980a).

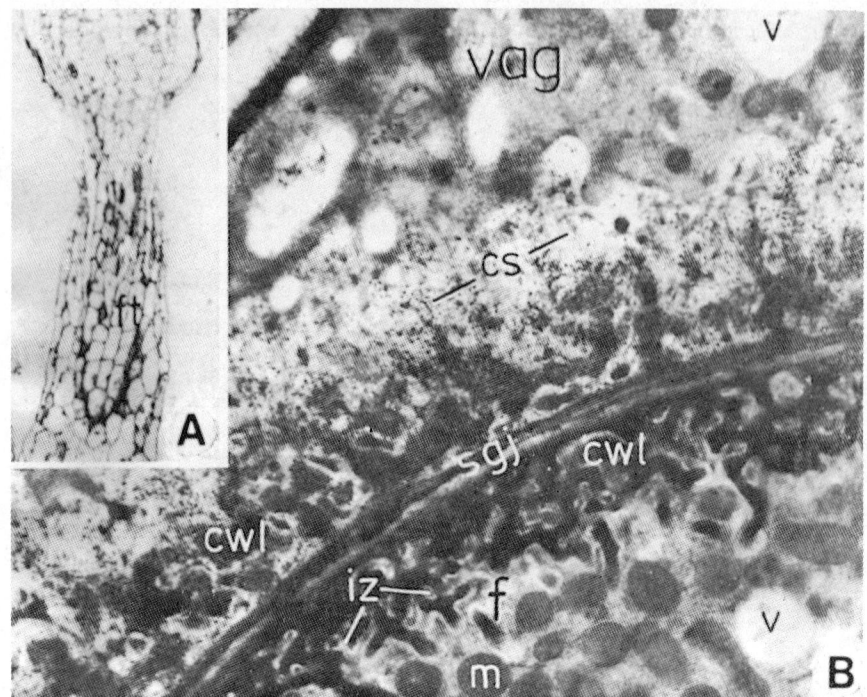

Plate 7.17 A, B. Sporophyte-gametophyte junction of *Physcomitrium cyathicarpum*.
A. Light micrograph of longitudinal section of haustorial region showing the disposition of foot (ft) and the surrounding vaginula (vag). **B.** Electron micrograph of a portion of the sporophyte-gametophyte junction (sgj) at high magnification to show the distension of the clear space (cs) and larger cell wall labyrinth (cwl) in the vaginula cells.
f—foot cell; iz—interfacial zone; m—mitochondrion; v—vacuole.
(After Lal & Chauhan, 1981).

The ultrastructural studies on *Physcomitrium cyathicarpum* have revealed the presence of extensive wall labyrinths in the peripheral foot cells as well as in the cells of the vaginula abutting the haustorium. These cells show ultrastructure typical of transfer cells (Plate 7.17A,B; Lal & Chauhan, 1981).

Recently, Chauhan and Lal (1984) have reported that in *P. cyathicarpum* haustorial tip has a well developed wall-membrane apparatus and typical transfer cells. It remains persistent and healthy throughout the life of the sporophyte. The haustorial tip consists of a quadrate of four broad cells which possess starch-filled plastids and appears to be involved in apoplastic transport. The authors suggest that the tip cells may even play an important role in secretion. Cytochemical studies of the haustorial foot also reveal a cluster of PAS-positive bodies, and a high concentration of nucleic acids and total proteins in the cells of the terminal quadrate.

SETA

In the recent past significant ultrastructural studies have been made on the structural and functional aspects of the conducting tissues of seta. Among the taxa studied are: *Mnium orthorhynchium* (Bassi & Favali, 1973), *Polytrichum commune* (Favali & Bassi, 1973), *Tortula muralis* (Favali & Gianni, 1974), *Pogonatum aloides* (Favali & Gianni, 1975), and *Leucophanes candidum* (Favali & Bassi, 1978).

In *Polytrichum commune* the seta has three main regions: (1) cortex, (2) parenchymatic ring, and (3) a central core. The cells of the cortical zone contain starch-filled plastids with hardly recognizable thylakoid membranes similar to chloroamyloplasts. The cells are pierced by a few plasmodesmata. In the parenchymatic or middle zone the cells are pierced by a greater number of plasmodesmata. In this region, two zones can be distinguished: an outer multilayered zone with wide intercellular spaces, and an inner zone made of a single layer of cells forming a continuous ring around the central core. In the central core as well, two different regions are visible: (1) an outer sheath of tightly packed cells with thin walls and cytoplasm rich in mitochondria, and (2) an inner cylinder with two types of cells: empty cells (hydroids) and cells containing cytoplasmic remnants (leptoids) which are interconnected by highly perforated parenchyma cells. All cells of the central core function as a conducting system (Favali & Bassi, 1973). Similar structures have also been reported in *Pogonatum aloides* (Favali & Gianni, 1975), but they are different from those observed in the seta of *Tortula muralis* (Favali & Gianni, 1974). In *Tortula* the seta shows a central region made of cells with very thin walls and narrow lumen.

In the mature seta of *Dawsonia superba* the central strand is anchored within an inner lacunar cortex (Hebant, 1975). This strand consists of a solid core of hydroids, surrounded by cells which by their general appearance seem to be the counterpart of the gametophytic leptoids. The hydroids are elongated cells devoid of living protoplasm at maturity. The leptoids are elongated elements with more or less oblique end walls perforated by numerous plasmodesmata. The central strand of the seta penetrates the lower portion of the capsule and terminates just below the columella. Conduction in the rest of the capsule is performed by parenchyma cells, and not by highly specialized tissues (viz. hydrom & leptom).

During elongation growth of *Pellia* setae, the thinning of cell walls (a consequence of their expansion) is compensated in part by formation of new wall material. Newly deposited microfibrils are arranged more or less circularly in the wall, i.e. normal to the direction of elongation. During wall extension the microfibrils reorient and become parallel with the long axis of the cell (Schnepf & Deichgraber, 1979).

HISTOENZYMOLOGICAL STUDIES

LOCALIZATION OF ENZYMES IN THE HAUSTORIAL FOOT

In addition to the ultrastructural investigations of the haustorial foot, interesting studies have also been made on the activities of some enzymes in the zone of transfer between the gametophyte and the sporophyte.

Maier and Maier (1972) investigated the distribution of two groups of membrane-bound enzymes, viz. Mg^{2+}-activated adenosine triphosphatase and beta-glycerophosphatase (acid phosphatase) in the haustorial foot and the surrounding gametophytic tissue of *Polytrichum piliferum*. Their observations revealed that only the cells with extensive wall labyrinths were the sites of intense enzyme activity. Precipitates of the reaction products of both acid phosphatase and Mg^{2+} activated adenosine triphosphatase were localized in the epidermis of the haustorium. The activities of the two enzymes were observed to be proportional to the state of differentiation of the labyrinth, thereby stressing the role of foot epidermis as an absorptive epithelium. Hébant and Suire (1974) demonstrated the activities of cytochrome oxidase, succinic dehydrogenase, peroxidase, and acid phosphatase in the peripheral cells of the foot as well as in the abutting cells of the vaginula (gametophyte) in *Dawsonia papuana*, *Dicranum scoparium*, *Polytrichum commune*, and *Scapania undulata*. Their study indicated that the external cells were involved in intensive transport of solutes, and showed strongly increased respiratory and phosphatase activities. They also recorded a variability in the intensity of reaction products in different zones of the haustoria of the various genera and concluded that all these enzyme activities in the foot decreased sharply at the time of spore dispersal, and upon withering of the seta. Beta-glycerophosphatases have also been reported in the conducting tissues of bryophytes (Hébant, 1973, 1977). These studies support the view that phosphatases may intervene in the storage and mobilization of organic substances, especially the carbohydrates. Thomas et al. (1978) reported that *Phaeoceros laevis (Anthoceros laevis)* shows intense acid phosphatase activities in the transfer cells of the foot.

Chauhan (1981) observed high density of the reaction products for four hydrolases viz. alkaline phosphatase, acid phosphatase, esterase, and Mg^{2+} and Ca^{2+}-activated adenosine triphosphatases in the sporophyte-gametophyte junction of *Physcomitrium cyathicarpum*. The reaction products for these hydrolases were observed in the peripheral cells of the foot, and the cells of the vaginula abutting the haustorium. These findings suggest that both of these regions are the seat of intense metabolic activity.

REFERENCES

ALLEN, C.E., *Am. J. Bot.*, 8, 456.(1916).
BASSI, M. and M.A. FAVALI, *Nova Hedwigia*, 24, 337 (1973).
BARBIER, C., *C.r. hebd. Séanc. Acad. Sci., Paris (Sér. D)*, 274, 3222 (1972).
BARBIER, C., *Bull. Soc. bot. Fr. Lett.*, 126, 507 (1979).
BELL, P.R., "Nucleocytoplasmic interaction in the maturing egg of *Marchantia polymorpha*." In H.Y. Mohan Ram, J.J. Shah. and C.K. Shah, Eds, *Form, Structure, and Function in Plants*. Sarita Prakashan, India, 1975.
BONNOT, E.J., *C.r. hebd. Séanc. Acad. Sci., Paris (Ser. D)*, 265, 958 (1967).
BROWN, R.C. and B.E. LEMMON, *Am. J. Bot.* 67, 918 (1980a)
BROWN, R.C. and B.E. LEMMON, *Bryologist*, 83, 137 (1980b).
BROWN, R.C. and B.E. LEMMON, *Bryologist*, 83, 153 (1980c).
BROWN, R.C. and B.E. LEMMON, *Protoplasma*, 110, 23 (1982).
BROWN, R.C., B.E. LEMMON, and Z.B. CAROTHERS, *Can. J. Bot.*, 60, 2394 (1982).
BROWNING, A.J. and B.E.S. GUNNING, *J. exp. Bot.*, 30, 1233 (1979).
BUTTERFASS, T., *Patterns of Chloroplast Reproduction—A Developmental Approach to Protoplasmic Plant Anatomy*, Springer-Verlag, Wien, New York, 1979.
CAROTHERS, Z.B. and G.L. KREITNER, *J. Cell Biol.*, 36, 603 (1968).
CASTALDO, R. and V.D. MARTINO, *Instituto E Orto Botanico Dell' Universita Di Napoli*, 10-11, 63 (1968-69).
CASTALDO, R., R. LIGRONE, and R. GAMBARDELLA, *Caryologia*, 32, 121 (1979a).
CASTALDO, R., R. LIGRONE, and R. GAMBARDELLA, *Rev. Bryol. Lichénol.*, 45, 345 (1979b).
CHAUHAN, E., "Ultrastructural, cytochemical and histoenzymological studies on some aspects of sporophyte development in the moss *Physcomitrium cyathicarpum* Mitt." Ph.D. Thesis, Univ. Delhi, India, 1981.
CHAUHAN, E. and M. LAL, *Cytobios*, 41, 85 (1984).
CHEVALLIER, D., F. NURIT, and H. PESEY, *Ann. Bot.*, 41, 527 (1977).
DeMAGGIO, A.E. and D.A. STETLER, *Am. J. Bot.*, 64, 449 (1977).
DENIZOT, J., *C.r. hebd. Séanc. Acad. Sci., Paris (Sér. D)*, 272, 2769 (1971).
DIERS, L., *Planta*, 66, 165 (1965a).
DIERS, L., *Z. Naturforsch.*, B20, 795 (1965b).
DIERS, L., *J. Cell Biol.*, 28, 527 (1966).
DIERS, L., "Origin of Plastids: cytological results and interpretations including some genetical aspects." In P.L. Miller, Ed., *Control of Organelle Development*, 24th Symposium of the Society for Experimental Biology, Cambridge, 1970. Pp. 129-145.
DILL, F.J., *Science*, 144, 541 (1964).
DUCKETT, J.G., *J. Bryol.*, 7, 405 (1973).
ESCHRICH, W. and M. STEINER, *Planta*, 74, 330 (1967).

ESCHRICH, W. and M. STEINER, *Planta*, **82**, 33 (1968).
EYME, J. and C. SUIRE, *C.r. hebd. Séanc. Acad. Sci., Paris (Sér. D)*, **265**, 1788 (1967).
EYMÉ, J. and C. SUIRE, *C.r. hebd. Séanc. Acad. Sci., Paris (Sér. D)*, **268**, 290 (1969).
EYMÉ, J. and C. SUIRE, *Botaniste*, **54**, 109 (1971).
FAVALI, M.A. and M. BASSI, *Nova Hedwigia*, **25**, 451 (1973).
FAVALI, M.A. and M. BASSI, *Nova Hedwigia*, **29**, 147 (1978).
FAVALI, M.A. and F. GIANNI, *Ost. Bot. Z.*, **122**, 323 (1974).
FAVALI, M.A. and F. GIANNI, *Giorn. Bot. Ital.*, **109**, 375 (1975).
GALATIS, B. and P. APOSTOLAKOS, *Can. J. Bot.*, **55**, 772 (1977).
GAMBARDELLA, R., R. CASTALDO, and R. LIGRONE, *Caryologia*, **32**, 125 (1979).
GAMBARDELLA, R., R. LIGRONE, and R. CASTALDO, *Cryptog. Bryol. Lichénol.*, **1**, 415 (1980).
GENEVÉS, L., *C.r. hebd. Séanc. Acad. Sci., Paris (Sér. D)*, **262**, 2215 (1966).
GENEVÉS, L., *C.r. hebd. Séanc. Acad. Sci., Paris (Sér. D)*, **267**, 849 (1968).
GENEVÉS, L., *C.r. hebd. Séanc. Acad. Sci., Paris (Sér. D)*, **275**, 197 (1972).
GOEBEL, K., *Organography of Plants—Especially of the Archegoniatae and Spermophyta* (English translation), Clarondon Press, Oxford, 1905
GÓRSKA-BRYLASS, A., *Bull. Acad. Pol. Sci. Cl. II, Ser. Sci. biol.*, **17**, 549 (1969).
GULLVAG, B.M., *J. Palynol.*, **1**, 49 (1967).
GUNNING, B.E.S., *Sci. Prog. Oxf.*, **64**, 539 (1977).
GUNNING, B.E.S. and J.S. PATE, *Protoplasma*, **68**, 107 (1969).
GUNNING, B.E.S. and J.S. PATE, "Transfer cells." In A.W. Robards, Ed., *Dynamic Aspects of Plant Ultrastructure*, McGraw-Hill Co., London, New York, 1974. Pp. 441-480.
GUNNING, B.E.S., J.S. PATE, and L.G. BRIARTY, *J. Cell Biol.*, **37**, 7 (1968).
HAUSMANN, M.K. and D.J. PAOLILLO, *Bryologist*, **80**, 143 (1977).
HAUSMANN, M.K. and D.J. PAOLILLO, *Am. J. Bot.*, **65**, 646 (1978).
HÉBANT, C., *Phytomorphology*, **20**, 390 (1970).
HÉBANT, C., *Protoplasma*, **77**, 231 (1973).
HÉBANT, C., *J. Hattori bot. Lab.* No. 37, 229 (1974).
HÉBANT, C., *J. Hattori bot. Lab.* No. 39, 235 (1975).
HÉBANT, C., *J. Hattori bot. Lab.* No. 40, 221 (1976).
HÉBANT, C., *The Conducting Tissues in Bryophytes*, J. Cramer, A.R. Gantner, Verlag K G, Vaduz, 1977.
HÉBANT, C., "Conducting tissues in bryophyte systematics." In G.C.S. Clarke and J.G. Duckett, Eds., *Bryophyte Systematics*, Academic Press, London and New York, 1979. Pp. 365-383.
HÉBANT, C. and E.J. BONNOT, *Z. PflPhysiol.*, **72**, 213 (1974).
HÉBANT, C. and C. SUIRE, *Rev. Bryol. Lichénol.*, **40**, 171 (1974).
HECKMAN, C., *Grana*, **10**, 109 (1970).
HORNER, H.T. JR., N.R. LERSTEN and C.C. BOWEN, *Am. J. Bot.*, **53**, 1048 (1966).
IDZIKOWSKA, K. and A. SZWEYKOWSKA, *Acta Soc. Bot. Pol.*, **67**, 1 (1978).
JARVIS, L.R., *Proc. VIII Int. Congr. el. Micr. Canberra*, **2**, 620 (1974).
JENSEN, K.G. and R.L. HULBARY, *Am. J. Bot.*, **65**, 823 (1978).
KELLEY, C., *J. Cell Biol.*, **41**, 910 (1969).
KREITNER, G.L., *Am. J. Bot.*, **64**, 57 (1977a).
KREITNER, G.L., *Am. J. Bot.*, **64**, 464 (1977b).
KREITNER, G.L. and Z.B. CAROTHERS, *Am. J. Bot.*, **63**, 545 1976).
LAL, M. and P.R. BELL, *Biol. J. Lin. Soc.*, **7**, 85 (1975).
LAL, M. and P.R. BELL, *Ann. Bot.*, **41**, 127 (1977).
LAL, M. and N.N. BHANDARI, *Bryologist*, **71**, 11 (1968).
LAL, M. and E. CHAUHAN, *Protoplasma*, **107**, 79 (1981).
LAL, M., G. KAUR, and E. CHAUHAN, *New Phytol.*, **92**, 441 (1982a).
LAL, M., G. KAUR, and E. CHAUHAN, *J. Bryol.*, **12**, 293 (1982b).
LAL, M., G. KAUR, and E. CHAUHAN, *Cryptog. Bryol. Lichénol.*, **6**, 51 (1985a).
LAL, M., G. KAUR, and E. CHAUHAN, *Cryptog. Bryol. Lichénol.*, **6**, 135 (1985b).
LAMBERT, A.M., *Bull. Soc. bot. Fr. Lett.*, **121**, 93 (1974).

LAMBERT, A.M., *Bryophytorum Bibliotheca*, 13, 113 (1977).
LIGRONE, R., R. GAMBARDELLA, and R. CASTALDO, *Cryptog. Bryol. Lichénol.*, 1, 115 (1980a).
LIGRONE, R., R. GAMBARDELLA, and R. CASTALDO, *Cryptog. Bryol. Lichénol.*, 1, 239 (1980b).
LÜTTGE, U. and G. KRAPF, *Planta*, 81, 132 (1968).
LÜTTGE, U. and N. HIGINBOTHAM, *Transport in Plants*, Springer-Verlag, New York, Heidelberg, Berlin, 1979.
MAIER, K., *Planta*, 77, 108 (1967).
MAIER, K. and U. MAIER, *Protoplasma*, 75, 91 (1972).
MANTON, I., *J. exp. Bot.*, 8, 382 (1957).
McCLYMONT, J.W. and D.A. LARSON, *Am. J. Bot.*, 51, 195 (1964).
MŁODZIANOWSKI, F., *Acta Soc. Bot. Pol.*, 39, 45 (1970).
MOGENSEN, G.S., *Can. J. Bot.*, 56, 1032 (1978).
MONROE, J.H., *Bot. Gaz.*, 129, 247 (1968).
MOSER, J.W., "An ultrastructural study of spermatogenesis in *Phaeoceros laevis* subsp. *carolinianus*," Ph.D. Thesis, Univ. Illinois, Urbana, U.S.A., 1970.
MOSER, J.W., J.G. DUCKETT, and Z.B. CAROTHERS, *Am. J. Bot.*, 64, 1097 (1977).
MUELLER, D.M.J., *Bryologist*, 75, 63 (1972).
MUELLER, D.M.J., *Am. J. Bot.*, 61, 525 (1974).
NEIDHART, H.V., "Elektronenmikroskopische und Physiologische Untersuchungen zur Sporogenese von *Funaria hygrometrica* Sibth," Ph.D. Thesis, Technische Universität, Hannover, 1975.
NURIT, F., *Bull. Soc. bot. Fr. Lett.*, 121, 169 (1974).
OLESEN, P. and G.S. MOGENSEN, *Bryologist*, 81, 494 (1978).
ÖPIK, H., "Mitochondria." In A.W. Robards, Ed., *Dynamic Aspects of Plant Ultrastructure*, McGraw-Hill Co., London, New York, 1974. Pp. 52-83.
PAOLILLO, D.J., JR., *Protoplasma*, 58, 667 (1964).
PAOLILLO, D.J., JR., *Can. J. Bot.*, 43, 669 (1965).
PAOLILLO, D.J., JR., *Trans. Am. microsc. Soc.*, 86, 428 (1967).
PAOLILLO, D.J., JR., *Cytologia*, 34, 133 (1969).
PAOLILLO, D.J., JR., *New Phytol.*, 74, 287 (1975).
PAOLILLO, D.J., JR. and M. CUKIERSKI, *Bryologist*, 79, 466 (1976).
PAOLILLO, D.J., JR. and J.A. REIGHARD, *Bryologist*, 70, 61 (1967).
PAOLILLO, D.J., JR., G.L. KREITNER, and J.A. REIGHARD, *Planta*, 78, 226 (1968a).
PAOLILLO, D.J., JR., G.L. KREITNER, and J.A. REIGHARD, *Planta*, 78, 248 (1968b).
PATE, J.S. and B.E.S. GUNNING, *Ann. Rev. Pl. Physiol.*, 23, 173 (1972).
RIPETSKY, R.T. and V.I. MATASOV, *Tsitol. Genet.*, 9, 307 (1975).
SAPĚHIN, A.A., *Ber. dt. bot. Ges.*, 29, 491 (1911).
SCHEIRER, D.C., *Planta*, 115, 37 (1973).
SCHEIRER, D.C., *Bryologist*, 78, 113 (1975).
SCHEIRER, D.C., *Protoplasma*, 89, 323 (1976).
SCHEIRER, D.C., *Am. J. Bot.*, 64, 369 (1977).
SCHEIRER, D.C., *Am. J. Bot.*, 70, 987 (1983).
SCHMIEDEL, G. and E. SCHNEPF, *Protoplasma*, 100, 367 (1979).
SCHMIEDEL, G. and E. SCHNEPF, *Planta*, 147, 405 (1980).
SCHNEPF, E. and G. DEICHGRÄBER, *Z. PflPhysiol.*, 94, 283 (1979).
SEABURY, F., JR., "Sporogenesis in selected genera of the Musci," Ph.D. Thesis, Texas A and M University College Station, Texas, 1975.
SILER, M.B., *Bot. Gaz.*, 95, 563 (1934).
STEINKAMP, M.P., "Spore wall ultrastructure in the Hepaticae with special reference to the Sphaerocarpales," Ph.D. Thesis, Univ. California, Santa Cruz, 1973.
STEINKAMP, M.P. and W.T. DOYLE, *Am. J. Bot.*, 66, 546 (1979).
STEINKAMP, M.P. and W.T. DOYLE, *Am. J. Bot.*, 68, 395 (1981).
SUIRE, C., *Botaniste*, 53, 125 (1970).

SUN, C.N., *Protoplasma*, **58**, 663 (1964).
THIERY, J.P., *J. Microscopie (Paris)*, **6**, 987 (1967).
THOMAS, R.J., D.S. STANTON, D.H. LONGENDROFER, and M.E. FARR, *Bot. Gaz.*, **139**, 306 (1978).
VALANNE, N., *Ann. Bot. Fenn.*, **3**, 1 (1966).
VALANNE, N., *Can. J. Bot.*, **49**, 547 (1971).
VALANNE, N., S. TOIVONEN, and R. SAARINEN, *Bryologist*, **79**, 188 (1976).
VIAN, B., C. BARBIER, and D. REIS-CREPIN, *C.r. hebd. Séanc. Acad. Sci., Paris (Sér. D)*, **271**, 1918 (1970).
WEIER, T.E., *Cellule*, **40**, 261 (1931).
WIENCKE, C. and D. SCHULZ, *Protoplasma*, **86**, 107 (1975).
ZINSMEISTER, D.D. and Z.B. CAROTHERS, *Am. J. Bot.*, **61**, 499 (1974).

8 Chemical Constituents of Bryophytes

IN recent times considerable information has been obtained on the chemical constituents of bryophytes. Among the organic compounds that have attracted attention are terpenoids, flavonoids, lignins, growth substances, antibiotics, lipids, and sterols. The terpenoids and flavonoids are very significant chemosystematic markers at different taxonomic levels, and the rich and unique chemistry of bryophytes offers taxonomists a vast, largely untapped supply of alternative taxonomic characters.

ANTIBIOTICS

Watt (1891) referred to medicinal uses of *Marchantia, Conocephalum, Jungermannia, Riccia,* and *Anthoceros,* and Wren (1956) mentioned about *Polytrichum juniperinum.* According to Hartwell (1971) extracts of *Marchantia polymorpha, M. stellata, Polytrichum commune,* and another unidentified species of *Polytrichum* possess antitumor elements. An interesting feature of bryophytes is that they are relatively free from attack by parasitic microorganisms. This may be due to their immunological properties or antimicrobial activity (see Banerjee & Sen, 1979). Madsen and Pates (1952), and Pates and Madsen (1955) studied eight bryophytes of which *C. conicum, Dumortiera hirsuta, Sphagnum portoricense,* and *S. strictum* were active. The first two were active against the fungus *Candida albicans,* and both the species of *Sphagnum* inhibited growth of the bacteria, *Staphylococcus aureus* and *Pseudomonas aeruginosa.* Ramaut (1959) demonstrated that extracts of *Sphagnum* inhibited growth of the bacterium *Sarcina lutea.* Antiseptic property of bogs has been ascribed to the prevailing antibiotic activity and the anaerobic conditions. Aqueous, alcoholic, and ether extracts of *Sphagnum* gave fractions which inhibited proliferation of *Sarcina.* McCleary et al. (1960) examined 12 species of mosses, of which *Anomodon rostratus* and *Orthotrichum rupestre* inhibited growth of bacteria like *Micrococcus flavus, M. rubens, Streptococcus pyogenes,* and a fungus *Candida albicans.* Ethanolic extracts of *Mnium cuspidatum* contained substances active against *Micrococcus flavus.* Subsequently,

McCleary and Walkington (1966) reported their findings on the activity of 50 species of mosses belonging to 33 genera against *Gaffkya tetragena* and *Staphylococcus aureus*. Eighteen species showed moderate to strong activity against one or both bacteria; seven others showed a slight but positive activity. Pronounced effect was shown by the genera *Atrichum*, *Mnium*, *Polytrichum*, and *Sphagnum*. Two reasons were suggested for this: (1) accumulation of non-ionized organic acids, with little correlation with pH of the extracts, and (2) the presence of certain compounds, probably polyphenolics, which may act in a manner similar to that of substances reported from vascular plants. Gupta and Singh (1971) reported antibacterial activity of petroleum ether extracts of *Barbula* and *Timmiella* against 33 bacterial species belonging to gram +ve, gram −ve, and acid fast bacteria. Banerjee and Sen (1979) studied antibiotic activity of 52 species belonging to 40 genera against some microorganisms, including three gram +ve, a penicillin sensitive, and penicillin resistant strain, five gram −ve, one acid fast bacterium, and three fungi. The plants were extracted in water, ethanol, methanol, ether, and acetone. Out of the 52 species, 29 (56%) were active against at least one of the test bacterium, but none possessed antifungal property. Among the mosses, 14 out of 31 species (41%) were inhibitory to at least one bacterium. Thirteen out of 19 species of hepatics (68%) tested yielded positive results. Both the hornworts tested were active against gram −ve bacteria. The moss *Brachythecium procumbens* and the liverworts *Asterella sanguinea* and *Marchantia paleacea* showed the broadest spectrum of antibiotic activity.

The antibiotic activity of bryophytes, as in other plant groups, varies from species to species. It also depends on the age of the plant, season of collection, and the ecological niche (Banerjee & Sen, 1979). Thus, *Conocephalum conicum* collected from the Rambi forest (2200 m) possessed antibiotic activity, but the same species collected from the Lebong Road (1650 m) in Darjeeling district of West Bengal was completely inactive, although both of them were collected in autumn. Madsen and Pates (1952) reported that *Conocephalum conicum* growing in Florida was active, whereas Hayes (1947) detected no activity in the same species growing in Ohio. Banerjee and Sen (1979) collected *Hyophila involuta* from two different altitudes and habitats. The extract of the sample from the Carron tea estate (800 m) showed antibiotic activity against *Vibrio cholerae*. The specimens from a hill top (1100 m) were effective against *Salmonella typhi* and *Pseudomonas aeruginosa*, but did not inhibit growth of *V. cholerae*. Even the species of *Marchantia* growing in association with *H. involuta* at the hill top showed activity against *S. typhi* and *P. aeruginosa*. Similarly, methanolic extracts of *Plagiochasma appendiculatum* collected from the Mall Road, Darjeeling, in October 1964, caused inhibition of *Bacillus subtilis*, *Mycobacterium phlei*, *S. typhi*, *V. cholerae*, *Staphylococcus aureus*, *Sarcina lutea*, and *Klebsiella pneumoniae*. Whereas, the plants collected from the Camel's Back Road, Mussoorie, in October 1966, showed antibiotic activity against the first four bacteria mentioned above, but did not inhibit growth of the last three. However, extracts of *Plagiochasma* inhibited growth of *Pseudomonas aeruginosa* which was not affected by the extract of plants collected from Darjeeling.

Apart from the above factors, the contradictions may arise due to the use of different solvents for extraction. Since the chemical properties of the active principles are not known, a large number of solvents should be tried. A common criticism of studies on antibiotic activity of plant extracts is that the effect observed may well be due to pH. Banerjee and Sen (1979) reported that pH of the aqueous extract was never below four or above seven. According to McCleary and Walkington (1966), pH was much lower than four in the extracts of *Sphagnum cuspidatum, S. palustre,* and *Polytrichum commune.* In several instances, they reported positive effects even after neutralization of the extracts to pH seven. According to Banerjee and Sen (1979), extracts with organic solvents, particularly alcohols, yield better results than with water or other solvents. A comparison of activity in different solvents together with a careful analysis of antimicrobial spectra reveal the occurrence of a variety of antibiotic substances. Two active principles were isolated from *Marchantia paleacea,* one of which was active against the penicillin-resistant strains of *Staphylococcus aureus, Salmonella typhi, Vibrio cholerae, Klebsiella pneumoniae,* and *Mycobacterium phlei.*

The taxa *Campylopus laetus, Asterella sanguinea, Plagiochasma appendiculatum,* and *Reboulia hemisphaerica* were moderately active against the gram –ve bacteria, *S. typhi* and *V. cholerae.* Growth of *Klebsiella pneumoniae,* resistant to many drugs, was inhibited by extracts of *Conocephalum conicum, Marchantia palmata, Reboulia hemisphaerica,* and *Targionia hypophylla. Anthoceros erectus, Brachythecium procumbens, Asterella sanguinea,* and *Plagiochasma appendiculatum* were particularly active against *Pseudomonas aeruginosa.* Infections of *P. aeruginosa* are at present best treated by polymixin (an antibiotic substance produced by the bacterium *Bacillus polymyxa*), which unfortunately also affects the kidney and is thus used rather sparingly. Growth of the acid fast bacterium, *Mycobacterium phlei,* was suppressed considerably by *Mnium* sp., *Brachythecium procumbens, Oxystegus cylindricus, Marchantia palmata, Asterella angusta, A. sanguinea, Plagiochasma appendiculatum,* and *Reboulia hemisphaerica.* Of these, the activity of *Asterella sanguinea* was most striking (Banerjee & Sen, 1979). Asakawa et al. (1980f) reported that bibenzyls from *Radula* species show antimicrobial activity.

Bryophytes also have antifungal property. For example *Dumortiera hirsuta, Sphagnum portoricense,* and *Orthotrichum rupestre* were active against *Candida albicans* (Madsen & Pates, 1952; McCleary et al., 1960). Wolters (1964) studied antifungal activity of 18 species of bryophytes, two belonging to Jungermanniales and 16 to Musci. *Pogonatum aloides, Plagiothecium denticulatum,* and *Diplophyllum albicans* were particularly remarkable in this respect. Canonica et al. (1969) isolated cinnamolide from *Porella* and *Makinoa* species and reported that this substance shows antifungal activity against dermatophytes like *Trichophyton rubrum, T. mentagrophytes,* and *Microsporum gypseum.* Lunularin and lunularic acid showed antifungal activity against spore germination of *Alternaria brassicola, Botrytis cinerea, Septoria nodorum,* and *Uromyces fabae* (Pryce. 1972b).

GROWTH SUBSTANCES

The endogenous hormonal substances of plants are classified into five major groups: auxins, gibberellins, cytokinins, abscisic acid, and ethylene (Gross, 1975). In general, the first three act as growth promoters and the other two as inhibitors. Besides these, some substances affecting growth are of restricted distribution. The following account deals with the specified and some non-specified growth regulators in bryophytes.

SPECIFIED GROWTH SUBSTANCES

Auxins

Narayanaswami and LaRue (1955) reported that gemmae of *Lunularia cruciata* normally do not germinate while within the gemma cup, but do so when the tissue around the apical notch is removed. They demonstrated that inhibition of gemma germination in situ caused by the apical notch is replaceable by lanolin paste containing an auxin. On the basis of chromatographic studies and biological activity, these investigators concluded that the auxin produced in the thallus apex of the liverwort may be IAA. von Maltzahn (1959) reported that apical dominance in the moss *Splachnum ampullaceum* could be replaced by IAA application. Schenider and Sharp (1962) also observed that gemmae failed to germinate while contained in the gemma 'cups' of *Tetraphis pellucida.* However, the first positive report of the occurrence of IAA in bryophytes was that of Schneider et al. (1967), who extracted a compound from the thalli of *Marchantia polymorpha* having Rf (retention factor) value similar to that of authentic IAA (Fig. 8.1).

Using paper chromatograms and thin-layer chromatograms, Schwabe and Valio (1970a) detected an indolic compound in extracts of thalli of *Lunularia cruciata.* They suggested that this compound may be identical with IAA.

Maravolo (1981) reported that extracts of *Marchantia polymorpha* contain substances which, in vitro, strongly inhibit or enhance indole acetic acid oxidase activity. The levels of these compounds in different regions of the thallus vary. Transport of ^{14}C-IAA in thallus explants is basipetal and is localized in the midrib region.

Recently, Ashton et al. (1985) demonstrated the presence of IAA in the

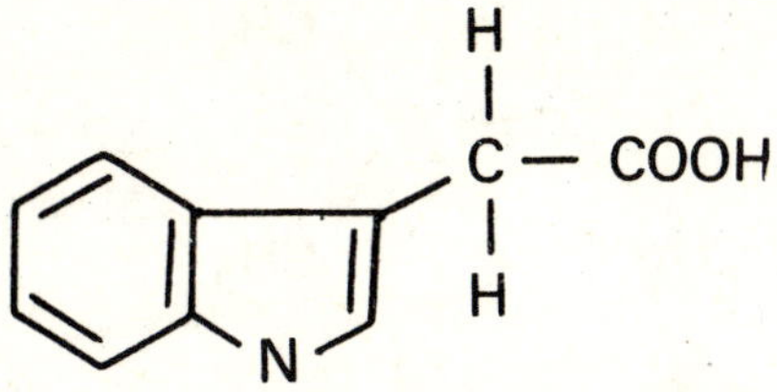

Fig. 8.1.

gametophytes of *Physcomitrella patens* by gas chromatography–selected ion monitoring-mass spectrometry using an isotope-dilution method.

Gibberellins

Endogenous gibberellin-like substances have been reported in *Polytrichum commune* (Muromtsev et al., 1964). Presence of three gibberellin-like substances has been demonstrated in *Marchantia polymorpha* (Melstrom et al., 1974). The concentration of these substances is altered in response to photoperiod. An increase of photoperiod from 12 to 18 h results in increased thallus elongation and orthogeotropic response. Gibberellin antagonists like AMO 1618 and CCC retard both thallus elongation and vertical growth.

There is evidence that gibberellin may promote auxin-mediated elongation, possibly through the inactivation of auxin oxidase or an auxin oxidase co-factor (Melstrom et al., 1974; Reynolds & Maravolo, 1973).

Cytokinins

An endogenous cytokinin (bryokinin) was isolated from callus cells of the hybrid sporophyte *Funaria hygrometrica* × *Physcomitrium pyriforme* by Bauer (1966). Bryokinin also replaces kinetin as a growth factor in tissue cultures of *Nicotiana tabacum*, and checks yellowing of leaves in *N. rustica*.

In mosses bryokinin is physiologically active at several stages of development. At the caulonema stage it promotes bud formation; in the phase immediately before sexual maturation it supports apogamous sporogonium formation in *Splachnum ovatum*; in the phase of sexual maturation it promotes archegonium differentiation in *S. ovatum* (Bauer, 1966).

Beutelmann (1973) observed that when callus cells derived from the hybrid sporophyte are supplied with adenine-8-^{14}C, they produce a labelled cytokinin which was later characterized by Beutelmann and Bauer (1977) as N^6-γ, γ- (Δ^2-isopentenyl) adenine (2iP) by means of gas-liquid chromatography and mass spectrometry (Fig. 8.2). The concentration of this compound in the culture medium and within the cells is about the same (10^{-6}M).

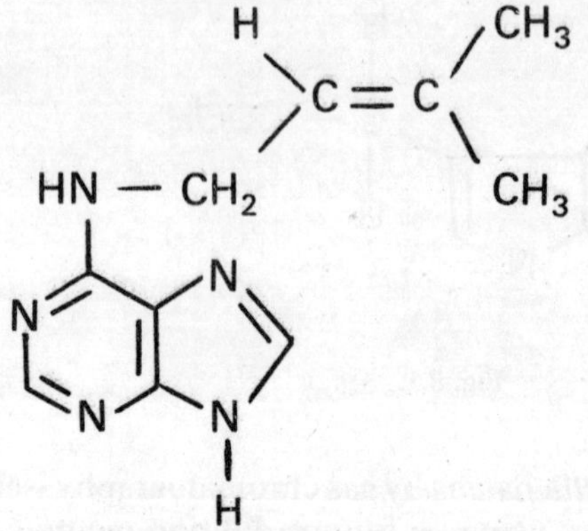

Fig. 8.2.

Lunularic Acid

In many plants incidence of natural dormancy has been correlated with the levels of an internal inhibitor. Studies on *Lunularia cruciata* (Schwabe & Nachmony-Bascomb, 1963) revealed that continuous light treatment leads to the production of an endogenous growth inhibitor which prevents further growth of thalli.

Increased growth of bisected gemmae as compared with intact controls suggested that self-inhibition also occurs. This indicated that the growing tips of gemmae are the locus of inhibitor production and they have some similarity to apical dominance in higher plants (Schwabe & Valio, 1970b). These investigators postulated that in short days (optimum growth) less of inhibitor is produced than in long days (dormancy inducing). This inhibitor is normally accumulated in the gemmae, and treatments which allow diffusion of the inhibitor from the gemmae into the medium promote growth. This endogenous growth inhibitor belongs to the group of chemical compounds termed stilbenes, and is commonly known as lunularic acid (Fig. 8.3). It was first isolated and characterized by Valio et al. (1969), and later, by using thin-layer chromatography and gas-liquid chromatography, it has been shown to be 3,4'-dihydroxybibenzyl-2-carboxylic acid or dihydro-hydrangeic acid (Gorham, 1977; Pryce, 1971a, 1972a,b,c; Pryce & Linton, 1974; Valio & Schwabe, 1970).

Lunularic acid is reported in all the liverworts and algae so far examined, but has not been detected in the hornworts, mosses, and pteridophytes. Among higher plants, it is known to be present only in *Hydrangea macrophylla*. Algae and liverworts lack ABA, which is the ubiquitous growth inhibitor in mosses, ferns, and higher plants. Thus, it seems that lunularic acid fulfils the same growth regulating functions as ABA (Gross, 1975).

Valio and Schwabe (1970) reported that in the liverwort *Lunularia*

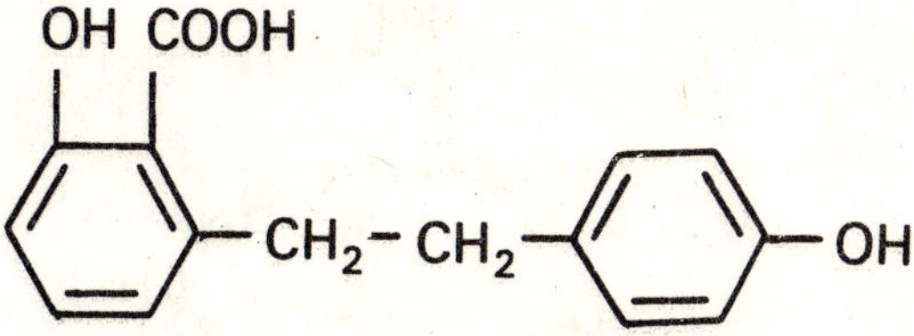

Fig. 8.3.

cruciata ABA could not be detected even though its growth is inhibited by applied ABA. This observation is of some phylogenetic significance. Although the active plant growth inhibitory agent, lunularic acid or ABA, may have changed during evolution, its site of action seems to have remained essentially unaltered. Two facts support this hypothesis: firstly, both promote dormancy and inhibit growth; and secondly, they have structural similarity.

The level of lunularic acid is highly variable in different taxa. For example, *Conocephalum conicum* has 66 µg/g fresh wt.; *L. cruciata* 5 µg/g fresh wt.; *Marchantia polymorpha* 50 µg/g fresh wt., and *Riccia fluitans* has 1µg/g fresh wt. Even species of the same genus differ in the lunularic acid content: *Solenostoma triste* has 11 µg/g fresh wt., whereas *S. crenulatum* has 1 µg/g fresh wt. (Pryce, 1971b).

Lunularic acid has some ecological significance as well. Its presence makes thalli drought-resistant, and the degree of resistance depends upon the period of long-day treatment. It also plays an important role in chemical control over the environment by inter- and intra- specific inhibition.

Cyclic AMP

Cyclic AMP mediates a multitude of responses in animals and microorganisms. It is also known to occur in fungi, algae, and possibly in mosses. Handa and Johri (1977) isolated a factor from the protonema of *Funaria hygrometrica*. They claim it to be indistinguishable from cAMP, since it stimulates the activity of protein kinase from rabbit skeletal muscle, and co-chromatographs with authentic cAMP (Fig. 8.4). They have also measured its intracellular level, which is four to seven fold higher in chloronema cells as compared with that in caulonema cells. They demonstrated that application of cAMP enhances the formation of chloronema cells (Handa & Johri, 1976). The chloronema cells contain both IAA and cAMP and their balance may, in fact, underlie protonemal differentiation.

Recently, Kaul and Sachar (1982) reported that the ^{14}C-labelled putative cAMP, isolated from *Funaria hygrometrica* tissue, showed no resemblance to authentic (^{3}H)-cAMP, and their studies implied that the tissue may contain authentic cAMP below picomole levels, and as such can not be

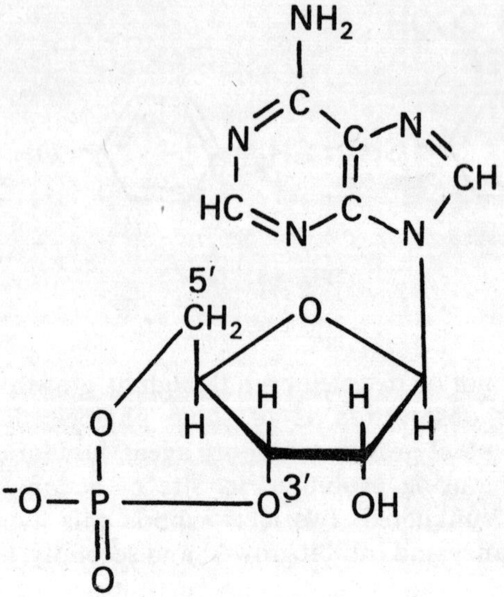

Fig. 8.4.

physiologically significant. Latest studies of Bhatla and Chopra (1984) have indicated the presence of adenylate cyclase in the shoot apices of the moss *Bryum argenteum*. This enzyme is responsible for the synthesis of cAMP.

Acetylcholine

Acetylcholine (Fig. 8.5), a neurohormone, has been detected in animals and its mode of action is very well known. Its presence in moss callus (regenerated from the seta of the hybrid sporophyte *Funaria hygrometrica* × *Physcomitrium pyriforme*; Bauer, 1963) was demonstrated by Hartmann in

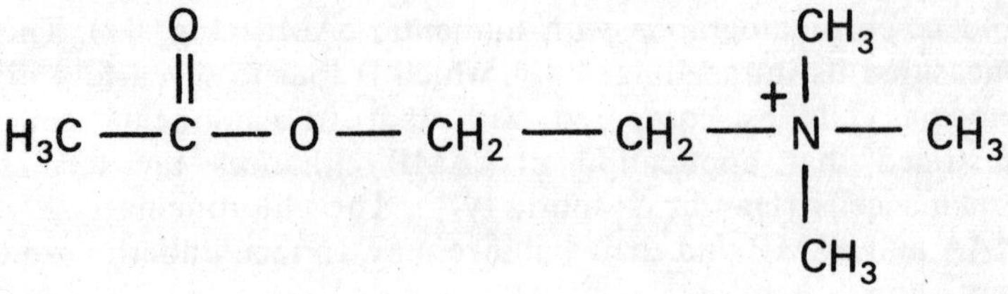

Fig. 8.5.

1971 by means of pharmacological experiments on the heart of frog and by chromatography.

Hartmann and Kilbinger (1974) identified and quantified acetylcholine from the moss callus by gas-liquid chromatography and reported that its concentration is regulated by phytochrome.

Ethylene

Production of ethylene by thalli of *Marchantia* (DeGreef et al., 1981) and setae of *Pellia* (Thomas et al., 1983) has been demonstrated. Recently, Rohwer and Bopp (1985) reported the production of ethylene in the protonema of *Funaria hygrometrica*. Exogenously applied ethylene precursor—ACC (1-aminocyclopropane-1-carboxylic acid) and IAA enhance ethylene formation. According to them ethylene may act as a senescence hormone in mosses as well.

NON-SPECIFIED GROWTH SUBSTANCES

Factor F and H

In 1959, Bopp observed that moss protonemata, especially those of *Funaria hygrometrica*, which are planted onto agar about 1 cm or more apart, do not grow over each other, whereas a single protonema can grow to a diameter of several centimetres. On the other hand, protonemata arising from spores lying only 3 mm apart grow completely together to form a common patch. It has been demonstrated that this inhibition is caused by substances excreted by the protonemata (Bauer, 1956; Bopp, 1959, 1961; Wolters, 1960).

The protonema of *Funaria hygrometrica* produces two different substances which diffuse into the substrate. In the chloronema, less than eight days of age, a thermolabile substance, Factor F, is formed. This factor promotes the growth of caulonema, but inhibits bud formation (Bopp, 1963). In the caulonema, after about 10 days, another substance, Factor H, is produced which is thermostable and is responsible primarily for bud formation. It also causes inhibition of caulonemal growth, but this is its secondary effect (Klein, 1967).

The investigation of Kockel (1967) revealed that Factor H is not limited to *Funaria hygrometrica*. The diffusates of protonemata of *Leptobryum pyriforme, Atrichum undulatum, Bryum argenteum, Pohlia nutans*, and *Dicranella heteromalla* as also of the liverwort *Marchantia polymorpha* affect *Funaria* as does *Funaria*-diffusate.

Eltz (1975) provided some more details about Factor H. It is soluble in organic solvents like ether, chloroform, dichloromethane, ethyl acetate, n-butanol, and petroleum ether. The molecular weight of this factor is in the range of 100 to 300 dalton. This factor is not identical with any of the known phytohormones (auxins, cytokinins, or gibberellins), and thin layer chromatographic studies have indicated that this factor contains two active ingredients: steroids and terpenes.

Gemma Factor

Rawat and Chopra (1976) reported that the secondary protonema of *Bryum klinggraeffii* releases a morphoregulatory substance in the medium during gemma formation. This substance inhibits protonemal growth and causes the induction of gemmae on fresh protonema before it attains the critical size required for gemma formation. It has therefore been named 'gemma factor'. The area of the test-protonema (under the influence of the diffusate) at the time of producing gemmae is 15 mm^2, whereas protonema in the control experiment grows to an area of 142 mm^2 before forming gemmae. Autoclaved protonemal diffusate retained its ability to accelerate gemma formation but did not inhibit protonemal growth. Thus it seems to contain at least two fractions, one of which is heat labile and the other is heat stable. The heat labile fraction inhibits protonemal growth but does not affect gemma induction. The heat stable fraction, on the other hand, does not inhibit protonemal growth but stimulates gemma initiation.

Kumra (1977) observed that high light level (3500 lux), 18 C temperature, and a pH range of 5 to 5.8 are most favourable for the production of gemma factor in *Bryum klinggraeffii*. Studies on the combined effect of protonemal diffusate and some known growth regulators (IAA, GA_3, & kinetin), on protonemal growth and gemma/bud formation suggest that protonemal diffusate has auxin-like properties (Chopra & Kumra, 1978). However, diffusate does not affect bud initiation in *Funaria hygrometrica* (Kumra, 1977) and *Timmiella anomala* (Rekhi, 1978). These findings indicate that the endogenous growth regulator/s are specific to *B. klinggraeffii*.

Sporogon Factor

Bauer (1959) obtained aposporous protonema from the hybrid sporogonium *Funaria hygrometrica* × *Physcomitrium pyriforme*. This protonema produced apogamous sporophytes only when in organic union with the parent sporophyte. He, therefore, postulated that a 'sporogon factor' emanating from the sporophyte is translocated into the protonema, where it induces apogamous sporophytes. Later, Lazarenko (1960) also pointed out that there is a factor in the diploid protonema of mosses which on attaining maturity results in the differentiation of apogamous sporophytes. Further studies by Bauer (1963) indicated that the factor is a labile one.

Menon and Lal (1977) reported the occurrence of a factor for apogamy in *Physcomitrium pyriforme*. They opine that this factor accumulates in the leaves where it initiates de novo differentiation of sporophytes. High concentration of sugar in the medium and dry conditions favour the production of this factor, whereas high light level (5000 to 6000 lux) is inhibitory. They also concluded that sucrose may be exercising some 'hormone-like' control on the production of sporogonial factor, possibly by interaction with the endogenous growth substances.

LIPIDS

Lipids are a heterogenous group of substances which yield fatty acids on hydrolysis. They are readily soluble in organic solvents, but sparingly so in water. There are two main classes of lipids: (1) the neutral lipids (the triglycerides, wax, & steryl esters), and (2) the polar lipids (the phospho- & glycolipids). The true fat molecule or triglyceride has two parts: an alcohol (usually glycerol), and a fatty acid. Phospholipids are complex in structure and contain a phosphate group, glycerol, and two fatty acids attached to it by ester linkages. Phosphatidyglycerol is a major component of phospholipids and is located particularly in the chloroplasts and mitochondria. Glycolipids may be defined as lipids with hydroxyl group of glycerol linked to a sugar by glycosidic bond. Most important glycolipids are monogalactosyl and digalactosyl diglycerides. These are highly surfactant molecules and play a role in chloroplast metabolism.

Considerable amounts of triglyceride, steryl, and wax esters occur in green moss shoots, spores, and protonemata (Gellerman et al., 1975; Karunen & Liljenberg, 1978; Karunen & Mikola, 1980; Karunen et al., 1980a,b; Liljenberg & Karunen, 1978; Swanson et al., 1976).

ALKANES

Bryophytes contain a wide range of alkanes (Table 8.1). In these alkanes the ratio of hydrocarbons (R) with an odd number of carbons to those with an even number varies from 1.03 to 31.7. Among the liverworts n-paraffins range from C_{15} to about C_{35}, and, in general, the n-paraffin homologues are distributed in two groups at C_{17}-C_{18} and C_{29}-C_{33}. These hydrocarbons are important constituents of the epicuticular waxes which act as protective coatings. It has been suggested that in many plants the composition of the hydrocarbon fraction is characteristic of the species and can thus be used as a taxonomic aid (Matsuo et al., 1974b).

TABLE 8.1.
Major alkanes met with in bryophytes.

Taxa	Major alkanes	Investigator/s
A. HEPATICAE		
Bazzania pompeana	C_{33}	Matsuo et al. (1974a)
Calypogeia integristipula	C_{27},C_{29}	Benešová et al. (1972)
Conocephalum conicum	C_{23},C_{25},C_{26}, C_{27}	Stránsky et al. (1967)
Gymnocolea inflata	C_{31},C_{33}	Huneck and Klein (1970); Benešová et al. (1972)
Isotachis japonica	C_{29},C_{31}	Matsuo et al. (1972a, 1974a)

(Table contd.)

Taxa	Major alkanes	Investigator/s
Jungermannia sphaerocarpa	C_{27},C_{29}	Benešová et al. (1972)
Macrodiplophyllum plicatum	C_{23}	Matsuo et al. (1974a)
Mylia taylorii	C_{25},C_{31}	Benešová et al. (1972); Stránsky et al. (1967)
Pellia fabbroniana	$C_{18},C_{27},$ C_{29},C_{31}	Matsuo et al. (1974a); Stránsky et al. (1967); Benešová et al. (1972)
P. epiphylla	$C_{26}C_{27}C_{29}$	Stránsky et al. (1967)
Porella platyphylla	C_{29}	—do—
Scapania parvitexta	C_{29},C_{31}	Matsuo et al. (1974a)
B. MUSCI		
Brachythecium rivulare	C_{29}	Catalano et al. (1976)
Campylopus introflexus	C_{31}	—do—
Ctenidium molluscum	C_{27}	Schuster (1966)
Hypnum cupressiforme ssp. imponens	C_{31}	Marsili et al. (1972)
Leucobryum glaucum	C_{27}	Stránsky et al. (1967)
Neckera crispa	C_{29},C_{31}	Marsili et al. (1972)
Pseudoscleropodium purum	C_{31}	Marsili et al. (1971)
Rhacomitrium lanuginosum	C_{29}	Catalano et al. (1976)
Rhytidiadelphus sp.	C_{29},C_{31}	Marsili et al. (1972); Stránsky et al. (1967)
Scleropodium toureti	C_{27},C_{33}	Catalano et al. (1976)
Sphagnum fuscum	C_{25}	Corrigan et al. (1973)
S. magellanicum	—do—	—do—
S. recurvum	C_{23}	—do—
S. rubellum	C_{23},C_{25}	—do—
S. teres	C_{27},C_{29},C_{31}	Marsili et al. (1972)
S. palustre	C_{23}	Caldicott and Eglinton (1976)
S. cuspidatum	—do—	—do—
Thuidium abietina (=Abietinella abietina)	C_{31}	Marsili et al. (1972)

Benesová et al. (1972) noticed compounds other than straight chain homologues in some liverworts. For example, in *Calypogeia meylanii* iso- and anteiso-isomers predominate in the C_{26}-C_{32} region. Similar hydrocarbons are present in traces in *Mylia taylorii*..

FATTY ACIDS

Moss lipids contain, besides fatty acids common for plant lipids, acids with four and five double bonds, like the following:

Acids	Abbreviations
eicosatetraenoic (arachidonic)	20:4 w6
eicosapentaenoic	20:5 w3
9,12,15-octadecatrien-6-ynoic (linolenic)	18:3 w3
9,12-octadecadien-6-ynoic (linoleic)	18:2 w6
11,14-eicosadien-8-ynoic	20:2

The ability to synthesize arachidonic acid and more highly unsaturated fatty acids is a major biochemical difference between mosses and seed plants. These acids bestow peculiar physiological characteristics on mosses. Acids 20:4 w6 and 20:5 w3 occur in complex lipids which are components of subcellular structures, and as such they may contribute to the adaptability of moss tissues to extreme habitat conditions (Gellerman et al., 1975). These investigators hypothesize that 20:4 w6 acid in monogalactosyldiacylglycerol, and phosphatidyl ethanolanine contributes to the survival of moss chloroplasts and other membranes. The high levels of 20:5 w3 acid in monogalactosyl diacylglycerol and digalactosyldiacylglycerol suggests that its role is correlated to photosynthesis, possibly in part substituting for linolenic acid. Thus, one can conclude that the highly unsaturated acids have more specific functions than merely substituting their precursors, linoleic and linolenic acids.

Information on the fatty acids in liverworts is available only for *Asterella, Conocephalum, Diplophyllum, Lophocolea, Marchantia, Mylia,* and *Pellia* (Table 8.2). Matsuo et al. (1971a) detected C_{10} to C_{18} fatty acid methyl esters in the steam distillate of *P. fabbroniana*, and these constituted 75.3 percent of the oil. Methyl esters of fatty acids, although unusual, are known to be present in both lower as well as higher plants.

In the thalli of *M. polymorpha*, 18:3 w3 acid was predominant, but significant amounts of 20:4 w6 and 20:5 w3 acids were also present (Gellerman et al., 1972). In the setae of *L. heterophylla*, 20:4 w6 and 20:5 w3 were present along with 18:2 w6 and 18:3 w3 acids. Glycerolipids and sterol esters were predominant in the unelongated setae. Phospho- and glycolipids increased dramatically with respect to the total lipid content during seta elongation and this occurred largely at the expense of diglyceride (Thomas, 1975).

Among mosses, fatty acids have been isolated from the following genera: *Abietinella, Anomodon, Bartramia, Campylium, Ceratodon, Climacium, Cratoneuron, Dicranum, Distichum, Drepanocladus, Fontinalis, Hedwigia, Hygrohypnum, Hylocomium, Hypnum, Mnium, Neckera, Plagiothecium, Pleurozium, Polytrichum, Pseudoscleropodium, Rhacomitrium, Rhytidiadelphus, Sphagnum, Splachnum, Taylonia, Thamnobryum, Thuidium,* and *Tortula* (Huneck, 1983).

A considerable information is also available on the fatty acid composition of mosses. Gellerman et al. (1975) studied the distribution of fatty acids among the various lipid classes of four mosses: *Mnium cuspidatum, M. medium, Hylocomium splendens,* and *Pleurozium schreberi*. All of these contained about 30 to 40 percent of 20:4 w6 and 20:5 w3 acids in their total lipids. However, the lipids from the last two species contained 75 percent neutral lipids (triglycerols, steryl, & wax esters), whereas the lipids of the first two species had only 20 percent or less of the neutral lipids. The amounts of wax esters in the samples described were low, but these contained a high percentage of phytyl and phytenoyl moieties. The content of 20:5 w3 in *Mnium medium* was somewhat less

TABLE 8.2.
Fatty acids met with in Hepaticae.

Taxa	Fatty acid/s	Investigator/s
Asterella	10, 15-Dihydroxyhexadecanoic acid, hexadecanoic acid, 15-hydroxyhexadecanoic acid cis-9, cis-12, cis-15-octadecatrienoic acid.	Caldicott and Eglinton (1976)
Conocephalum	Eicosanoic acid, eicosenoic acid, fatty acid ethyl ester (n-C_{14}-C_{24}), hexadecanoic acid, 2-hydroxyhexadecanoic acid, 2-hydroxyoctadecanoic acid, 2-hydroxytetracontanoic acid, 2-hydroxytriacontanoic acid, octadecanoic acid, tetracontanoic acid, tetracosanoic acid, triacontanoic acid	Caldicott and Eglinton (1976); Benešová et al. (1969a); Matsuo et al. (1980)
Diplophyllum	n-C_{36}-C_{50} esters	Benešová et al. (1975)
Lophocolea	Diglycerides, eicosanoic acid, 5,8,11,14,17-eicosapentaenoic acid, 5,8,11,14-eicosatetraenoic acid, glycolipids, hexadecanoic acid, 9-hexadecenoic acid, monoglycerides, cis-9,·, cis-12-, cis-15-octadecatrienoic acid, cis-9-octadecenoic acid, phospholipids, tetradecanoic acid, triglycerides.	Thomas (1975)
Marchantia	Docosanoic acid, 5,8,11,14,17-eicosapentaenoic acid, 5,8,11,14-eicosatetraenoic acid, hexadecanoic acid, 9-hexadecenoic acid, cis-12-octadecadienoic acid, octadecanoic acid, cis-9-octadecenoic acid.	Gellerman et al. (1972)
Mylia	n-C_{38}-C_{52} esters	Benešová et al. (1971)
Pellia	Fatty acid methyl ester (n-C_{10}-C_{16}, C_{18})	Matsuo et al. (1971a)

than in *M. cuspidatum*. These investigators (Gellerman et al., 1972) also reported that the above acids are less abundant in *Polytrichum juniperinum*, in which the major fatty acid is linolenic (18:3 w3).

Fatty acid composition of the lipids in *Fontinalis antipyretica* reveals the presence of linoleic (18:2 w6), linolenic (18:3 w3), and 11,14-eicosadien-8-ynoic (20:2) acids as the major components of triglycerides (Jamieson & Reid, 1976). The acetylenic acids accounted for 62.6 percent of the acids from the triglyceride fraction, the $C_{18}:C_{20}$ ratio being 3.2:1. Anderson et al. (1975) observed 75 percent acetylenic acids in the triglycerides of *F. antipyretica* and the ratio of $C_{18}:C_{20}$ was 0.9:1. Swanson et al. (1976) correlated the occurrence of cytoplasmic droplets in the gametophores of some mosses with the presence of triglycerides. They noticed that the droplets were more when the triglyceride content was high. Among the investigated mosses, *Mnium punctatum* and *Hygrohypnum luridium* have the least amount of highly unsaturated fatty acids.

Koskimies and Simola (1979) studied the fatty acid composition of total lipids in the gametophores of *Sphagnum fimbriatum*, *S. majus*, *S. magellanicum*, and *S. nemoreum*. These species had low lipid concentration because of the large number of dead, empty cells. The fatty acids of total lipids in several *Sphagnum* species were mainly palmitic (16:0), linoleic, and linolenic (Corrigan et al., 1976). The amount of 20:4 w6 acid was high in the steryl and methyl esters, and 20:5 w3 was abundant in the triglycerides. The membrane lipids of *Sphagnum* species contained only minor amounts of polyunsaturated C_{20} fatty acids, although they were rather abundant in some mosses like *Mnium cuspidatum* (Gellerman et al., 1975).

Karunen et al. (1979) studied the lipid content of *Sphagnum fuscum*, *S. angustifolium*, and *S. papillosum* at different ages (one to five years). The total lipid content was slightly higher in *S. angustifolium* and *S. papillosum* than in *S. fuscum*. This is in agreement with the observations of Pakarinen and Vitt (1974), according to whom the highest lipid values occur in hydrophilic species. The lipid content was maximum in the youngest portion of the moss shoots, ranging from 4.8 to 5.2 percent of dry weight, and decreased with increasing age to about 40 to 60 percent of the original value. In the crown (0 to 1 cm) of the shoot of *S. fuscum*, which was richest in lipids, the more polar lipids (79.3% of total lipid) formed the main fraction, whereas the steryl and wax esters, and triglycerides were present only in minor amounts (9.4 & 5.8%, respectively). Segment 1 to 2 cm contained about 25 percent less lipid than the crown, whereas the segment 3 to 6 cm contained about 55 percent less lipids. In the older segments (6 to 24 cm) there was no marked change in the total lipid content. However, a distinct decrease of the steryl and wax fraction was noticed. Karunen and Salin (1980) observed that in *Sphagnum fuscum* increased age resulted in a shift towards more saturated and long chain (20.2 to 26.0) fatty acids, mainly due to a preferential breakdown of unsaturated fatty acids in senescent shoots.

Regarding the role of lipids in adaptation of *Dicranum elongatum* to the subarctic environment, two main lipid classes: triglycerides and, steryl and wax esters have been considered (Karunen & Mikola, 1980; Karunen et al., 1980a,b). In this moss, triglyceride is separated into two fractions: the common and acetylenic triglyceride. The major fatty acids were 18:2 w6 (20.5%), 18:3 w3 (32.6%), and 20:4 w6 (13.9%) (Karunen & Mikola, 1980). These investigators suggested that the brown parts of the moss contain storage lipids which are formed in early spring and utilized during summer and winter. Possibly the storage lipids also serve as an energy source for the onset of regeneration process and subsequent formation of new gametophytes. The major fatty acids of steryl and wax esters of *D. elongatum* are palmitic (16:0), stearic (18:0), linoleic (18:2 w6), linolenic (18:3 w3), and arachidonic (20:4 w6) (Ekman & Karunen, 1980). In the green parts the major fatty acids of the steryl esters are polyunsaturated (18:2 w6, 18:3 w3, & 20:4 w6), whereas those of the wax esters are saturated (16:0 & 18:0). Polytenic acid is exclusively found in the wax esters (Karunen et al., 1980b). The wax esters of *D. elongatum* fall into two groups with physiologically distinct functions. The first group comprises fatty alcohol with l-octadecanol as the major alcohol. They are concentrated in the surface wax layers of moss shoots, and are esterified primarily with saturated fatty acid. The second group, the isoprenoid esters, are located in the cells and are components of both alkyl (phytol & geranyl-geranoil) and acyl (phytenic & phytanic) moieties of the esters (Karunen et al., 1980a).

The fatty acid composition of the steryl esters of *D. elongatum* does not change much with increased shoot age. However, slight changes are noticed in the proportions of w3 and w6 acids. The fatty acid composition of the wax esters changes in response to shoot age. These are slightly more unsaturated in the older segments (Karunen et al., 1980a,b). The changes in the fatty acid pattern of the wax esters in response to shoot age may be connected with the proportional changes in the alkyl moieties of the wax esters (Karunen et al., 1980a).

Karunen (1981) investigated the influence of temperature and light on the quality and quantity of triglycerides and, steryl and wax esters in an attempt to elucidate the feasible interaction between the green and senescent parts of the moss *Dicranum elongatum*. The influence of temperature and increased light level on the quantitative changes in the steryl and wax ester fraction (9% increase at 1 C) and triglyceride fraction (13% increase at 1 C) indicated that at low temperature the energy obtained from photosynthesis is primarily conserved in the triglycerides and not in the steryl esters. According to Gellerman et al. (1975) steryl esters may have a function of holding polyunsaturated fatty acids in mosses. Similarly, they may have the function of holding phytosterols or their precursors in reserve to be used for building up membranes during regeneration of brown parts, and also contribute to the adaptation of *D. elongatum* to harsh environmental conditions such as low temperature and desiccation

(Karunen et al., 1980a). According to Karunen (1981) their level decreases in response to elevated temperature, and the rapidly growing moss shoots contain the lowest level of steryl esters. Karunen (1975) identified saturated and monoenoic long chain fatty acids in the triglyceride fraction of *Polytrichum commune* spores. The saturated methyl esters were straight chained and even numbered, with C-number ranging from 12 to 26 or odd numbered with C-number ranging from 13 to 25. Later, in 1977, he separated the various isomers' of fatty acid methyl esters by gas chromatography. It gives a good separation of C 16:1 w7 and w9; C 18:1 w9 and w7; and C 20:1 w9 from C 18:3 w3 and C 18:2 w6.

Germinating spores of *P. commune* contained 5.61 ± 0.52 mg steryl and wax esters including volatile compounds per 100 mg dry weight (Karunen & Liljenberg, 1978). The content of steryl and wax esters increased slightly during the first 6 h of germination, but decreased thereafter. The major fatty acids of steryl and wax ester fraction of dry spores and germinating spores as well as of protonemata were palmitic, oleic, linolenic, and linoleic. Phytanic and phytenic acids were present in small amounts.

CUTICULAR COMPONENTS

Cutin acids, although not fat components, are closely related in structure and may be synthesized from lipid fatty acids by chain elongation before being deposited in the leaf cutin. It appears that cutin lipids principally consist of hydroxy and epoxyhydroalkanoic acids of C_{16} and C_{18} chain length.

Caldicott and Eglinton (1976) studied two mosses (*Sphagnum palustre* & *S. cuspidatum*) and two liverworts (*Asterella lindenbergiana* & *Conocephalum conicum*). The cutin acids of *S. palustre* were similar to those of higher plants, except for the higher ratio of 16-hydroxy-hexadecanoic to dihydroxyhexadecanoic acid. In *S. palustre* and *S. cuspidatum* cutin acids were predominantly of the C_{16}-type. The cutin of *Conocephalum* contained lipids which were frequently associated with wax (Kolattukudy & Walton, 1972; Martin & Juniper, 1970). Whether they are, in fact, chemical constituents of cutin or are simply in very close physical association with it, remains to be determined.

TERPENOIDS

Terpenoids are made up of two or more isoprene molecules $CH_2=C(CH_3)-CH=CH_2$, and these compounds are classified on the basis of the number of such C_5 units they contain. The terpenoids range from the essential oil components, the volatile mono- and sesquiterpenes (C_{10} &

C_{15}), through the less volatile diterpene (C_{20}) to the involatile triterpenes, sterols (C_{30}), and carotenoid pigments (C_{40}).

Isoprene, the basic unit of terpenoids, is derived from isopentenyl pyrophosphate (IPP). In biosynthesis, two IPP molecules are linked to give geranyl pyrophosphate (C_{10}), the key intermediate in monoterpene formation. Geranyl pyrophosphate and IPP are, in turn, linked to give farnesyl pyrophosphate (C_{15}), the key intermediate of sesquiterpene synthesis, and by dimerisation give rise to triterpene (C_{30}). Condensation of farnesyl pyrophosphate with a further molecule of IPP gives rise to geranyl-geranyl pyrophosphate (C_{20}), the immediate precursor of diterpenes.

Terpenoids appear to be ubiquitous in bryophytes. In the hepatics there have been a few surveys for monoterpenes; a considerable work has been done in elucidating the sesquiterpenoids; and limited but accurate information is available on the di- and triterpenoids. Up till now, mono- and sesquiterpenoids have been identified only in the liverworts which have oil bodies. Steroids and triterpenoids occur in both mosses and liverworts. Of all the diterpenoids only one is present in mosses. Terpenoids or aromatic compounds are very significant chemosystematic markers in bryophytes (Asakawa, 1981).

MONOTERPENOIDS

Monoterpenes are reported only in a few bryophytes. The intense fragrance after crushing some species of the Jungermanniales and Marchantiales is due to the presence of monoterpenoids. So far, 29 monoterpenoids have been reported in the Hepaticae and these are as follows: borneol, bornyl acetate, bornyl ferulate, bornyl 2-methoxy -4-hydroxycinnamate, camphene, camphor, carvacrol, p-cymene, geraniol, limonene, linalool, linalyl acetate, myrcene, myrtenal, ocimene, α-phellandrene, β-phellandrene, α-pinene, β-pinene, β-sabinene, terpinen-4-ol, α-terpinene, γ-terpinene, α-terpineol, terpinolene, thujanol, thymol, thymyl acetate, tricyclene (Asakawa, 1983). Suire (1970) noticed camphene, and α- and β-pinene in *Radula complanata*. Svensson (1974) detected α-pinene, camphene, myrcene, α-terpinene, limonene, γ-terpinene, terpinolene, and p-cymene in two species of *Jungermannia*. In *J. cordifolia*, camphene is the major component, whereas in *J. obovata*, terpinolene and limonene predominate.

Hörster and Wiermann (1976) analyzed *Conocephalum conicum* and tentatively identified 12 of the 39 volatile components: α-pinene, camphene, sabinene, myrcene, α- and γ-terpinene, limonene, p-cymene, terpinen-4-ol, lineoyl acetate, bornyl acetate, and α-terpineol. Suire and Bourgeois (1977) reported that the monoterpenes of *Frullania tamarisci* consist almost entirely of α- and β-pinene and camphene, and *Porella platyphylla* contains α- pinene, camphene, sabinene, α-terpinene (major component), limonene, β-phellandrene, p-cymene, and terpinolene.

Asakawa et al. (1976a) demonstrated that the major monoterpenoid and volatile component in *Conocephalum conicum* is (+)-bornyl acetate (Fig. 8.6). It is present in high quantity (0.6%). The most significant difference in the chemical constitution of the thallus and the female receptacle is that (+)-bornyl acetate is not detected in the latter (Asakawa, 1980). The same phenomenon has been observed in *Wiesnerella denudata*. Asakawa et al. (1981a) observed that *Conocephalum supradecompositum* is chemically quite different from *C. conicum* and *W. denudata*. The latter two species elaborate a large amount of monoterpene hydrocarbons and bornyl acetate, and the chemical constitution of the female and male gametophytes is different, whereas the female and male gametophytes of *C. supradecompositum* elaborate the same components. Furthermore, the monoterpenoid content in *C. supradecompositum*, is considerably less than that of *C. conicum*. On the basis of above results, it is suggested that *C. supradecompositum* is more primitive than *C. conicum* and *W. denudata*.

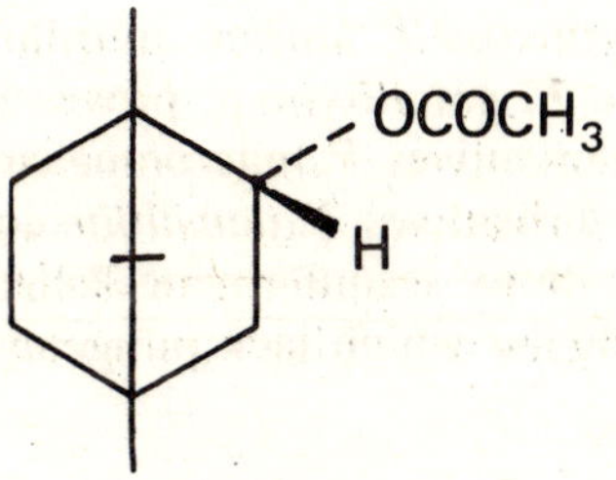

Fig. 8.6.

SESQUITERPENOIDS

Information about the sesquiterpenoids is limited to liverworts. Their presence was initially recognized by Fujita et al. (1956) through a study of the essential oil in *Bazzania pompeana*. The first positive identification of a sesquiterpene was that of drimenol (Fig. 8.7) from *Bazzania trilobata* by Huneck (1967). In the same year, Huneck and Klein discovered that *Scapania undulata* contains (−)-longifolene and (−)-longiborneol. The sesquiterpenoids are generally located in the oil bodies and there is a direct correlation between the abundance and size of the oil bodies and the amount of essential oils obtained from the thalli of different species.

Drimanes

The simplest member of this group, drimenol, was isolated from *B. trilobata* (Huneck, 1967). A number of other drimanes, all with the same absolute configuration as drimenol, have been isolated more recently.

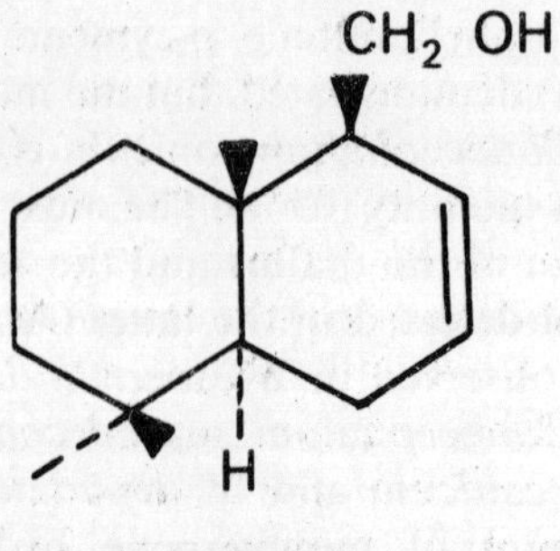

Fig. 8.7.

Asakawa et al. (1976b) have proved that drimane derivatives occur quite commonly in *Porella* species. They demonstrated that the pungent component of *P. vernicosa* is the drimane dial, (+)-tadeonal (Fig. 8.8). They distinguished six species of *Porella* on the basis of sesquiterpenoid composition. *P. vernicosa* contains drimane (including the pungent dialdehyde, tadeonal), aromadendrane, and pinguisane derivatives. *P. gracillissima* has an identical composition, and this confirms that this species is closely related to *P. vernicosa*. *P. faurieri*, morphologically quite similar to *P. vernicosa*, but not to *P. gracillissima*, proves to be quite distinct as it contains only drimane derivatives. *P. macroloba* can be distinguished by the lack of aromadendrane derivatives. *P. densifolia* contains only pinguisanes. *P. perottetiana* lacks the above sesquiterpene skeletons, but contains two, α- and β-unsaturated aldehydes which lack pungency.

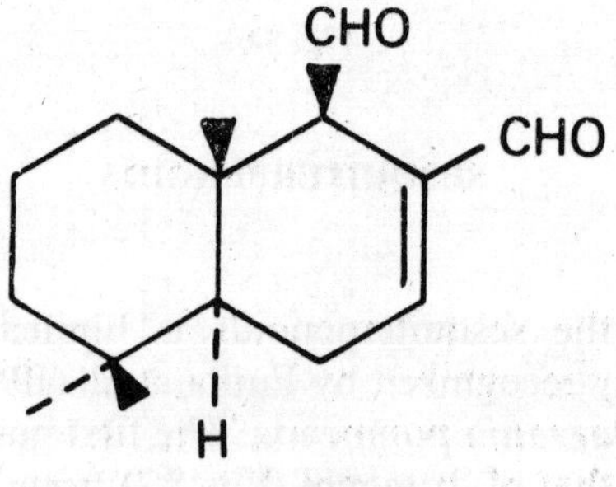

Fig. 8.8.

Bisabolanes

These are the largest group of sesquiterpenoids met with in liverworts. Structurally the simplest representatives of this family isolated from *Scapania undulata*, are the cis- and trans-isomers of α-bisabolene (Andersen et al., 1977). Matsuo et al. (1971b) detected cuparene (Fig. 8.9a) in *Bazzania pompeana*. Later, Hopkins and Perold (1974) isolated (S)-2-hydroxycuparene [(−)-δ-cuparenol] (Fig. 8.9b) from *Marchantia polymorpha*.

Connolly et al. (1972, 1974) isolated a completely new sesquiterpene alcohol from *Gymnomitrion obtusum*. It was named gymnomitrol (Fig. 8.10a), and was accompanied by its parent hydrocarbon gymnomitrene (Fig. 8.10b). Andersen et al. (1973a) isolated a number of sesquiterpenes from four species of *Barbilophozia*. Among the components were two unknown tricyclic isomeric sesquiterpenes ∝- and β-barbatene. Matsuo et al. (1973a) isolated a novel tricyclic isomeric sesquiterpene, ∝-pompene, from *Bazzania pompeana*. In 1975, Matsuo et al. observed that structure of ∝-pompene is similar to ∝-barbatene. Asakawa et al. (1981b) isolated ∝-barbatene from *Riccardia jackii*. Hayashi et al. (1969) isolated a sesquiterpene hydrocarbon with a new carbon skeleton, bazzanene (Fig. 8.11a) from *Bazzania pompeana*. From the same plant, bazzanenol (Fig. 8.11b), an alcohol, was isolated by Hayashi and Matsuo (1970). Andersen and Huneck (1973) reported bazzanene in *B. trilobata*. This compound is also present in traces in *Scapania undulata* (Andersen et al., 1977).

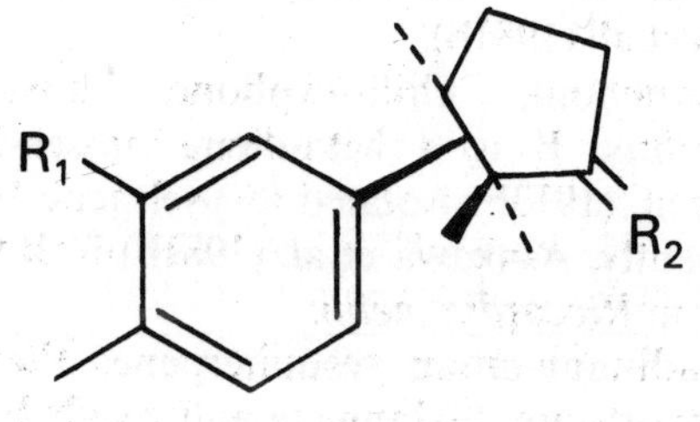

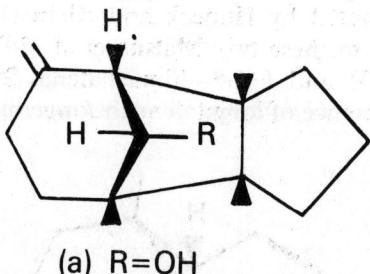

(a) $R_1=H$; $R_2=H_2$
(b) $R_1=OH$; $R_2=H_2$

Fig. 8.9a,b.

(a) R=OH
(b) R=H

Fig. 8.10a,b.

Cadinanes

Benešova et al. (1969a) reported the presence of δ-cadinene (Fig. 8.12) in

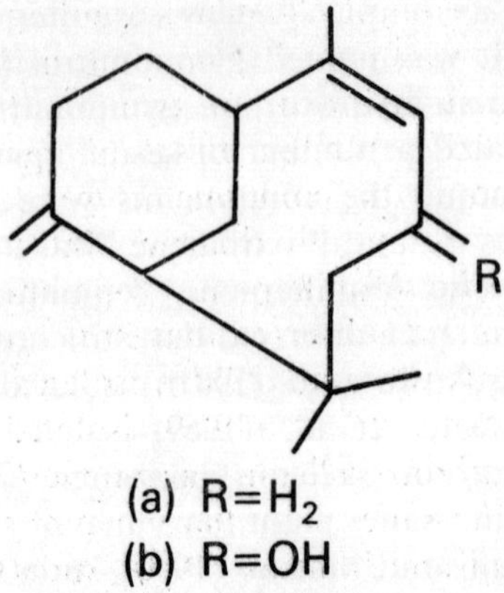

(a) R=H₂
(b) R=OH

Fig. 8.11a,b.

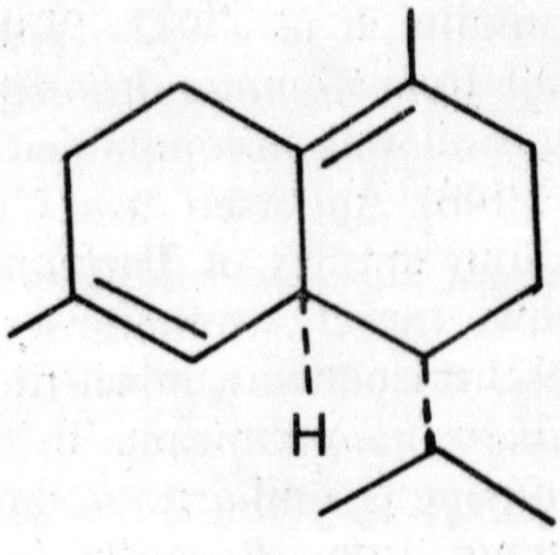

Fig. 8.12.

Conocephalum conicum. The presence of other cadalenes (10-epi-zonarene, calamenene, & α-calcorene) was indicated in *Riccardia multifida* by Asakawa et al. (1981b).

Another sesquiterpenoid, chiloscyphone, was isolated from *Chiloscyphus polyanthus.* It is a keto-diene possessing the cadinane skeleton. Andersen et al. (1973b) isolated (+)-selinene from *Chiloscyphus polyanthus.* More recently, Asakawa et al. (1981b) isolated (+)-α-selinene and (−)-β-selinene from *Riccardia jackii.*

The only other cadinane-group sesquiterpenes that have tentatively been detected in liverworts are: β-ylangene and p-cubebene (Matsuo et al., 1973b), α-copaene, and α- and β-muurolene (Matsuo et al., 1976a).

Humulanes and Caryophyllanes

There are a group of sesquiterpenes containing either nine or an 11 membered ring resembling the essential oil of *Pinus longifolia.* The first isolation of sesquiterpenes of this type, (−)-longifolene and (−)-longiborneol was reported by Huneck and Klein (1967) from *Scapania undulata.* In addition to these two, Matsuo et al. (1973c) isolated (−)-α-longipinene (Fig. 8.13) and (+)-α-himachalene. Svensson and Bendz (1972) detected the presence of longifolene in *Jungermannia cordifolia* and

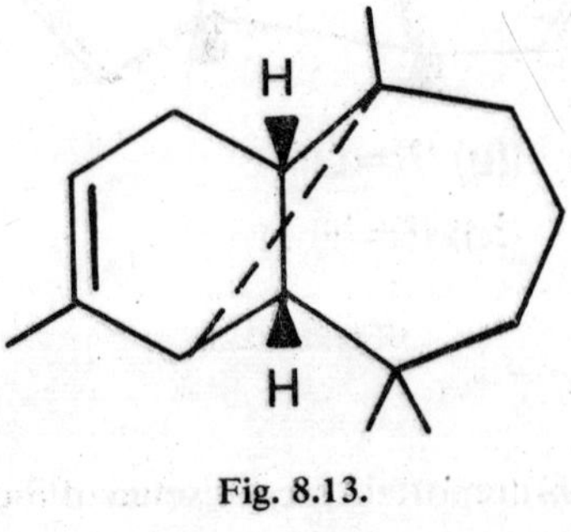

Fig. 8.13.

Scapania nemorosa, and isolongifolene in *S. subalpina*, *S. uliginosa*, and *S. undulata*. The simplest patent hydrocarbon of this series, caryophyllene, is present in the male, female, and sterile gametophytes of *Conocephalum supradecompositum* (Asakawa et al., 1981a).

Germacranes and Elemanes

The only genuine germacranes isolated are costunolides, and in this chapter these have been dealt with under sesquiterpene lactones. However, (−)-bicyclogermacranes were isolated from the liverworts *Plagiochila asplenioides* and *P. semidecurrens* (Asakawa et al., 1980a, 1981b). Ent-bicyclogermacrane has also been isolated from two species of *Porella* (Asakawa et al., 1978) and *Trocholejeunea sandvicensis* (Asakawa et al., 1980b). The β-elemene has been identified in *Marchantia polymorpha*, *Riccardia multifida*, and *Conocephalum supradecompositum* (Asakawa et al., 1981a,b; Gleizes et al., 1973-1974). *Dumortiera hirsuta* is also reported to contain β-, δ- and γ-elemene (Matsuo et al., 1976a).

Maalianes and Aromadendranes

These sesquiterpenoids are derived from bicyclogermacrene. A number of ent-sesquiterpenoids in this class have been isolated. The first two with established structures (−)-maaloxide (Fig. 8.14) and (+)-cyclocolorenone (Fig. 8.15) are from *Plagiochila acanthophylla* ssp. *japonica* (Matsuo et al., 1974c).

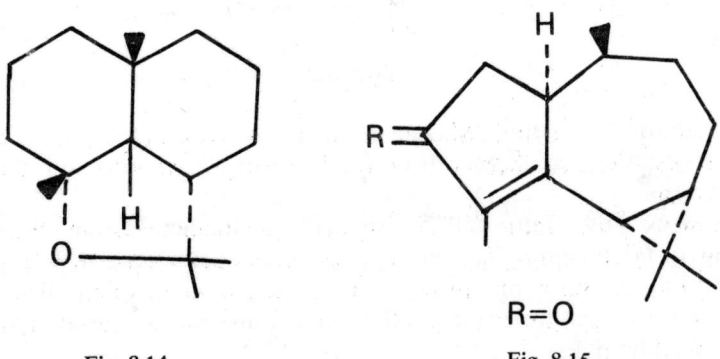

Fig. 8.14.   Fig. 8.15.

Asakawa et al. (1976c, 1978) detected a number of aromadendrane sesquiterpenoids in *Porella vernicosa* and *P. gracillissima*. These include, in addition to ent-cyclocolorenone, the closely related (+)-α-gurjunene, the previously unknown (+)-β-gurjenene, and a hydroxylated ent-cyclocolorenone.

Asakawa et al. (1979a) have reported four novel secoaromadendrane type sesquiterpene hemiacetals from the liverwort *Plagiochila asplenioides*: plagiochiline C, D, E, and F, together with the previously known (−)-bicyclogermacrene. Plagiochiline A and C have been isolated from *P. semidecurrens* along with (−)-bicyclogermacrene and its related hydrocarbons. Asakawa et al. (1980a) described the distribution of ent-2,3-secoaromadendrane, ent-aromadendrane, and ent-maaliane type sesquiterpenes in 14 species of *Plagiochila*. The intense pungent smell of some species of this genus is due to an ent-2,3,-secoaromadendrane type sesquiterpene hemiacetal, plagiochiline A. On the basis of presence or absence of plagiochiline A, the 14 species of *Plagiochila* are divided into two groups. The ent-secoaromadendrane type sesquiterpenoids and their related ent-sesquiterpenoids are one of the significant chemosystematic markers in the Plagiochilaceae.

The liverwort *Mylia taylorii* has a rich sesquiterpenoid chemistry and contains at least 24 components (Benešová et al., 1973), the outstanding among which is a tetracyclic sesquiterpene alcohol, myliol (Fig. 8.16). Subsequently, Andersen et al. (1977) isolated a closely related hydrocarbon anastreptene from *Anastrepta orcadensis*. This hydrocarbon is closely related to myliol and is present in high proportions in the essential oil, but is susceptible to oxidation.

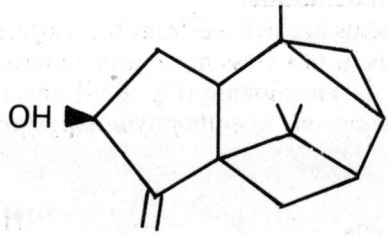

Fig. 8.16.

In addition to myliol, Matsuo et al. (1974b) isolated a novel carbon skeletal sesquiterpene ketone from the liverwort *M. taylorii*, and named it as taylorione.

Bourgeois and Suire (1977) isolated guaiazulene from the steam distillate of *Pellia epiphylla*. These investigators distinguished *P. epiphylla* and *P. fabbroniana* by the presence or absence of guaiazulene. The earlier distinction between these species (presence or absence of sporophytes) was of not much help.

Pinguisanes

The first representative of this group, the ketone pinguisone (Fig. 8.17), was isolated from *Aneura pinguis* (Benešová et al., 1969b). This compound lacks β-methyl group in the furan ring which occurs in most furanoid sesquiterpenoids (Corbella et al., 1974). Krutov et al. (1973) isolated the second member of this group, deoxypinguisone or pinguisane from *Ptilidium ciliare*. Asakawa et al. (1976b,d) isolated five more compounds related to pinguisane from *Porella vernicosa* and *P. densifolia*. The major sesquiterpenoid was the tertiary alcohol; α-pinguisene is also isolated from *Porella* species (Asakawa et al., 1978).

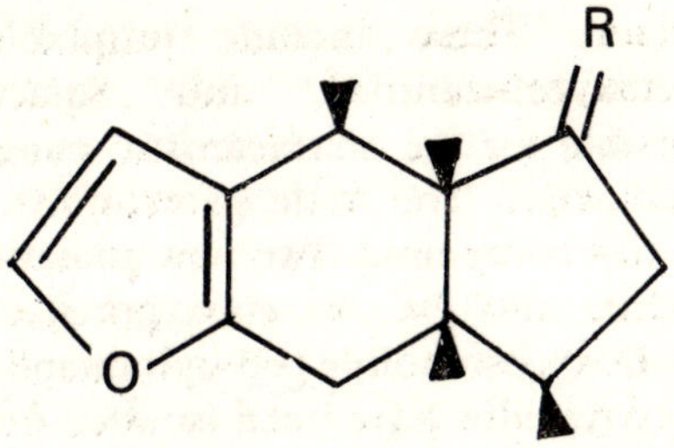

R=O

Fig. 8.17.

Asakawa et al. (1979b) reported three new pinguisane type sesquiterpenes: pinguisanin, pinguisanolide, and β-pinguisenediole from *P. platyphylla*. Dehydropinguisanin, dehydropinguisenol, and pinguisenal together with the previously known pinguisanin, pinguisanolide, and bicyclogermacrene were isolated from *Trocholejeunea sandvicensis* (Asakawa et al., 1980d).

Sesquiterpene Lactones

Knoche et al. (1969) studied the structure of the active principle in *Frullania tamarisci*. The compound was named (−)-frullanolide (Fig. 8.18). Connolly and Thornton (1973) isolated three more lactones from *F. tamarisci*: γ-cyclocostunolide, ∝-cyclocostunolide, and costunolide. Asakawa et al. (1976e) isolated ent-arbusculin β, (+)-dihydrofrullanolide, (+)-5β-hydroxyfrullanolide, and (+)-cis-β-cyclocostunolide from *Frullania dilatata*. These investigators also reported two more lactones: (+)-eremofrullanolide and (+)-dihydroeremofrullanolide from the same liverwort. Both these possess the ent-eremophilene skeleton with an ∝-isopropyl group. Asakawa et al. (1976f) tested the sensitivity of patients to *Frullania* sesquiterpene lactones using patch tests with one percent solution of the above compounds isolated from *F. dilatata*. All patients were sensitive to at least one of the compounds, but the actual pattern of reaction varied widely.

Asakawa and Takemoto (1979) isolated five sesquiterpene lactones from

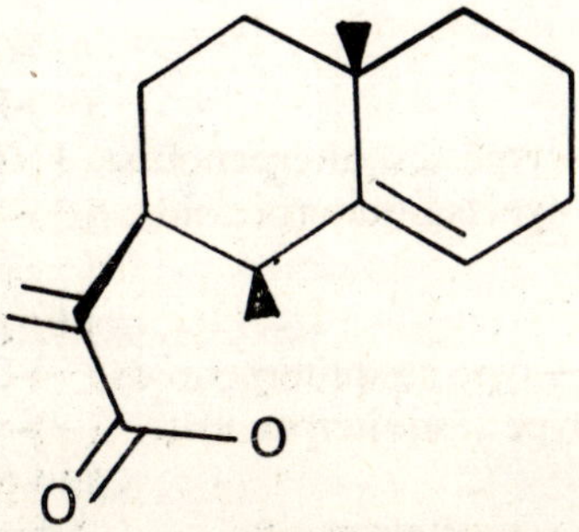

Fig. 8.18.

Conocephalum conicum. These include tulipinolide, zaluzanin C, zaluzanin D, 8∝-acetoxyzaluzanin C, and 8α-acetyloxyzaluzanin D. Tulipinolide is responsible for the characteristic pungency of the female gametophytes of *C. conicum*. The male gametophyte has no pungency because of the lack of this compound. Two new guaiane type sesquiterpene dilactones, porelladiolide and 3∝, 4∝-epoxyporelladiolide, and a new germacranolide, 3β-hydroxycostunolide [=3-epitamaulipin β] together with isoeremanthin and eregoyazidin have been isolated from *Porella japonica* (Asakawa et al., 1981d).

Recently, sesquiterpenoids have been classified on the basis of their skeletons into the following groups (Huneck, 1983).

(1) Acyclic sesquiterpenoids: Farnesene
(2) Monocarbocyclic sesquiterpenoids
    (a) Bisabolane—type sesquiterpenoids: Ar-curcumene
                                                                 ∝-Bisabolene
                                                                 (−)-β-Bisabolene
    (b) Elemane—type sesquiterpenoids: Dehydrosaussurea lactone
                                                                   ∝-Elemene
                                                                   β-Elemene
                                                                   δ-Elemene
                                                                   γ-Elemene
                                                                   Elemol
                                                                   Saussurea lactone
    (c) Germacrane—type sesquiterpenoids: Costunolide
                                                                   (+)-Dihydrocostunolide
                                                                   3β-Hydroxycostunolide
                                                                   Parthenolide
                                                                   (+)-Tulipinolide
    (d) Humulane—type sesquiterpenoids: ∝-Humulene
(3) Bicarbocyclic sesquiterpenoids
    (a) Cuparane—type sesquiterpenoids: (−)-(S)-Cuparene
                                                                   γ-Cuprenene
                                                                   (−)-Cuprenenol
                                                                   Cyclopentanone(−)-(R)-
                                                                       ∝-cuparenone
                                                                   (--)-(S)-2-Hydroxycuparene
                                                                   (−)-Rosulantol
    (b) Trichothecane—type sesquiterpenoids: (+)-Trichodiene
    (c) Himachalane—type sesquiterpenoids: (+)-∝-Himachalene
                                                                  (−)-β-Himachalene
                                                                  (+)-γ-Himachalene
    (d) Caryophyllane—type sesquiterpenoids:(−)-Caryophyllene
    (e) Chamigrane—type sesquiterpenoids: (+)-∝-Chamigrene
                                                                  (+)-β-Chamigrene
    (f) Drimane—type sesquiterpenoids: Albicanol
                                                                   Confertifolin

Chemical Constituents of Bryophytes 201

	cis-Dihydroconfertifolin
	Drimenin
	Drimeninol
	(−)-Drimenol
	iso-Drimeninol
	(−)-Tadeonal
(g) Carotane—type sesquiterpenoids:	γ-Lactone hercynin
(h) Eremophilane—type sesquiterpenoids:	(+)-Dihydroeremofrullanolide
	(+)-Eremofrullanolide
	Eremophilene
(i) Selinane—type sesquiterpenoids:	Cyclocostunolide
	Frullanolide
	α-Helmiscapene
	β-Helmiscapene
	α-Methylene-γ-lactones
	ent-Selina-4,11-diene
	(−)-α-Selinene
	β-Selinene
	δ-Selinene
	Sibirene
	6α-Hydroxyselina-4 (14) 11-Diene-13-aldehyde
	(−)-α-Eudesmol
	Diplophyllin and its derivatives
(j) Cadinane—type sesquiterpenoids:	(+)-γ-Bulgarene
	(−)-δ-Cadinene
	ent-γ-Cadinene
	β-Calcorene
	Calamenene
	Muurolene
	γ-Muurolene
	(+)-γ$_1$-Cadinene
	(+)-ent-Epicubenol
	(−)-Chiloscyphone
	α-Copaene
	β-Cubebene
(k) Acorane—type sesquiterpenoids:	Acoradiene
	(−)-α-Alaskene
(l) Guaiane—type sesquiterpenoids:	8α-Acetoxyzaluzanin C
	8α-Acetoxyzaluzanin D
	Desacetoxydehydromatricarin
	Desacetoxymatricarin
	3α,4α-Epoxyporelladiolide

	Eregoyazidin
	Guaianolide
	α-Guaiene
	β-Guaiene
	Isoeremanthin
	iso-α-Gurjunene B
	Porelladiolide
	Zaluzanin C
	Zaluzanin D
(m) Bicyclogermacrane—and bicyclohumulane—type sesquiterpenoids:	(−)-3α-Acetoxybicyclogermacrene
	3β-Acetoxybicyclogermacrene
	(−)-Bicyclogermacrene
	(+)-Bicyclohumulenone
	(−)-Isobicyclogermacrenal
(n) Santalane—and bergamotane-type sesquiterpenoids:	β-Bergamotene
	β-Santalene
(o) Azulene — type sesquiterpenoids:	1,4-Dimethylazulene
	Guaiazulene
	1-Methoxycarbonyl-4-methylazulene
(p) Pinguisane—type sesquiterpenoids:	Dehydropinguisanin
	Dehydropinguisanol
	Deoxopinguisone
	5-Hydroxypinguis-10-ene
	Methyl deoxopinguisanoate
	Pinguisanene
	Pinguisanin
	Pinguisanol
	α-Pinguisene
	Pinguisone
	Pinguisonemethylester
	Methyl-3-oxonorpinguisonate
	Methyl norpinguisonate
	Norpinguisone
	β-Pinguisenediol
	Pinguisanolide
	Isopinguisanolide
(q) Seco-aromadendrane—type sesquiterpenoids:	Acetoxyplagiochiline C
	Acetoxyovalifoliene
	Furanoplagiochilal
	(+)-Hanegokidial

(+)-Hanegokitrial
(+)-Ovalifolienalone
(+)-Ovalifolienal
Plagiochilide
Plagiochiline (A-I)
(−)-Taylorione
Methoxyplagiochilines A$_1$, A$_2$ and C

(4) Tricarbocyclic sesquiterpenoids
  (a) Maaliane–type sesquiterpenoids:
    (+)-Maalian-5-ol
    β-Maaliene
    ent-Maaliol
    (−)-Maalioxide
  (b) Aromadendrane–type sesquiterpenoids:
    Alloaromadendrene
    (+)-Cyclocolorenone
    ent-Globulol
    l-Hydroxycyclocolorenone
    ent-α-Gurjunene
    ent-β-Gurjunene
    ent-Ledene
    ent-Spathulenol
  (c) Cedrane–type sesquiterpenoids:
    α-Cedrene
    β-Cedrene
  (d) Gymnomitrane–type sesquiterpenoids:
    α-Gymnomitrene
    β-Gymnomitrene
    (−)-β-Gymnomitrene
    (+)-Gymnomitrol
    (+)-Gymnomitrol acetate
    (+)-5-Hydroxygymnomitrol diacetate
    (−)-8,12-Epoxygymnomitrol acetate
    5,11-Diacetoxy-8,12-epoxygymnomitrane
    12-Hydroxy-11-acetoxygymnomitr-8-ene
  (e) Longifolane-, longibornane-, longipinane-, longicyclane-, and isolongifolane–type sesquiterpenoids:
    (−)-Longiborneol
    (−)-Longifolene
    (−)-Longipinanol
    (−)-α-Longipinene
    (−)-β-Longipinene
    (+)-Marsupellol
    (−)-Marsupellone
    (+)-9-Acetoxymarsupellone
    (−)-12-Acetoxymarsupellone
    Longicyclene
    Isolongifolene
  (f) Sativane–type sesquiterpenoids: (−) Sativene

(g) Ylangane-type sesquiterpenoids: α-Ylangene
   β-Ylangene
(h) Cycloguaiane–and cyclopseudoguaiane–type
   sesquiterpenoids: β-Bourbonene
   Caespitenone
(i) 6 (5→1)-Abeo-aromadendrane–type sesquiterpenoids: (+)-Vitrenal
(5) Tetracarbocyclic sesquiterpenoids: Anastreptene
   (−)-Dihydromylione A
   (−)-Myliol

DITERPENOIDS

Nilsson and Martensson (1971) reported ent-16-kaurenol (Fig. 8.19) in the moss: *Saelania glaucescens*, where it is considered to contribute to the bluish tinge of the plant. Earlier, Huneck and Vevle (1970) identified the same compound as the main component of waxy coatings on two liverworts, *Anthelia juratzkana* and *A. julacea*.

Connolly and Thornton (1972), Matsuo et al. (1976a), and Benes et al. (1977) isolated ent-kauranes from *Solenostoma triste*, *Jungermannia infusca*, and *J. sphaerocarpa*, respectively.

Matsuo et al. (1972b) detected another series of compounds i.e. a diterpene alcohol (−)-manool, in *Jungermannia torticalyx*. This established the presence of ent-labdane diterpenoids in liverworts. Later, another member of this series, a new ent-labdane diol, jungermanool, was isolated from this species (Matsuo et al., 1976b). Matsuo et al. (1976c) isolated three compounds of a third series, ent-pimaranes, from *Jungermannia thermarum*. These included ent-pimar-δ-(14), 15-dien-19-oic acid, ent-pimar-δ-(14), 15-dien-19-ol, and ent-pimar-15-en-δ, 19-diol. The last compound has been named thermarol.

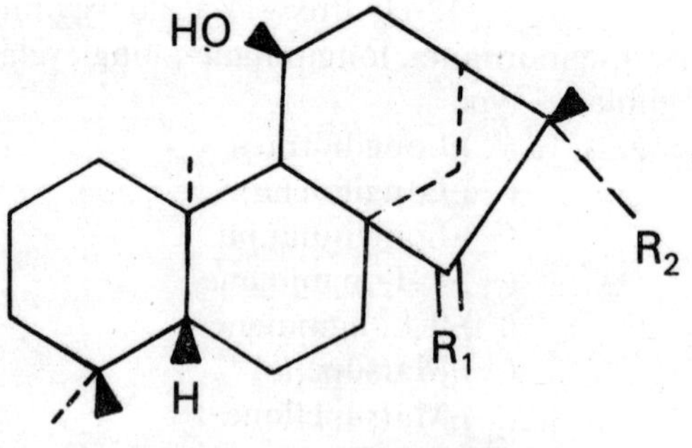

$R_1 = H; R_2 = OH$

Fig. 8.19.

Asakawa et al. (1976f, 1977) isolated two diterpene dialdehydes, sacculatal and isosacculatal, from *Trichocoleopsis sacculata* and *Porella vernicosa*. Sacculatal is also reported in *Porella japonica* (Asakawa et al., 1981b).

Huneck and Overton (1971) observed the occurrence of diterpenoids in *Anastrepta orcadensis*, *Scapania undulata*, *Barbilophozia barbata*, *B. floerkei*, *B. lycopodioides*, and *Gymnocolea inflata*.

Diterpenoids from bryophytes include open chain di-, tri-, and tetra-carbocyclic derivatives (Huneck, 1983).

(1) Open chain diterpenoids: Geranylgeraniol
Phytol

(2) Bicarbocyclic diterpenoids
  (a) Labdane–type diterpenoids: (−)-Jungermanool
    (−)-(5R, 7S, 8S, 9S, 10S)-labda-12, 14-dien-7, 8 diol
    (−)-Manöol
    Polyhydroxylated labdane derivatives
    Scapanin
  (b) Clerodane—type diterpenoids: Gymnocolin
  (c) Dolabellane—type diterpenoids: Barbilycopodin
    Deoxybarbilycopodin
  (d) Sacculatane–type diterpenoids:
    9-Hydroxysacculatal
    18-Hydroxysacculatal
    19-Hydroxysacculatal
    3-Hydroxysacculatanolide
    Isosacculatal
    Perrottetianal A,B
    Sacculatal
    Sacculatanolide

(3) Tricarbocyclic diterpenoids:
    15-Dien-19-oic-acid
    15-Dien-19-ol
    ent-Pimara-δ (14)
    (−)-Thermarol

(4) Tetracarbocyclic diterpenoids
  (a) Kaurane–type diterpenoids:
    ent-11α-Acetoxy-15-oxo-(16R)-kaurane
    ent-11α-Acetoxy-15-oxokaur-16-ene
    ent-11α, 15α-Diacetoxykaur-16-ene
    ent-11α, 15α-Dihydroxykaur-16-ene
    ent-11α-Hydroxy, 15α-acetoxykaur-16-ene
    ent-11α-Hydroxy-15-oxokaur-16-ene
    ent-15α-Hydroxykaur-16-ene
    ent-18-Hydroxykaur-16-one
    (16R)-ent-11α-Hydroxykauran-15-one

(16R)-ent-18-Hydroxykauran-15-one
(−)-ent-Kauran-16β-ol
(16R)-ent-Kauran-15-one
ent-15-Oxokaur-16-ene

(b) Verrucosane- and neoverrucosane—type
diterpenoids: (−)-2β-Acetoxy-11α-hydroxyverrucosane
(−)-2β-, 9α-Dihydroxyverrucosane
(−)-2β-Hydroxy-9α-acetoxyverrucosane
(−)-2β-Hydroxy-11α-acetoxyverrucosane
(−)-2β-Hydroxy-9-oxoverrucosane
(−)-5β-Hydroxyneoverrucosane
(−)-2β-Hydroxyverrucosane

TRITERPENOIDS AND STEROLS

Of major taxonomic interest was the discovery by Marsili and Morelli (1968, 1970) of triterpene hydrocarbons such as the hopenes and fernenes in mosses. These are quite rare in liverworts. In 1968, these investigators reported the presence of 22 (29)-hopene (Fig. 8.20) in *Thamnium alopecurum*. It is also reported in *Thuidium tamariscifolium* (Marsili & Morelli, 1970). In the latter species, 7-fernene, 9(11)-fernene, and 21-hopene were also present.

Sixteen types of triterpenoids have been detected in bryophytes and these are as follows: α-amyrin, fern-7-ene, fern-9(11)-ene, 3a-friedelanol, 3e-friedelanol, friedeline, hop-17(21)-ene, hop-21-ene, hop-22(29)-ene, lupeol, 21α-methoxy-serrat-14-en-3-one, neohop-13(18)-ene, squalene, taraxerol, taraxerone, and ursolic acid (Huneck, 1983).

The steroids found in bryophytes are given in Table 8.3.

To sum up, so far 14 steroids have been isolated from bryophytes. Sitosterol and stigmasterol occur in all the investigated taxa. Campesterol is more common in liverworts, and rest of the steroids occur mainly in mosses.

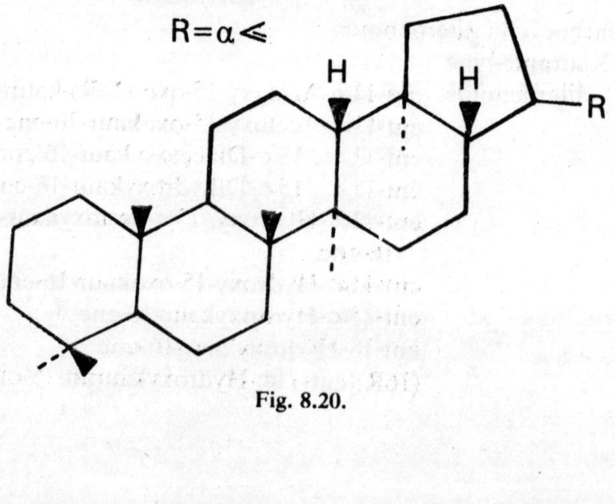

Fig. 8.20.

TABLE 8.3.
Steroids met with in bryophytes.

Taxa	Steroid/s	Investigator/s
MOSSES		
Abietinella	Campesterol, sitosterol, stigmasterol	Huneck (1971); Marsili et al. (1972)
Andreaea	31-Norcyclolaudenol, sitosterol. stigmasterol	Huneck et al. (1973)
Anomodon	Sitosterol	Catalano et al. (1976)
Brachythecium	Cycloeucalenol, cyclolaudenol, ergosterol. 31-norcyclolaudenol. sitosterol. stigmasterol	—do—
Breutelia	Sitosterol	Huneck et al. (1972)
Campylopus	Campesterol, cycloeucalenol, cyclolaudenol, ergosterol, sitosterol, stigmasterol	Catalano et al. (1976)
Climacium	Campesterol, sitosterol, stigmasterol	Marsili et al. (1972)
Ctenidium	Ergosterol. sitosterol. stigmasterol	Catalano et al. (1976)
Dicranum	Campesterol, cholesterol, cycloartenol, 24-methylenecycloartanol, sitosterol, stigmasterol	Ekman and Karunen (1980); Karunen et al. (1980a)
Hookeria	Sitosterol	Huneck et al. (1973)
Hypnum	Campesterol, 31-norcyclolaudenol, sitosterol, stigmasterol	Marsili et al. (1972)
Leucobryum	Sitosterol	Huneck et al. (1973)
Neckera	Campesterol, cyclolaudenol, sitosterol, stigmasterol	Marsili et al. (1972)
Platyhypnidium	Ergosterol, sitosterol	Huneck et al. (1973)
Polytrichum	Campesterol, cyclolaudenol, sitosterol, stigmasterol	Marsili et al. (1972)
Pseudoscleropodium	Campesterol, cyclolaudenol, 31-norcyclolaudenol, sitosterol, stigmasterol	Marsili et al. (1971)
Rhacomitrium	Campesterol, cycloeucalenol, cyclolaudenol, ergosterol. 31-norcyclolaudenol, obtusifoliol, sitosterol, stigmasterol	Huneck et al. (1973); Catalano et al. (1976)
Rhytidiadelphus	Campesterol, cyclolaudenol, 31-norcyclolaudenol, sitosterol, stigmasterol	Marsili et al. (1972); Huneck et al. (1973)
Scleropodum	Campesterol. cycloeucalenol. cyclolaudenol. sitosterol. stigmasterol	Catalano et al. (1976)
Sphagnum	Campesterol, cyclolaudenol, ergosterol. 31-norcyclolaudenol. obtusifoliol, sitosterol, stigmasterol	Black et al. (1955); Ives and O'Neil (1958); Marsili et al. (1972); Huneck et al. (1973)

(*Table contd.*)

Taxa	Steroid/s	Investigator/s
Thamnium	Ergosterol, sitosterol, stigmasterol	Marsili and Morelli (1968)
Thuidium	Campesterol, 24-methylenecycloartanol, sitosterol, stigmasterol	Marsili and Morelli (1970)
Tortella	Sitosterol	Huneck (1971)
LIVERWORTS		
Adelanthus	Campesterol, sitosterol, stigmasterol	see Asakawa (1983)
Anastrepta	Sitosterol	Huneck and Overton (1971)
Aneura	—do—	Asakawa et al. (1981c)
Anoplolejeunea	Campesterol, sitosterol, stigmasterol	see Asakawa (1983)
Anthelia	—do—	Asakawa et al. (1980c)
Anthoceros	Sitosterol	Asakawa et al. (1981c)
Archilejeunea	Campesterol, sitosterol, stigmasterol	see Asakawa (1983)
Asterella	—do—	—do—
Barbilophozia	—do—	Asakawa et al. (1981c)
Bazzania	—do—	Asakawa et al. (1979a)
Blasia	Brassicasterol, campesterol, sitosterol, stigmasterol	Asakawa et al. (1980b)
Brachiolejeunea	Campesterol, sitosterol, stigmasterol	see Asakawa (1983)
Bryopteris	—do—	—do—
Calycularia	—do—	—do—
Calypogeia	Campesterol, campesteryl behenate, sitosterol, stigmasterol	Meuche and Huneck (1966); Huneck (1974); Beneš and Kuzmiakova (1980); Asakawa et al. (1981c)
Cavicularia	Brassicasterol	Asakawa et al. (1980b)
Chiloscyphus	Brassicasterol, campesterol, cholesterol, stigmasterol	Matsuo et al. (1973d)
Clasmatocolea	Brassicasterol, campesterol, sitosterol, stigmasterol	Gradstein et al. (1981)
Conocephalum	Campesterol, sitosterol, stigmasterol	Benešová et al. (1969a); Asakawa (1980); Asakawa et al. (1980c)
Cryptochila	Brassicasterol, campesterol, sitosterol, **stigmasterol**	Gradstein et al. (1981)
Dicranolejeunea	Campesterol, sitosterol, stigmasterol	see Asakawa (1983)
Frullania	—do—	Asakawa et al. (1980b,c, 1981c)
Gongylanthus	Brassicasterol, campesterol, sitosterol	Gradstein et al. (1981)

(Table contd.)

Chemical Constituents of Bryophytes 209

Taxa	Steroid/s	Investigator/s
Gymnocolea	Campesterol, sitosterol, stigmasterol	Asakawa et al. (1981c)
Gymnomitrion	—do—	Asakawa et al. (1980b)
Haplomitrium	—do—	Asakawa et al. (1979a)
Herbertus	Brassicasterol, campesterol, sitosterol, stigmasterol	—do—
Heteroscyphus	Brassicasterol, campesterol, cholesterol, sitosterol, stigmasterol	Matsuo et al. (1973a); Asakawa et al. (1981c)
Hygrobiella	Campesterol, sitosterol, stigmasterol	Asakawa et al. (1980b)
Isotachis	Brassicasterol, campesterol, stigmasterol	Asakawa et al. (1979a, 1980b)
Jackiella	Stigmasterol	—do—
Jamesoniella	—do—	Asakawa et al. (1980b)
Jubula	Campesterol, sitosterol, stigmasterol	Asakawa et al. (1979a, 1980c, 1981c).
Jungermannia	Brassicasterol, campesterol, cycloartenol, sitosterol, stigmasterol	Matsuo et al. (1973d); Asakawa et al. (1980b,c, 1981c)
Lejeunea	Campesterol, sitosterol, stigmasterol	Asakawa et al. (1979a, 1981c)
Lepicolea	Campesterol, stigmasterol	see Asakawa (1983)
Lepidozia	Campesterol, sitosterol, stigmasterol	Asakawa et al. (1981c)
Leucolejeunea	—do—	see Asakawa (1983)
Lophocolea	—do—	Asakawa et al. (1980b)
Lophozia	Campesterol, phytosterols, sitosterol, stigmasterol	see Asakawa (1983)
Lunularia	Campesterol, sitosterol, stigmasterol	Asakawa et al. (1980b,c)
Macrodiplophyllum	Brassicasterol, campesterol, cholesterol, sitosterol, stigmasterol	Matsuo et al. (1973d)
Macvicaria	Campesterol, sitosterol, stigmasterol	Asakawa et al. (1979a)
Makinoa	—do—	Asakawa et al. (1980b)
Marchantia	Brassicasterol, campesterol, sitosterol, stigmasterol	Hopkins and Perold (1974); Asakawa et al. (1981c)
Marchesinia	Campesterol, sitosterol, stigmasterol	Asakawa et al. (1980c)
Marsupella	—do—	—do—
Megaceros	Brassicasterol, sitosterol, stigmasterol	Asakawa et al. (1979a)
Metzgeria	Campesterol, sitosterol, stigmasterol	Asakawa et al. (1981c)
Mylia	Sitosterol	Benešová et al. (1973)

(Table contd.)

Taxa	Steroid/s	Investigator/s
Neohattoria	Campesterol, sitosterol, stigmasterol	see Asakawa (1983)
Nowellia	—do—	Asakawa et al. (1980b)
Omphalanthus	—do—	see Asakawa (1983)
Pallavicinia	—do—	—do—
Pedinophyllum	—do—	Asakawa et al. (1981c)
Pellia	Campesterol, sitosterol, stigmasterol	Asakawa et al. (1980b, 1981c)
Phaeoceros	—do—	Asakawa et al. (1980c)
Plagiochila	Brassicasterol, campesterol, cholesterol, sitosterol, stigmasterol	Matsuo et al. (1973d); Asakawa et al. (1979a, 1981c)
Plagiochilion	Campesterol, sitosterol, stigmasterol	Asakawa et al. (1981c)
Porella	—do—	Huneck and Schreiber (1975); Asakawa et al. (1979a, 1980b,c)
Ptilidium	—do—	Asakawa et al. (1980b)
Ptychanthus	—do—	—do—
Radula	—do—	Asakawa et al. (1980b,c)
Reboulia	—do—	Asakawa et al. (1981c)
Riccardia	Campesterol, phytosterols, sitosterol, stigmasterol	see Asakawa (1983)
Riccia	Brassicasterol, campesterol, sitosterol, stigmasterol	Huneck et al. (1972); Asakawa et al. (1979b, 1980b)
Ricciocarpos	Sitosterol	Huneck et al. (1972)
Scapania	Brassicasterol, campesterol, cholesterol, sitosterol, stigmasterol	Matsuo et al. (1973d); Asakawa et al. (1980b, 1981c)
Spruceanthus	Campesterol, sitosterol, stigmasterol	Asakawa et al. (1980c)
Symbicczidium	—do—	see Asakawa (1983)
Syzygiella	Campesterol, stigmasterol	—do—
Takakia	Campesterol, sitosterol, stigmasterol	Asakawa et al. (1979a,c)
Trichocoleopsis	—do—	Asakawa et al. (1979a)
Tricholejeunea	—do—	Asakawa et al. (1980b,d)
Tuzibeanthus	Stigmasterol	Asakawa et al. (1980b)
Wiesnerella	Campesterol, sitosterol, stigmasterol	Asakawa (1980); Asakawa et al. (1979a)

FLAVONOIDS

Flavonoids are one of the most numerous and widespread groups of natural phenolic products in all major categories of green plants. These compounds, based upon the C_6-C_3-C_6 flavone nucleus, are classified according to the structure of the connecting C_3 portion of the molecule. Their biosynthesis involves a C_6-C_3 phenylpropanoid molecule, which then condenses with another aromatic ring.

The earliest report of the existence of flavonoids in bryophytes appears to be that of Paul (1908) who noticed anthocyanin-like pigments in a number of *Sphagnum* species. Molisch (1911) detected saponarin, a flavonoid-C-glycoside, from *Madotheca platyphylla*. Kozlowski (1921) observed this compound in the moss *Mnium*. In 1962, three rare flavonoids: 3-deoxyanthocyanin, luteolinidin 5-O-glycoside, and 5-O-diglucoside, were isolated by Bendz et al. from cell sap of the moss *Bryum cryophilum*. Markham et al. (1969) presented the first substantial proof of flavonoids in liverworts, and isolated two apigenin di-C-glycosides from *Hymenophyton flabellatum*.

Flavonoids seem to be more frequent in Hepaticae than in Musci. They are also not equally distributed within each group. Nearly all investigated species of Marchantiales possess flavonoids, whereas in Jungermanniales these compounds are not common. Among mosses the subclass Bryidae appears to have the most flavonoids. The study of flavonoid chemistry in liverworts have been very extensive and systematic. The predominant type of flavonoids in bryophytes are flavones, but other groups and some derivatives seem to occur fairly occasionally. Flavonoid composition is species specific and is a promising taxonomic tool for this group of plants (Markham and Porter, 1978a).

FLAVONES

Flavone glycosides are the most frequent flavonoids in bryophytes and the structure of nearly 50 such compounds has been determined.

Apigenin O-glycosides

The biosynthetically simplest flavone oxygenation pattern is represented by apigenin (Fig. 8.21). It is reported in the liverworts *Marchantia polymorpha* (Markham & Porter, 1974a), *M. berteroana* (Markham & Porter, 1975a), *Plagiochila asplenioides* (Mues & Zinsmeister, 1976), and the moss *Pleurozium schreberi* (Vandekerkhove, 1980). Apigenin has been isolated as various glycosides, mainly from liverworts. The most common is apigenin 7-O-glucuronide (Fig. 8.22) which is widespread in the Marchantiales and occurs almost exclusively in this order. Several apigenin derivatives listed in

212   Biology of Bryophytes

Fig. 8.21. $R_1 = R_2 = OH$
Fig. 8.22. $R_1h = GlcA-O-$; $R_2 = OH$

Table 8.4 are of interest because they are unique to bryophytes. The 4'-O-glucuronide, 7-O-glucuronide-4'-O-rhamnoside, 7-O-diglucuronide-4'-O-glucuronide, 7-O-, [2,4-di-O-(α-2-rhamnosyl)]-β-D-glucoside, and the acylated derivatives of apigenin 7- and 7,4'-O-glucosides are included in this group.

Apigenin derivatives also include the acacetin derivatives and the C-

TABLE 8.4.
Apigenin O-glycosides met with in bryophytes.

Apigenin glycoside	Taxa	Investigator/s
7-O-β-D-glucuronide	Conocephalum	Markham et al. (1976a)
	Hymenophyton	Markham et al. (1976b)
	Marchantia	Markham and Porter (1973, 1974b, 1975a,b)
	Ricciocarpos	Markham and Porter (1975c)
	Riccia	—do— Vandekerkhove (1978a)
	Riella	Markham et al. (1976c)
7-O-galacturonide	Reboulia	Markham et al. (1972)
7-O-β-D-glucoside	Haplomitrium	Markham (1977)
	Marchantia	Markham and Porter (1973)
	Plagiochasma	Schier (1974)
	Riccia	—do—
7,4'-di-O-β-D-glucoside (also acylated)	Haplomitrium	Markham (1977)
7-O-[2,4-di-O-(α-L-rhamnosyl)]-β-D-glucoside	Dicranum	Nilsson et al. (1973)
7,4'-di-O-β-D-glucuronide	Conocephalum	Markham et al. (1976a)
	Marchantia	Markham and Porter (1974a)
7-O-β-D-glucuronide-4'-O-α-L-rhamnoside	Conocephalum	—do—
7-O-β-D-diglucuronide-4'-O-β-D-glucuronide	—do—	—do—
7-O-neohescridoside	Hylocomium	Vandekerkhove (1977a);
	Pleurozium	Vandekerkhove (1980)

glycosylated apigenins. Acacetin was detected for the first time in *Marchantia berteroana* by Markham et al. (1978a). From *Reboulia hemisphaerica*, the previously unknown flavone O-glycoside, acacetin 7-O-rhamnosylgalacturonide, was described by Markham et al. (1972). This compound was accompanied by O-glycoside of an acacetin 8-C-glycoside. Acacetin 7-O-galacturonide was also detected in this plant and it has since been isolated together with 7-O-glucuronide from the liverwort *M. berteroana* (Markham et al., 1978a). It is possible that acacetin 7-O-glucoside is present in the moss *Sphagnum magellanicum*, since a flavone glucoside with the approximate molecular formula $C_{22}H_{35}O_{17}$ was reported by Manskaja and Drozdova (1955).

Apigenin C-glycosides

C-Glycosides of apigenin occur frequently in liverworts and mosses, but in most instances the nature of the sugar moieties has not been firmly established. Apigenin 6, 8-di-C-glycosides (vicenin-2) (Fig. 8.23) is reported in *Marchantia foliacea* (Markham & Porter, 1973). The dominant C-glycosides of *Hymenophyton* sp. are an isomeric pair of apigenin di-C-pentosides. Other C-glycosylated apigenins from *Hymenophyton* sp. include the schaftoside-isoschaftoside isomers, apigenin 6-C-glucoside-8-C-arabinoside, and 6-C-arabinoside-8-C-glucoside (Markham et al., 1976c).

The presence of isovitexin and its 7-O-glucoside, saponarin, in liverworts has been firmly established since the very early work of Molisch (1911). Nilsson (1969a) and Tjukavkina et al. (1970) independently provided confirmatory evidence for the presence of isovitexin and saponarin in *Porella platyphylla*. Nilsson (1969a, 1973a) noticed the presence of saponarin in a number of other *Porella* species including *P. cordaeana*, *P. platyphylloidea*, *P. bolanderii*, *P. capensis*, and *P. baueri*. Nilsson (1973b) also reported apigenin 6, 8-di-C-glucoside (vicenin-2) in *P. platyphylla*. Saponarin and vicenin-2 are also present in *Mnium undulatum* (Vandekerkhove, 1978a).

Melchert and Alston (1965) reported the occurrence of apigenin 8-C-glucoside (vitexin), apigenin 6,8-di-C-glycoside, and O-glucosides in the moss *Mnium affine*. These possibly occur with saponarin in *M. cuspidatum*

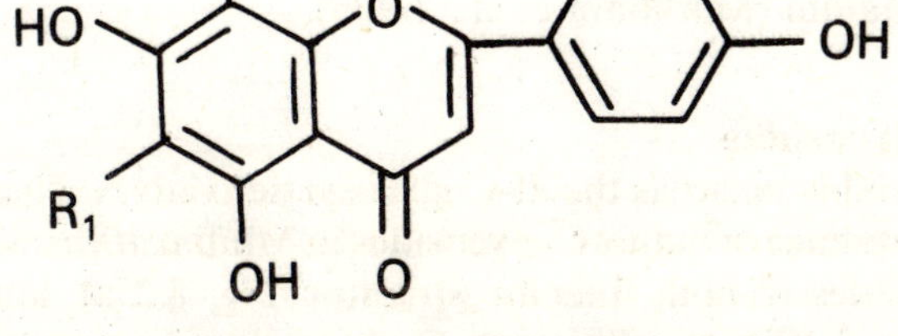

$R_1 = R_2 = \beta\text{-D-glucopyranosyl}$

Fig. 8.23.

as reported by Kozlowski (1921), and Melchert and Alston (1965). Mues and Zinsmeister (1976) observed the occurrence of 6,8-di-C-hexosylapigenin in the liverwort *Plagiochila asplenioides*.

Luteolin O-glycosides

Luteolin (Fig. 8.24) and its derivatives occur widely in liverworts and are also represented in mosses. Luteolin is met with in *Marchantia berteroana*

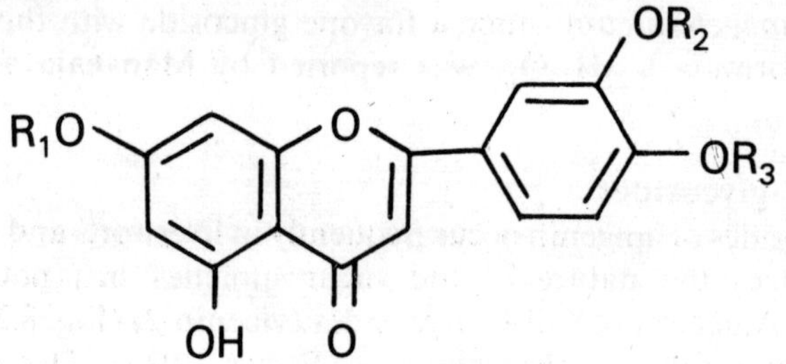

$R_1 = R_2 = R_3 = H$

Fig. 8.24.

(Markham & Porter, 1975a), *Riccia fluitans* (Vandekerkhove, 1978b), and in the sporophytic tissue of *Ceratodon purpureus* (Vandekerkhove, 1978c). Several luteolin derivatives are listed in Table 8.5.

The luteolin 3',4'- and 7,3',4'-glycosylation pattern were reported in *Marchantia polymorpha* and *Lunularia cruciata* by Markham and Porter (1974a,b). Glycuronide-rhamnoside derivatives of luteolin occur only in *Conocephalum conicum*, and on the basis of presence of luteolin 7-O-glucuronide-3',4'-di-O-rhamnoside the European geographic race is distinguished from the North American race which lacks this compound (Markham et al., 1976a).

A methylated derivative of luteolin, chrysoeriol, has been detected in *Marchantia foliacea* (Markham & Porter, 1973). Chrysoeriol in its O-glycosylated form has been reported as the 7-O-glucuronide in species of *Riella* (Markham et al., 1976b), *Marchantia* (Markham & Porter, 1973, 1975b), and *Conocephalum* (Markham et al., 1976b).

Luteolin C-glycosides

These invariably occur as the di-C-glycosyl derivatives. There is only one report of the presence of mono-C-glycosides in *Mnium affine* which contains the 8-C-glucosides of both luteolin (orientin Fig. 8.25a) and chrysoeriol (scoparin Fig. 8.25b) together with O-glycosyl derivatives of these compounds (Melchert & Alston, 1965).

Several luteolin 6,8-di-C-glycosides have been reported in bryophytes, the most common of which is luteolin 6,8-di-C-β-D-glucopyranoside (lucenin-2).

TABLE 8.5.
Luteolin O-glycosides met with in bryophytes.

Luteolin glycoside	Taxa	Investigator/s
7-O-β-D-glucuronide	Conocephalum	Markham et al. (1976a)
	Marchantia	Markham and Porter (1974a, 1975a,b)
	Riccia	Markham and Porter (1975c)
	Riella	Markham et al. (1976b)
	Sphaerocarpos	—do—
7-O-galacturonide	Marchantia	Markham and Porter (1975a)
	Reboulia	Markham et al. (1972)
3'-O-β-D-glucuronide	Lunularia	Markham and Porter (1974b)
	Marchantia	Markham and Porter (1974a, 1975a,b)
	Riccia	Markham and Porter (1975c)
	Ricciocarpos	—do—
	Riella	Markham et al. (1976c)
3'-O-galacturonide	Marchantia	Markham and Porter (1975a)
7,3'-di-O-galacturonide	—do—	—do—
7,3'-di-O-β-D-glucuronide	Conocephalum	Markham et al. (1976a)
	Marchantia	Markham and Porter (1974a, 1975a,b)
	Riccia	Markham and Porter (1975c)
	Ricciocarpos	—do—
7,4'-di-O-β-D-glucuronide	Conocephalum	Markham et al. (1976a)
	Marchantia	Markham and Porter (1974a, 1975a)
	Sphaerocarpos	Markham et al. (1976a)
7,4'-di-O-galacturonide	Marchantia	Markham and Porter (1975a)
7-O-β-D-glucuronide-4'-O-α-rhamnoside	Conocephalum	Markham et al. (1976a)
3,4'-di-O-β-D-glucuronide	Lunularia	Markham and Porter (1974b)
	Marchantia	Markham and Porter (1974a, 1975a)
3,4'-di-O-galacturonide	—do—	Markham and Porter (1974a, 1975a)
	Riccia	Markham and Porter (1975c)
7-O-β-D-glucuronide-3',4'-di-O-α-L-rhamnoside	Conocephalum	Markham et al. (1976a)
7-O-glucuronide-3'-O-rhamnoside	Riccia	Vandekerkhove (1978b)
7-O-glucuronide-3'-mono (trans) ferulyl-glycoside	—do—	Markham et al. (1978b)

This was isolated initially from *Marchantia foliacea* (Markham & Porter, 1973) and has subsequently been detected in *Conocephalum conicum* (Markham et al., 1976a), *Plagiochila asplenioides* (Mues & Zinsmeister, 1976), and *Riccia fluitans* (Vandekerkhove, 1978b).

Tricetin Derivatives

Tricetin (Fig. 8.26a) O-glycosides are rare in plants. The only related O-glycoside reported in bryophytes is the 7-O-glucuronide of tricin (Fig. 8.26b)

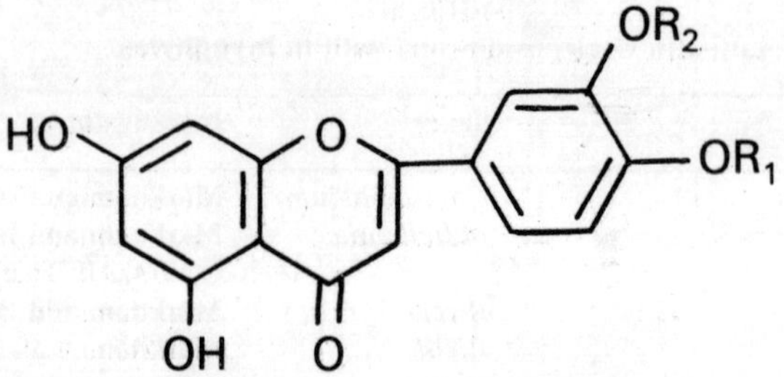

(a) $R_1 = H$; $R_2 = \beta$-D-glucopyranosyl
(b) $R_1 = CH_3$; $R_2 = \beta$-D-glucopyranosyl

Fig. 8.25a,b.

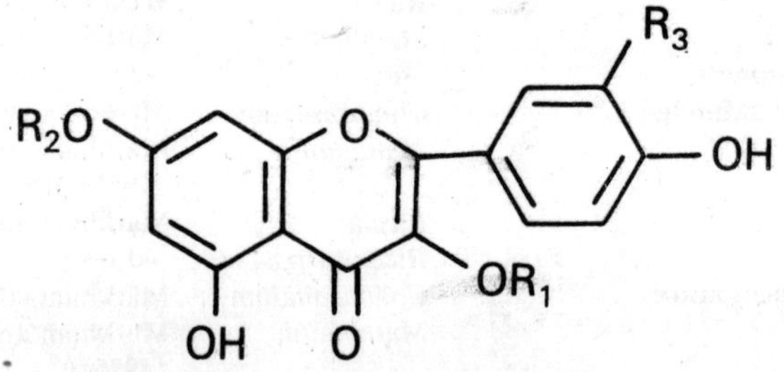

(a) $R_1 = R_2 = R_3 = H$
(b) $R_1 = R_2 = H$; $R_3 = CH_3$

Fig. 8.26a,b.

in *Marchantia foliacea* (Markham & Porter, 1973), where it co-occurs with 7-O-rhamnosylglucuronide.

Tricetin itself is represented in bryophytes by the series of unique di-C-glycosides, which have been isolated from *Plagiochila asplenioides* (Mues & Zinsmeister, 1976), *Takakia* sp. (Markham & Porter, 1979), *Apometzgeria pubescens* (Theodar et al., 1980, 1981a), and *Metzgeria* sp. (Theodar et al., 1981b). From *Metzgeria* sp., a new di-C-glucoside, tricin 6-C-xyloside-8-C-hexoside has been reported (Theodar et al., 1981b).

6- and 8-Hydroxylated Flavone Derivatives

Nilsson (1969b) isolated a flavone with additional 6- or 8-oxygenation from the moss *Bryum weigelii*. Scutellarein 7-O-glucoside (Fig. 8.27) was also identified in this plant.

Other examples of 6- and 8-hydroxylated flavones are confined to liverworts. Markham (1977) isolated an acylated isoscutellarein (8-hydroxy-apigenin)-7-O-glucoside from *Haplomitrium gibbsiae*. Isoscutellarein was

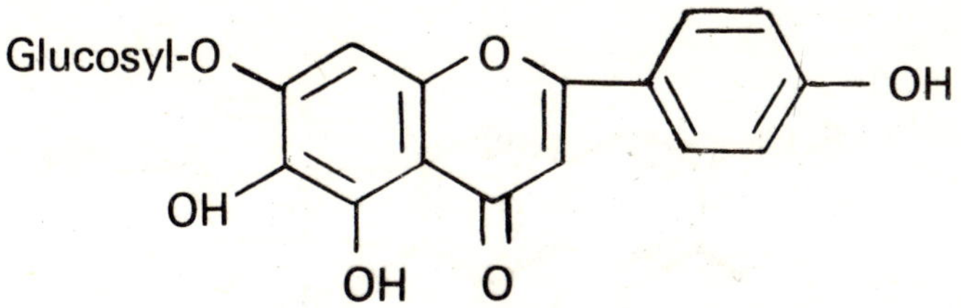

Fig. 8.27.

first isolated from *Marchantia berteroana* (Markham & Porter, 1975a) where it occurs as the 8-O-glucuronide, and probably also as its 8,4'-di-O-glucuronide. Hypolaetin (8-hydroxyluteolin)-8-O-glucuronide and 8,4'-di-O-glucuronide were also isolated from the same liverwort.

8-Methoxyluteolin has been detected in the giant liverwort *Monoclea forsteri* (Markham, 1972). It is glycosylated at the 7- and 4'-position with an oligosaccharide or oligosaccharides such that its molecular weight is approximately 3200, and is accompanied by at least two similar compounds, which probably contain a C-linked oligosaccharide. Recently, Markham and Mues (1983) identified a new flavonoid 5,7-dihydroxy-8,4'-dimethoxyflavone (bucegin) 7-O-β-D-glucuronide from *Bucegia romanica* of Marchantiales.

ISOFLAVONES

Anhut et al. (1984) reported isoflavonoids in bryophytes. They reported isoflavones, orobol and pratensein, as the 7-O-glucosides and 7-(6''-malonylglucosides), in *Bryum capillare*.

FLAVONOLS

The first evidence of flavonols in bryophytes was given by Reznik and Wiermann (1966) who reported kaempferol (Fig. 8.28a) and quercetin (Fig. 8.28b) in *Corsinia coriandrina*. This was substantiated by Markham (*cited in* Markham & Porter, 1978a) who observed that a Spanish sample of this plant contained two kaempferol 3-O-di-, or tri-glycosides and at least three quercetin 3-O-glycosides, one of which is thought to be the 3-O-galacturonide. Quercetin 3-O-rutinoside is also identified in *C. coriandrina* and *Reboulia hemisphaerica* (Schier, 1974).

Markham et al. (1976c) noticed four kaempferol glycosides: kaempferol 3-O-rhamnoglucoside; kaempferol 3,7-di-O-rhamnoside; kaempferol-3, O-glucoside-7-O-rhamnoside; and kaempferol 3-O-rhamnoglucoside-7-O-rhamnoside from *Hymenophyton leptopodum*. These investigators identified this species from the closely related *H. flabellatum* on the basis of

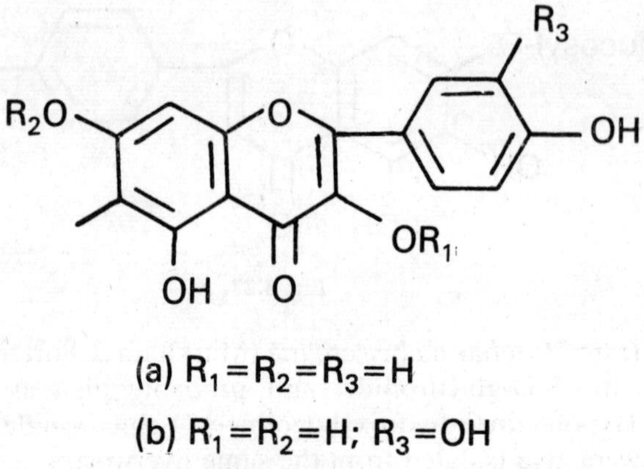
R3
R2O
O
OH
OR1
OH O
(a) R1=R2=R3=H
(b) R1=R2=H; R3=OH

Fig. 8.28a,b.

presence of these compounds. Markham (1980) reported 3-O-glycuronide derivatives from *Carrpos sphaerocarpos* and *Corsinia coriandrina*. *Carrpos* produced biosynthetically simple flavone 7-O-glucuronides and aureusidin 6-O-glucuronide.

Flavonols are normally absent in mosses. The only available report is that of a quercetin 3-O-diglycoside in *Mnium affine* by Melchert and Alston (1965).

DIHYDROFLAVONOIDS AND BIFLAVONYLS

Dihydroflavones are of common occurrence in bryophytes. Naringenin (Fig. 8.29a) and its 7-O-glucoside (Fig. 8.29b) were isolated from *Riccia crystallina* (Markham & Porter, 1975c). These two compounds comprised about 60

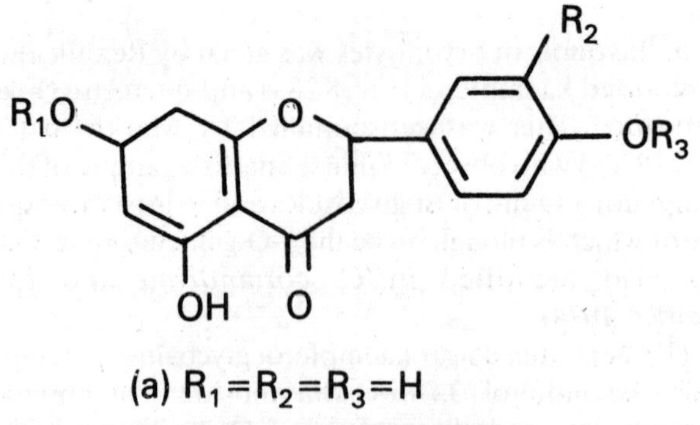
R2
R1O
O
OR3
OH O
(a) R1=R2=R3=H
(b) R1=β-D-glucosyl; R2=R3=H

Fig. 8.29a,b.

percent of the total flavonoids in the plant *Georgia pellucida* and *Riccia bicarinata* contain dihydroflavones of unknown structures (Vandekerkhove, 1977b; Spencer, 1979).

The only example of the occurrence of a biflavone in a non-vascular plant is that of the new 5'. 8"-biluteolin which Lindberg et al. (1974) isolated from *Dicranum scoparium*.

AURONES AND CHALCONES

Aurones occur both in liverworts and mosses, but chalcones have been reported only from Hepaticae. The antheridiophores of *Marchantia polymorpha*, *M. berteroana*, and *Conocephalum supradecompositum* contain the aurone, aureusidin 6-O-glucuronide (Fig. 8.30; Markham & Porter, 1978b). The aurone bracteatin (Fig. 8.31) has been identified in the moss *Funaria hygrometrica* (Weitz & Ikan, 1977).

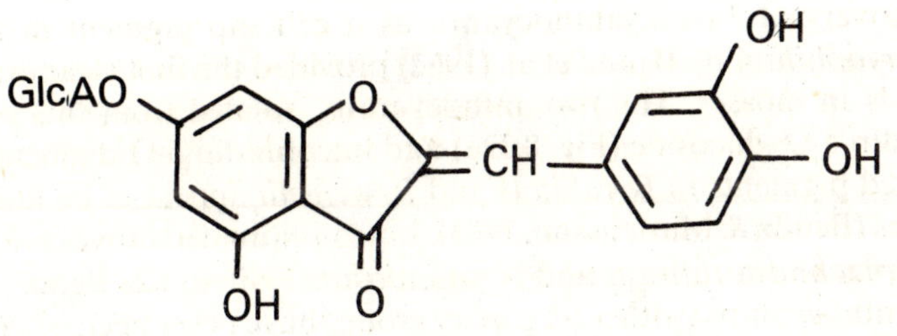

Fig. 8.30.

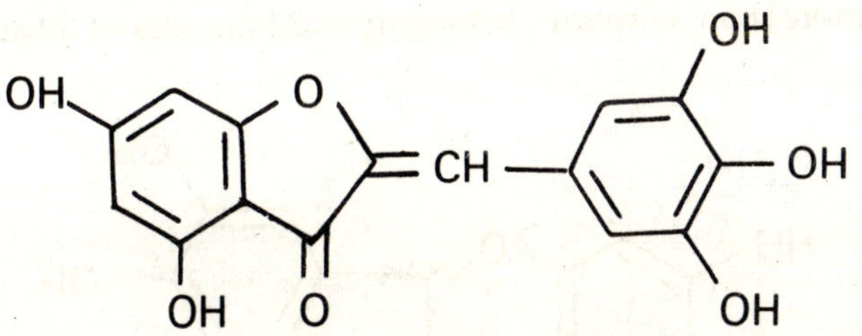

Fig. 8.31.

Chalcones of unknown structures have been reported from *Plagiochasma peruvianum*, *P. rupestre*, and *P. tenue* (Schier, 1974). Dihydrochalcone glycoside has been reported from *Hymenophyton leptopodum*, and is the first example of a chalcone C-glycoside in plants (Markham et al., 1976c). It has been assigned the structure 4,2',4',6'-tetrahydroxydihydrochalcone 3',5'-di-C-glucoside.

ACYLATED FLAVONOIDS

Acylated flavonoids have only recently been shown to occur in bryophytes. The liverwort *Haplomitrium gibbsiae* contains, three major flavonoid glucosides: apigenin 7-O-glucoside, apigenin 7,4'-di-O-glycoside, and isoscutellarein 7-O-glucoside. All of these are thought to contain at least one acyl group per sugar (Markham, 1977).

In *Riccia fluitans*, luteolin 7-O-glucuronide-3'-O-ferulyl-glucoside occurs in both acylated and unacylated forms (Markham et al., 1978b).

The only other example of acylation of flavonoids in the liverworts is the recent finding of an acylated flavone C-glycosides in a species of *Metzgeria* (Theodar et al., 1981b). In this plant the isoschaftoside (apigenin 6-C-arabinoside-8-C-glucoside) is acylated with one or more ferulyl units.

ANTHOCYANINS AND PROANTHOCYANIDINS

The discovery of 3-deoxyanthocyanin as a cell sap pigment in the moss *Byrum cryophilum* by Bendz et al. (1962) provided the first clear evidence of flavonoids in mosses. The two anthocyanins isolated from this plant are: luteolinidin 5-O-glucoside (Fig. 8.32a) and luteolinidin 5-O-diglucoside (Fig. 8.32b). Red pigments in *B. rutilans* and *B. weigelii* appear to be luteolinidin glucosides (Bendz & Martensson, 1963). Luteolinidin derivatives are also present in *Splachnum rubrum* and *S. vasculosum* (Nilsson & Bendz, 1973).

True anthocyanins (with a 3-hydroxy group) have never been isolated from liverworts. Schier (1974) observed anthocyanin-like spots in a number of Marchantiales by a paper chromatographic survey.

An extensive investigation by Bendz et al. (1966), and Nilsson and Bendz (1973) of more than 30 species belonging to 22 families of all subclasses of

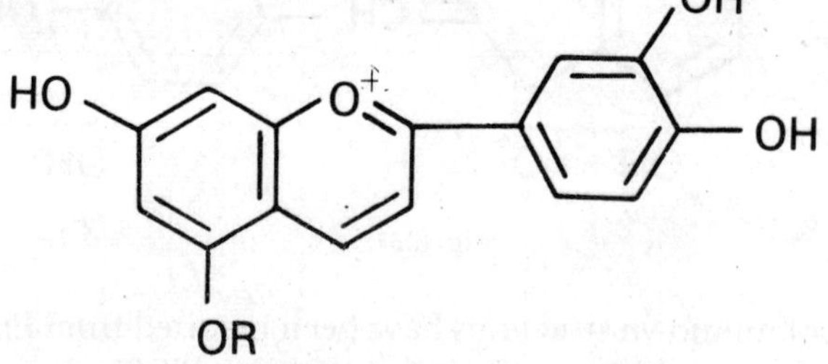

(a) R=glucosyl
(b) R=diglucosyl

Fig. 8.32 a,b.

Musci, for proanthocyanidins did not yield positive results. The existence of proanthocyanidins in liverworts is assumed on the basis of a chromatographic study (Schier, 1974).

SPHAGNORUBINS

An entirely new type of flavonoid derivative is the reddish-violet pigment of *Sphagnum magellanicum* named sphagnorubin (Fig. 8.33) by Rudolph and Vowinkel (1969). This pigment has been shown to be anthocyanin derived. In contrast to the anthocyanins, this compound is fixed to the constituents of the cell wall, and its molecular formula is $C_{23}H_{14}O_6$. The structure established by Vowinkel (1975) proves it to be 8, 11-dihydroxy-2-(3,4-dihydroxyphenyl)-9H-phenanthro·(2,1-b) pyran-9-one.

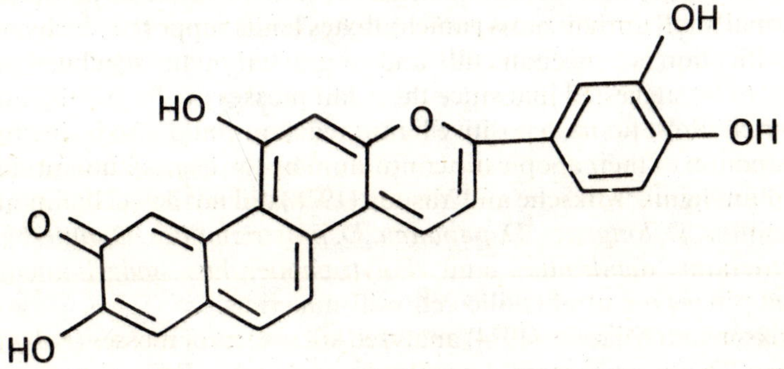

Fig. 8.33.

The biosynthesis of sphagnorubin clearly involves both the shikimate and acetate-malonate pathways. Vowinkel (1975) has suggested two possible routes: one via an intermediate involving one cinnamoyl residue; and the other via an intermediate involving two cinnamoyl residues. The first route would require the loss of two hydroxyl groups and the introduction of one, whereas the second provides for direct oxidation.

LIGNINS

Lignin is the polyphenolic C_6-C_3 polymer which acts as a water-resistant cement essential for formation of cells such as tracheids, fibers, and vessels. It is ultimately associated with the cell-wall polysaccharides and tends to undergo a variety of secondary reactions when attempts are made to extract it. For this reason, no complete structure of any naturally occurring lignin has yet been determined, although partial model structures involving known inter-monomer bond types have been assembled (Freudenberg & Neish, 1968). It has been suggested that lignin could have arisen as a detoxification product by conversion of phenolic compounds to an insoluble form by oxidative polymerization (Markham & Porter, 1978a).

Studies using Maule test for the detection of syringyl groups in lignin failed to show lignin in *Marchantia* and *Funaria* species, but it was detected in vascular plants such as *Psilotum, Lycopodium, Equisetum,* and *Selaginella* (Towers & Gibbs, 1953). In 1962, Siegel examined the dioxan soluble lignin of a number of lower plants and found evidence for the presence of traces of lignin-like materials in the moss *Polytrichum* (peristome teeth tissue) and in the liverwort *Marchantia polymorpha.* However, in a later communication Siegel (1969) commented that lignin is absent during all growth stages in liverworts of the genera *Marchantia, Riccia,* and *Jungermannia,* and in the gametophytes of *Bryum, Rhodobryum, Mnium, Polytrichum,* and *Funaria.* But, he reported lignin in the gametophyte axes of the giant New Zealand mosses, *Dawsonia* and *Dendroligotrichum.* The presence of lignin in these exceptionally tall upright moss gametophytes lends support to the hypothesis that lignification is a mechanically and/or gravitationally regulated process. Hébant (1974) suggested that since these tall mosses are frequently found in humus-rich forest floors, a positive lignin reaction could also be the result of contamination of their apoplastic continuum by the degradation products of higher plant lignin. Miksche and Yasuda (1978) did not detect lignin in *Dawsonia grandis, D. longiseta, D. papuana, D. polytrichoides, D. superba, Dendroligotrichum dendroides,* and *Polytrichadelphus magellanicus,* but observed other type of phenolic cell wall materials.

Erickson and Miksche (1974) analyzed six species of mosses (*Polytrichum commune, Dicranum bergeri, Leptobryum pyriforme, Ptilium crista-castrensis, Pogonatum urnigerum,* & *Sphagnum palustre*) and two of liverworts (*Plagiochila asplenioides* & *Scapania undulata*) for the presence of lignin. Oxidative degradation in each case yielded minor amounts of methyl veratrate (Fig. 8.34a), dimethyl isohemipate (Fig. 8.34b), and dimethyl metahemipate (Fig. 8.34c) which were shown to be derived from catechol-containing products. Oxidation products also gave methyl 4,7, 9-trimethyl-2-dibenzofurancarboxylate (Fig. 8.35a) and the derived related ester (Fig. 8.35b). One half of the precursor of these possesses a phenylpropane skeleton and the other probably originates from acetates. There were no oxidation products typical of guaiacyl and guaiacyl-syringyl lignins, such as dimethyl 2′,5,6-trimethoxydiphenyl-ether-3,4-dicarboxylate. It was therefore concluded that none of these plants contains lignin, not even *Sphagnum* which was earlier reported to contain a p-hydroxyphenyl type of lignin (Bland et al., 1968). For comparative purposes, a synthetic equivalent of p-hydroxyphenyl lignin was prepared, and the oxidative degradation products of this lignin were indeed quite different from those obtained from *Sphagnum.* Two of the studied bryophytes, *Ptilium crista-castrensis* and *Plagiochila asplenioides,* however, yielded large quantities of dimethyl 5,5-dehydrodianisate, one of the four significant degradation products from the

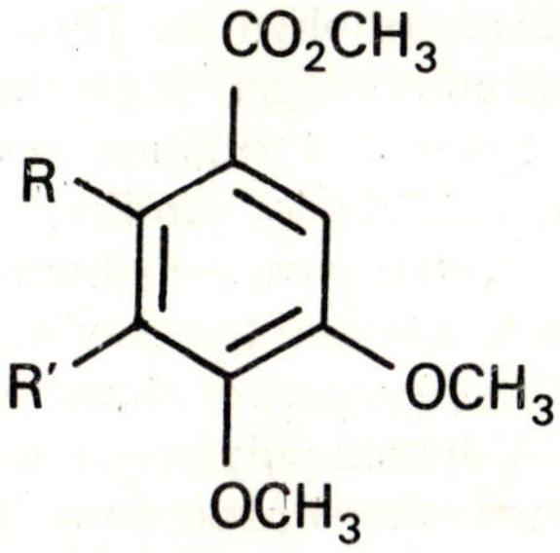

CO_2CH_3

(a) R=R'=H
(b) R=H; R'=CO_2CH_3
(c) R=CO_2CH_3; R'=H

Fig. 8.34a,b,c.

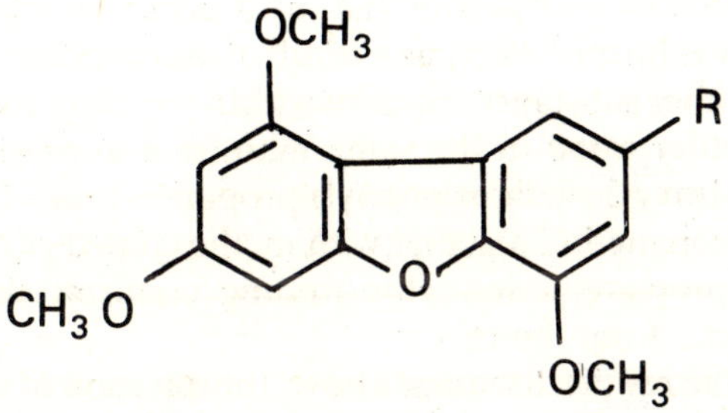

(a) R=CO_2CH_3
(b) R=$CH_2CH_2CO_2CH_3$

Fig. 8.35a,b.

synthetic lignin. This does not imply that these species contain true lignin. The investigators interpreted this as an evidence for the presence in these plants of a constituent rich in 2,2'-dihydroxy-4,4'-dialkylbiphenyl units rather than a p-hydroxyphenyl lignin which they considered would also have yielded the other three degradation products. This interpretation, however, is based on the observed behaviour of a synthetic lignin which was prepared from p-coumaryl alcohol with peroxidase enzyme in a cell-free environment.

Contrary to the conclusions based on oxidative degradation other recent work supports earlier studies which indicated that *Sphagnum* contains a lignin built chiefly of p-hydroxyphenyl units. For example, Bland et al. (1968) used a vibratory-ball milling technique and solvent extraction to isolate a low yield (0.9%) of a presumed lignin from pre-extracted *Sphagnum*. Elemental analysis indicated its molecular formula to be $C_9H_{8.62}O_2$

$(H_2O)_{0.62}(OCH_3)_{0.27}$ after removal of carbohydrate. This substance was low in methoxyl aromatic hydrogen and ether oxygen. It gave an acetate containing 36.5 percent $COCH_3$ and nitrobenzene. Oxidation of this acetate yielded small amounts of p-hydroxybenzaldehyde, vanillin, and syringaldehyde. Permanganate oxidation of the acetate gave p-hydroxybenzoic acid, small amounts of vanillic and syringic acids, and traces of 6-hydroxyisophthallic and 5-carboxyvanillic acids. These properties show that the substance is basically a highly condensed C-C-linked polymer of methoxyl-free units of structure with some methoxylated ether linked units. This substance may therefore reasonably be classified as a lignin. Characteristics of lignin isolated from *Sphagnum* and from plants belonging to other classes support the concept that lignins are p-hydroxyphenyl propane polymers. The different degrees of condensation in different lignins appear to be determined by the degree of substitution of the phenyl propane units. Reznikov and Novitskii (1975; *cited in* Markham & Porter, 1978a) studied the dioxan lignin of *Sphagnum medium* by the technique of reductive degradation with Na/liquid ammonia. The structure of the isolated dioxan lignin was insignificantly different from that of protolignin (i.e. as it occurs in plants), and the phenolic fraction was formed solely as a result of decomposition of the lignin and not from any other substances present within the plant tissue. Reductive degradation of conifer wood in the same manner also produced products such as 1-(4-hydroxy-3-methoxyphenyl)-propane and 1-(4-hydroxy-3-methoxyphenyl)-propane-1-ol together with methoxylated equivalents of the p-hydroxyphenyl propane series. This finding confirms that such compounds are produced from lignin.

On the basis of the results discussed above, the question of the existence of lignin in bryophytes is still not clearly answered. Confusion results partly from the lack of a sound chemical definition of lignin, and partly from the variety of isolation techniques and degradative procedures used. However, if one accepts that lignin is a phenolic cell-wall polymer of shikimate derived C_6-C_3 subunits, then it is highly probable that such a lignin is present at least in some mosses. The lignin is confined to specialized structures, particularly stems and is composed primarily of p-hydroxyphenyl propane subunits, the methoxyl content always being low. Evidences for the type of intermonomer linkages present is sketchy, although in *Sphagnum*-lignin there are good indications that both C-C and ether bonds are involved. Indeed, the similarity of the structural features deduced from a study of the reductive degradation products of both *Sphagnum* and conifer-lignins, has led Reznikov and Novitskii (1975; *cited in* Markham & Porter, 1978a) to conclude that there is a definite genetic relationship between the lignins of higher and lower plants.

OTHER CONSTITUENTS

In addition to the major constituents described earlier, some relatively less important ones are being dealt with in the following pages.

CAROTENOIDS

Although the presence of carotenoids in bryophytes was established as early as 1902 by Kohl, very little is as yet known about their quantitative and qualitative occurrence in this group. The bryophytes analyzed for carotenoids are listed in Table 8.6. Douin (1956, 1958) employed paper chromatography for investigating 40 species of Bryales and some species of Sphagnales and Andreaeales as well as 20 species of Marchantiales and Jungermanniales. He observed α- and β- carotene, lutein, and 5,6-epoxylutein in all the investigated taxa and concluded that the carotenoid distribution seems to be fairly uniform in bryophytes, the difference being mainly of a quantitative nature.

Some qualitative variations have also been noticed by Freeland (1957) and Bendz et al. (1968). Freeland (1957) detected α - and β-carotene, lutein, violaxanthin, and zeaxanthin in both gametophytes and sporophytes of five moss species. Two species (*Entodon seductrix* & *Ditrichum vaginas*) also had neoxanthin, and in one of them (*Entodon seductrix*) cryptoxanthin was identified in both the generations. Bendz et al. (1968) isolated ten carotenoids from *Fontinalis antipyretica*, eight of which have been tentatively identified as α- and β-carotene, neo-β-carotene U, lutein, 5,6-epoxylutein, violaxanthin, neoxanthin, and auroxanthin or auroxanthin-like pigment. Czeczuga (1972) used column and thin-layer chromatography and reported the presence of α-carotene, cryptoxanthin, lutein, epoxylutein, zeaxanthin, violaxanthin, and neoxanthin in *Bryum ventricosum*. β-Carotene was not identified in this moss.

Karunen and Ihantola (1977a) studied the quantitative composition of carotenoids in *Polytrichum commune* spores, and reported the presence of carotenes (555.98 $\pm$ 11.6 µg/g spores), violaxanthin (77.37 $\pm$ 1.9), lutein (+ antheraxanthin) (622.49 $\pm$ 12.4), neoxanthin (313.88 $\pm$ 12.3), and zeaxanthin (185.25 $\pm$ 7.3). The ratio of carotenes and xanthophylls is about the same as that in green tissues of higher plants. During the germination process the formation of xanthophylls is more pronounced than that of carotenes, and the increase is greatest in the contents of violaxanthin and zeaxanthin (Karunen & Ihantola, 1977b).

CARBOHYDRATES

The reports of specific carbohydrates in bryophytes are listed in Table 8.7. Maass and Craigie (1964) reported the presence of sucrose, fructose, glucose, and a series of fructosides in *Sphagnum* species and the liverwort *Scapania*

TABLE 3.6.
Carotenoids met with in bryophytes.

Taxa	Carotenoid/s	Investigator/s
MOSSES		
Aulacomnium	α- and β-Carotene, lutein, neo-viola-, and zeaxanthin	Strain (1958)
Bryum	α-Carotene, epoxylutein, lutein, neo-, viola-, zea-, and cryptoxanthin	Czeczuga (1972)
Chamberlainia	α- and β-Carotene, lutein, viola-, and zeaxanthin	Freeland (1957)
Dawsonia	α- and β-Carotene, lutein, neo-, viola-, and zeaxanthin	Stránský et al. (1967)
Ditrichum	—do—	Freeland (1957)
Entodon	α- and β-Carotene, crypto-, neo-, viola-, and zeaxanthin, lutein	—do—
Fontinalis	Auro-, neo-, and violaxanthin, α- and β-carotene, epoxylutein, lutein, neo-β-carotene U	Bendz et al. (1968)
Hygrohypnum	α- and β-Carotene, lutein, neo-, viola-, and zeaxanthin	Strain (1958)
Mnium	α- and β-Carotene, lutein, viola- and zeaxanthin	Freeland (1957)
Philonotis	α- and β-Carotene, lutein, neo-, viola-, and zeaxanthin	Strain (1958)
Pohlia	β-Carotene, lutein, neo-, viola-, and zeaxanthin	—do—
Polytrichum	Anthera-, viola-, and zeaxanthin, α- and β-carotene, lutein, neoxanthin, neo A	Karunen and Ihantola (1977a,b)
Sphagnum	α- and β-Carotene, lutein, neo-, viola-, and zeaxanthin	Strain (1958)
Thuidiopsis	α- and β-Carotene, lutein, neo-, viola-, and zeaxanthin	Stránský et al. (1967)
Thuidium	α- and β-Carotene, lutein, viola-, and zeaxanthin	Freeland (1957)
LIVERWORTS		
Aneura	β-Carotene	Benešová et al. (1969a)
Asterella	α- and β-Carotene, lutein, neo-, viola-, and zeaxanthin	Strain (1958); Stránský et al. (1967)

(Table contd.)

Taxa	Carotenoid/s	Investigator/s
Conocephalum	—do—	Strain (1958)
Lophocolea	Anthera-, neo-, viola-, and zeaxanthin, β-carotene, lutein	Taylor et al. (1972); Mues et al. (1973)
Marchantia	Anthera-, neo-, viola-, and zeaxanthin, β-carotene, lutein, phytochrome	Strain (1958); Fredericq and DeGreef (1968); DeGreef and Fredericq (1969)
Pellia	Neo- and violaxanthin	Lichtenthaler (1968)
Riccardia	β-Carotene	Benešová et al. (1969a); Huneck and Schreiber (1975)
Ricciocarpos	β-Carotene	Huneck and Schreiber (1975)
Sphaerocarpos	Anthera-, neo-, viola-, and zeaxanthin, α- and β-Carotene, lutein	Herrmann (1968)

TABLE 8.7.
Carbohydrates met with in bryophytes.

Taxa	Carbohydrate/s	Investigator/s
MOSSES		
Camptothecium	Maltose, melibiose, sucrose	Margaris and Kalaitzakis (1974)
Homalothecium	Glucose, maltose, mannose, sucrose	—do—
Platyhypnidium	Glucose, maltose, melibiose, sucrose	—do—
Rhyncostegium	Glucose, maltose, mannose, sucrose	—do—
Sphagnum	Arabinose, fructose, galactose, glucose, mannose, uronic acids, xylose	Theander (1954)
Tortula	Deoxyribose, fructose, glucose, maltose, mannose, melibiose, sucrose	Margaris and Kalaitzakis (1974)
LIVERWORTS		
Aerobolbus	Inuline	Rancken (1914)
Apometzgeria	Polyfructoside	Quillet (1956a,b)
Asterella	Amidon, mannitol	Rao and Das (1968)
Balantoipsis	Inuline	Rancken (1914)
Balbilophozia	Polyfructoside	Quillet (1956a)
Bazzania	Arabane, maltose, methyl pentosane, sedoheptulose, sorbose, volemitol, xylane	Ono and Yoshimura (1958); Ono (1958a,b); Walland and Kinzel (1966); Suire (1975)
Blasia	Fructose, glucose, sucrose	Ono (1958b)
Calypogeia	—do—	—do—
Cephalozia	Inuline	Rancken (1914)
Chiloscyphus	Fructose, glucose, sucrose	Ono (1958b); Holligan and Drew (1971)
Conocephalum	Amidon, hexitol, maltose, arabinose, fucose, galactose, glucose, mannose, rnannose, xylose, uronic acids	Quillet (1956b); Walland and Kinzel (1966); Holligan and Drew (1971); Taylor and Kaufman (1976)
Diplophyllum	Raffinose	Allsopp (1951)
Dumortiera	Fructose, glucose, sucrose	Ono (1958b)
Exormotheca	Mannuronic acid	Rao and Das (1968)
Fossombronia	Sucrose	Allsopp (1951)

(Table contd.)

Chemical Constituents of Bryophytes

Taxa	Carbohydrate/s	Investigator/s
Frullania	Glycerol, hexitol	Holligan and Drew (1971)
Gymnocolea	D-Mannitol	Asakawa et al. (1980e)
Heteroscyphus	Fructose, glucose, sucrose	Ono (1958b)
Jamesoniella	Hexitol, mannitol, myoinositol	Lewis (1970)
Jubula	Fructose, glucose, sorbose, sucrose	Ono (1958b)
Lepidozia	Sedoheptulose, sorbose, volemitol	Ono (1958b); Lewis (1971)
Lethocolea	Inuline	Rancken (1914)
Lophocolea	Mannitol, volemitol, arabinose, fucose, galactose, glucose, mannose	Lewis (1971); Thomas (1977)
Marchantia	Amidon, hexitol	Quillet (1956b); Holligan and Drew (1971)
Metzgeria	Fructose, glucose, sucrose	Quillet (1956b)
Monoclea	Inuline	Rancken (1914)
Mylia	Arabane, mannitol, methyl pentosane, volemitol, xylane	Lewis (1971); Suire (1975)
Nardia	Glucose, sucrose	Suire (1975)
Pallavicinia	Amidon, mannuronic acid	Rao and Das (1968)
Pellia	Amidon, maltose, polyfructoside, raffinose	Allsopp (1951); Quillet (1956b); Walland and Kinzel (1966)
Plagiochasma	Mannuronic acid	Rao and Das (1968)
Plagiochila	Arabinose, fucose, galactose, hexitol, D-mannitol, marinitol, mannose, myoinositol, rhamnose, volemitol, xylose	Lichtenthaler (1968); Lewis (1970, 1971); Suleiman (1972); Suire (1975); Asakawa et al. (1980e)
Plectocolea	Glucose, fructose, sucrose	Ono (1958b)
Porella	—do—	—do—
Radula	—do—	—do—
Reboulia	Fructose, glucose	Rao and Das (1968)
Riccardia	L-Arabinose, D-galactose, mannuronic acid, fucose, glucose, mannose, rhamnose, xylose	Das and Rao (1966, 1967); Taylor and Kaufman (1976)
Saccogyna	Glucose, sucrose	Suire (1975)
Scapania	Arabinose, fructose, galactose, inuline, mannose, polyfructoside, rhamnose, xylose	Rancken (1914); Quillet (1956a); Maass and Craigie (1964)
Trichocolea	Fructose, glucose, sucrose	Quillet (1956b); Ono (1958a,b)

undulata. Sucrose was crystallized from *S. balticum* by Theander (1954), and the presence of fructosides in protonemata and gametophytes was indicated by paper chromatography (Black et al., 1955; Theander, 1954). Lewis (1970) recorded that certain leafy liverworts contain considerable amounts of free sugars including a high proportion of acyclic sugar alcohols. Two species of *Plagiochila* (*P. asplenioides* var. *major* & *P. carringtonii*) contained two unusual acyclic sugar alcohols in high proportion. One of these was identified as volemitol, a C_7 straight-chain sugar, and the second component was a branched chain C_6 acyclic polyol. This finding is of taxonomic value, since volemitol is present in the genus *Plagiochila*, but is absent in the two investigated species of *Jamesoniella*. *P. carringtonii* was originally placed in the genus *Jamesoniella*, but this result suggests that it should best be considered as a species of *Plagiochila*.

In addition to the known sugars from mosses (sucrose, fructose, glucose), presence of mannose, melibiose, and deoxyribose is also reported in *Tortula princeps, Rhynchostegium* sp., *Platyhypnidium riparioides, Homalothecium* sp., and *Camptothecium* sp. (Margaris & Kalaitzakis, 1974). Rao et al. (1975) noticed that *Riccia plana* is rich in carbohydrates and proteins, but fat is less abundant.

ORGANIC ACIDS

The occurrence and metabolism of organic acids in liverworts had not been investigated until a chromatographic survey by Allsopp (1951). He failed to detect any such acid in the eight investigated genera. Das (1961) reported trans-aconitic acid in *Riccia*, a genus not studied by Allsopp. Later, Das and Rao (1963) also detected the presence of trans-aconitic and malic acids in *Riccardia, Plagiochasma*, and *Riccia*. In addition, mannuronic acid is present in *Riccardia* and *Plagiochasma*. The presence of cis-aconitic and malic acids, both being Krebs cycle intermediates, makes it probable that their role in intermediary metabolism of liverworts may be similar to that in higher plants. As for the appearance of aconitic acid mostly in the trans-position, it was argued by Das (1961) that the instability of cis-isomer might have caused its transformation into the former. Maass and Craigie (1964) reported the occurrence of citric, malic, fumaric, and succinic acids in *Sphagnum* species.

Several examples of cinnamic acid derivatives occurring in bryophytes have been reported, and it has been suggested by Nilsson (1973a) that these are probably common constituents of bryophytes. *Sphagnum magellanicum* also yielded a cinnamic acid derivative, 4-hydroxy-β-carboxymethylcinnamic acid (Tutschek et al., 1973). Amongst the liverworts, several instances of hydroxy-cinnamic acids occurring esterified with the sugar moieties of flavonoid glycosides are on record. Cinnamic acid esters are also known in *Anthoceros* species (Méndez & Sanz-Cabanilles, 1979). The so far reported organic acids from bryophytes have been listed in Table 8.8.

TABLE 8.8.
Organic acids met with in bryophytes.

Taxa	Carboxylic acid/s	Investigator/s
Bazzania	cis-Aconitic acid, citric acid, malic acid, shikimic acid	Kinzel and Walland (1966); Walland and Kinzel (1966); Takeda et al. (1980)
Blasia	Calcium oxalate	Suire (1975)
Cavicularia	—do—	—do—|
Conocephalum	cis-Aconitic acid, calcium oxalate, citric acid, malic acid, shikimic acid	Kinzel and Walland (1966); Walland and Kinzel (1966); Suire (1975)
Pellia	cis-Aconitic acid, citric acid, malic acid, shikimic acid	Kinzel and Walland (1966); Walland and Kinzel (1966)
Plagiochasma	trans-Aconitic acid, malic acid	Das and Rao (1963)
Plagiochila	Citric acid, fumaric acid, glycolic acid, malic acid, malonic acid, succinic acid	Walland and Kinzel (1966); Suleiman (1972)
Riccardia	cis-Aconitic acid, trans-aconitic acid, malic acid	Das and Rao (1963)
Riccia	—do—	—do—
Sphagnum	Cinnamic acid, citric acid, fumaric acid, malic acid, succinic acid	Maass and Craigie (1964); Tutschek et al. (1973)

DIHYDROSTILBENES

These are phenolic compounds with basic skeleton C_6-C_2-C_6. An important growth inhibitor, lunularic acid, is a dihydrostilbene carboxylic acid. Lunularin, a dihydrostilbene, is met with in *Marchantia polymorpha* (Hopkins & Perold, 1974) and *Conocephalum conicum* (Pryce & Linton, 1974). Lunularin also occurs as its monomethyl ether in *Frullania pedicellata* (Asakawa et al., 1976e).

Pellepiphyllin (Fig. 8.36) was the first dihydrostilbene to be isolated from a bryophyte thought to be *Pellia epiphylla* (Benešová & Herout, 1970). Further study, however, revealed that the species was *P. neesiana* (Benešová & Herout, 1972). In fact, *P. epiphylla* does not contain pellepiphyllin. The structure of pellepiphyllin has been confirmed by Huneck (1976). Pryce (1971b) suggested that pellepiphyllin could be derived from lunularic acid by methylation and oxidative decarboxylation. However, apart from its co-occurrence with lunularic acid in liverworts *Lunularia cruciata*, *P. endiviaefolia*, and *Marchantia polymorpha* there is no evidence to support this.

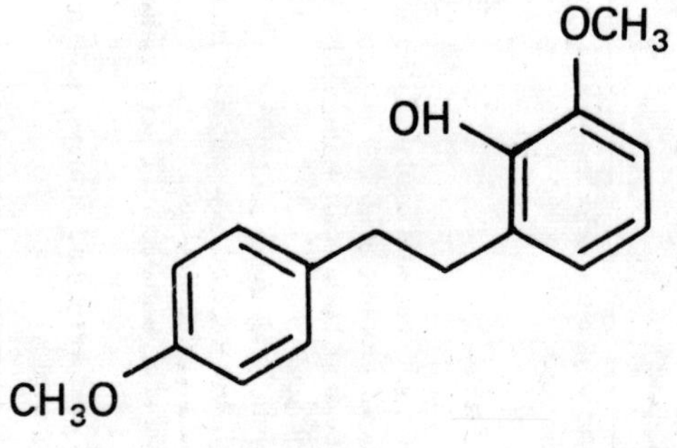

Fig. 8.36.

From the essential oil of the liverwort *Radula complanata*, 3-methoxydihydrostilbene (Fig. 8.37) has been separated by Suire (1970). Two new dihydrostilbenes, brittonin-A (Fig. 8.38a) and brittonin-B (Fig. 8.38b)

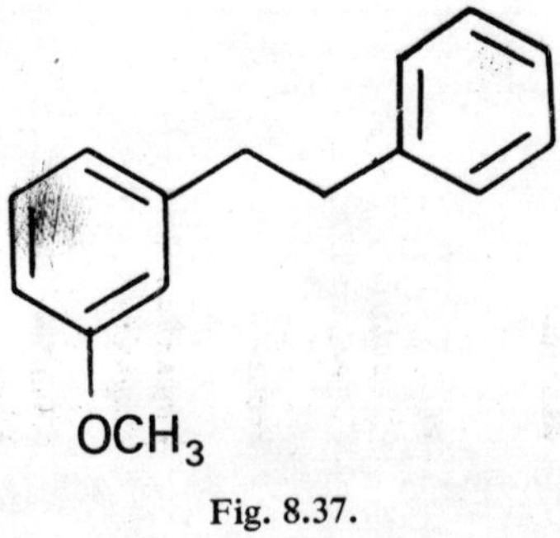

Fig. 8.37.

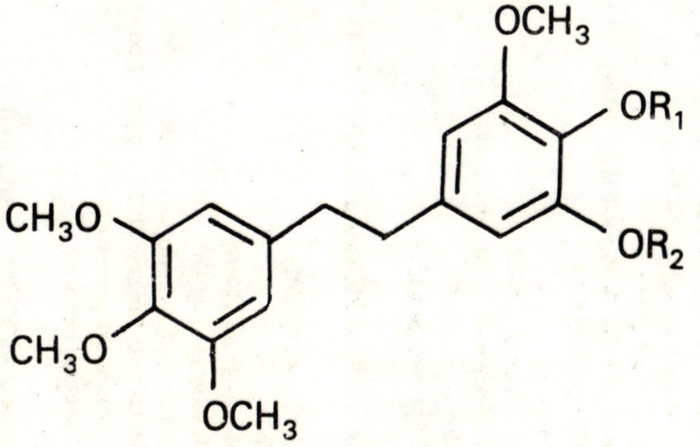

(a) $R_1 = R_2 = CH_3$
(b) $R_1 = R_2 = CH_2$

Fig. 8.38a,b.

have been isolated and characterized from the Japanese liverwort *Frullania brittoniae* ssp. *truncatifolia*. European species of this genus have well known allergenic properties which are absent in this species (Asakawa et al., 1976b).

ENZYMES

The bryophytes analyzed for enzymes are listed in Table 8.9.

AMINO ACIDS AND QUINONES

Only a few members of Hepaticae have been analyzed for amino acids (Table 8.10) and quinones (Table 8.11).

Black et al. (1955) and Maass and Craigie (1964) analyzed several *Sphagnum* species and reported presence of the following amino acids: aminobutyric acid, arginine, citrulline, glycine, glutamine, histidine, leucine, lysine, methionine, ornithine, phenylalanine, proline, tyrosine, and valine. Hartmann and Geissler (1973) isolated allantoine and allantoic acids from *Funaria hygrometrica*.

INORGANIC COMPOUNDS

Bryophytes have been analyzed for the following elements: nitrogen,

TABLE 8.9.
Enzymes met with in bryophytes.

Taxa	Enzyme/s	Investigator/s
MOSSES		
Abietinella	Oxalic acid oxidase	Houget et al. (1928); Datta (1956)
Atrichum	Allantoicase, allantoinase	Brunel and Capelle (1947)
Aulacomnium	Dehydrogenase	Taylor et al. (1970)
Brachythecium	β-Fructosidase, oxalic acid oxidase	Mason (1916); Niekerk-Blom (1946)
Campylium	Dehydrogenase	Taylor et al. (1970)
Catoscopium	—do—	—do—
Catharinea	Cinnamyl alcohol dehydrogenase	Mansell et al. (1974)
Ceratodon	Transfer RNA-nucleotidyl transferase	Schneider and Schneider (1977)
Cratoneuron	Dehydrogenase	Taylor et al. (1970)
Dawsonia	Acid phosphatase, cytochrome oxidase, peroxidase, succinate dehydrogenase	Hébant and Suire (1974)
Dicranum	Acid phosphatase, **cytochrome** oxidase, β-fructosidase, oxalic acid oxidase, peroxidase, succinate dehydrogenase	Mason (1916); Hébant and Suire (1974)
Distichum	Dehydrogenase	Taylor et al. (1970)
Ditrichum	—do—	—do—
Eurhynchium	Dehydrogenases, oxalic acid oxidase	Niekerk-Blom (1956); Taylor et al. (1970)
Fontinalis	Carbonic acid anhydrase, oxalic acid oxidase	Steeman and Kristiansen (1949); Datta and Meeuse(1955); Datta (1956)
Funaria	Phenylalanin ammonium lyase	Meyer and Angerman (1973)
Homalothecium	Dehydrogenases	Taylor et al. (1970)
Hylocomium	—do—	—do—
Hypnum	Oxalic acid oxidase	Houget et al. (1928); Franke and Hasse (1937)
Leucobryum	—do—	Houget et al. (1928)
Mnium	Catalase, oxalic acid oxidase	Houget et al. (1928); Datta (1956); Suzuki and Meeuse (1965)
Philonotis	Dehydrogenases	Taylor et al. (1970)

(Table contd.)

Taxa	Enzyme(s)	Investigator(s)
Pleurozium	—do—	—do—
Pogonatum	cis-3-Hexenal synthetase, n-hexenal synthetase	Hatanaka et al. (1978)
Polytrichum	Acid phosphatase, allantoicase, allantoinase, amylase, cytochrome oxidase, dehydrogenases, β-fructosidase, maltase, oxalic acid oxidase, peroxidase, phosphatase, succinate dehydrogenase	Mason (1916); Houget et al. (1928); Niekerk-Blom (1946); Brunel and Capelle (1947); Datta (1956); Taylor et al. (1970); Maier and Maier (1972); Hébant and Suire (1974)
Ptilium	Dehydrogenases	Taylor et al. (1970)
Rhytidiadelphus	Dehydrogenases, oxalic acid oxidase	Datta and Meeuse (1955); Datta (1956); Taylor et al. (1970)
Sphagnum	Ascorbic acid oxidase, cinnamyl alcohol dehydrogenase, β-fructosidase, peroxidase, phenolase	Mason (1916); Mattison (1961); Mansell et al. (1974); Tutschek (1979)
Thuidium	Dehydrogenases, β-fructosidase	Mason (1916); Taylor et al. (1970)
Tomenthypnum	Dehydrogenase	Taylor et al. (1970)
Tortula	—do—	
LIVERWORTS		
Asterella	Amylase, butyrase, catalase, invertase, laccase, Lipase, maltase, a proteolytic enzyme, urease	Udar and Chandra (1960a)
Chiloscyphus	Urease	Jones et al. (1973)
Conocephalum	Lunularic acid decarboxylase	Pryce and Linton (1974)
Frullania	Oxalic acid oxidase	Houget et al. (1928)
Lophocolea	Urease	Jones et al. (1973)
Lunularia	Cinnamic acid, 4-hydroxylase, lunularic acid decarboxylase, phenylalanin ammonium lyase	Pryce (1972b); Gorham (1977)
Marchantia	Esterase, indole-3-acetic acid oxidase, phosphatase, urease	Maravolo et al. (1967); Jones et al. (1973); Spaeth and Maravolo (1973)
Plagiochasma	Amylase, butyrase, catalase, invertase, laccase, lipase, maltase, a proteolytic enzyme, urease	Udar and Chandra (1960a)

(Table contd.)

Taxa	Enzyme/s	Investigator/s
Riccia	Acid phosphatase, aldolase, amylase, ascorbic acid oxidase, butyrase, catalase, de-aminase, β-fructosidase, invertase, laccase, lipase, maltase, peroxidase, phosphorylase, polyphenol oxidase, a proteolytic enzyme, ribonuclease, urease	Udar and Chandra (1960a,b,c); Rao et al. (1969)
Scapania	Acid phosphatase, cytochrome oxidase, peroxidase, succinate dehydrogenase	Hébant and Suire (1974)
Sphaerocarpos	Urease	Jones et al. (1973)

TABLE 8.10.
Amino acids met with in Hepaticae.

Taxa	Amino acid/s	Investigator/s
Diplophyllum	Allantoic acid	Touffet and Villeret (1958)
Marchantia	Tryptamine, tryptophan	Schneider et al. (1967)
Pallavicinia	Tryptophan	—do—
Pellia	Allantoin, allantoic acid	Touffet and Villeret (1958)
Plagiochila	Allantoin, alanine, allantoic acid, asparagine, aspartic acid, glutamic acid, glutamine, glycine, leucine, lysine, phenylalanine, proline, serine, threonine, valine	—do— Suleiman (1972)

TABLE 8.11.
Quinones met with in Hepaticae.

Taxa	Quinone/s	Investigator/s
Conocephalum	Plastoquinone, α-tocoquinone	Bucke et al. (1966)
Lunularia	Phylloquinone K, plastohydroquinone-9, plastoquinone-9, α-tocopherol, α-tocoquinone	Lichtenthaler et al. (1977)
Marchantia	Plastohydroquinone, plastoquinone, α-tocoquinone	Lichtenthaler (1968)
Pellia	Phylloquinone K_1, plastohydroquinone-9, plastoquinone-9, α-tocopherol, α-tocoquinone	Lichtenthaler et al. (1977)

phosphorus, silicium, sulphur, copper, aluminium, silver, iron, zinc, lead, sodium, potassium, barium, nickle, manganese, magnesium, and strontium (see Chapter 9). Among the bryophytes which prefer substrates rich in copper are: *Merceya latifolia, M. ligulata, Mielichhoferia macrocarpa, M. mielichhoferi, Oligotrichum hercynicum, Cephaloziella massalongi, C. phyllacantha.* and *Gymnocolea acutiloba* (Persson, 1956; Schatz, 1955; Shacklette, 1965a,b, 1967).

MISCELLANEOUS

Bendz and Svensson (1971) made a study of steam distillate of *Fontinalis antipyretica* and detected 40 compounds. The major constituent and odour compound being hexanal; some others were ethanal, ethyl acetate and formate, ethanol, 2-heptanone, ethyl hexanoate and heptanoate, l-hexyl acetate and 2-octanoate. Tetracosanoic acid has also been isolated from this moss.

Matsuo et al. (1971c) studied the non-polar constituents of the leafy liverwort *Isotachis japonica*. Six major non-terpenoid components were present in the fraction. Benzyl benzoate was in highest proportion and among the others were benzyl and β-phenylethyl cinnamates. The rest of the constituents were all esters but were not isolated in sufficient quantity. Terpenoid constituents were not detected in *I. japonica*.

In a study of the non-polar fraction of *Bazzania pompeana*, Matsuo et al. (1971b) isolated, in addition to the sesquiterpene bazzanene, the secondary alcohol nonacosan-10-ol. Matsuo et al. (1972a) detected a series of sec-alcohols from the hexane extract of *Isotachis japonica*. A homologous series (C_{20}-C_{34}) was observed with the most abundant alcohol being C_{25} (28%). The ratio of odd to even homologues was higher (16.5) for the alcohols than for the alkanes (3.9). The alkanes (2%), sec-alcohol (3%) fractions were minor compounds with the earlier reported aromatic-esters (67%) fraction.

Benešová et al. (1969c) isolated two simpler isoprenylated indoles, 7-(3-methyl-2-butenyl)-indole (Fig. 8.39) and 6- (3-methyl-2-butenyl)-indole (Fig. 8.40) from *Riccardia sinuata*.

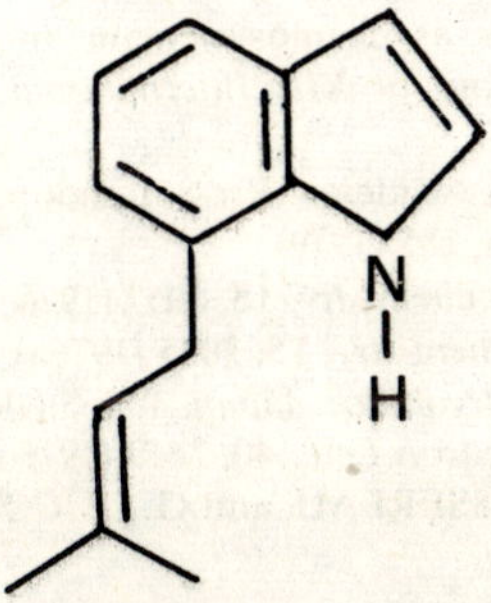

Fig. 8.39.

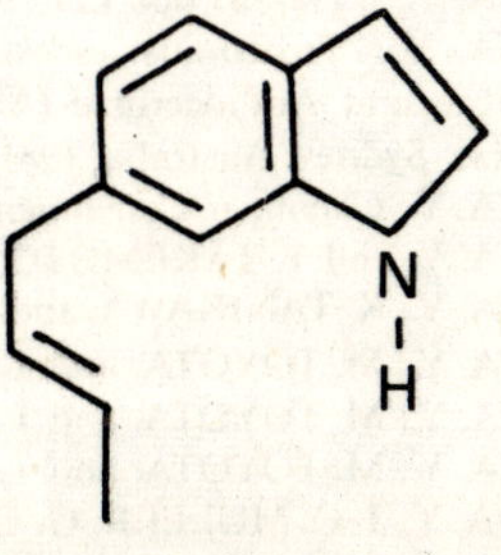

Fig. 8.40.

ANTITUMOUR ACTIVITIES

Belkin et al. (1952/53) observed that the ethanolic extract of *Polytrichum juniperinum* is active against Sarcoma 37. Diplophyllin isolated from *Diplophyllum* sp. showed activity against human epidermoid carcinoma (Ohta et al., 1977). Frullanolide, tulipinolide, and diplophyllin indicated cytotoxicity against KB cells. These three compounds are major components of *Conocephalum, Wiesnerella,* and *Diplophyllum* (Asakawa et al., 1978). Zaluzanin C isolated from a higher plant showed antitumour activity against P-388 lymphocytic leukemia (Dominguez et al., 1975; Jolad et al., 1974).

ALLERGENIC ACTIVITIES

Some skin allergies are caused by epiphytic liverworts, especially *Frullania* and *Radula* species. Knoche et al. (1969) and Perold et al. (1972) recognized the sesquiterpene lactone frullanolide as active allergenic principle. Asakawa (1981) reported that eudesmanolides from *Chiloscyphus* and *Diplophyllum* species; germacranolide and guainolides from species of *Porella, Frullania, Conocephalum,* and *Wiesnerella* are all potential allergenic agents.

REFERENCES

ALLSOPP, A., *J. exp. Bot.*, 2, 121 (1951).
ANDERSEN, N.H. and S. HUNECK, *Phytochemistry*, 12, 1818 (1973).
ANDERSEN, N.H., C.R. COSTIN, C.M. KRAMER, Y. OHTA, and S. HUNECK, *Phytochemistry*, 12, 2709 (1973a).
ANDERSEN, N.H., B. SHUNK, and C.R. COSTIN, *Experientia*, 29, 645 (1973b).
ANDERSEN, N.H., P. BISSONETTE, C.-B. LIU, B. SHUNK, Y. OHTA, C.-LIW TSENG, A. MOORE, and S. HUNECK, *Phytochemistry*, 16, 1731 (1977).
ANDERSON, W.H., J.L. GELLERMAN, and H. SCHLENK, *Lipids*, 10, 501 (1975).
ANHUT, S., H.D. ZINSMEISTER, R. MUES, W. BARZ, K. MACKENBROCK, J. KÖSTER, and K.R. MARKHAM, *Phytochemistry*, 23, 1073 (1984).
ASAKAWA, Y., *J. Hattori bot. Lab. No.* 48, 277 (1980).
ASAKAWA, Y., "Terpenoids and aromatic compounds as chemosystematic indicators in Hepaticae and Anthocerotae (Abstr.)". In *Proceedings of XIII International Botanical Congress*, Sydney, Australia, 1981. Pp. 148.
ASAKAWA, Y., *Chemical Constituents of the Hepaticae*, Academic Press, London, 1983.
ASAKAWA, Y. and T. TAKEMOTO, *Phytochemistry*, 18, 285 (1979).
ASAKAWA, Y., K. TANIKAWA, and T. ARATANI, *Phytochemistry*, 15, 1057 (1976c).
ASAKAWA, Y., M. TOYOTA, and T. ARATANI, *Phytochemistry*, 15, 2025 (1976a).
ASAKAWA, Y., M. TOYOTA, and T. ARATANI, *Proc. Bryol. Soc. Japan*, 1, 155 (1976b).
ASAKAWA, Y., M. TOYOTA, and T. ARATANI, *Tetrahedron Lett.*, 40, 3619 (1976d).
ASAKAWA, Y., J.-C. MULLER, G. OURISSON, J. FOUSSEREAU, and G. DUCOMBS, *Bull Soc. Chim. France*, 1976, 1465 (1976e).
ASAKAWA, Y., T. ARATANI, and G. OURISSON, *Misc. Bryol. Lichenol.*, 7, 99 (1976f).

ASAKAWA, Y., T. TAKEMOTO, M. TOYOTA, and T. ARATANI, Tetrahedron Lett., 16, 1407 (1977).
ASAKAWA, Y., M. TOYOTA, and T. TAKEMOTO, Phytochemistry, 17, 457 (1978).
ASAKAWA, Y., N. TOKUNAGA, M. TOYOTA, T. TAKEMOTO, S. HATTORI, M. MIZUTANI, and C. SUIRE, J. Hattori bot. Lab. No. 46, 67 (1979a).
ASAKAWA, Y., S. HUNECK, M. TOYOTA, T. TAKEMOTO, and C. SUIRE, J. Hattori bot. Lab. No 46, 163 (1973b).
ASAKAWA, Y., N. TOKUNAGA, M. TOYOTA, T. TAKEMOTO, and C. SUIRE, J. Hattori bot. Lab. No. 45, 395 (1969c).
ASAKAWA, Y., H. INOUE, M. TOYOTA, and T. TAKEMOTO, Phytochemistry, 19, 2623 (1980a).
ASAKAWA, Y., C. SUIRE, M. TOYOTA, N. TOKUNAGA, T. TAKEMOTO, S. HATTORI, and M. MIZUTANI, J. Hattori bot. Lab. No. 48, 285 (1980b).
ASAKAWA, Y., N. TOKUNAGA, T. TAKEMOTO, S. HATTORI, M. MIZUTANI, and C. SUIRE, J. Hattori bot. Lab.No. 47, 153 (1980c).
ASAKAWA, Y., M. TOYOTA, M. KANO, and T. TAKEMOTO, Phytochemistry, 19, 2651 (1980d).
ASAKAWA, Y., M. TOYOTA, and T. TAKEMOTO, Phytochemistry, 19, 2141 (1980e).
ASAKAWA, Y., M. TOYOTA, and T. TAKEMOTO, Phytochemistry, 20, 257 (1981d).
ASAKAWA, Y., M. TOYOTA, T. TAKEMOTO, H. FUJIKI, and T. SUGIMURA, Planta Medica, 39, 233 (1980f)
ASAKAWA, Y., R. MATSUDA, and R. TAKEDA, Phytochemistry, 20, 1423 (1981a).
ASAKAWA, Y., J.C. MULLER, M. TOYOTA, S. HATTORI, and G. OURISSON, Phytochemistry, 20, 2187 (1981b).
ASAKAWA, Y., R. MATSUDA, S. HATTORI, M. MIZUTANI, H. INOUE, C. SUIRE, and S. HUNECK, J. Hattori bot. Lab.No. 50, 107 (1981c).
ASHTON, N.W., A. SCHULZE, P. HALL, and R.S. BANDURSKI, Planta, 164, 142 (1985).
BANERJEE, R.D. and S.P. SEN, Bryologist, 82, 141 (1979).
BAUER, L., Planta, 46, 604 (1956).
BAUER, L., Naturwissenschaften, 46, 154 (1959).
BAUER, L., J. Linn. Soc. (Bot.), 58, 343 (1963).
BAUER, L., Z. PflPhysiol., 54, 241 (1966).
BELKIN, M., D.B. FITZGERALD, and M.D. FELIX, J. Natl. Cancer Inst., 13, 741 (1952/53).
BENDZ, G. and O. MÅRTENSSON, Acta chem. Scand., 17, 266 (1963).
BENDZ, G. and L. SVENSSON, Phytochemistry, 10, 3283 (1971).
BENDZ, G., O. MARTENSSON, and L. TERENIUS, Acta chem. Scand., 16, 1183 (1962).
BENDZ, G., O. MARTENSSON, and E. NILSSON, Acta chem. Scand., 20, 277 (1966).
BENDZ, G., L. G. LOOF, and MÅRTENSSON, Acta chem. Scand., 22, 2215 (1968).
BENES, I. and N. KUZMIAKOVA, "Isolation of steryl esters from the liverworts of the genus Calypogeia and their chemotaxonomic significance." First Conference on Organic and Bioorganic Chemistry of Young Scientists, Bechyne/CSSR (Abstr.), 1980. Pp. 102.
BENES, I., V. BENEŠOVÁ, and V. HEROUT, Coll. Czech. chem. Comm., 42, 1229 (1977).
BENEŠOVÁ, V. and V. HEROUT, Coll. Czech. chem. Comm., 35, 1926 (1970).
BENEŠOVÁ, V. and V. HEROUT, Coll. Czech. chem. Comm., 37, 1764 (1972).
BENEŠOVÁ, V., V. HEROUT, and F. ŠORM, Coll. Czech. chem. Comm., 34, 1810 (1969a).
BENEŠOVÁ, V., Z. SAMEK, V. HEROUT, and F. ŠORM, Coll. Czech. chem. Comm., 34, 582 (1969b).
BENEŠOVÁ, V., Z. SAMEK, V. HEROUT, and F. ŠORM, Coll. Czech. chem. Comm., 34, 1807 (1969c).
BENEŠOVÁ, V., P. SEDMERA, V. HEROUT, and F. ŠORM, Tetrahedron Lett., 1971, 2679 (1971).
BENESOVÁ, V., M. STREIBEL, H.M. CHÂU, I. BENES, and K. KONECNY, Coll. Czech. chem. Comm., 37, 3090 (1972).
BENEŠOVÁ, V., H.M. CHAU, and V. HEROUT, Phytochemistry, 12, 211 (1973).

BENEŠOVÁ, V., Z. SAMEK, and S. VAŠIČKOVA, *Coll. Czech. chem. Comm.*, **40**, 1966 (1975).
BEUTELMANN, P., *Planta*, **112**, 181 (1973).
BEUTELMANN, P. and L. BAUER, *Planta*, **133**, 215 (1977).
BHATLA, S.C. and R.N. CHOPRA, *Ann. Bot.*, **54**, 195 (1984).
BLACK, W.A.P., W.J. CORNHILL, and F.N. WOODWARD, *J. appl. chem.*, **5**, 484 (1955).
BLAND, D.E., A. LOGAN, M. MENSHUN, and S. STERNHELL, *Phytochemistry*, **7**, 1373 (1968).
BOPP, M., *Planta*, **53**, 178 (1959).
BOPP, M., *Biol. Rev.*, **36**, 237 (1961).
BOPP, M., *J. Linn. Soc. (Bot.)*, **58**, 305 (1963).
BOURGEOIS, G. and C. SUIRE, *Rev. Bryol. Lichénol.*, **43**, 343 (1977).
BRUNEL, A. and G. CAPELLE, *Bull. Soc. Chim. Biol.*, **29**, 427 (1947).
BUCKE, C., R.M. LEECH, M. HALLAWAY, and R.A. MORTON, *Biochem. Biophys. Acta*, **112**, 19 (1966).
CALDICOTT, A.B. and G. EGLINTON, *Phytochemistry*, **15**, 1139 (1976).
CANONICA, L., A. CORBELLA, P. GARIBOLDI, G. JOMMI, J. KREPINSKY, G. FERRARI, and C. CASAGRANDE, *Tetrahedron*, **25**, 631 (1969).
CATALANO, S., A. MARSILI, I. MORELLI, and M. PACCHIANI, *Phytochemistry*, **15**, 1178 (1976).
CHOPRA, R.N. and P.K. KUMRA, *Phytomorphology*, **28**, 298 (1978).
CONNOLLY, J.D. and I.M.S. THORNTON, *J. chem. Soc. Perkin Trans.*, **1**, 736 (1972).
CONNOLLY, J.D. and I.M.S. THORNTON, *Phytochemistry*, **12**, 631, (1973).
CONNOLLY, J.D., A.E. HARDING, and I.M.S. THORNTON, *J. chem. Soc. chem. Comm.*, **20**, 1320 (1972).
CONNOLLY, J.D., A.E. HARDING, and I.M.S. THORNTON, *J. chem. Soc. Perkin Trans.*, **1**, 2487 (1974).
CORBELLA, A., P. GARIBOLDI, G. JOMMI, F. ORSINI, A. DeMarco and A. IMMIRZI, *J. chem. Soc. Perkin Trans.*, **1**, 1875 (1974).
CORRIGAN, D., C. KLOOS, C.S. O'CONNOR, and R.F. TIMONEY, *Phytochemistry*, **12**, 213 (1973).
CORRIGAN, D., C. KLOOS, C.S. O'CONNOR, and R.F. TIMONEY, *Planta*, **29**, 26 (1976).
CZECZUGA, B., *Planta*, **103**, 87 (1972).
DAS, V.S.R., *Indian J. Pl. Physiol.*, **4**, 60 (1961).
DAS, V.S.R. and M.P. RAO, *Nature*, **198**, 970 (1963).
DAS, V.S.R. and M.P. RAO, *Curr. Sci.*, **35**, 20 (1966).
DAS, V.S.R. and M.P. RAO, *Indian J. Exp. Biol.*, **5**, 193 (1967).
DATTA, P.K., "A study of moss oxalic acid oxidase," Ph.D. Thesis, Univ. Washington, U.S.A., 1956.
DATTA, P.K. and B.J.D. MEEUSE, *Biochem. Biophys. Acta.*, **17**, 602 (1955).
DeGreef J. and J. FREDERICQ, *Physiol. Plant.*, **22**, 462 (1969).
DeGreef, J., M. DePROFT, F. VEROUSTRAETE, and H. FREDERICQ, "Case studies of ethylene release in higher and lower plant systems." In B. Jeffcoat, Ed., *Aspects and Prospects of Plant Growth Regulators*. Wessex, Oxfordshire, 1981. Pp. 9-18.
DOMINGUEZ, X.A., J. MARRIOQUIN, and E. CARDENAS, *Planta Medica*, **28**, 89 (1975).
DOUIN, F., *C.r. hebd. Séanc. Acad. Sci., Paris (Ser.D)*, **243**, 1051 (1956).
DOUIN, F., *C.r. hebd. Séanc. Acad. Sci., Paris (Sér.D)*, **246**, 1248 (1958).
EKMAN, R. and P. KARUNEN, *Phytochemistry*, **19**, 1243 (1980).
ELTZ, H. V., *Nativer Morphoregulator (Faktor H) aus Moosprotonemen, Staatsexamensarbeit*, Heidelberg, 1975.
ERICKSON, M. and G.E. MIKSCHE, *Phytochemistry*, **13**, 2295 (1974).
FRANKE, W. and K. HASSE, *Z. Physiol. Chem.*, **249**, 231 (1937).
FREDERICQ, H. and J. DeGreef, *Physiol. Plant.*, **21**, 346 (1968).
FREELAND, R.O., *Pl. Physiol.*, **32**, 64 (1957).
FREUDENBERG, K. and A.C. NEISH, *Constitution and Biosynthesis of Lignin*, Springer-

Verlag, New York, 1968.
FUJITA, Y., T. UEDA, and T. ONO, *Nippon Kagaku Zasshi*, **77**, 400 (1956).
GELLERMAN, J.L., W.H. ANDERSON, and H. SCHLENK, *Bryologist*, **75**, 550 (1972).
GELLERMAN, J.L., W.H. ANDERSON, D.G. RICHARDSON, and H. SCHLENK, *Biochem Biophys. Acta*, **388**, 277 (1975).
GLEIZES, M., G. PAULY, and C. SUIRE, *Le Botaniste*, **56**, 209 (1973-74).
GORHAM, J., *Phytochemistry*, **16**, 249 (1977).
GRADSTEIN, S.R., R. MATSUDA, and Y. ASAKAWA, *J. Hattori bot. Lab.* No. 50, 231 (1981).
GROSS, D., *Phytochemistry*, **14**, 2105 (1975).
GUPTA, K.G. and B. SINGH, *Res. Bull. Panjab Univ.*, **22**, 237 (1971).
HANDA, A.K. and M.M. JOHRI, *Nature*, **259**, 480 (1976).
HANDA, A.K. and M.M. JOHRI, *Pl. Physiol.*, **59**, 490 (1977).
HARTMANN, E., *Planta*, **101**, 159 (1971).
HARTMANN, E. and G. GEISSLER, *Biochem. Physiol. Pflanzen*, **164**, 614 (1973).
HARTMANN, E. and H. KILBINGER, *Biochem. J.*, **137**, 249 (1974).
HARTWELL, J.L., *Lloydia*, **34**, 410 (1971).
HATANAKA, A., J. SEKIYA, and T. KAJIWARA, *Phytochemistry*, **17**, 869 (1978).
HAYASHI, S. and A. MATSUO, *Experientia*, **26**, 347 (1970).
HAYASHI, S., A. MATSUO, and T. MATSUURA, *Experientia*, **25**, 1139 (1969).
HAYES, L.E., *Bot. Gaz.*, **108**, 408 (1947).
HÉBANT, C., *J. Hattori bot. Lab.* No. 38, 365 (1974).
HÉBANT, C. and C. SUIRE, *Rev. Bryol. Lichénol.*, **40**, 171 (1974).
HERRMANN, R.G., *Z. Naturforsch.*, **23B**, 1496 (1968).
HOLLIGAN, P.M. and E.A. DREW, *New Phytol.*, **70**, 271 (1971).
HOPKINS, B.J. and G.W. PEROLD, *J. chem. Soc. Perkin Trans.*, 1, 32 (1974).
HÖRSTER, H. and R. WIERMANN, *Nova Hedwigia*, **27**, 183 (1976).
HOUGET, J., A. MAYER, and L. PLANTEFOL, *Ann. Physiol. Physicochim. Biol.*, **4**, 123 (1928).
HUNECK, S., *Z. Naturforsch.*, **22B**, 462 (1967).
HUNECK, S., *Phytochemistry*, **10**, 3283 (1971).
HUNECK, S., *Phytochemistry*, **31**, 1289 (1974).
HUNECK, S., *Tetrahedron*, **32**, 109 (1976).
HUNECK, S., "Chemistry and biochemistry of bryophytes." In R.M. Schuster, Ed., *New Manual of Bryology* Vol. I, The Hattori Botanical Laboratory, Nichinan, Miyazaki, Japan, 1983. Pp. 1-116.
HUNECK, S. and E. KLEIN, *Phytochemistry*, **6**, 383 (1967).
HUNECK, S. and E. KLEIN, *J. Hattori bot. Lab.* No. 33, 1 (1970).
HUNECK, S. and K.H. OVERTON, *Phytochemistry*, **10**, 3279 (1971).
HUNECK, S. and K. SCHREIBER, *J. Hattori bot. Lab.* No. 39, 215 (1975).
HUNECK, S. and G. VEVLE, *Z. Naturforsch.*, **25B**, 227 (1970).
HUNECK, S., R. GROLLE, and G. VEVLE, *J. Hattori bot. Lab.* No. 36, 93 (1972).
HUNECK, S., K. SCHREIBER, and S. JANICKE, *Phytochemistry*, **12**, 2533 (1973).
IVES, D.A.J. and A.N. O'NEIL, *Can. J. Chem.*, **36**, 434 (1958).
JAMIESON, G.R. and E.H. REID, *Phytochemistry*, **15**, 1731 (1976).
JOLAD, S.D., R.M. WIEDHOPH, and J.R. COLE, *J. Pharm. Sci.*, **60**, 1321 (1974).
JONES, R.A., M.J. MONTAGUE, and J. TAYLOR, *Phytochemistry*, **12**, 1675 (1973).
KARUNEN, P., *Physiol. Plant.*, **33**, 98 (1975).
KARUNEN, P., *Physiol. Plant.*, **40**, 239 (1977).
KARUNEN, P., *Can. J. Bot.*, **59**, 1902 (1981).
KARUNEN, P. and A. IHANTOLA, *Bryologist*, **80**, 88 (1977a).
KARUNEN, P. and A. IHANTOLA, *Bryologist*, **80**, 313 (1977b).
KARUNEN, P. and C. LILJENBERG, *Physiol. Plant.*, **44**, 417 (1978).
KARUNEN, P. and H. MIKOLA, *Phytochemistry*, **19**, 319 (1980).
KARUNEN, P. and L.M. SALIN, *Kamia-Kami*, **7**, 500 (1980).

KARUNEN, P., H. MIKOLA, H. LINKE, and E.K. EURANTO, *Can. J. Bot.*, **57**, 1335 (1979).
KARUNEN, P., H. MIKOLA, and R. EKMAN, *Physiol. Plant.*, **48**, 554 (1980a).
KARUNEN, P., H. MIKOLA, and R. EKMAN, *Physiol. Plant.*, **49**, 351 (1980b).
KAUL, R. and R.C. SACHAR, *Biochem. Biophys. Res. Comm.*, **104**, 126 (1982).
KINZEL, H. and A. WALLAND, *Z. PflPhysiol.*, **54**, 371 (1966).
KLEIN, B., *Planta*, **73**, 12 (1967).
KNOCHE, H., G. OURISSON, G.W. PEROLD, J. FOUSSEREAU, and J. MALEVILLE, *Science*, **166**, 239 (1969).
KOCKEL, H., *Vorkommen und Wirkung des Faktor H bei Verschiedenen Laubmoosprotonemen*, Staatsexamensarbeit, Hannover, 1967.
KOHL, F.G., *Untersuchungen über das Carotin*, Borntraeger, Leipzig, 1902.
KOLATTUKUDY, P.E. and T.J. WALTON, "The biochemistry of plant cuticular lipids." In R.T. Holman, Ed., *Progress in the Chemistry of Fats and other Lipids*, Pergamon Press, Oxford, 1972. Pp. 121-175.
KOSKIMIES, K. and L.K. SIMOLA, *Can. J. Bot.*, **58**, 259 (1979).
KOZLOWSKI, A., *C.r. hebd. Séanc. Acad. Sci., Paris (Sér. D)*, **173**, 429 (1921).
KRUTOV, S.M., Z. SAMEK, V. BENEŠOVÁ, and V. HEROUT, *Phytochemistry*, **12**, 1405 (1973).
KUMRA, P.K., "Experimental studies on some Delhi mosses." M.Sc. Thesis, Univ. Delhi, India, 1977.
LAZARENKO, A.S., *Dokl. Akad. nauk. SSSR*, **134**, 1240 (1960).
LEWIS, D.H., *Trans. Br. bryol. Soc.*, **6**, 108 (1970).
LEWIS, D.H., *Trans. Br. bryol. Soc.*, **6**, 391 (1971).
LICHTENTHALER, H.K., *Planta*, **81**, 140 (1968).
LICHTENTHALER, H.K., P. KARUNEN, and K.H. GRUMBACH, *Physiol. Plant.*, **40**, 105 (1977).
LILJENBERG, C. and P. KARUNEN, *Physiol. Plant.*, **44**, 369 (1978).
LINDBERG, G., B.G. OSTERDAHL, and E. NILSSON, *Chemica Scripta*, **5**, 140 (1974).
MAASS, W.S.G. and J.S. CRAIGIE, *Can. J. Bot.*, **42**, 805 (1964).
MADSEN, G.C. and A.L. PATES, *Bot. Gaz.*, **113**, 293 (1952).
MAIER, K. and U. MAIER, *Protoplasma*, **75**, 91 (1972).
VON MALTZAHN, K.E., *Nature*, **183**, 60 (1959).
MANSELL, R.L., G.G. GROSS, J. STOCKIGT, H. FRANKE, and M.H. ZENK, *Phytochemistry*, **13**, 2427 (1974).
MANSKAJA, S.M. and T.V. DROZDOVA, *Dokl. Akad. nauk. SSSR*, **102**, 789 (1955).
MARAVOLO, N.C. "The control of endogenous auxin levels in hepatics (Abstr.)." In *Proceedings of XIII International Botanical Congress*, Sydney, Australia, 1981. Pp. 145.
MARAVOLO, N.C., E.D. GARBER, and P.D. VOTH, *Am. J. Bot.*, **54**, 1113 (1967).
MARGARIS, N.S. and J. KALAITZAKIS, *Bryologist*, **77**, 470 (1974).
MARKHAM, K.R., *Phytochemistry*, **11**, 2047 (1972).
MARKHAM, K.R., *Phytochemistry*, **16**, 617 (1977).
MARKHAM, K.R., *Biochemical Systematics and Ecology*, **8**, 11 (1980).
MARKHAM, K.R. and R. MUES, *Phytochemistry*, **22**, 143 (1983).
MARKHAM, K.R. and L.J. PORTER, *Phytochemsitry*, **12**, 2007 (1973).
MARKHAM, K.R. and L.J. PORTER, *Phytochemistry*, **13**, 1937 (1974a).
MARKHAM, K.R. and L.J. PORTER, *Phytochemistry*, **13**, 1553 (1974b).
MARKHAM, K.R. and L.J. PORTER, *Phytochemistry*, **14**, 1593 (1975a).
MARKHAM, K.R. and L.J. PORTER, *Phytochemistry*, **14**, 1641 (1975b).
MARKHAM, K.R. and L.J. PORTER, *Phytochemistry*, **14**, 199 (1975c).
MARKHAM, K.R. and L.J. PORTER, "Chemical constituents of the bryophytes." In L. Reinhold, J.B. Harborne, and T. Swain, Eds., *Progress in Phytochemistry*, Pergamon Press, New York, 1978a. Pp. 181-272.
MARKHAM, K.R. and L.J. PORTER, *Phytochemistry*, **17**, 159 (1978b).
MARKHAM, K.R. and L.J. PORTER, *Phytochemistry*, **18**, 611 (1979).
MARKHAM, K.R., L.J. PORTER, and B.G. BREHM, *Phytochemistry*, **8**, 2193 (1969).

MARKHAM, K.R., T.J. MARBY and J.E. AVERETT, *Phytochemistry*, 11, 2875 (1972).
MARKHAM, K.R., L.J. PORTER, R. MUES, H.D. ZINSMEISTER, and B.G. BREHM, *Phytochemistry*, 15, 147 (1976a).
MARKHAM, K.R., L.J. PORTER, E.O. CAMPBELL, J. CHOPIN, and M.-L. BOUILLANT, *Phytochemistry*, 15, 1517 (1976b).
MARKHAM, K.R., L.J. PORTER, and N.G. MILLER, *Phytochemistry*, 15, 151 (1976c).
MARKHAM, K.R., N.A. MOORE, and L.J. PORTER, *Phytochemistry*, 17, 911 (1978a).
MARKHAM, K.R., H.D. ZINSMEISTER, and R. MUES, *Phytochemistry*, 17, 1601 (1978b).
MARSILI, A. and I. MORELLI, *Phytochemistry*, 7, 1705 (1968).
MARSILI, A. and I. MORELLI, *Phytochemistry*, 9, 651 (1970).
MARSILI, A., I. MORELLI, and A.M. IORI, *Phytochemistry*, 10, 432 (1971).
MARSILI, A., I. MORELLI, C. BERNARDINI, and M. PACCHIANI, *Phytochemistry*, 11, 2003, (1972).
MASON, T.G., *Scient. Proc. Roy. Dublin Soc.*, 15, 13 (1916).
MARTIN, J.T. and B.E. JUNIPER, *The Cuticle of Plants*, Edward Arnold, London, 1970.
MATSUO, A., M. NAKAYAMA, and S. HAYASHI, *Phytochemistry*, 10, 430 (1971a).
MATSUO, A., M. NAKAYAMA, Y. ISHIDA, and S. HAYASHI, *Z. Naturforsch.*, 26B, 32 (1971b).
MATSUO, A., M. NAKAYAMA, and S. HAYASHI, *Z. Naturforsch.*, 26B, 1023 (1971c).
MATSUO, A., M. NAKAYAMA, S. HAYASHI, and S. NISHIMOTO, *Phytochemistry*, 11, 3313, (1972a).
MATSUO, A., M. NAKAYAMA, J. ONO, and S. HAYASHI, *Z. Naturforsch.*, 27B, 1437 (1972b).
MATSUO, A., T. MAEDA, M. NAKAYAMA, and S. HAYASHI, *Tetrahedron Lett.*, 42, 4131 (1973a).
MATSUO, A., M. NAKAYAMA, and S. HAYASHI, *Bull. chem. Soc. Japan*, 46, 1010 (1973b).
MATSUO, A., M. NAKAYAMA, and S. HAYASHI, *Chemistry Lett.*, 1973, 769 (1973c).
MATSUO, A., M. NAKAYAMA, M. MAEDA, and S. HAYASHI, *Phytochemistry*, 12, 2413 (1973d).
MATSUO, A., M. NAKAYAMA, H. GOTO, and S. HAYASHI, *Phytochemistry*, 13, 957 (1974a).
MATSUO, A., S. SATO, M. NAKAYAMA, and S. HAYASHI, *Tetrahedron Lett.*, 42, 3681 (1974b).
MATSUO, A., M. NAKAYAMA, S. SATO, T. NAKAMOTO, S. UTO, and S. HAYASHI, *Experientia*, 30, 321 (1974c).
MATSUO, A., M. NAKAYAMA, M. MAEDA, Y. NODA, and S. HAYASHI, *Phytochemistry*, 14, 1037 (1975).
MATSUO, A., T. NAKAMOTO, M. NAKAYAMA, and S. HAYASHI, *Proceedings of 34th Meeting of the Chemical Society of Japan*, 1976a. Pp. 661.
MATSUO, A., T. NAKAMOTO, M. NAKAYAMA, and S. HAYASHI, *Experientia*, 32, 966 (1976b).
MATSUO, A., S. UTO, M. NAKAYAMA, S. HAYASHI, K. YAMASAKI, R. KASAI, and D. TANAKA, *Tetrahedron Lett.*, 28, 2451 (1976c).
MATSUO, A., M. NAKAYAMA, S. HAYASHI, and K. NAKAI, *Phytochemistry*, 19, 1848 (1980).
MATTISON, N.L., "Enzymes des Sphaignes." Kompleken. Izuchen. Fisiol. Aktivn. Veshchestv. Nizshikh Rasten., Akad. nauk SSSR, Bot. Inst., 1961. Pp. 102-112.
McCLEARY, J.A. and D.L. WALKINGTON, *Rev. Bryol. Lichénol.*, 34, 309 (1966).
McCLEARY, J.A., P.S. SYPHERD, and D.L. WALKINGTON, *Science*, 131, 108 (1960).
MELCHERT, T.E. and R.E. ALSTON, *Science*, 150, 1170 (1965).
MELSTROM, C.E., N.C. MARAVOLO, and J.R. STROEMER, *Bryologist*, 77, 33 (1974).
MÉNDEZ, J. and F. SANZ-CABANILLES, *Phytochemistry*, 18, 1409 (1979).
MENON, M.K.C. and M. LAL, *Ann. Bot.*, 41, 1169 (1977).
MEUCHE, D. and S. HUNECK, *Chem. Ber.*, 99, 2669 (1966).
MEYER, M.W. and C. ANGERMAN, *Bryologist*, 76, 361 (1973).

MIKSCHE, G.E. and S. YASUDA, *Phytochemistry*, **17**, 503 (1978).
MOLISCH, H., *Ber. dt. bot. Ges.*, **29**, 487 (1911).
MUES, R. and H.D. ZINSMEISTER, *Phytochemistry*, **15**, 1757 (1976).
MUES, R., E. EDELBLUTH, and H.D. ZINSMEISTER, *Öst. Bot. Z.*, **122**, 177 (1973).
MUROMTSEV, G.S., V.N. AGNISTIKOVA, L.M. LUPOVA, L.P. DUBOVAYA, and T.A. LEKAREVA, *Invest. Akad. nauk. SSSR Ser. Biol. Moscow*, **5**, 727 (1964).
NARAYANASWAMI, S. and C.D. LaRue, *Phytomorphology*, **5**, 356 (1955).
NIEKERK-BLOM, C.J., *Proc. Kon. Ned. Akad. Wetensch.*, **49**, 1096 (1946).
NILSSON, E., *Acta chem. Scand.*, **23**, 2910 (1969a).
NILSSON, E., *Arkiv. för Kemi*, **31**, 475 (1969b).
NILSSON, E., "Abstracts of Uppsala dissertation," Fac. Sci., Uppsala, 1973a. Pp. 239.
NILSSON, E., *Phytochemistry*, **12**, 722 (1973b).
NILSSON, E. and G. BENDZ, "Flavonoids in bryophytes." In G. Bendz and J. Santesson, Eds, *Chemistry in Botanical Classification*, Academic Press, New York, 1973. Pp. 117-120.
NILSSON, E. and O. MÅRTENSSON, *Acta chem. Scand.*, **25**, 1486 (1971).
NILSSON, E., G. LINDBERG, and B.G. OSTERDAHL, *Chemica Scripta*, **4**, 66 (1973).
OHTA, Y., N.H. ANDERSON, and C.-B. LIU, *Tetrahedron*, **33**, 617 (1977).
ONO, M., *J. Jap. Bot.*, **33**, 285 (1958a).
ONO, M., *J. Jap. Bot.*, **33**, 267 (1958b).
ONO, M. and H. YOSHIMURA, *J. Jap. Bot.*, **33**, 60 (1958).
PAKARINEN, P. and D.H. VITT, *Can. J. Bot.*, **52**, 1151 (1974).
PATES, A.L. and G.C. MADSEN, *Bot. Gaz.*, **116**, 250 (1955).
PAUL, H., *Mitt. Bayer bot. Ges. (Moork-Anst)*, **2**, 63 (1908).
PEROLD, G.W., J.C. MULLER, and G. OURISSON, *Tetrahedron*, **28**, 5797 (1972).
PERSSON, H., *J. Hattori bot. Lab.* No. **17**, 1 (1956).
PRYCE, R.J., *Planta*, **97**, 354 (1971a).
PRYCE, R.J., *Phytochemistry*, **10**, 2679 (1971b).
PRYCE, R.J., *Phytochemistry*, **11**, 872 (1972a).
PRYCE, R.J., *Phytochemistry*, **11**, 1355 (1972b).
PRYCE, R.J., *Phytochemistry*, **11**, 1759 (1972c).
PRYCE, R.J. and L. LINTON, *Phytochemistry*, **13**, 2497 (1974).
QUILLET, M., *C.r. Acad. Sci., Paris*, **242**, 2475 (1956a).
QUILLET, M. *C.r. Acad. Sci., Paris*, **242**, 2656 (1956b).
RAMAUT, J.L., *Les Naturalistes Belges*, **40**, 9 (1959).
RANCKEN, H., *Acta Soc. F. et Fl. Fenn.*, **39**, 1 (1914).
RAO, M.P. and V.S.R. DAS, *Z. PflPhysiol.*, **59**, 87 (1968).
RAO, K.R., N. M. KONDAIAH, and K.V.R. RAMANA, *Z. PflPhysiol.*, **60**, 168 (1969).
RAO, K.R., C.S.P. NAIDU, and N.M. KONDAIAH, *Bryologist*, **78**, 216 (1975).
RAWAT, M.S. and R.N. CHOPRA, *Z. PflPhysiol.*, **78**, 372 (1976).
REKHI, ANITA, "Studies on protonemal differentiation and bud formation in *Timmiella anomala*," M.Sc. Thesis, Univ. Delhi, India, 1978.
REYNOLDS, A.C. and N.C. MARAVOLO, *Am. J. Bot.*, **60**, 406 (1973).
REZNIK, H. and R. WIERMANN, *Naturwissenschaften*, **53**, 230 (1966).
ROHWER, F. and M. BOPP, *J. Plant Physiol.*, **117**, 331 (1985).
RUDOLPH, H. and R. VOWINKEL, *Z. Naturforsch.*, **24B**, 1211 (1969).
SCHATZ, A., *Bryologist*, **58**, 113 (1955).
SCHIER, W., *Nova Hedwigia*, **25**, 549 (1974).
SCHNEIDER, Z. and J. SCHNEIDER, *Biochem. Physiol. Pflanzen.*, **171**, 239 (1977).
SCHNEIDER, M.J. and A.J. SHARP, *Bryologist*, **65**, 154 (1962).
SCHNEIDER, M.J., R.F. TROXLER, and P.D. VOTH, *Bot. Gaz.*, **128**, 174 (1967).
SCHUSTER, R.M., *Hepaticae and Anthocerotae of North America*, Vol. 1, Columbia University Press, New York, 1966. Pp. 381.
SCHWABE, W.W. and S. NACHMONY-BASCOMB, *J. exp. Bot.*, **14**, 353 (1963).
SCHWABE, W.W. and I.F.M. VALIO, *J. exp. Bot.*, **21**, 112 (1970a).
SCHWABE, W.W. and I.F.M. VALIO, *J. exp. Bot.*, **21**, 122 (1970b).

SHACKLETTE, H.T., *Geol. Surv. Bull.*, **1198C**, 1 (1965a).
SHACKLETTE, H.T., *Geol. Surv. Bull.*, **1198D**, 1 (1965b).
SHACKLETTE, H.T., *Geol. Surv. Bull.*, **1198G**, 1 (1967).
SIEGEL, S.M., *The Plant Cell Wall*, Pergamon Press, New York, 1962. Pp. 69.
SIEGEL, S.M., *Am. J. Bot.*, **56**, 175 (1969).
SPAETH, S.C. and N.C. MARAVOLO, *Bot. Gaz.*, **134**, 274 (1973).
SPENCER, K.C., *Phytochem. Bull.*, **12**, 4 (1979).
STEEMAN NIELSEN, E. and J. KRISTIANSEN, *Physiol. Plant.*, **2**, 325 (1949).
STRAIN, H.H., *Annual Priestley Lectures. Penn. State Univ.*, **32**, 1 (1958).
STRÁNSKY, K., M. STREIBL, and V. HEROUT, *Coll. Czech. chem. Comm.*, **32**, 3213 (1967).
SUIRE, C., *Le Botaniste*, **53**, 125 (1970).
SUIRE, C., *Rev. Bryol. Lichénol.*, **41**, 105 (1975).
SUIRE, C. and G. BOURGEOIS, *Phytochemistry*, **16**, 284 (1977).
SULEIMAN, A.A.A., "The carbohydrates of the leafy liverwort, *Plagiochila asplenioides*, and aspects of their metabolism." Ph.D. Thesis, Univ. Sheffield, 1972.
SUZUKI, Y. and B.J.D. MEEUSE, *Pl. Cell Physiol.*, **6**, 25 (1965).
SVENSSON, L., *Phytochemistry*, **13**, 651 (1974).
SVENSSON, L. and G. BENDZ, *Phytochemistry*, **11**, 1172 (1972).
SWANSON, E.S., W.H. ANDERSON, J.L. GELLERMAN, and H. SCHLENK, *Bryologist*, **79**, 339 (1976).
TAKEDA, R., Y. OHTA, and Y. HIROSE, "Studies on chemical constituents of *Ptychanthus striatus* (3)." In *100th Annual Meeting of Pharm. Soc. Japan, Tokyo*, Symposium Papers, 1980. Pp. 195.
TAYLOR, I.E.P., W.B. SCHOFIELD, and A.M. ELLIOT, *Can. J. Bot.*, **48**, 367 (1970).
TAYLOR, J. and P.B. KAUFMAN, *J. Hattori Bot. Lab.* No. **40**, 115 (1976).
TAYLOR, J., R.J. THOMAS, and J.G. OTERO, *Bryologist*, **75**, 36 (1972).
THEANDER, O., *Acta chem. Scand.*, **8**, 989 (1954).
THEODAR, R., H.D. ZINSMEISTER, R. MUES, and K.R. MARKHAM, *Phytochemistry*, **19**, 1695 (1980).
THEODAR, R., K.R. MARKHAM, R. MUES, and H.D. ZINSMEISTER, *Phytochemistry*, **20**, 1457 (1981a).
THEODAR, R., H.D. ZINSMEISTER, R. MUES, and K.R. MARKHAM, *Phytochemistry*, **20**, 1851 (1981b).
THOMAS, R.J., *Phytochemistry*, **14**, 623 (1975).
THOMAS, R.J., *Pl. Physiol.*, **59**, 337 (1977).
THOMAS, R.J., M.A. HARRISON, J. TAYLOR, and P.B. KAUFMAN, *Pl. Physiol.*, **73**, 395 (1983).
TJUKAVKINA, N.A., V. BENEŠOVA, and V. HEROUT, *Coll. Czech. chem. Comm.*, **35**, 1306 (1970).
TOUFFET, J. and S. VILLERET, *Bull. Soc. bot. Fr.*, **105**, 312 (1958).
TOWERS, G.H.N. and R.D. GIBBS, *Nature*, **172**, 25 (1953).
TUTSCHEK, R., *Phytochemistry*, **18**, 1437 (1979).
TUTSCHEK, R., H. RUDOLPH, P.H. WAGNER, and R. KREHER, *Biochem. Physiol. Pflanzen.*, **164**, 461 (1973).
UDAR, R. and S. CHANDRA, *Curr. Sci.*, **29**, 104 (1960a).
UDAR, R. and S. CHANDRA, *J. Hattori bot. Lab.* No. **23**, 85 (1960b).
UDAR, R. and S. CHANDRA, *Bryologist*, **63**, 173 (1960c).
VALIO, I.F.M. and W.W. SCHWABE, *J. exp. Bot.*, **21**, 138 (1970).
VALIO, I.F.M., R.S. BURDON, and W.W. SCHWABE, *Nature*, **223**, 1176 (1969).
VANDEKERKHOVE, O., *Z. PflPhysiol.*, **85**, 135 (1977a).
VANDEKERKHOVE, O., *Z. PflPhysiol.*, **82**, 455 (1977b).
VANDEKERKHOVE, O., *Z. PflPhysiol.*, **86**, 135 (1978a).
VANDEKERKHOVE, O., *Z. PflPhysiol.*, **86**, 217 (1978b).
VANDEKERKHOVE, O., *Z. PflPhysiol.*, **86**, 279 (1978c).

VANDEKERKHOVE, O., Z. PflPhysiol., **100**, 369 (1980).
VOWINKEL, E., Chem. Ber., **108**, 1166 (1975).
WALLAND, A. and H. KINZEL, Flora, **156A**, 597 (1966).
WATT, G., A Dictionary of the Economic Products of India, Part V, W.H. Allen, 1891, Pp. 85.
WEITZ, S. and R. IKAN, Phytochemistry, **16**, 1108 (1977).
WOLTERS, B., Arch. Mikrobiol., **37**, 293 (1960).
WOLTERS, B., Planta, **62**, 88 (1964).
WREN, R.W., Potter's New Cyclopedia of Botanical Drugs and Preparations, R.C. Wren, Sir Issac Pitman and Sons Ltd., London, 1956.

9 Bryophytes as Indicators of Pollution

POLLUTION of air, water, and soil caused mostly by increasing industrialization, is a matter of great concern. Pollutants may be defined as substances which exist in such concentrations as to cause unwanted effects. A primary pollutant is one which is emitted as such into the atmosphere, whereas a secondary pollutant is formed as a result of a chemical reaction in the atmosphere. Pollutants may be gaseous such as carbon monoxide, fluorides, hydrocarbons, hydrogen sulphide, nitrogen oxides, ozone, and sulphur dioxide. Among the particulate pollutants may be mentioned dust, particles of metallic oxides, coal, fly ash, cement, liquid particles, and smoke. In this chapter only the pollutants important in respect of bryophytes have been considered.

Pollutants inhibit sexual reproduction in bryophytes (De Sloover & LeBlanc, 1970). They also reduce photosynthesis and growth of plants and may eventually cause their death. Some species of bryophytes have become rare, whereas others have become extinct as a result of increasing pollution (Barkman, 1969; Daly, 1970; Devosalle et al., 1969; LeBlanc & De Sloover, 1970; LeBlanc & Rao, 1973a; Rao, 1982). Among the factors which influence the degree of damage are distance from the source of pollution, the exposure factor (concentration × time), nature of the substrate, shelter, pH, precipitation, stage of development, reproductive potential, and life-form of the species. Increasing precipitation results in increase in the heavy metal content of bryophytes (Tyler, 1971). In general the rough mats, tall turfs, large cushions, and leafy liverworts are least resistant to pollutants (Gilbert, 1970a). The protonemata of mosses are more sensitive than the mature gametophores (Gilbert, 1968). Species which produce large number of spores and/or gemmae and show fast vegetative propagation survive in polluted areas (Gilbert, 1971).

Bryophytes can play a significant role as indicators of environmental pollution (LeBlanc & Rao, 1975; Taoda, 1973a). These plants either independently or together with lichens, can be employed in developing an index of atmospheric purity (IAP) which is based on the number, frequency-coverage, and resistance factor of species, and this index can provide a fair picture of the long-range effects of pollution in a given area (see De Sloover & LeBlanc, 1970; LeBlanc et al., 1974; Rao, 1982).

There are two types of bryophytes. The first are very sensitive to pollution and show visible symptoms of injury even in the presence of minute quantities of pollutants. Such types, therefore, serve as good bio-indicators of the degree of pollution and also of the nature of pollutants. The second type have the capacity to absorb and retain pollutants in quantities much higher than those absorbed by other plant groups growing in the same habitat. Thus, they trap and prevent recycling of such pollutants in the ecosystem for different periods of time. An analysis of such plants can give a fair idea about the degree of metal pollution. Studies using the herbarium specimens in different regions of the world have indicated the relative importance of such pollutants during an extended period of time. This approach has added a new vista to the study of pollution.

HEAVY METALS

Heavy metals constitute a very important class of pollutants. The most significant among these are: lead, cadmium, zinc, mercury, arsenates, and chromium. Recently the effects with nickel and vanadium have also been studied on some liverworts. Bryophytes, and mosses in particular, have the capacity to absorb and accumulate one or more of these, in concentrations toxic for other groups of plants (Rao et al., 1977; Rühling & Tyler, 1971). Czarnowska and Rejment-Grochowska (1974) reported that iron content in the investigated mosses was ten times higher than in grasses. Bryophytes even concentrate some of the rare earth elements which are below the detection level in the substrate (Shacklette, 1965a). Heavy metals are absorbed either from the atmosphere or from the substrate or from both the sources. Very often the older tissues of the plant have higher concentrations of the metallic ions as compared to the younger portions (Shimwell & Laurie, 1972).

Rejment-Grochowska (1976) exposed four mosses (*Atrichum undulatum, Ceratodon purpureus, Polytrichum piliferum,* & *P. juniperinum*) to polluted air at three locations in the city of Warsaw. The fourth location was in the Botanical Garden. It has been suggested that the ability of mosses to accumulate heavy metals depends upon the total leaf surface, and the number of thin walled parenchymatous cells. Among the above mosses *A. undulatum* is highly sensitive to air pollution and proves best as a bio-indicator, whereas *C. purpureus* is not a good indicator because its leaves have a small surface and contain many thick walled cells. For *A. undulatum* the maximal concentrations of Pb- 400 ppm, Fe- 2500 ppm, and Cu- 211 ppm were observed at location I, and those of Zn- 1150 ppm and Mn- 2010 ppm at location II. In the Botanical Garden the content of lead, copper, and iron was considerably lower than in other three locations. The samples of mosses grown in the centre of the town showed inhibited growth, but resumed growth after transfer to Botanical Garden. Rejment-Grochowska (1976) suggests that tolerance of mosses to polluted air depends to a greater extent on the ratio of metals taken up than on the absolute metal level. For mosses the main role in complex accumulation of metals seems to be played by lead and the levels of other metals depend on it, as also reported by Ruhling and Tyler (1969).

LEAD

Lead is one of the most toxic metals known. Widespread lead pollution takes place from automobile exhaust. Lead is absorbed through precipitation or is supplied as dust, the constituents of which are released due to action of organic acids (Svensson & Ledén, 1965). Therefore, lead tends to accumulate more in the ectohydric' mosses like *Dicranella varia*, which lack cuticle and thus are able to absorb water from all over their body, than in the 'myxohydric' and 'endohydric' mosses which have a cuticle-like covering to minimize surface absorption.

It has been reported that in *Grimmia doniana* lead is ionically bound to the cell wall, thus preventing toxic amounts of lead from penetrating the cytoplasm (Brown & Bates, 1972). Skaar et al. (1973) observed electron dense lead inclusions in the nuclei, and damaged nuclear envelopes in leaf cells of *Rhytidiadelphus squarrosus* from lead polluted area and also in cultured plants watered with lead acetate solution. No such inclusions were noticed within cells of leaves of plants from unpolluted area. These authors have suggested that binding of lead within the nuclear membrane keeps the concentration of diffusible lead in the cytoplasm below a level that would be toxic to the mitochondrial and other lead-sensitive functions of the cytoplasm. Some mosses have a very simple way of getting rid of excess of heavy metals absorbed by them. In *Dicranella varia* (an ectohydric moss) lead and zinc are excreted in the form of a powder of the metal sulphates from the leaf tips during summer drought (Shimwell & Laurie, 1972). Such precipitates comprise high concentrations of the metallic ions, and an analysis of one sample showed the value of lead to be as high as 60,450 ppm. The crust on the leaf surface rapidly dissolves in rain water and is thus lost.

Ruhling and Tyler (1968) analyzed the specimens of three mosses. *Hylocomium splendens*, *Pleurozium schreberi*, and *Hypnum cupressiforme* from herbarium specimens in Lund (Sweden) which had been collected at intervals from 1860 to 1968 from the same location. They reported that lead content in the specimens over the years 1860 to 1875 was only about 20 ppm (dry wt), in the next 25 years it became almost double. From the year 1900 to 1950 there was not much change, but in the next ten years there was a rapid increase and this brought the average of lead content to 80 to 90 ppm. In a similar study by Lee and Tallis (1973) in Britain, based on living and herbarium specimens of *Hypnum cupressiforme*, it was reported that in respect of lead pollution the peak occurred in the eighteenth and nineteenth centuries, and in the last 100 years lead had declined in its importance as a pollutant. Lee (1972) studied the effect of pollution from a factory manufacturing anti-knock compounds, on mosses like *Ceratodon purpureus* and *Eurhynchium praelongum*. In the first 400 m a rapid decline was recorded in lead concentration in the vegetation away from the source of pollution. In the next 400 m the decline was slower, but even at 800 m the amount of lead in the mosses remained up to three times that in the same species collected from an area remote from industry. In *Ceratodon purpureus* lead content was 320 ppm at a distance of 2 m, and 185 ppm at a distance of 20 m from a road in central Manchester.

Lead tolerance can also play as a factor in natural selection, as indicated by the work of Briggs (1972). He grew different clones of *Marchantia polymorpha* in cultures and reported that such of the clones which were subjected to high lead pollution in their natural habitats proved tolerant to lead in cultures. Whereas, others which were growing in areas exposed to less lead contamination were sensitive to lead. Therefore, the tolerant species were successful in urban industrial environment.

CADMIUM

Mosses can take up airborne cadmium as well as absorb it from the substrate. Nash (1974) reported that in cultures bryophytes show sensitivity to this element at concentrations as low as 10^{-8} M. Simola (1977a) recorded differences in two species of *Sphagnum* (*S. fimbriatum* & *S. nemoreum*) in respect to cadmium tolerance. The indications of damage to the plants are noticeable in the pigmentation and growth rate. In some plants low concentrations of cadmium have been reported to promote growth. Lepp and Roberts (1977) observed that growth of *Marchantia polymorpha* gemmalings was more with 1 and 5 ppm cadmium as compared to cadmium-free controls or those receiving high cadmium levels. This enhancement in growth possibly occurs because of activation of respiratory enzymes. Cadmium at 1 ppm also significantly enhanced the elongation of germ tubes in *Funaria hygrometrica*, but at higher concentrations (10 ppm) spore germination was markedly reduced. Cadmium is known to accumulate in mosses in rather great amounts. Concentrations as high as 30 mg/kg dry weight have been recorded in mosses in a polluted area of Sweden (Tyler, 1972). Rasmussen (1977) surveyed the cadmium content in living and herbarium specimens in an attempt to know the levels of airborne metals in rural areas of Denmark between the years 1951 to 1975. The content of cadmium remained almost unchanged excepting a slight increase of two percent during the years 1973 to 1975.

ZINC

The uptake of zinc is also related to the water economy of mosses (Shimwell & Laurie, 1972). The 'ectohydric' ones like *Dicranella varia* have much higher zinc content (2420 ppm) than the 'myxohydric' ones like *Philonotis fontana* (297 ppm). In *Marchantia polymorpha* and *Funaria hygrometrica*, zinc concentrations more than 50 ppm result in complete inhibition of spore germination and germ tube extension. The protonema of *Funaria hygrometrica* reacts to this pollutant in cultures containing more than 10 ppm by producing rounded 'brood' cells. This possibly is a defence mechanism, since the surface area of protonema in contact with zinc gets reduced (Coombes & Lepp, 1974). The effect of distance on the concentration of zinc in moss tissues has been reported in *Eurhynchium praelongum* (Burkitt et al., 1972), *Hypnum cupressiforme*, and *Hylocomium splendens* (Rühling & Tyler, 1969, 1970). In the first moss collected from distances of three miles and six miles from an industrial area in Bristol (England) the concentration of zinc was 1315 ppm and 876 ppm, respectively. Rühling and Tyler (1969) analyzed herbarium specimens collected in southern Sweden during 1870 to 1943. An increase was recorded between 1870 to 1919, but not later. Robitaille and LeBlanc (*cited in* Rao et al., 1977) collected some interesting data in respect of five metals including zinc in some mosses from Mount Royal in Montreal (Canada) during a 66-year period (1905 to 1971). The concentration of zinc showed a steady increase as a result of heavy metal pollution due to industrialization. The extent of increase in zinc content ranged from 1.2 to 5 times in different moss species. The highest increase was recorded in *Plagiomnium cuspidatum* from 88 to 440 ppm. In rural areas of Denmark, the content of zinc in bryophytes increased from 23 to 76 percent in the period 1951 to 1975, and from 42 to 108 percent in the urban areas during this period (Rasmussen, 1977). Barclay-Estrup and Rinne (1978) reported that zinc content in *Pleurozium schreberi* and *Hylocomium splendens* was more in urban areas than in rural areas of Canada.

MERCURY

Mercury is collected by plants through rain water or dust particles. Mosses and lichens tolerate high concentrations of mercury, and have often been used as indicators of atmospheric mercury. Near the heavy mercury mining activity in Boston the average levels of mercury in soil were 100 to 200 ppb, whereas in the mosses they were as high as 200 to 2000 ppb (Yeaple, 1972). Mondano and Smith (1974) recorded 4.34 ppm mercury in *Dicranella heteromalla* at the urban end of a transect, and only 0.24 ppm in the rural areas. An estimation of mercury values in mosses and conifer twigs and needles indicated that bryophytes give a more reliable indication of mercury fall-out. Siegel et al. (1975) studied the mercury content of 10 mosses from Alaska and Hawaii and reported a range between 0.1 and 120 µg/100 g material, with an average of 13.0 µg. Wallin (1976) reported the mercury values in the carpet-forming moss, *Hypnum cupressiforme*, to be between 9 to 15 µg/100 g. In Norway *Hylocomium splendens* was collected from 43 sampling sites and the average mercury concentration was 12 µg (Steinnes, 1977). Solberg and Selmer-Olsen (1978) determined the mercury level of 11 species of mosses. The average in 10 of these was 16.6 µg/100 g. In the eleventh, *Tomenthypnum nitens*, the content was as high as 94 µg/100 g.

ARSENATES

Arsenicals are relatively more toxic inorganics in the environment. Not much is known about the effect of arsenic pollution on bryophytes. Simola (1977b) studied the effect of arsenate on *Sphagnum nemoreum* grown in cultures. The highest concentration tested (1 mM) was lethal. At the lowest level the shoots turned yellowish-green. At 0.1 mM the cytoplasm degenerated, chloroplast membranes were broken, and the thylakoids swelled. The chloroplasts contained more starch and in addition had plastoglobuli.

CHROMIUM

Shacklette (1965a) determined the chromium level in 38 species of mosses and liverworts, and reported the average to be 79 µg/g ash weight. In another study based on 16 species of bryophytes the chromium content was observed to be as high as 99 µg/g (Lounamaa, 1956). Lee et al. (1977) recorded a very interesting instance of the epiphytic moss *Aerobryopsis longissima* which accumulates chromium from the bark of the host (*Homalium guillaini*) and also from the water dripping from its leaves. The level of chromium in the moss is 5000 µg/g, it being the highest record for bryophytes. The host had a concentration of chromium nearly 20 times that present in the moss.

NICKEL

Lepp and Hockenhull (1983) cultivated gemmalings of *Marchantia polymorpha* in liquid medium in the presence of nickel. At 0.15 ppm Ni (as $NiSO_4$) resulted in significant growth stimulation, but had marked toxic effects at levels in excess of 0.25 ppm. $NiCl_2$ and $Ni(NO_3)_2$ elicited different responses. The former produced significant growth stimulation at 0.25 ppm, whereas the latter was tolerated even at 0.5 ppm. Growth stimulation was noticed with 1 ppm Ni as Ni-EDTA, and no significant growth retardation was observed even with 10 ppm Ni-EDTA or the Ni-glycine complex. Furthermore, the gemmalings from urban populations of *M. polymorpha* showed no growth reduction at 0.5 ppm Ni (as $NiSO_4$), whereas those from the glasshouse grown plants were less tolerant.

VANADIUM

Lepp and Lawson (1984) have recently demonstrated that urban and rural populations of *Marchantia polymorpha* showed differential responses to VO_3^- (metavanadate ion) and VO^{2+} (vanadyl cation). Gemmalings from urban populations showed tolerance to VO^{2+}, whereas those from the rural population showed tolerance to VO_3^-. Excessive vanadium levels in both forms resulted in reduced growth, necrosis, chlorosis, and reduction/inhibition of rhizoid development. In *Lunularia cruciata* the light brown oil bodies turned black at higher vanadium levels.

STABILITY PATTERN OF METAL IONS

Metal ions are retained by plants either as a result of simple ion exchange or by chelation with specific organic groups within the cell. If one metal complex is more stable than another of the same type, it will displace the latter from a less stable complex. The stability of organic chelates with divalent metal ions was observed to follow the average sequence: Pb > Cu > Ni > Co > Zn > Cd > Fe > Mn > Mg (*see* Rao et al., 1977). Using *Hylocomium splendens*, the absorption and retention of heavy metals was reported by Rühling and Tyler (1970) to be in the order: Cu > Pb > Ni > Co > Zn, Mn. In the mosses *H. splendens* and *Pleurozium schreberi* LeBlanc et al. (1974) noticed the absorption of the metals to be in the order: Pb > Cu > Zn > Cd > As.

As to what determines the order in the series is not very clear. The nature of the bond (dsp_2) influences stability. Since lead and copper form stronger bonds they head the list. It also appears that the more basic the metal is, the less stable complex it forms (*see* Brown, 1984).

METAL TOLERANCE

Very little is known about the maximum tolerance levels of bryophytes for individual metals and combination of cations. Marked differences in various species in this respect are on record (Noeske et al., 1970). It has been suggested that the majority of mosses have a tendency for the accumulation of some metals like iron, lead, zinc, nickel from the substrate and the atmosphere. Certain mosses like *Weissia controversa*, *Grimmia tenera*, and *Rhacomitrium lanuginosum* can successfully grow on soil derived from serpentine rock which contains excessive amounts of nickel (4000 ppm) and chromium (900 ppm) (Shacklette, 1965b). In their behaviour towards certain metals, mosses generally resemble higher plants rather than lichens (Lounamaa, 1956).

In the moss *Atrichum undulatum* Czarnowska and Rejment-Grochowska (1974) recorded that the concentrations of iron, manganese, zinc, and copper were 2000, 97, 45, and 4.5 ppm in the gametophyte, whereas in the sporophyte they were 215, 93, 76, and 7.4 ppm, respectively.

COPPER MOSSES

The majority of plants are extremely sensitive to copper concentration. Some species of mosses serve as indicators of high copper concentration in the substrate and are known as 'copper mosses'. Geissler (1982) has suggested that the copper concentration tolerated by the 'copper mosses' is lethal to phanerogams, and on such sites bryophytes have no competition. The important copper mosses are: (1) *Mielichhoferia elongata*, (2) *M. mielichhoferi*, (3) *M. macrocarpa*, (4) *Merceya ligulata* [= *Scopelophila ligulata* (Spruc.) Spruc.], and (5) *Dryptodon stratus* (Shacklette, 1961, 1966). Among the liverworts whose principal substrata are copper ores are: (1) *Cephaloziella phyllacantha*, (2) *C. massalongi*, (3) *Gymnocolea acutiloba*, and (4) *G. inflata*. Warncke (1968) reported that *Marchantia alpestris* is generally restricted to copper-rich areas in Scandinavia, and suggested that it should be added to the list. Some authorities prefer to call these species as sulphur mosses' (Schatz, 1955), because they often occur on substrata rich in reduced forms of sulphur, like copper, iron, and zinc sulphides, and also in hot sulphur springs where the content of copper and iron is fairly low, but that of reduced and elementary sulphur is very high. These mosses assimilate carbon dioxide with hydrogen sulphide as the reducing agent, and accumulate free sulphur intracellularly (Nakamura, 1938). Such species seem to constitute the missing link between anaerobic sulphur bacteria and aerobic green plants.

Mårtensson and Berggren (1964) analyzed the substrata of *Mielichhoferia elongata* and *Dryptodon stratus*. Copper values of 320 to 770 ppm were recorded, indicating that these mosses are resistant to or may prefer a substrate containing copper in amounts about 100 times higher than present in ordinary soils. Brassard (1969) analyzed the moss *M. elongata* and reported that it accumulates several elements like copper, barium, lanthanum, lead, nickel, strontium, and zinc. These mosses also prefer very low pH (2.7 to 4.2), and avoid lime stone rocks which are alkaline. In the sulphide ores, which are a substrate for cuprophilic bryophytes, pH is lowered as a result of oxidation of sulphites and formation of sulphuric acid.

Tiwari (1984) has reported a species of *Mielichhoferia* from the

Kumaun region of North-West Himalaya. The plants occur on a polymetallic copper contaminated substrate (Ca, 0.02%; Mg, 0.39%; Fe, 3.30%; Cu, 0.39%; Zn, 0.46%).

An extensive survey conducted by Shacklette (1967) showed that many copper mosses grow on substrata low in sulphur, and others on substrata with relatively high pH. It is, therefore, likely that the presence of high amounts of copper control the distribution of such mosses. However, it is still not known whether a specific requirement of this element exists. Differences in the copper tolerance and pH requirement have been recorded by Gams (1966) in different species of the genus *Mielichhoferia*.

Although copper mosses occur on soils rich in copper, and also show morphological changes typical of certain types of mineralization, it is difficult to use them as plant indicators in exploring for copper. This is because such mosses are very rare and are also difficult to identify (*see* Hartman, 1969).

PEAT MOSSES

Peat mosses have the ability to accumulate heavy metals and serve as indirect indices of air pollution. Like most other mosses they have haploid gametophytes, but have an added advantage due to their extraordinary capacity for vegetative multiplication. In *Sphagnum* a single mutated vegetative cell in the plant may give rise to a new tolerant clone. Therefore, bog mosses are good systems for studies on adaptation of plants to pollution.

Studies by Rühling and Tyler (1968) in Sweden have indicated that the amount of lead and cadmium in *Sphagnum* species are much higher (42 to 49 ppm) in polluted than in unpolluted areas (5 to 12 ppm). A similar trend in the level of lead, nickel, chromium, and iron has been reported in Canada by Pakarinen and Tolonen (1976). Simola (1977a,b) used the aseptic culture technique for studies on *Sphagnum* species and reported that they are tolerant to cadmium, lead, arsenate, and fluoride. This tolerance is due to the high exchange capacity of *Sphagnum*.

Species of *Sphagnum* are now being used by most workers in the 'moss bag technique' for measuring the level of heavy metal pollution (*see* Brown, 1984).

GASEOUS POLLUTANTS

The effect of three gaseous pollutants is being considered: sulphur dioxide, fluorides, and ozone.

SULPHUR DIOXIDE

The main source of sulphur dioxide (SO_2) in the atmosphere is the burning of coal and oil. This gas is also produced in the iron sintering plants or thermal power plants. In the presence of moisture a major part of sulphur dioxide gets converted to sulphurous acid, and this in turn is oxidized to sulphuric acid, which is about 20 times more toxic than sulphur dioxide (Gartrell et al., 1963). However, sulphuric acid soon gets converted to sulphates, like ammonium sulphate and calcium sulphate, which are less toxic. Thus, one of the good indicators of sulphur dioxide pollution is the concentration of soluble sulphates in water and soil samples (Gordon & Gorham, 1963).

As a result of exposure to sulphur dioxide the sensitive species gradually lose green colour. The leafy liverwort *Radula complanata* changes its colour within ten minutes and its chloroplasts disintegrate at 120 ppm sulphur dioxide (Coker, 1967). Chlorophyll 'a' is degraded to phaeophytin 'a' with the release of magnesium ions:

$$\text{Chlorophyll 'a'} + 2H^+ \longrightarrow \text{Phaeophytin 'a'} + Mg^{2+}$$

The level of pollution can also be determined by the amount of water soluble magnesium liberated during the process of reduction of chlorophyll 'a' to phaeophytin 'a'. As a result of permanent damage to chloroplasts at high sulphur dioxide concentration, photosynthesis fails to take place even after the source of pollution is removed. Damage due to sulphur dioxide is also caused because of irreversible plasmolysis of the cell contents, and this eventually results in the death of plants. Sulphur dioxide is highly toxic for bryophytes, and most of them can not survive when its average concentration exceeds 0.017 ppm (Daly, 1970; Taoda, 1972). Taoda (1973b) reported that most epiphytic bryophytes were injured by 0.8 ppm sulphur dioxide in 10 to 40 h, or by 0.4 ppm in 20 to 80 h of total exposure. At 0.2 ppm growth retardation was noticed on prolonged exposure (100 h or more). Normally the limiting concentration of sulphur dioxide is below 5 ppm, but the tolerant species can withstand up to 10 ppm for short duration. *Ctenidium molluscum* is known to be the most resistant moss. Among the other toxiphilous species (tolerant to high sulphur dioxide pollution) are *Bryum argenteum, Ceratodon purpureus,* and *Tortula muralis* (Daly, 1970; Gilbert, 1970b). Among the six bryophytes tested by Syratt and Wanstall (1969) *Dicranoweisia cirrata* was able to withstand higher levels of sulphur dioxide than the other species. All sulphur dioxide-resistant bryophytes have fast growth rate. Among the factors known to modify sensitivity to sulphur dioxide are: sulphur dioxide concentration, humidity, pH, shelter, nature of the substrate, and relative percentages of sulphur dioxide derivatives (bisulphite ions & sulphurous acid) (Comeau & LeBlanc, 1971; Robitaille et al., 1977). Species growing on tree trunks are generally more sensitive than those growing on other substrata (see Daly, 1970; Gilbert, 1970b). At lower humidity levels bryophytes show less damage than at higher levels (see Syratt & Wanstall, 1969; Taoda, 1973b). The degree of ionization of sulphur dioxide in solution is influenced by pH. It also affects the oxidation of sulphur dioxide to less toxic sulphates (Gilbert, 1968). The rate of oxidation is independent of pH between 8.8 and 8.2. At low pH values (5.9 to 3.2) the rate of oxidation decreases (Joslyn & Braverman, 1954). This is perhaps why bryophytes disappear from the more acidic habitats first and the sensitive species survive at high pH values. Rao and LeBlanc (1967) reported that in Wawa (Canada) there was a marked reduction of epiphytes, both in number and abundance of species, with a rise in the concentration of sulphur dioxide in the air and of sulphate in water and soil. Studies conducted in Denmark by Johnsen and Søchting (1976) also revealed that distribution of epiphytic lichens and bryophytes was well-correlated with sulphur dioxide emission and pH of the bark. Shelter can be provided by topographic barriers, grassland, or even by fissures in bark, and can greatly reduce sulphur dioxide levels (Gilbert, 1970a,b). Emission of sulphur dioxide during rain (acid rain) enhances the solubility of metal salts and increases their uptake by bryophytes. It also lowers the pH and brings about other marked changes in the chemical properties of epiphytes as well as their substrata (Robitaille et al., 1977).

Gilbert (1968) cultured two sensitive species (*Hypnum cupressiforme* & *Polytrichum commune*) and two resistant ones (*Ceratodon purpureus* & *Bryum argenteum*), and exposed them to sulphate and sulphite ions. Even at the relatively high concentration of 600 ppm, neither of them was damaging. In contrast, bisulphite ions proved toxic to the gametophores of sensitive species and to all protonemata, which survived only at concentrations lower than 20 ppm. The gametophores of resistant species tolerated up to about 200 ppm of bisulphite.

Among the methods employed for ascertaining the resistance of mosses to sulphur dioxide are fumigation and plasmolysis. Fumigation with sulphur dioxide tests their total resistance (Coker, 1967). For determining their plasmotic resistance mosses are submerged in a 0.15 mM solution of sodium thiosulphate. Viability is measured by the capacity of gametophores for carbon dioxide exchange. Hackemesser (1981) tested six sulphur dioxide concentrations (varying from 0.5 to 5.5 mg/m^3) over an exposure of 48 h on some selected moss species. While low concentrations partly caused a slight increase in the activity of acid-phosphatase, higher levels resulted in considerable decrease. According to the author the change in activity of this enzyme is of limited value for determining sulphur dioxide damages in mosses. In yet another method bryophytes are transplanted along with their substrate in ecologically more or less similar, but sulphur dioxide-polluted sites. LeBlanc and Rao (1973b) transplanted bark discs bearing *Orthotrichum obtusifolium* and *Pylaisiella polyantha*, in the region of Sudbury (Ontario) where the environment is polluted with sulphur dioxide. After exposure for one year the epiphytes were either greatly damaged or were dead, depending upon the level of pollution in the area (*see* LeBlanc et al., 1976). The common symptoms of injury were plasmolysis and chlorophyll degradation (*see* Daly, 1970).

Field surveys and pollution mapping techniques are being profitably used for determining the degree of pollution in a given area and for delimiting pollution zones (Granger, 1972; Rao, 1982; Taoda, 1972). Maps prepared on the basis of index of atmospheric purity (IAP) values are more or less similar to those based on sulphur dioxide levels (LeBlanc et al., 1972). Turk and Wirth (1975) conducted a comparative study of sulphur dioxide tolerance in mosses, and arranged them in the following decreasing order: *Ctenidium molluscum, Rhytidiadelphus triquetrus, Hypnum cupressiforme, Grimmia pulvinata,* and *Mnium undulatum.* An estimation of sulphur content of the species resistant to sulphur dioxide pollution (*Ceratodon purpureus* & *Bryum argenteum*) revealed that their gametophores contain more sulphur than those of sensitive ones. Mosses in general seem to be more sensitive to sulphur dioxide treatment than liverworts. Among mosses, acrocarps are more resistant than pleurocarps (Gilbert, 1968).

FLUORIDES

Fluorides are generally released into the atmosphere as gaseous hydrogen fluoride (HF) or as volatile fluorides (Na_3AlF_6 & SiF_4). The main source of this pollution is aluminium factories although manufacture of phosphate fertilizers, brick plants, pottery, and ferroenamel works also contribute to fluoride pollution. With an increase in aluminium industry, fluoride pollution has become a serious problem throughout the world. Bryophytes are very sensitive to hydrogen fluoride and show symptoms of injury at concentrations as low as 0.001 to 0.1 ppm Experiments performed by LeBlanc et al. (1971) with bark discs bearing the mosses *Pylaisiella polyantha* and *Orthotrichum obtusifolium* deserve a special mention. The discs were transplanted onto trees at varying distances from an aluminium industry in Quebec, Canada. On the transplanted discs in polluted areas *P. polyantha* changed its colour from the normal golden green to brown, dark brown, or to burnt appearance, depending upon the length of exposure, the distance from the source of pollution, and the direction from the factory. Fluoride pollution brings about plasmolysis in the cell and also disintegrates chlorophyll. However, the exact mode of injury is not yet well understood. In contrast to the damage done by sulphur dioxide, no magnesium is released during chlorophyll degeneration due to fluoride injury.

OZONE

Action of sunlight on nitrogen dioxide and on certain hydrocarbons results in the production of ozone (O_3) and, therefore, it is a secondary pollutant. Ozone pollution brings about acute injury and senescence (Heggestad, 1968), but not much is known about the precise mode of its phytotoxicity.

Preliminary experiments on *Funaria hygrometrica* using the tissue culture technique (Comeau & LeBlanc, 1971) revealed that ozone stimulated regenerative ability of leaves when administered at low concentrations for short periods. The percentage of regeneration was inversely proportional to the exposure factor.

RADIONUCLIDES (RADIO ISOTOPES)

Angiosperms exhibit a mean count rate of only 63 per min, whereas mosses and lichens, which hold the fall-out strongly, have 152 and 183 mean count rates, respectively. This difference is quite significant (Gorham, 1959).

CESIUM

Cesium is very tightly bound in the clay minerals, and, therefore, the chief means of entry of ^{137}Cs in plants is through foliar absorption. Such studies are especially important in the arctic zone where mosses and lichens form the first link in the food chain which leads to man (Hoffman, 1972).

For terrestrial bryophytes in a forest, ^{137}Cs is available mainly through the raindrip, whereas the epiphytic ones receive their supply chiefly from the stemflow (Hoffman, 1972). It has been suggested that cryptogams should be used as meters for radioactive fall-out from nuclear bomb tests in the atmosphere, since they have high absorption and retention capabilities for radionuclides (Svensson & Ledén, 1965). Terrestrial mosses in forests are reported to have higher amounts of this element than do other undergrowth plants or litter (Richard, 1967; Svensson, 1967). Based on laboratory experiments it has been suggested by Witkamp and Frank (1967) that the floor in *Pinus verginiana* forest intercepted 93 percent of the total ^{137}Cs which reached it, with mosses and lichens accounting for two-third of the total amount intercepted. Loss of accumulated cesium is more rapid from dead plants than from live ones. The efficiency of mosses and lichens in absorbing this element is high. *Cladonia subtenuis* absorbed 88 percent of the ^{137}Cs input, whereas *Dicranum scoparium* absorbed 95 percent (Witkamp & Frank, 1967).

According to Steere (1970) leafy hepatics are more sensitive to ionizing radiations than mosses, and *Fissidens garberi* appears to be the most resistant because it can survive within 10 m of a cesium source.

Hoffman (1972) tagged a 500 m^2 *Liriodendron tulipifera* forest with 467 mc (millicounts) of ^{137}Cs. The terrestrial bryophytes such as *Dicranum scoparium, Eurhynchium hians, Mnium affine,* and *Rhynchostegium serrulatum* accumulated 1.16 mc; the tree-base mosses like *Anomodon rostratus, Leucobryum albidum, Thuidium delicatum,* and *Hypnum curvifolium* accumulated 0.37 mc; the epiphytic lichens on main stems had 2.52 mc and the canopy lichens accumulated 0.16 mc. The sum of 4.21 mc accumulated by cryptogams is less than one percent of the total introduced into the forest, but even this amount was greater on a unit dry weight basis than that in the tree foliage. Experimentally determined uptake efficiencies of ^{137}Cs were 92 percent for the terrestrial bryophytes mat (living as well as dead), and 67 percent for the vertically oriented epiphytic lichens, most of which were foliose Tree-base mosses and their substrate bark absorbed 90 percent of the ^{137}Cs supply. Hoffman (1972) suggested that even though lichens and bryophytes are minor components of forest ecosystems, they represent an important sink in cycling and accumulation of ^{137}Cs. Their slow growth rates and long lives increase the time during which they might hold elements out of more rapid circulation.

STRONTIUM

Strontium-90 is an isotope of great biological interest. The occurrence of small quantities of strontium in plants under natural conditions has been known for many years. *Buxbaumia aphylla* is reported to concentrate more strontium than that found in its substrate (Hancock & Brassard, 1974). Radioactive strontium was supplied to *Grimmia orbicularis* under field and laboratory conditions (Hebrard et al., 1972, 1974). Maximum ^{90}Sr uptake in the field coincided with period of maximum rainfall. Brown and Buck (1978) reported that in *Brachythecium rutabulum* strontium is more concentrated at the base of the shoot than calcium. In higher plants as well, more strontium is retained in the roots. Strontium and calcium are mainly bound physically rather than biologically. Brown and Buck (1978) confirmed that potassium is mainly held intracellularly, whereas calcium and strontium are mostly bound extracellularly.

A comparison of pleurocarpous and acrocarpous mosses revealed very little difference in calcium content between the two types, but the average strontium values were lower in pleurocarpous species (Brown & Buck, 1978). Although calcium and strontium are chemically similar, the latter is more firmly bound to the anionic exchange sites.

URANIUM

Terrestrial bryophytes accumulate uranium, but aquatic bryophytes are usually employed for chemical analysis, because they are constantly washed by stream water from which they sorb and concentrate trace elements.

Bryophytes usually have simple ion exchange capacity, and a life time up to several decades. They, therefore, might afford of give an indication of the mean uranium content of stream water over a long period of time.

Whitehead and Brooks (1960) obtained data for the uranium content of water by placing peat adsorbers in the streams. The uranium content of the ash of peat adsorbers and of the bryophytes growing in the stream was directly related.

RADIATIONS

The most commonly investigated radiations are x-rays, gamma-rays, and alpha-rays. These ionizing radiations activate plant cell molecules and thus quicken chemical reactions. Injury results when the chemically altered molecules are essential components of protoplasm. In nature the quantities of ionizing radiations are negligible and therefore no injury is caused to the plants.

Most plants tolerate much higher doses of radiation than those fatal to animals. The average lethal dose for man is about 500 R (Roentgen), whereas some plants tolerate as much as 20,000 R (Kuzin et al., 1958). Among the bryophytes the dose of alpha-irradiation which produced partial or complete killing in 24 h was 1460 mc/sec for the liverwort *Plagiochila asplenioides*, whereas for the moss *Hookeria lucens* it was 21,900 mc/sec (Biebl & Url, 1963). Bryophytes and other non-vascular plants are less sensitive to radiation than are the vascular plants (Pullum & Erbisch, 1972).

Most morphogenetic effects of gamma radiation are non-genetic, and seem to be due to their effect on hormone balance (Sax, 1963). Some of the effects are instantaneous whereas others are delayed (Gunckel & Sparrow, 1961). There are numerous factors which seem to influence the degree of effect. Some of the external factors are: type of radiation, dose rate, temperature, day length, humidity, nature of substrate in which the plant grows, time of the day, and the season. The internal factors are: stage of life cycle, type of tissue, portion of the cell hit by radiation, age of the cell, nutritional state of the plant, water content of cells, metaphase length, chromosome size, and nuclear volume. Sparrow (1962) and Steere (1970) consider nuclear volume to be more important. A comparison among different species indicates that smaller nuclear volumes generally confer greater resistance to irradiation.

Snyder (1961) studied the effects of ten specific doses of x-ray radiation on spore germination in *Polytrichum commune*. The dosage ranged from 100 to 5000 R. There was no inhibition of spore germination up to 300 R, and the maximum decrease in germination and stunting of protonemal growth occurred at 1000 R and above: 5000 R was lethal. Grossman and Hillson (1974) gave four levels of gamma radiation to *Polytrichum commune*: 1000, 1600, 2200, and 2800 R. For the first 10 to 15 weeks 1000 R had a stimulatory effect. However, a delayed inhibitory response was observed after a few months.

The life cycle features of bryophytes may be of some help in knowing whether nuclear volume or chromosome size is a more accurate index of radiosensitivity, since mosses have diploid and haploid generations growing together. One would expect that with an increase in ploidy, nuclear volume would increase. Sparrow (1964a,b) compared the nuclear volume and interphase chromosome volume ratio of many polyploid and diploid series, and reported that polyploids may have larger nuclei, but the chromosome volumes do not increase correspondingly. In many instances the polyploid chromosomes are even smaller. Moutschen (1959) studied the radiosensitivity of *Brachythecium rutabulum* and reported that the survival rates initially vary inversely with ploidy, but the subsequent rates are the same in tissues of different ploidy.

Because of their marked capacity for cellular proliferation bryophytes are very suitable for studies on cell survival within an irradiated multicellular organism. Miller and Sparrow (1964) observed that in the thalli of *Marchantia polymorpha* nuclei of the apical cells are about twice the volume of the nuclei of other vegetative cells, and these are nearly twice as sensitive to radiations, at lower levels of exposure. However, at the lethal level there is no difference in the tolerance. Similar observations were made by Miller and Sparrow (1965) in their experiments on *Marchantia* gemmae. Increased exposure to gamma-rays decreased the number of functional apical cells, but there was an initial increase in the number of non-apical outgrowths per gemma. The greatest number of non-apical outgrowths occurred at 19.4 kr exposure. Survival of the gemmae at this exposure was 99 percent, but only 65 percent of the apical cells survived. However, the lethal exposure for both types of cells was approximately the same.

Mnium cuspidatum showed a marked diurnal and annual rhythm in its resistance to α-and UV-rays (Biebl & Hofer, 1965). Maximum resistance to α-radiation was registered at mid-night, and minimum between 12 noon and 4 p.m. Seasonwise, bryophytes have maximum resistance in June and minimum in December.

Bryophytes, especially the sensitive epiphytic taxa, hold a great potential for studies on atmospheric pollution. However, the amounts of pollutants present within the investigated plants depend upon the capacity of the species to sorb and accumulate these elements, and also on many environmental factors. These data, therefore, should not be taken as an exact index of the degree of pollution in a given area.

Practically all the work on these lines has so far been done in developed countries. There is an urgent need for such investigations to be conducted in developing countries, in order to assess the effect of increasing industrialization.

In recent years the tissue culture technique has been employed for investigating the response of various species and clones of a particular species to different pollutants. With an extended use of this technique very useful information can be collected, and pollution resistant strains can be developed.

REFERENCES

BARCLAY-ESTRUP. P. and R.J.K. RINNE. *Oikos*. 30. 106 (1978).
BARKMAN. J.J.. "The influence of air pollution on bryophytes and lichens". In *Air Pollution: Proceeding of the First European Congress on the Influence of Air Pollution on Plants and Animals*. Wageningen. 1969. Pp. 197-209.
BIEBL. R. and K. HOFER. *Radiat. Bot.*. 6, 225 (1965).
BIEBL. R. and W. URL. *Protoplasma*. 57, 84 (1963).
BRASSARD. G.R.. *Nature*. 222, 584 (1969).
BRIGGS. D.. *Nature*. 238, 166 (1972).
BROWN, D.H., "Uptake of mineral elements and their use in pollution monitoring". In A.F. Dyer and J.G. Duckett, Eds. *The Experimental Biology of Bryophytes*, Academic Press, New York, 1984. Pp. 229-255.
BROWN. D.H. and J.W. BATES. *J. Bryol.*. 7, 187 (1972).
BROWN. D.H. and G.W. BUCK. *J. Bryol.*. 10, 199 (1978).
BURKITT. A.. P. LESTER. and G. NICKLESS. *Nature*. 238, 327 (1972).
COKER, P.D., *Trans. Br. bryol. Soc.*, 5, 341 (1967).
COMEAU, G. and F. LeBlanc. *Natur. Can.*, 98, 347 (1971).
COOMBES, A.J. and N.W. LEPP, *Bryologist*, 77, 447 (1974).
CZARNOWSKA. K. and I. REJMENT-GROCHOWSKA. *Acta Soc. Bot. Pol.*. 43, 39 (1974).
DALY, G.T., *Proc. N.Z. Ecol. Soc.*, 17, 70 (1970).
DEVOSALLE. L.. F. DEMARET, J. LAMBINON, and A. LAWALREE. *Serv. Res. Nat. dom. Cons. Nat.*, 4, 1 (1969).
DE SLOOVER. J. and F. LeBlanc. *Bull. Acad. Soc. Lorraines Sci.*, 9, 82 (1970).
GAMS. H., *Vereins zum Schutze der Alpenpflanzen und Tiere Jahrb.*, 31, 1 (1966).
GARTRELL, F.E., F.W. THOMAS, and S.B. CARPENTER. *Amer. Indust. Hyg. Assoc. Jour.*, 24, 113 (1963).
GEISSLER, P., "Alpine Community". In A.J.E. Smith, Ed., *Bryophyte Ecology*, Chapman and Hall, London and New York, 1982. Pp. 167-189.
GILBERT, O.L., *New Phytol.*, 67, 15 (1968).
GILBERT, O.L., *New Phytol.*, 69, 605 (1970a).
GILBERT, O.L., *New Phytol.*, 69, 629 (1970b).
GILBERT, O.L., *Lichenologist*, 5, 26 (1971).
GORDON, A.G. and E. GORHAM. *Can. J. Bot.*, 41, 1063 (1963).
GORHAM. E., *Can. J. Bot.*, 37, 327 (1959).
GRANGER. J.M., *Sarracenia*, 5, 43 (1972).
GROSSMAN, H.H. and C.J. HILLSON, *Bryologist*, 77, 142 (1974).
GUNCKEL, J.E. and A.H. SPARROW, "Ionizing radiations: biochemical, physiological, and morphological aspects of their effects on plants". In W. Rühland, Ed., *Handbuch der Pflanzenphysiologie*, Gottingen, Germany, 1961. Pp. 555-611.
HACKEMESSER. H., *Angew. Botanik.*, 55, 83 (1981).
HANCOCK, J.A. and G.R. BRASSARD, *Can. J. Bot.*, 52, 1861 (1974).
HARTMAN, E.L., *Bryologist*, 72, 56 (1969).
HEBRARD, J.P., L. FOULQUIER, and A. GRAUBY, *Radioprotection*, 7, 157 (1972).
HEBRARD, J.P., L. FOULQUIER, and A. GRAUBY. *Bull. Soc. bot. Fr.*, 121, 235 (1974).
HEGGESTAD, H.E., *Phytopathology*, 58, 1089 (1968).
HOFFMAN. G.R., *Bot. Gaz.*, 133, 107 (1972).
JOHNSEN, I. and U. SØCHTING, *Bryologist*, 79, 86 (1976).
JOSLYN, M.A. and J.B.S. BRAVERMAN, *Adv. Fd. Res.*, 5, 97 (1954).
KUZIN, A.M., S. CHI, and G.N. SAENKO, *Biofizika*, 3, 308 (1958).
LeBlanc, F. and J. DeSloover, *Can. J. Bot.*, 48, 1485 (1970).
LeBlanc, F. and D.N. RAO, *Bryologist*, 76, 1 (1973a).
LeBlanc, F. and D.N. RAO, *Ecology*, 54, 612 (1973b).
LeBlanc, F. and D.N. RAO, "Effects of air pollution on lichens and bryophytes". In J.B. Mudd and T.T. Kozlowski, Eds. *Responses of Plants to Air Pollution*, Academic Press, New York, 1975. Pp 237-272.

LeBlanc, F., G. COMEAU, and D.N. RAO, *Can. J. Bot.* **49**, 1691 (1971).
LeBlanc, F., D.N. RAO, and G. COMEAU, *Can. J. Bot.*, **50**, 519 (1972).
LeBlanc, F., G. ROBITAILLE, and D.N. RAO, *J. Hattori bot. Lab.* No., 38, 405 (1974).
LeBlanc, F., G. ROBITAILLE, and D.N. RAO, *J. Hattori bot. Lab.* No., 40, 27 (1976).
LEE, J.A., *Nature*, **238**, 165 (1972).
LEE, J.A. and J.H. TALLIS, *Nature*, **245**, 216 (1973).
LEE, J.A., R.R. BROOKS, and R.D. REEVES, *Bryologist*, **80**, 203 (1977).
LEPP, N.W. and Y. HOCKENHULL, *Bryologist*, **86**, 342 (1983).
LEPP, N.W. and J.A. LAWSON, *Bryologist*, **87**, 37 (1984).
LEPP, N.W. and M.J. ROBERTS, *Bryologist*, **80**, 533 (1977).
LOUNAMAA, K.J., *Ann. Bot. Soc. Zool. Fenn.*, **29**, 1 (1956).
MARTENSSON, O. and A. BERGGREN, *Oikos*, **5**, 99 (1964).
MILLER, M.W. and A.H. SPARROW, *Nature*, **204**, 596 (1964).
MILLER, M.W. and A.H. SPARROW, *Radiat. Res.*, **25**, 219 (1965).
MONDANO, M. and W.H. SMITH, *Environ. Conserv.*, **1**, 201 (1974).
MOUTSCHEN, J., *Radiobiol. Latina*, **2**, 105 (1959).
NAKAMURA, H., *Acta Phytochim., Japan*, **10**, 271 (1938).
NASH III, T.H., "The effect of air pollution on other plants, particularly vascular plants". In B.W. Ferry, M.S. Baddeley, and D.L. Hawksworth, Eds. *Air Pollution and Lichens*, London, 1974. Pp. 192-223.
NOESKE, P., A. LACHLI, O.L. LANGE, G.H. VIEWEG, and H. ZIEGLER, *Vort. Bot. Ges. (Dtsch. Bot. Ges.) N.F.*, **4**, 67 (1970).
PAKARINEN, P. and K. TOLONEN, *Ambio*, **5**, 38 (1976).
PULLUM, P.A. and F.H. ERBISCH, *Bryologist*, **75**, 48 (1972).
RAO, D.N., "Responses of bryophytes to air pollution". In A.J.E. Smith, Ed., *Bryophyte Ecology*, Chapman and Hall, London and New York, 1982. Pp. 445-471.
RAO, D.N. and F. LeBlanc, *Bryologist*, **70**, 141 (1967).
RAO, D.N., G. ROBITAILLE, and F. LeBlanc, *J. Hattori bot. Lab.* No., 42, 213 (1977).
RASMUSSEN, L., *Environ. Pollut.*, **14**, 37 (1977).
REJMENT-GROCHOWSKA, I., *J. Hattori bot. Lab.* No., 41, 225 (1976).
RICHARD, W.H., "Accumulation of ^{137}Cs in litter and understorey plants of forest stands from various climatic zones of Washington". In B. Aberg and F.P. Hungate, Eds, *Proceedings International Symposium on Radioecological Concentration Processes*, Stockholm, 1966, Pergamon, New York, 1967. Pp. 527-531.
ROBITAILLE, G., F. LeBlanc, and D.N. RAO, *Rev. Bryol. Lichénol*, **43**, 53 (1977).
RÜHLING, Å. and G. TYLER, *Bot. Notiser.*, **121**, 321 (1968).
RÜHLING, Å. and G. TYLER, *Bot. Notiser.*, **122**, 248 (1969).
RÜHLING, Å. and G. TYLER, *Oikos*, **21**, 92 (1970).
RÜHLING, Å. and G. TYLER, *J. appl. Ecol.*, **8**, 497 (1971).
SAX, K., *Radiat. Bot.*, **3**, 179 (1963).
SCHATZ, A., *Bryologist*, **58**, 113 (1955).
SHACKLETTE, H.T., *Bryologist*, **64**, 1 (1961).
SHACKLETTE, H.T., *U.S. Geol. Surv. Bull.*, **1198-D**, 1 (1965a).
SHACKLETTE, H.T., *U.S. Geol. Surv. Bull.*, **1198-C**, 1 (1965b).
SHACKLETTE, H.T., *U.S. Geol. Surv. Bull.*, **1198-G**, 1 (1966).
SHACKLETTE, H.T., *U.S. Geol. Surv. Bull.*, **1198-G**, 18 (1967).
SHIMWELL, D.W. and A.E. LAURIE, *Environ. Pollut.*, **3**, 291 (1972).
SIEGEL, S.M., B.Z. SIEGEL, N. PUERNER, T. SPEITEL, and F. THORARINSSON, *Water, air, and soil pollution*, **4**, 9 (1975).
SIMOLA, L.K., *Ann. Bot. Fenn.*, **14**, 1 (1977a).
SIMOLA, L.K., *Can. J. Bot.*, **55**, 426 (1977b).
SKAAR, H., E.M. OPHUS, and B.M. GULLVÅG, *Nature*, **241**, 215 (1973).
SNYDER, C.R., *Am. Biol. Teacher*, **23**, 506 (1961).
SOLBERG, Y. and A.R. SELMER-OLSEN, *Bryologist*, **81**, 144 (1978).
SPARROW, A.H., "The role of the cell nucleus in determining radiosensitivity". In *Brookhaven Lect. Ser.* No. 17, BNL 766 (T-287), Upton, New York, 1962. Pp. 22.

SPARROW, A.H., "Relationship between chromosome volume and radiosensitivity in plant cells". In *Proceedings M.D. Anderson 18th Annual Symposium on Fundamental Cancer Research*, Huston, Texas, 1964a.

SPARROW, A.H., "Comparisons of the tolerance of higher plant species to acute and chronic exposures of ionizing radiation". In *Conference on Mechanism of the Dose Rate Effect at the Genetic and Cellular Level*, Oiso, Japan, Nov. 4-7, 1964b.

STEERE, W.C., "Bryophyte studies on the irradiated and control sites in the rainforest at El Verde". In H.T. Odum, Ed., *Tropical Rain Forest: A Study of the Irradiation and Ecology at El Verde, Puerto Rico*, U.S. Atomic Energy Commission, Washington, D.C., 1970. Pp. D213-D225.

STEINNES, E., Atmospheric deposition of trace elements in Norway studied by means of moss analysis. Kjeller Report KR-154, Kjeller Research Establishment, 1977.

SVENSSON, G.K., "The increasing ^{137}Cs level in forest moss in relation to the total ^{137}Cs fall-out from 1961 through 1965". In B. Aberg and F.P. Hungate, Eds, *Proceedings International Symposium on Radioecological Concentration Processes*, Stockholm, 1966, Pergamon, New York, 1967. Pp. 539-546.

SVENSSON, G.K. and K. LEDÉN, *Health Physics*, **11**, 1033 (1965).

SYRATT, W.J. and P.J. WANSTALL, "The effect of sulphur dioxide on epiphytic bryophytes" In *Air Pollution: Proceeding of the First European Congress on the Influence of Air Pollution on Plants and Animals*, Wageningen, 1969. Pp. 79-85.

TAODA, H., *Jap. J. Ecol.*, **22**, 125 (1972).

TAODA, H., *Hikobia*, **6**, 224 (1973a).

TAODA, H., *Hikobia*, **6**, 238 (1973b).

TIWARI, S.D., "Studies on bryophytes of Kumaun Himalaya", Ph.D. Thesis, Univ. Kumaun, Nainital, India, 1984.

TÜRK, R. and V. WIRTH, *Bryologist*, **78**, 187 (1975).

TYLER, G., "Moss analysis—a method of surveying heavy metal deposition". In M. Englund and W. Berry, Eds, *Proceedings of the Second International Clean Air Congress*, Academic Press, NewYork, 1971. Pp. 129-132.

TYLER, G., *Ambio*, **1**, 52 (1972).

WARNCKE, E., *Bot. Tidskr.*, **69**, 358 (1968).

WALLIN, TH., *Environ. Pollut.*, **10**, 101 (1976).

WHITEHEAD, M.E. and R.R. BROOKS, *Bryologist*, **72**, 501 (1960).

WITKAMP, M. and M.L. FRANK, *Health Physics*, **13**, 985 (1967).

YEAPLE, D.S., *Nature*, **235**, 229 (1972).

10 Protoplast Culture

THE isolated protoplasts are developmentally independent systems quite analogous to microbes which have so far been the favourite material for investigations by geneticists. Moreover, the isolated protoplasts can be made to fuse, and thus the genotypes of divergent taxa can be combined to produce a new genotype which can not be had by the usual breeding programmes. In addition, the fusion of protoplasts brings about mixing of cytoplasms, thus permitting an analysis of cytoplasmic inheritance. These naked 'cells' are also capable of direct incorporation of the desired traits and are thus a suitable system for genetic engineering.

One of the most promising aspects of plant protoplast research is production of somatic hybrids and selection of genetically altered lines. A pre-requisite for these studies is the successful regeneration of plants from protoplasts. Unfortunately, this has not been possible in many instances, probably because of complex nutritional requirements of the protoplasts of higher plants. By contrast, lower plants, such as mosses and liverworts, provide excellent developmental systems with extremely high regeneration potential, and simple nutritional requirement for differentiation. Of particular significance is the gametophytic moss plant which comprises a large population of haploid cells, each capable of regenerating into an entire plant. Investigations on protoplasts of bryophytes can also be of great significance in determining the differences between the developmental regulation mechanisms of haploid and diploid plants.

The term protoplast refers to the plasmolyzed contents of a cell without the cell wall. These naked free cells—which are considered to be totipotent—can be easily manipulated for plating, counting, and raising of mutants. The possible applications of protoplast technology are listed below:

(1) Parasexual hybridization.
(2) Study of plant morphogenesis from single isolated cells.
(3) Genetic modification of cells by uptake of foreign genome.
(4) Studies on mutation and selection of altered cell lines.
(5) Understanding the mechanisms of viral, bacterial, and fungal infection.
(6) Isolation of cell organelles and other constituents.
(7) Biosynthesis and composition of primary cell wall and its structural and functional properties.

(8) Study of membranes.
(9) Transport and co-transport of molecules across membranes.
(10) Study of hormone action at the membrane level.

ISOLATION OF PROTOPLASTS

Protoplasts are isolated by removal of cell wall in the presence of a suitable osmoticum, and this can be accomplished by one of the following two methods:

MECHANICAL METHOD

It involves random sectioning of plasmolyzed cells and results in the release of some undamaged protoplasts. Binding (1966) successfully isolated protoplasts from *Funaria hygrometrica* and *Sphaerocarpos donnellii*. Gay (1976) also employed this technique for *Polytrichum juniperinum*.

Binding (1966) plasmolyzed the cells for longer durations (3 h) in 1 M sucrose solution. The leafy stems were then minced with a razor blade on a glass slide and the fragments transferred to 2 ml of a 0.7 M sucrose solution. Swelling of the protoplasts continued for 1 h, and gentle shaking of the medium allowed the undamaged protoplasts to emerge from the severed cells. The suspension was then filtered through a cloth sieve (pore diameter 140 um) to eliminate fragments of tissues and cells. The filtrate contained only isolated protoplasts and cell organelles, especially chloroplasts.

ENZYMATIC METHOD

This method was developed by Cocking (1960). The protoplasts are released by enzymatic digestion of cell wall in an isolation medium which comprises pectin and cellulose-dissolving hydrolytic enzymes, and a suitable osmoticum which prevents the protoplasts from bursting.

Enzymes

A wide range of commercial enzyme preparations is now available, under different trade names. For example, the commonly used fungal cellulases and hemicellulases are cellulase 'Onozuka R 10', Driselase, Rhozyme HP 150, hemicellulase, β-glucuronidase, and Helicase. Amongst the various pectinases, Macerozyme R 10, polygalacturonase 'Rohament P', and pectic acid transeliminase (PATE) or pectic acid-acetyl transferase are employed. These enzymes have been used in the concentration range of 0.5 to 5 percent.

Methods of Enzymatic Isolation

The tissue can either be treated with two enzymes—Macerozyme and cellulase—in two steps, sequentially one after the other, or simultaneously in one-step.

In the two-step or sequential method, separation of cells is brought about by pectinases and it is followed by release of protoplasts by cellulases. In this method the cells remain in contact with enzymes for a shorter period as compared to the one-step method. In *Sphaerocarpos donnellii*, plants were placed in a water bath-shaker with 2.5 percent glucuronidase or Helicase in 0.55 M mannitol at 27 C. After 25 to 30 min this enzyme solution was discarded and the plant material subjected to digestion in two percent cellulase dissolved in 0.5 M mannitol (Schieder & Wenzel, 1972). Taylor (1979) also employed the same type of enzymatic treatment, but used higher concentration of cellulase (5%) and another enzyme, pectinase (0.75%), in his second step. Bopp et al. (1980) obtained best results for protoplast isolation of *Funaria hygrometrica* by employing the following procedure:

(1) Pre-incubation for plasmolysis in glycine (5% w/v) for 10 to 15 min.
(2) Pre-maceration in glycine (5%) and PATE (0.2% w/v) for 2 h.
(3) Maceration in glycine and cellulase 'Onozuka R 10' (5%) for 2 to 3 h.

In the one-step method the tissue is subjected to an enzyme mixture comprising Macerozyme and cellulase, and the chances of microbial contamination are minimized. This method has been employed for the following taxa: *Anoectangium thomsonii* (Saxena & Rashid, 1980), *Funaria hygrometrica* (Gwóźdź & Waliszewska, 1979; Saxena & Rashid, 1981), *Lunularia cruciata* (Thomas & Silcox, 1983), *Marchantia polymorpha* (Ono et al., 1979; Takeuchi et al., 1980), *Physcomitrella patens* (Stumm et al., 1975), *Plagiomnium vesicatum* (Maeda, 1979), and *Pottia intermedia* (Ripetsky, 1978).

In a few higher plants a combination of mechanical and enzymatic methods has been employed. To begin with, mechanical separation of cells is done by homogenizing the tissue, which is then subjected to enzyme treatment. Yet another modification is an intermediate process between the one-step and the two-step method. The tissue is treated with Macerozyme to loosen the cells, and before free cells are formed it is transferred to cellulase.

Osmoticum

For the successful release of protoplasts it is essential to have a solution of accurately determined osmolality. Otherwise, for want of pressure hitherto exerted by the cell wall, immediate lysis of protoplasts occurs. The following substances have been used as osmotica: $CaCl_2$, $Ca(NO_3)_2$, sugars, and sugar alcohols (mannitol & sorbitol). It is not essential that the osmotic stabilizer be metabolically inert. Thus, sucrose, glucose, and sorbitol have also been employed. A combination of sugar/s and mannitol is more desirable because the gradual decline in overall osmotic potential due to the utilization of sugar/s during the period of early growth of protoplasts helps to avoid a sudden change when the regenerated cells are transferred to osmoticum-free medium for continued growth. The optimal concentration of the osmoticum, as is expected, varies considerably among plant species and the organ employed for isolation of protoplasts. It generally varies from 0.3 to 0.7 M.

SOURCE MATERIAL FOR PROTOPLASTS

In mosses protoplasts can be isolated from protonemata as well as from gametophores. In liverworts protoplasts have been isolated from thalli as well as calli grown in suspension cultures (see Table 10.1).

It has been demonstrated that with the currently available enzymes only cells of protonema in *Physcomitrella patens* are susceptible to degradation (Bach, 1973; Stumm et al., 1975). In order to obtain reproducible results it is necessary to culture the protonema under controlled conditions (Stumm, 1974).

Taylor (1979) mentioned some advantages of the suspension culture for isolation of protoplasts. These are: (1) the growth rate of a tissue in liquid culture is much higher than on solidified media, (2) the cells are physiologically more uniform because all cells are equally in contact with the culture medium, and (3) the cells are better adapted to *in vitro* conditions prior to protoplast isolation. He also suggested that for high yields of protoplasts a rapidly growing suspension culture of the tissue is a prerequisite. It is therefore essential that the tissue be harvested in the exponential growth phase just prior to the onset of the stationary phase, and this is on the average, 24 days after the inoculation of the suspension culture.

Steps for the Isolation of Protoplasts

The steps involved for the isolation of protoplasts are diagrammatically represented represented on page 273.

FACTORS AFFECTING PROTOPLAST ISOLATION

Rinsing of the Tissue

Rinsing is useful in isolating sterile protoplasts from material grown under non-sterile conditions.

Pretreatment of the Tissue

Incubation of plants in weak illumination or in dark enhances the yield and viability of protoplasts. Treatment of the tissue with chemicals such as

TABLE 10.1.
Source of protoplast isolation in bryophytes.

Taxa	Source	Investigator/s
LIVERWORTS		
Lophocolea heterophylla	Suspension cultures raised from thallus	Taylor (1979)
Lunularia cruciata	Thallus	Thomas and Silcox (1983)
Marchantia polymorpha	Suspension cultures raised from thallus	Taylor (1979)
	Callus	Ono et al. (1979); Takeuchi et al. (1980); Sugawara et al. (1983); Mori et al. (1982, 1983)
Sphaerocarpos donnellii	Thallus	Binding (1966); Schieder and Wenzel (1972); Schieder (1974a,b); Wenzel and Schieder (1973)
	Suspension cultures raised from thallus	Taylor (1979)
Sphaerocarpos texanus	—do—	—do—
MOSSES		
Anoectangium thomsonii	Protonema	Saxena and Rashid (1980, 1981)
Funaria hygrometrica	Leafy stem	Binding (1966)
	Protonema	Gwóźdź and Waliszewska (1979); Bopp et al. (1980); Nowak and Młodzianowski (1980); Saxena and Rashid (1981); Batra and Abel (1981)
Physcomitrella patens	—do—	Bach (1973); Stumm (1974); Stumm et al. (1975); Grimsley et al. (1977a,b); Burgess and Linstead (1981); Jenkins and Cove (1983)
Physcomitrium sp.	Spore mother cells	Ripetsky (1975)
Plagiomnium vesicatum	Protonema	Maeda (1979)
Polytrichum juniperinum	Leafy stem	Gay (1976, 1980)
Pottia intermedia	Protonema, gametophore apices	Ripetsky (1978, 1979)

arginine, cycloheximide, L-lysine, kinetin, glycine, and L-cysteine also improves yield and quality of protoplasts. Bopp et al. (1980) used glycine (5%) in the maceration mixture to isolate protoplasts from *Funaria hygrometrica*.

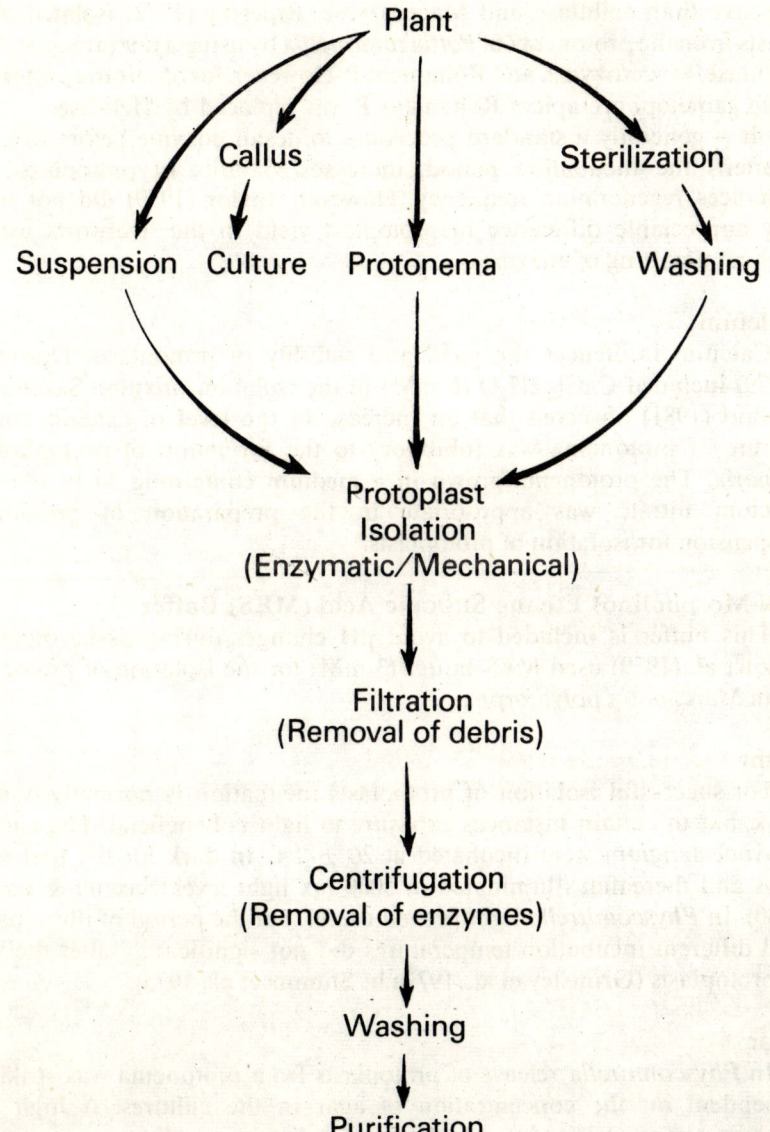

Enzyme/s

Effective isolation of protoplasts depends on: (1) the specific enzyme used, (2) the combination of enzymes, and (3) enzyme purification. These are determined empirically, for example Saxena and Rashid (1980) isolated protoplasts from preplasmolyzed protonemal filaments of *Anoectangium thomsonii* grown in suspension cultures, by using the enzymes cellulase and Macerozyme, or Driselase. Driselase was more effective than cellulase and Macerozyme, or Driselase. Driselase was more effective than cellulase and Macerozyme. Ripetsky (1978) isolated protoplasts from the protonema of *Pottia intermedia* by using a mixture containing cellulase, Macerozyme, and Rohament P. However, for obtaining protoplasts from gametophore apices Rohament P was replaced by Helicase.

It is generally a standard procedure to desalt enzyme before use, as it shortens the incubation period, increases viability of protoplasts, and enhances regeneration frequency. However, Taylor (1979) did not notice any appreciable difference in protoplast yield in the liverworts with or without desalting of enzymes.

Calcium

Calcium influences the yield and stability of protoplasts. Ono et al. (1979) included $CaCl_2.2H_2O$ (6 mM) in the isolation mixture. Saxena and Rashid (1981) observed that an increase in the level of calcium for the culture of protonema was inhibitory to the formation of protoplasts in *Funaria*. The protonema grown in a medium containing 50 to 100 mg/l calcium nitrate was appropriate in the preparation of protonemal suspension for isolation of protoplasts.

2(N-Morpholino) Ethane Sulfonic Acid (MES) Buffer

This buffer is included to avoid pH changes during tissue digestion. Ono et al. (1979) used MES buffer (3 mM) for the isolation of protoplasts from *Marchantia polymorpha*.

Light

For successful isolation of protoplasts incubation is normally done in dark, but in certain instances exposure to light is beneficial. The cultures of *Anoectangium* were incubated at 26 ± 2 C in dark for the first seven days and thereafter illuminated in 3000 lux light level (Saxena & Rashid, 1980). In *Physcomitrella* light proved useful, but the period of illumination and different incubation temperatures did not significantly alter the yield of protoplasts (Grimsley et al., 1977a,b; Stumm et al., 1975).

Agar

In *Physcomitrella* release of protoplasts from protonema was strikingly dependent on the concentration of agar in the cultures. A high agar concentration (1.3%) improved the quality as well as quantity of protoplasts (Stumm et al., 1975).

Wetting Agents

Another factor which increased disintegration of cell walls in a remarkable manner was pretreatment of protonema with non-ionic detergents, Tween-80, and Triton X-100. Stumm et al. (1975) reported that Tween-80 probably caused increased wettability of the protonema which enhanced the possibility for enzymes to attack cell walls resulting in higher protoplast yield.

Age

Age of the material can affect the composition of cell wall. Therefore, Saxena and Rashid (1981) employed actively growing protonemal suspension for isolation of protoplasts in *Anoectangium thomsonii*. Bopp et al. (1980) observed maximum yield of protoplasts from 16 to 20-day-old protonema. Gwóźdź and Waliszewska (1979) noticed that only a young protonema, six to seven days after spore germination, was susceptible to enzyme treatment.

pH and Osmolality

Some other factors which influence successful isolation of protoplasts are the initial pH (5.5 to 6.8), and osmolality (0.3 to 0.6 M) of the enzyme solution. Schieder and Wenzel (1972) observed that in *Sphaerocarpos donnellii* protoplasts can be isolated at pH as low as 4.0.

CULTURE OF PROTOPLASTS

The methods employed for culture of protoplasts are basically similar to those employed for tissue and cell culture. Some minor modifications are made in view of delicate nature of protoplasts.

LIQUID CULTURE

Protoplasts are cultured in the form of a suspension of appropriate density in a small volume of liquid medium. This method was used by Ono et al. (1979) and Takeuchi et al. (1980). They dispensed 2 to 5 ml of *Marchantia polymorpha* protoplast suspension in petri plates ranging from 3 to 9 cm in diameter.

In liquid cultures a serious problem is the bursting of protoplasts due to collision with the culture vessel. However, one distinct advantage is that such cultures allow the osmotic potential of the medium to be monitored, since a gradual decrease promotes rapid proliferation. For culture of protoplasts in flasks or petri dishes larger volumes of suspension are required. Liquid cultures of small volumes can be done by using one of the following methods:

Drop Culture

Small drops of protoplast suspension are placed on the inner side of lid of a petri dish, which when inverted brings the culture drops in hanging position (Gwóźdź & Waliszewska, 1979; Stumm et al., 1975).

Wenzel and Schieder (1973) used erect drops instead of hanging drops. They placed a drop (0.2 ml) of protoplast suspension of *Sphaerocarpos donnellii* in the centre of a petri dish and surrounded it with several drops of 60 percent sea water to give a moist chamber effect. The dish was incubated in a moist plastic box at 25 to 30 C in low light level or in dark. Fresh medium was added as and when required. This method is advantageous because it provides good aeration by increasing the surface to volume ratio.

Hanging Drop Culture in Cavity Slides

This method is not frequently used. In this procedure a drop of protoplast suspension is placed on a sterile cover glass, which is in turn inverted on a cavity slide and sealed with sterile mineral oil. The slides are then stored in sterile petri plates in dark.

Microchamber Culture

A microchamber is set up by fixing two cover glasses (22 mm × 22 mm) with mineral oil about 18 mm apart on a slide. A droplet of about 30 µl of nutrient medium containing protoplasts is placed between the cover glasses. This is covered by another cover glass which is made to rest on the other two. This method has an advantage over the hanging drop technique in that protoplasts or the fusion products can be kept under constant view and the entire development can be followed.

Multiple Drop Array (MDA) Technique

In this method, 49 drops, each of 20 µl and having about 400 protoplasts, are placed on the inside of the lid of a petri dish along a square grid. With this technique as many as 5000 media variations in two to three factor combinations can be tested with a single population of two million protoplasts.

Microdroplet Culture

By this method, cultures can be raised from protoplasts in volumes as small as 0.25 to 0.50 µl by using a special cuprak culture dish which has built-in wells capable of accommodating just the desired volume of culture medium. The regeneration frequency in microdroplet culture of protoplasts is comparable to that obtained with microcultures. By employing microdroplet culture it is envisaged that even single protoplasts or the hybrid cells can be cultured.

AGAR PLATING OF PROTOPLASTS

In this method protoplast suspension at an appropriate density is gently mixed with an equal volume of nutrient medium gelled with agar. It is then dispensed in petri dishes in a molten state at 45 C. The cultures are sealed with parafilm and kept upside down at 28 C. Plating of protoplasts enables not only the growth of a large population, but also allows a ready observation of the growth of individual protoplasts.

In the conventional plating technique described above protoplasts are inevitably exposed to a temperature shock. Therefore, instead of agar the use of sodium alginate has been proposed, which is easily gelled by the addition of Ca^{++}, and again liquified with a chelate (sodium citrate). This process can be carried out at room temperature, and recovery of regenerants from liquid is easier than from agar medium in which the colonies get embedded.

The plating technique has been employed by Gay (1976), Maeda (1979), and Saxena and Rashid (1980). Innovations of this technique allow growth of protoplasts even at a very low density.

REGENERATION OF PROTOPLASTS

The regenerative capacity of isolated protoplasts depends upon the isolation procedure and the plant system (see Table 10.2).

TABLE 10.2.
Percentage of regenerating protoplasts in bryophytes.

Taxa	Regenerative frequency	Investigator/s
Funaria hygrometrica	2-5%	Gwóźdź and Waliszewska (1979)
	50%	Bopp et al. (1980)
	40-60%	Saxena and Rashid (1981)
Marchantia polymorpha	90%	Sugawara et al. (1983)
Physcomitrella patens	5-10%	Stumm et al. (1975)
	80%	Grimsley et al. (1977a,b)
Polytrichum juniperinum	20%	Gay (1976)
Sphaerocarpos donnellii	70%	Wenzel and Schieder (1973)

The cultured viable protoplasts soon start regenerating new cell walls. The presence of cell wall can be confirmed on transfer of protoplasts to a deplasmolyzing solution which results in their bursting. In routine experiments cell wall synthesis can be tested at the level of light microscope by observing fluorescence on staining with Calcofluor White-ST (Plate 10.1; Mori et al., 1983), the selective stain for cellulose, or with Tinopal 'BOPT' (see Evans & Cocking, 1977). In *Anoectangium thomsonii* the protoplasts regenerated cell walls in three days, and divisions commenced after five to seven days (Saxena & Rashid, 1980). On the other hand, in

Funaria hygrometrica protoplasts developed cell walls within 36 h, and the cells 'germinated' in a typical spore-like fashion (Saxena & Rashid, 1981).

Mori et al. (1983) observed that in *Marchantia polymorpha* regeneration of cell walls began within a few hours of cultivation. New cell walls completely covered the surface of protoplasts within 48 h. Coumarin and 2,6-dichlorobenzonitrile (DCBN) treatment inhibited formation of new cell wall. They also studied the relationship between regeneration of cell walls and ultrastructural changes in protoplasts. In the initial stage of cell wall regeneration endoplasmic reticula developed remarkably close to plasma membrane in the protoplasts, but development of Golgi bodies was not observed at the same locus. This may suggest that the Golgi bodies do not play an active role in the cell wall formation, at least not in very early periods of cell wall regeneration. The development of endoplasmic reticula and an ultrastructural change of plasma membrane from smooth to rough may be important in cell wall formation of protoplasts.

After the protoplasts develop cell walls, they divide to form irregular groups of spherical cells which later show cell differentiation and organogenesis. Protonemal filaments or thalli may result depending upon the system. A small percentage of these protonemata can be demonstrated to be diploid or tetraploid (Binding, 1966; Knoop, 1978; Bopp et al., 1980). In *Sphaerocarpos donnellii*, isolated protoplasts regenerated entire plants (Binding, 1966). In *Physcomitrella patens* and *Funaria hygrometrica* (Plate 10.2A-C) protoplasts isolated from protonemata regenerated cells which again formed protonema (Grimsley et al., 1977a; Gwozdź & Waliszewska, 1979; Stumm et al., 1975), but in *Marchantia polymorpha* protonema formation was always preceded by development of callus tissues (Plate 10.3A-E; Ono et al., 1979).

Thus, the behaviour of the regenerated protoplasts in bryophytes can be summarized as follows:

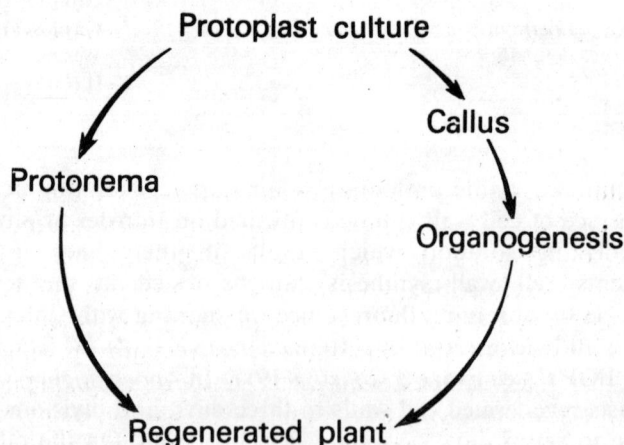

FACTORS AFFECTING REGENERATION

A perusal of literature reveals that apart from different parameters of isolation and growth conditions of donor tissue, regeneration of protoplasts critically depends on the physical (density, temperature, & light) and chemical (carbon source, ammonium nitrate, calcium salts, amino acids, vitamins, growth regulators etc.) environment. Recently, Saxena (1985), using *Dalbergia sissoo* as a system, reported that cold-conditioning of the source tissue prior to protoplast isolation has a marked promotory effect on protoplast regeneration.

Stumm et al. (1975) observed that the most important requirement for protoplast development was a screen of glass between the protoplast and the light source. This possibly resulted in the removal of harmful ultraviolet rays from the light source. They also reported that pretreatment of the protonema with the detergent 'Tween-80' promoted the regeneration process in *Physcomitrella patens*. The isolated protoplasts also readily regenerated cell walls in media enriched with divalent cations such as calcium and magnesium. The subsequent growth of these cells into germ tubes, protonema, and leafy gametophores required special media.

Gwóźdź and Waliszewska (1979) observed that in *Funaria hygrometrica* despite wall formation, the protoplasts failed to divide in an inorganic medium. Extensive experimentation showed that glucose at a concentration of 0.5 percent was essential for survival and subsequent divisions of protoplasts. The protoplasts of mosses do not require hormones, and in general proliferation can be achieved on mineral medium (Gay, 1976; Gwóźdź & Waliszewska, 1979; Maeda, 1979; Saxena & Rashid, 1980; Stumm et al., 1975). However, a marked stimulation of mitotic activity of *Funaria* protoplasts was possible on hormone-enriched medium (Saxena & Rashid, 1981). On a medium supplemented with 2,4-D and BAP, each at 1 mg/l, divisions could be observed as early as the second day of culture. Nowak and Mlodzianowski (1980) also studied the influence of kinetin on regeneration of protonema from isolated protoplasts of *Funaria hygrometrica*. No stimulating effect of kinetin was noticed on formation of cell wall around isolated protoplasts, although kinetin at concentration of 1 µM distinctly accelerated division of cells obtained from protoplasts and stimulated formation of protonema and gametophores. At higher kinetin concentration (5 µM) protoplasts, and later cells, increased their volume, and regeneration processes took place considerably more slowly than in the control.

Protoplasts were isolated from callus culture of *Marchantia polymorpha*. The rate of cell division of regenerated protoplasts was increased by an addition of activated charcoal to the culture medium. Partial but considerable enhancement of cell division was also observed with frequent washing of the protoplasts and replacement of the medium during the initial stage of culture (Sugawara et al., 1983). These workers suggested that certain substance(s) inhibitory to cell division is/are released from protoplasts and/or regenerated protoplasts, and these are sorbed by activated charcoal.

PROTOPLAST FUSION AND SOMATIC HYBRIDS

Fusion of protoplasts may be spontaneous or induced. Formation of multinucleate structures "fusion bodies" is a common feature during the process of enzymatic isolation of protoplasts. These result by coalescence of adjoining protoplasts, during cell wall degradation.

Fusion of protoplasts can be induced by adding fusogenic agents into the medium. Under normal circumstances the protoplast surface carries a negative charge. Therefore, the first pre-requisite for a good fusogenic agent is that it should lower or if possible neutralize the surface charge altogether, thereby preventing the protoplasts to repel one another. The commonly used fusogenic techniques/agents are: mechanical method (the protoplasts are brought into intimate contact using a partially blocked micropipette and compressing them by the flow of liquid); calcium nitrate; sodium nitrate; Ca^{++}; high pH and temperature; an agglutinating agent, polyethyleneglycol (PEG); and lectins. The general procedure for inducing fusion of protoplasts is given on page 281.

By these methods somatic hybrids between mutants have been produced successfully and used as a method of genetic analysis of non-fertile mutant strains (Ashton & Cove, 1977; Grimsley et al., 1977a,b, 1980; Schieder, 1974a, 1976). Wenzel and Schieder (1973) and Schieder (1974a) obtained fusion products of two non-allelic mutants nic_2 and pal_2 (a normal green, nicotinic acid-deficient female, nic_2; and a pale green, glucose-deficient male, pal_2) of *Sphaerocarpos donnellii*. Female plants of this liverwort possess seven autosomes together with a larger heterochromatic X-chromosome, whereas male plants possess, in addition to the seven autosomes, one smaller Y-chromosome. From the numerous fused protoplasts, only one autotrophic green plant developed. This plant had 14 autosomes and both X and Y sex chromosomes, proving it to be a somatic hybrid.

The genetic modification of cells requires long-term maintenance of protoplasts in the naked state. Takeuchi et al. (1980) isolated protoplasts from callus cells of *Marchantia polymorpha* which were frozen and stored in liquid nitrogen for four to six months. These protoplasts recovered and showed callus formation, which eventually differentiated into thalli. The cryoprotectants (chemicals used for the freeze preservation technique) used singly or in combination are glucose, sorbitol, mannitol, glycerol, dimethyl sulfoxide (DMSO), polyvinyl pyrrolidone, ethylene glycol, and hydroxyethyl starch. The highest degree of cryoprotection was afforded by five percent DMSO plus ten percent glucose.

Only one report is available on the use of bryophyte protoplasts for a study of hormone action at the membrane level. Recently, Thomas and Silcox (1983) incubated *Lunularia* protoplasts in IAA and fusicoccin (FC, a probable activator of membrane-bound ATPase), to test, first, for the possible proton efflux, and then for swelling and lysis. They also tested for possible hormone-stimulated uptake of the plasmolytic agent by incubating protoplasts.

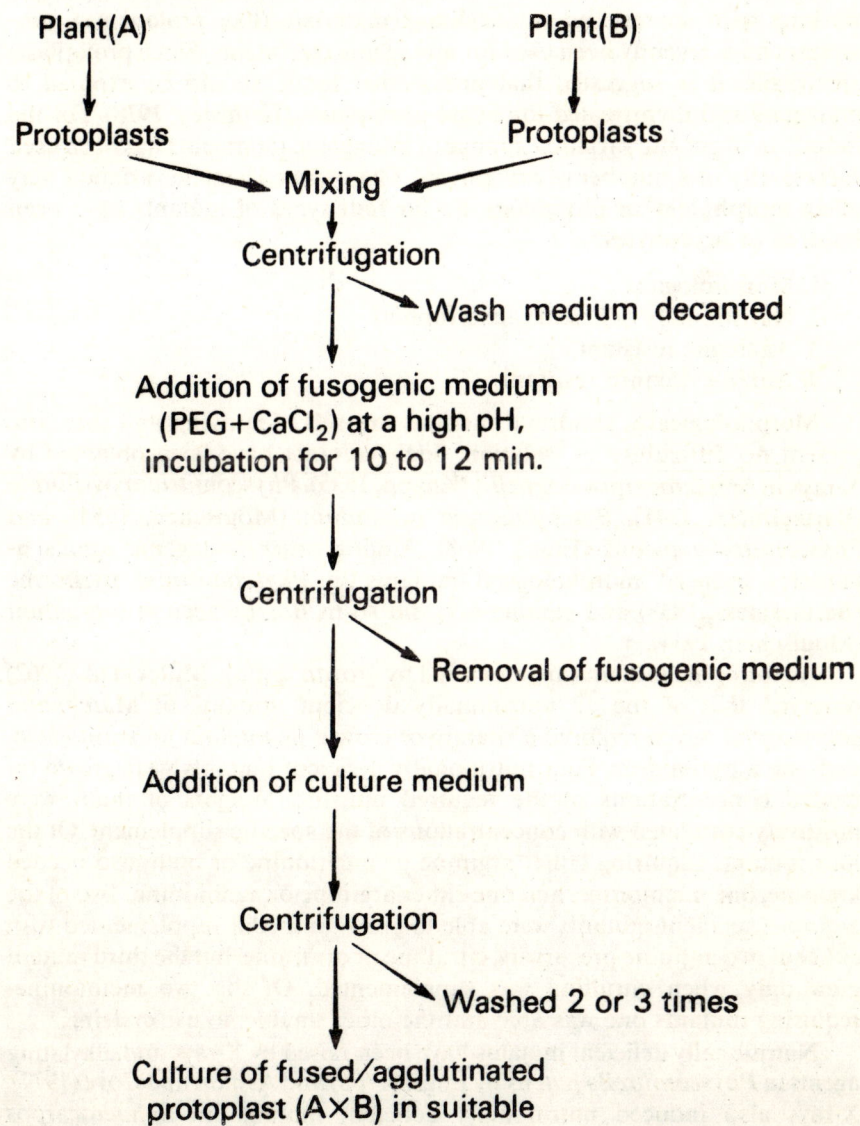

INDUCTION AND ISOLATION OF MUTANTS

The haploid spore is an ideal material for mutagenic treatment, and therefore spores have often been used for such investigations (Ashton & Cove, 1977; Engel, 1968; Schieder, 1973). Since many developmentally abnormal mutants are sexually sterile, a mutagenic treatment involving somatic tissue would be required for reversion studies (to subject already-mutant strains to further mutagenic treatment). To overcome some of the problems faced while working with spores and multicellular materials (like protonema), protoplasts have recently been used for mutagenic treatments. Since protoplasts are fragile, it is suggested that protonemal tissue should be exposed to mutagens and then treated to release protoplasts (Grimsley, 1978). For the induction of genetic variation a range of mutagenic treatments has been used successfully in a number of bryophytes. The induced genetic variation may affect morphology or physiology. So far four types of mutants have been induced in bryophytes:

1. Morphological
2. Nutritionally deficient (auxotrophic)
3. Analogue resistant
4. Auxin-cytokinin resistant

Morphologically abnormal mutants are self identifying and therefore present no difficulties in isolation. Such mutants have been obtained by X-rays in *Sphaerocarpos donnellii* (Knapp, 1935), *Physcomitrium pyriforme* (Barthelmess, 1941), *Brachythecium rutabulum* (Moutschen, 1954), and *Physcomitrella patens* (Engel, 1968). Among other mutagenic agents α-particles induced morphological mutants in *Physcomitrium pyriforme* (Barthelmess, 1938) and gamma-rays did so in *Brachythecium rutabulum* (Moutschen, 1954).

Auxotrophic mutants are identified by growth testing. Miller et al. (1962) observed that of the 12 nutritionally deficient mutants of *Marchantia polymorpha*, seven required a vitamin or growth factor, four an amino acid, and one a pyrimidine. Four nutritionally deficient mutants were grown on graded concentrations of the required nutrilite. Weights of thalli were positively correlated with concentrations of the specific supplement. Of the four mutants requiring either arginine or methionine or both, two needed arginine, one methionine, and one either arginine or methionine. Two of the arginine deficient mutants were able to grow on media supplemented with either of two arginine precursors, citrulline or ornithine, but the third mutant grew only when citrulline was supplemented. Of the two methionine-requiring mutants one was able and the other unable, to use cysteine.

Nutritionally deficient mutants have been raised by X-rays and alkylating agents in *Physcomitrella patens* by Engel (1968) and Ashton and Cove (1977). X-rays also induced nutritionally deficient mutants in *Sphaerocarpos donnellii* (Schieder, 1973).

Courtice et al. (1978) studied the fertility of vitamin auxotrophic strains of *Physcomitrella patens*. Sporophytes failed to develop on strains which were either self-sterile or cross-fertilized by strains containing non-complementary auxotrophic markers when they were grown on normal medium. Complementary strains, which could presumably manufacture their own vitamins, produced normal sporophytes. Sporophyte production could only be restored to the first two categories by a considerable increase in the vitamin supplement above the level necessary for normal gametophyte growth. From the above data, Courtice et al. (1978) concluded that at least for certain vitamins the sporophyte is independent of the gametophyte.

Mutants resistant to the amino acid analogues, p-fluorophenylalanine and D-serine; and purine analogue, 8-azaguanine; have been isolated in *Physcomitrella patens*. The amino acid analogue resistant mutants are isolated by growing mutagenized spores directly on analogue-containing medium; whereas to obtain mutants resistant to the purine analogue, it was found necessary to allow treated spores to germinate in the absence of the analogue and to add the analogue after seven to nine days of growth (Ashton & Cove, 1977).

Ashton and Cove (1977), Ashton et al. (1979), and Cove et al. (1980) obtained mutants of *Physcomitrella patens* that were resistant to the effects of high auxin or cytokinin concentrations in the medium, and many of these displayed enhanced chloronema and reduced caulonema production. The following types were isolated:

Type 1—unaffected by auxin or cytokinin, caulonema production blocked at some stages prior to that at which auxin has an effect,

Type 2—resistant to auxin but sensitive to cytokinin, fails to synthesize cytokinin, and

Type 3—sensitive to auxin, non-synthesizing strains produced no caulonema, which demonstrates the absolute requirement of auxin for caulonema production.

The investigations on protoplast culture of bryophytes are still in their inception, and most of the possibilities listed in the beginning of this chapter have yet to be tried.

REFERENCES

ASHTON, N.W. and D.J. COVE, *Mol. Gen. Genet.*, 154, 87 (1977).
ASHTON, N.W., N.H. GRIMSLEY, and D.J. COVE, *Planta*, 427, 337 (1979).
BACH, A., *Versuche zur Gewinnung von Protoplasten an verschiedenen Leber—und Laubmoosen*, Diplomarbeit, Heidelberg, 1973.
BARTHELMESS, A., *Z. Indukt. Abstamm. -u. VererbLehre*, 74, 479 (1938).
BARTHELMESS, A., *Z. Indukt. Abstamm. -u. VererbLehre*, 79, 153 (1941).
BATRA, A. and W. O. ABEL, *Pl. Sci. Lett.*, 20, 183 (1981).
BINDING, H., *Z. PflPhysiol.*, 55, 305 (1966).
BOPP, M., S. ZIMMERMANN, and B. KNOOP, *Protoplasma*, 104, 119 (1980).

BURGESS, J. and P.J. LINSTEAD, *Planta*, **151**, 331 (1981).
COCKING, E.C., *Nature*, **187**, 927 (1960).
COURTICE, G.R.M., N.W. ASHTON, and D.J. COVE, *J. Bryol.*, **10**, 191 (1978).
COVE, D.J., N.W. ASHTON, D.R. FEATHERSTONE, and T.L. WANG, "The use of mutant strains in the study of hormone action and metabolism in the moss *Physcomitrella patens*." In D.R. Davies and D.A. Hopwood, Eds, *Proc. 4th John Innes Symp.*, John Innes Charity, Norwich, 1980. Pp. 231-241.
ENGEL, P.P., *Am. J. Bot.*, **55**, 438 (1968).
EVANS, P.K. and E.C. COCKING, "Isolated plant protoplasts." In H.E. Street, Ed., *Plant Tissue and Cell Culture*, Blackwell Scientific Publications, Oxford, 1977. Pp. 130-135.
GAY, L., *Z. PflPhysiol.*, **79**, 33 (1976).
GAY, L., Etude de la régéneration du gamétophyte feuille des Polytrichacées. Thèse de l'Université Claude Bernard Lyon. 1980.
GRIMSLEY, N.H., "Genetic analysis of the moss *Physcomitrella patens* using protoplast fusion", Ph.D. Thesis, Univ, Cambridge, England, 1978.
GRIMSLEY, N.H., N.W. ASHTON, and D.J. COVE, *Mol. Gen. Genet.*, **154**, 97 (1977a).
GRIMSLEY, N.H., N.W. ASHTON, and D.J. COVE, *Mol. Gen. Genet.*, **154**, 103 (1977b).
GRIMSLEY, N.H., D.R. FEATHERSTONE, G.R.M. COURTICE, N.W. ASHTON, and D.J. COVE, "Somatic hybridization following protoplast fusion as a tool for the genetic analysis of development in the moss, *Physcomitrella patens*." In L. Ferenczy and G.L. Farkas, Eds, *Advances in Protoplast Research*, Pergamon Press, Oxford, 1980. Pp. 363-376.
GWÓŹDŹ, E.A. and B. WALISZEWSKA, *Pl. Sci. Lett.*, **15**, 41 (1979).
JENKINS, G.I. and D.J. COVE, *Planta*, **157**, 39 (1983).
KNAPP, E., *Z. Indukt. Abstamm. -u. VererbLehre*, **70**, 309 (1935).
KNOOP, B., *Protoplasma*, **94**, 307 (1978).
MAEDA, M., *Bot. Mag. Tokyo*, **92**, 105 (1979).
MILLER, M.W., E.D. GARBER, and P.D. VOTH, *Bot. Gaz.*, **124**, 94 (1962).
MORI, K., H. MATSUSHIMA, and M. TAKEUCHI, "A study of cell wall regeneration of protoplasts in *Marchantia polymorpha* L." In F. Akio, Ed., *Plant Tissue Culture*, Maruzen, Tokyo, 1982. Pp. 45-46.
MORI, K., H. MATSUSHIMA, and M. TAKEUCHI, *Bot. Mag. Tokyo*, **96**, 281 (1983).
MOUTSCHEN, J., *La Cellule*, **46**, 181 (1954).
NOWAK, U. and F. MŁODZIANOWSKI, *Acta Soc. Bot. Pol.*, **49**, 195 (1980).
ONO, K., K. OHYAMA, and O.L. GAMBORG, *Pl. Sci. Lett.*, **14**, 225 (1979).
RIPETSKY, R.T., *Cytologia & Genetica*, **5**, 447 (1975).
RIPETSKY, R.T., *Ukr. bot. Zh.*, **35**, 467 (1978).
RIPETSKY, R.T., *Cytologia & Genetica*, **13**, 347 (1979).
SAXENA, P.K., *J. Plant Physiol.*, **119**, 385 (1985).
SAXENA, P.K. and A. RASHID, *Protoplasma*, **103**, 401 (1980).
SAXENA, P.K. and A. RASHID, *Pl. Sci. Lett.*, **23**, 117 (1981).
SCHIEDER, O., *Z. PflPhysiol.*, **70**, 185 (1973).
SCHIEDER, O., *Z. PflPhysiol.*, **74**, 357 (1974a).
SCHIEDER, O., *Biochem. Physiol. Pflanzen*, **165**, 433 (1974b).
SCHIEDER, O., *Mol. Gen. Genet.*, **144**, 63 (1976).
SCHIEDER, O. and G. WENZEL, *Z. Naturforsch.*, **27b**, 479 (1972).
STUMM, I., Isolierung und Kulture von Protoplasten aus dem Laubmoose *Physcomitrella patens*, Staatsexamenarbeit, Heidelberg, 1974.
STUMM, I., Y. MEYER, and W.O. ABEL, *Pl. Sci Lett.*, **5**, 113 (1975).
SUGAWARA, Y., K. MORI, H. MATSUSHIMA, and M. TAKEUCHI, *Z. PflPhysiol.*, **109**, 275 (1983).
TAKEUCHI, M., H. MATSUSHIMA, and Y. SUGAWARA, *Cryo-Letters*, **1**, 519 (1980).
TAYLOR, J., *J. Hattori bot. Lab. No.*, **46**, 55 (1979).
THOMAS, R.J. and K.R. SILCOX, *Pl. Sci. Lett.*, **29**, 169 (1983).
WENZEL, G. and O. SCHIEDER, *Pl. Sci. Lett.*, **1**, 421 (1973).

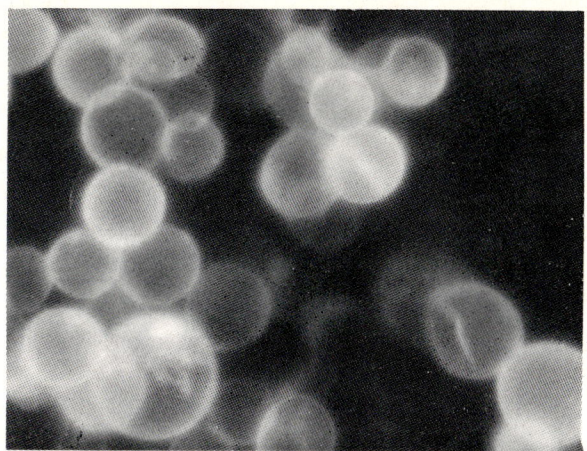

Plate 10.1. Protoplast culture of *Marchantia polymorpha*. Protoplasts regenerating cell walls after 48 h of cultivation. Mark the fluorescence which has resulted from staining cell walls with Calcofluor White-ST.
(After Mori et al., 1983).

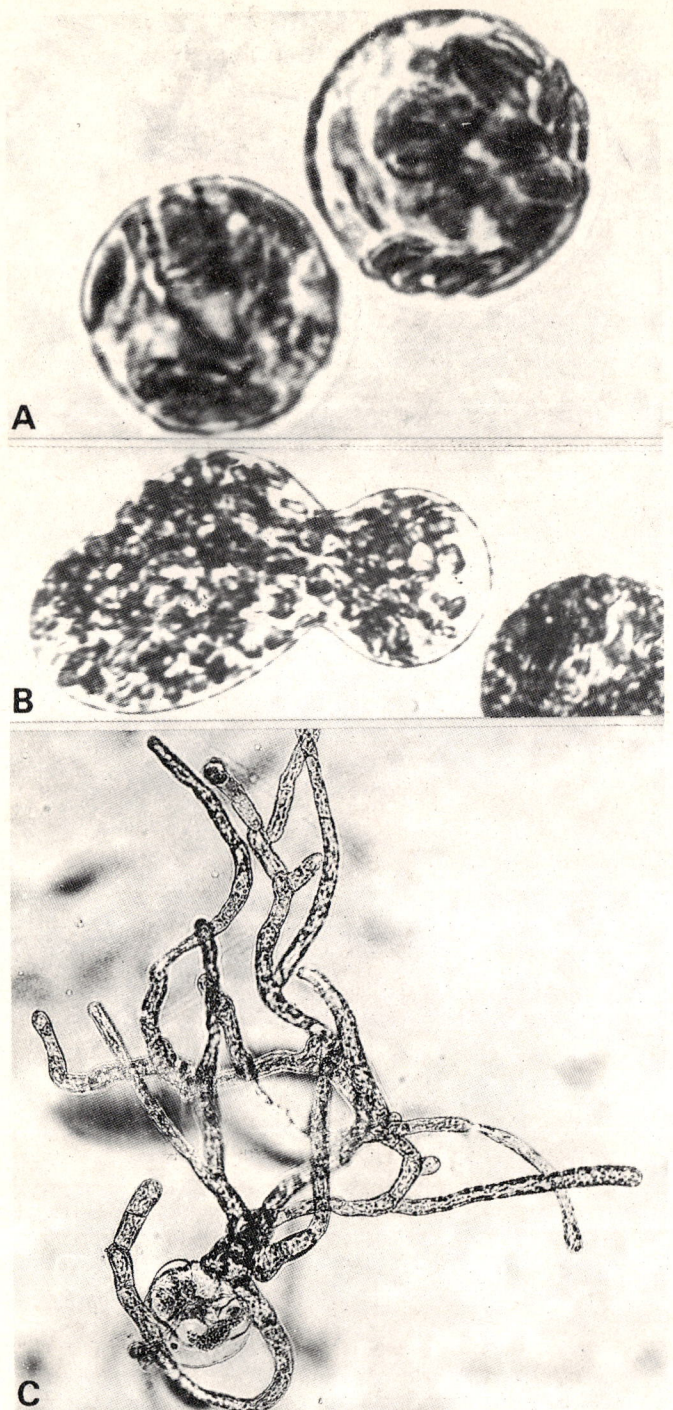

Plate 10.2 A-C. Protoplast culture of *Funaria hygrometrica*. **A.** Protoplasts shortly after isolation. **B.** Budding of a newly-formed cell after 12 days of culture. **C.** Protonema produced from a newly formed cell.
(After Gwóźdź & Waliszewska, 1979).

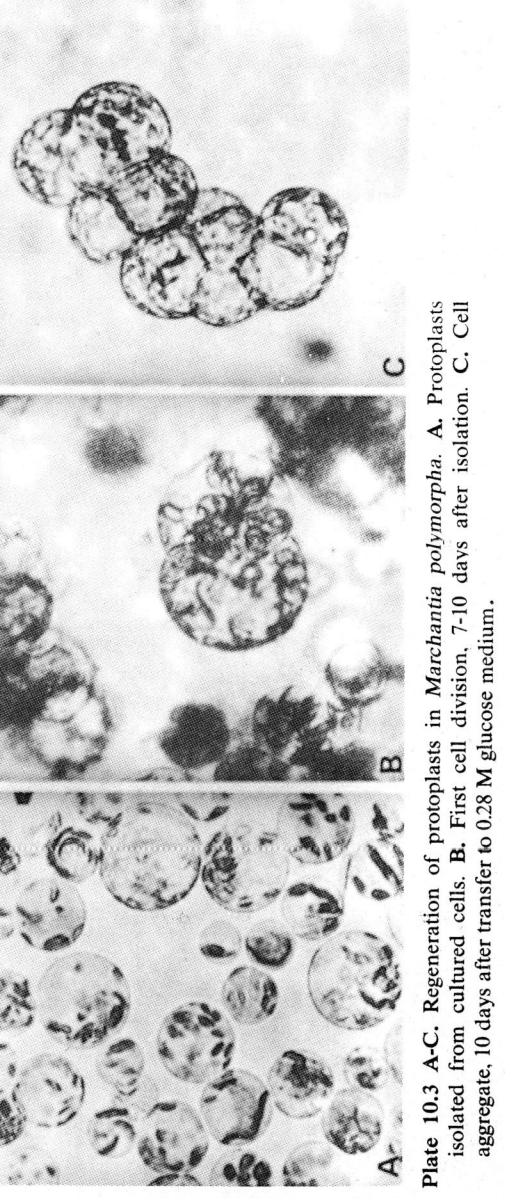

Plate 10.3 A-C. Regeneration of protoplasts in *Marchantia polymorpha*. **A.** Protoplasts isolated from cultured cells. **B.** First cell division, 7-10 days after isolation. **C.** Cell aggregate, 10 days after transfer to 0.28 M glucose medium.

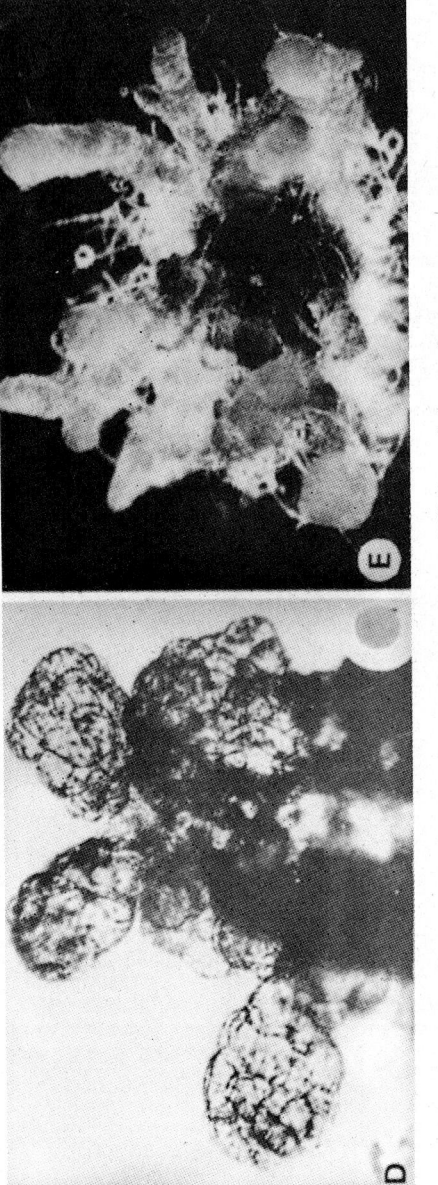

Plate 10.3 D, E. Young thalli arising from callus tissue, 15 days and 30 days after transfer to Knop's medium.
(After Ono et al., 1979).

11 Conduction in Bryophytes

LIFE on land requires many adaptations, conduction of water being one of them. Most bryophytes are terrestrial, but the majority lacks well developed internal conducting strands. This deficiency is made up by several characteristics:

(1) Bryophytes usually grow in places with abundant humidity, and their gametophytes absorb water from all over the surface. This reduces the need for internal conduction or even of external redistribution of water. Trachtenberg and Zamski (1979) have experimentally demonstrated that in *Polytrichum juniperinum* absorption of water through the aerial surface of gametophyte is more efficient than through the rhizome. In some forms like *Leucobryum* and *Sphagnum* surface absorption is more efficient because of the pores in the dead and empty cells in leaves and stems. The water sacs present in the gametophytes in some Jungermanniales like species of *Jubula*, *Polyota*, and *Frullania* are responsible for quick absorption and retention of water. Mucilage hairs at the apices of many forms also absorb water directly.

(2) The majority of bryophytes lie prostrate (all thalloid liverworts, the vast majority of leafy liverworts, & the pleurocarpous mosses). This habit increases the area of contact with the substrate.

(3) The bryophytes with erect gametophytes are in general small, so that the distance through which water has to move is not much.

(4) Lastly, effective external conduction occurs, for which there are many adaptations.

EXTERNAL CONDUCTION

GAMETOPHYTE

The gametophytes of several mosses grow in close tufts, and some like *Leucobryum glaucum* 'the pin cushion moss' form compact cushions with effective capillaries between the gametophytes.

In addition, there are several appendages, on the stem which help in producing capillary channels through which water moves.

Leaves

In many mosses leaves are closely placed on the stem (Fig. 11.1A), and in some taxa like *Polytrichum* they have sheathing bases. Dry leaves are curled and form more efficient capillaries (Bayfield, 1973). Lamellae on the leaves of some mosses like members of Polytrichaceae and Dawsoniaceae also help in external conduction.

Rhizoids

In the members of Marchantiales pegged rhizoids are normally arranged parallel to the flat surface, and make channels for effective water transport. In the Marchantiaceae the rhizoidal grooves in the archegoniophores serve the same purpose (Fig. 11.1B). In mosses, rhizoids are clustered in the lower portion of stem and in some instances like *Pogonatum* they twine together to form efficient capillary channels.

Scales

In the members of Marchantiales overlapping scales on the ventral surface of thallus are responsible for external transport of water (Fig. 11.2).

Paraphyllia

Some mosses have extensive growth of paraphyllia on stem and these help in conduction of water.

Branches

Closely placed branches help in external conduction.

SPOROPHYTE

The situation in the sporophytic generation is entirely different. Firstly, there are no appendages, and hence no capillaries for external transport. Secondly, in mature sporophytes of mosses the seta and capsule have thick walled cutinized epidermis, and thick walled hypodermis which prevent direct absorption of water from the atmosphere. However, during early stages of development some amount of external absorption occurs, and circulation of water is also possible in the narrow space between the capsule wall and calyptra.

In the sporophytes of liverworts the seta elongates only at maturity and this is followed by quick spore dispersal. Thus, there is no need for conduction.

SIGNIFICANCE OF EXTERNAL CONDUCTION

Since the gametophytes of only a few bryophytes have well developed conducting tissues, external conduction of water by capillary action has

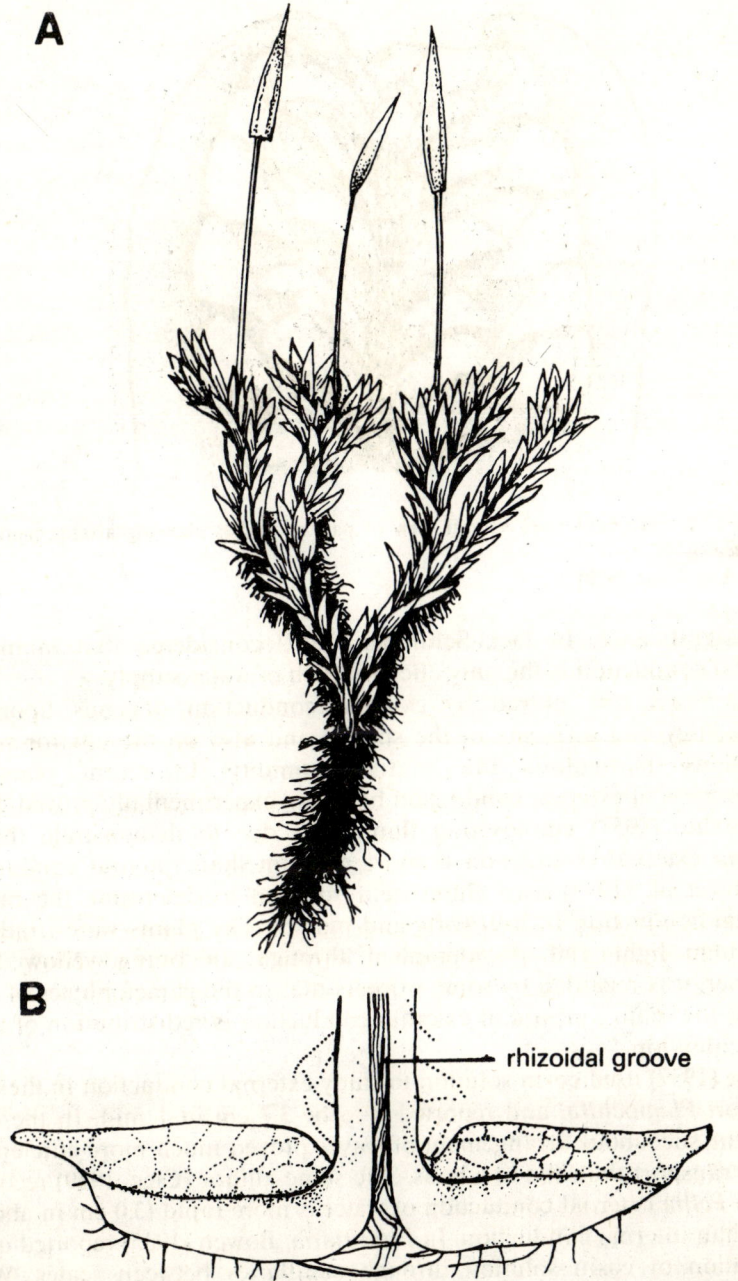

Fig. 11.1 A, B. External conduction in gametophyte. A. *Encalypta streptocarpa*. Gametophyte bearing sporophytes. The dense imbrication of leaves and numerous rhizoids provide capillary channels for external conduction. B. *Fimbriaria bleumeana*. Vertical section of thallus and basal part of archegoniophore. The latter shows the rhizoidal groove through which water and solutes travel up.
(A. After Bruch et al., 1836-55; B. After Bowen, 1935).

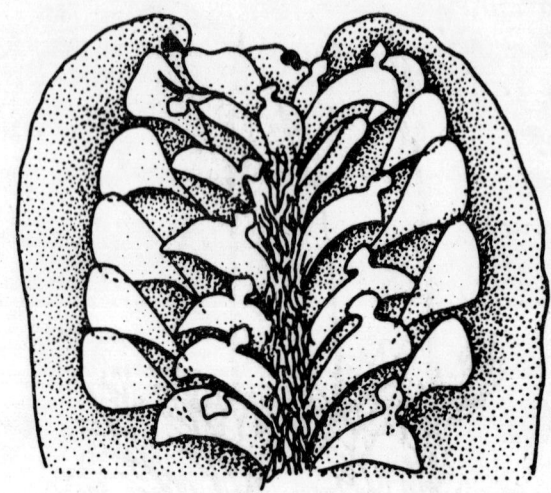

Fig. 11.2. *Marchantia* sp. Ventral view of part of thallus showing arrangement of scales.
(After Udar, 1970).

great significance. In fact, Schimper (1857) considered that in mosses external conduction is the only effective path of water supply.

The exact role played by external conduction depends upon the morphology and anatomy of the species and also on the environmental conditions, particularly the relative humidity. In recent years the effectiveness of external conduction has been experimentally proved. Bopp and Stehle (1957) employed a fluorescent dye to demonstrate that in *Funaria* external conduction is more efficient than internal conduction. Deloire et al. (1979) used fluorescein solution to determine the rate of external conduction in liverworts and mosses. The plants were irradiated with blue light and photographed through an orange-yellow filter. However, it is regarded by some workers that in the gametophytes of erect mosses the main purpose of external conduction is redistribution of water of precipitation.

Clee (1937) used eosin solution to study external conduction in the leafy liverwort *Plagiochila*, and reported it to be 3.7 cm in 1 min. In the leafy types the succubous arrangement of leaves proved much more efficient for water transport than the incubous. The same author (Clee, 1939) recorded that in *Pellia* external conduction of water is more rapid (3.0 cm in about 2 min) than internal conduction. In *Fimbriaria*, Bowen (1935) reported quick circulation of eosin solution through capillaries between scales. Water reaches the female receptacle through rhizoidal grooves of the archegoniophores. McConaha (1939) observed that in *Conocephalum* conduction of water along the entire length of thallus takes only 20 to 30 sec. In 1941, the same author reported that the rate of external conduction of *Conocephalum* and *Preissia* is about 1 mm/sec, whereas in *Lunularia* and *Reboulia* it is only 0.5 mm/sec.

Conduction in Bryophytes 289

Mägdefrau (1935-36) reported that in *Polytrichum* external conduction is sufficient to maintain turgidity at 90 percent relative humidity, but at 70 percent both internal and external conduction are necessary. According to this investigator, in the great majority of mosses external conduction is good whereas internal conduction is negligible or absent. Bayfield (1973) noticed that water conduction in *Polytrichum commune* is mostly internal under high evaporative flux, and is external under moderate flux. He considered that under many stress conditions both the modes are essential for the maintenance of optimum water balance.

INTERNAL CONDUCTION

CELLS INVOLVED IN CONDUCTION

All cells are basically capable of some conduction. But specialization for this function involves steps like: (1) increase in dimensions (length & breadth), (2) modifications in the end walls (development of plasmodesmata/greatly developed primary pit fields/small pores/ hydrolysis/degeneration), (3) thickening of side walls for support, and (4) degeneration of contents (for greater efficiency of conduction). Bryophytes show a wide variety in the structure of cells concerned with conduction of water, and various degrees of modifications are observed in different groups. The extent of development of the conducting cells is also influenced by the stage of development and the environmental conditions, especially humidity. The younger stages, and the plants growing under high humidity show poorly developed conducting strands.

Hornworts and Liverworts

In the gametophytes of hornworts and the majority of liverworts the cells responsible for conduction are ordinary parenchyma cells in the ventral region of thallus. These possess well developed primary pit fields in their walls (Fig. 11.3A; Proskauer, 1960; Siegel, 1962).

Cells in the centre of the stem of leafy liverworts, and the midrib of thalloid ones are usually elongated and have plasmodesmata in their walls, especially the end walls. Even callose has been observed in the species of *Madotheca* and *Conocephalum* (see Winkler, 1969). Such cells can be termed as conducting parenchyma.

Slightly more specialized water conducting cells occur in some liverworts like *Takakia, Haplomitrium, Moerckia, Pallavicinia, Symphyogyna,* and *Hymenophyton.* These liverworts have distinct conducting strands (Figs 11.3B to 11.5A), the cells of which are empty and dead at maturity. Among these the cells in *Haplomitrium, Moerckia,* and *Takakia* are relatively less specialized and possess the following features (Fig. 11.5B,C):

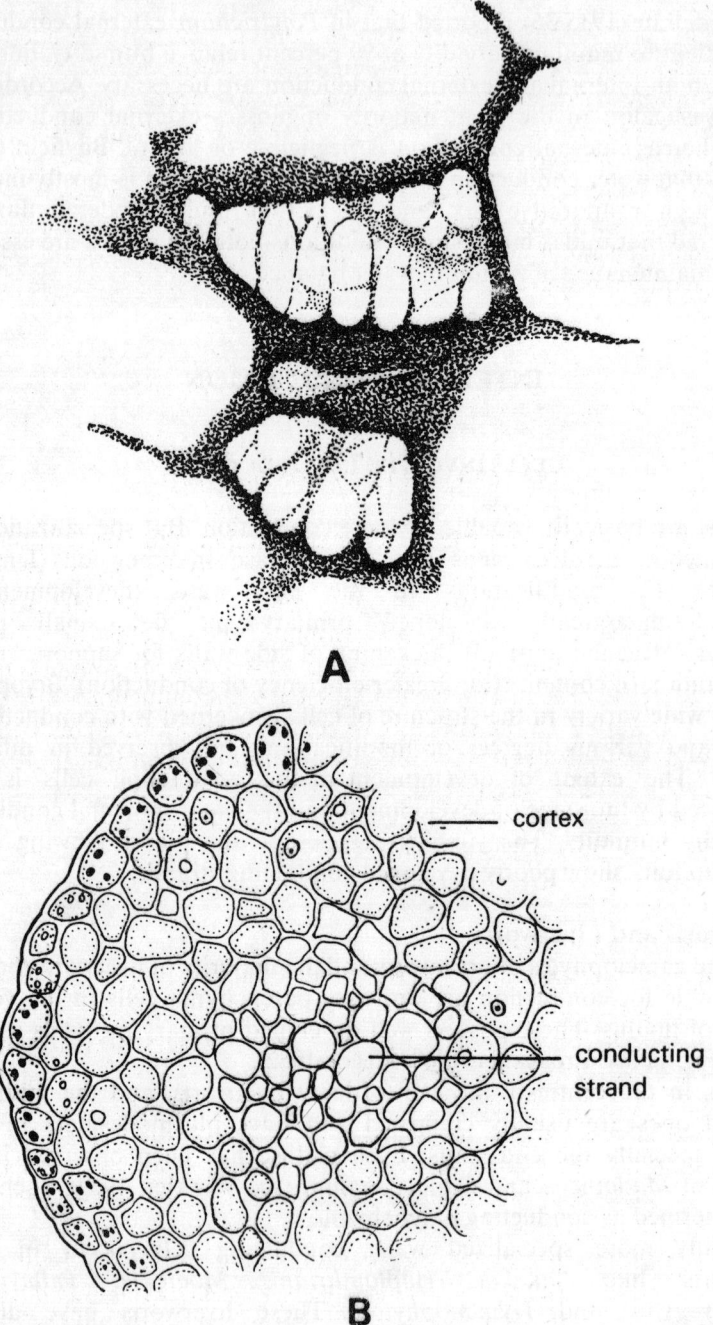

Fig. 11.3 A, B. Internal conduction in gametophyte. A. *Dendroceros crispus*. Cells from thallus showing large primary pit fields. B. *Takakia lepidozioides*. Transverse section of robust female leafy stem showing a central, solid true water conducting strand comprising dead and empty cells.
(A. After Proskauer, 1960; B. After Hébant, 1977).

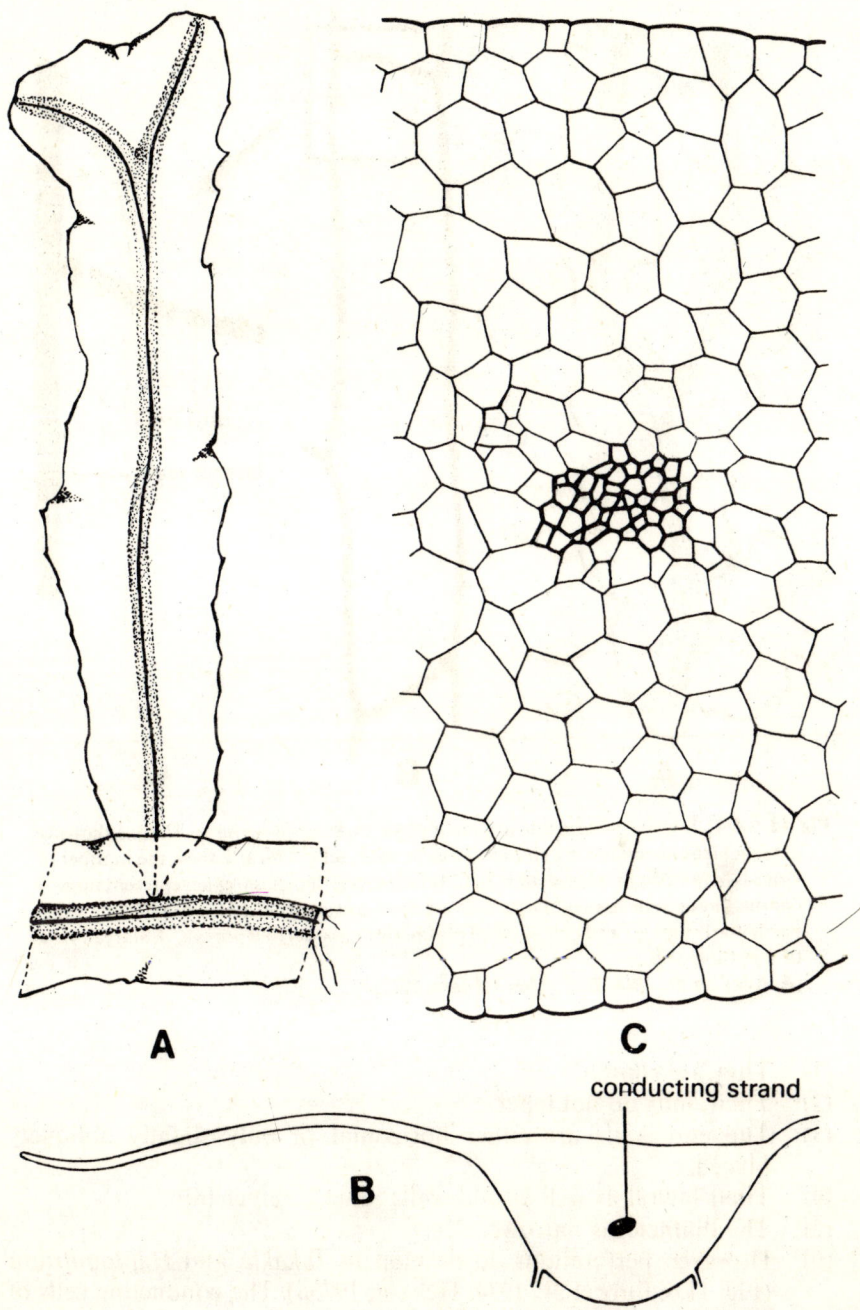

Fig. 11.4 A-C. *Pallavicinia lyellii.* **A.** Portion of thallus showing bifurcated branch arising on ventral surface of parent axis. **B.** Portion of mature thallus in cross section. **C.** Magnified view of a part of midrib in **B** to show details of conducting strand. (After Smith, 1966).

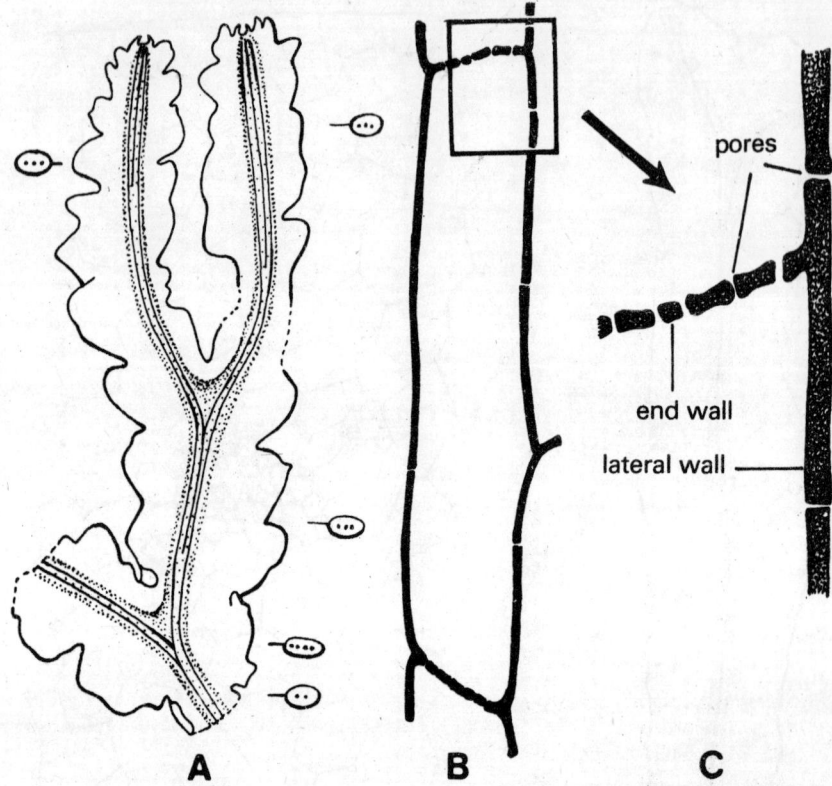

Fig. 11.5 A-C. Internal conduction in gametophyte. A. *Symphyogyna* sp. Thallus showing arrangement of conducting strands. Dots in ovals schematically show the number of strands in midribs in cross section. B, C. *Takakia* sp. B. Diagrammatic representation of conducting cells from stem. These cells are dead and empty at maturity and their walls, particularly the end walls, show small plasmodesmata-derived pores. C. Enlarged view of a portion of B.
(A. After Smith, 1966; B, C. After Hébant, 1977).

(1) They are shorter.
(2) Their ends do not taper.
(3) The end walls are either horizontal or only slightly obliquely placed.
(4) Their lateral as well as end walls remain rather thin.
(5) The diameter is narrower.
(6) However, perforations do develop in *Takakia* and *Haplomitrium* (Fig. 11.6; Burr et al., 1974; Hébant, 1973a). The conducting cells of *Moerckia* do not possess pores but have hydrolyzed walls (Hébant, 1979).

The other three genera have slightly more advanced type of conducting cells with the following characteristics (Fig. 11.7 A-C):
(1) They are greatly elongated.

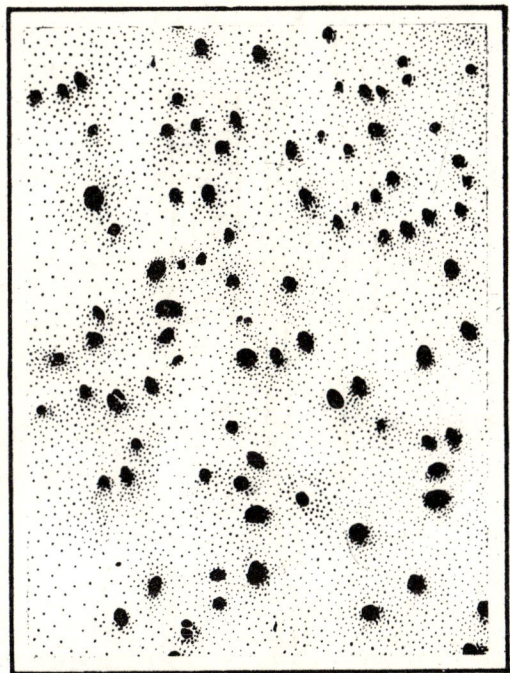

Fig. 11.6. *Haplomitrium gibbsiae*. Representation of scanning electron micrograph of end wall between two water-conducting cells in the central strand. It shows numerous plasmodesmata-derived pores.
(After Burr et al., 1974).

(2) They taper towards the ends.
(3) They have oblique end walls.
(4) Their lateral and terminal walls become conspicuously thick at maturity.
(5) Their diameter is larger.
(6) True perforations develop in their walls (especially end walls) from enlarged plasmodesmata (Campbell et al., 1975; Hébant, 1978, 1980; Smith, 1966).

Mosses

Conducting parenchyma is of common occurrence in stems, leaves, and sporogonia of mosses. Such cells have plasmodesmata, pits derived from them, and show callose (Hébant, 1968; Paolillo & Reighard, 1967).

Several mosses possess conducting strands. In addition to the specialized water conducting cells, food conducting cells may also be present. Water conducting cells of mosses are known as hydroids and these occur in stem, leaf, and seta. The hydroids collectively constitute the hydrom. The food conducting cells are called leptoids, and these are collectively known as the leptom (together with the accompanying parenchyma from which they are not very easily distinguishable).

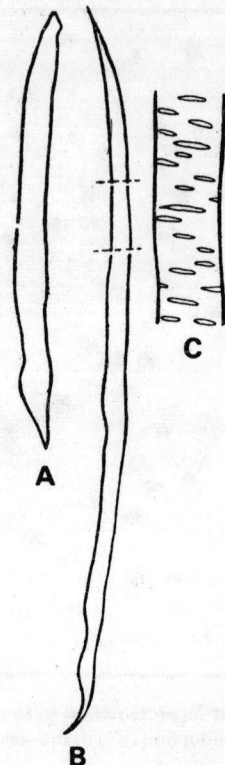

Fig. 11.7 A-C. *Pallavicinia lyellii*. A,B. Conducting cells isolated by maceration of thallus. C. Magnified view of portion marked in B to show pits in the side wall. (After Smith, 1966).

In addition to hydroids and leptoids the leaf nerves and stem axes may have thick walled, elongated cells which are known as stereids. Stereids often retain their contents even at maturity. Stereids are collectively known as the stereome.

Water conducting and food conducting strands may be present in one or both the generations of mosses (see Table 11.1).

In the stem, water conducting cells (hydroids) may form a well developed central strand as in a number of acrocarpous mosses (Fig. 11.8), which have an erect habit, especially so in Polytrichales (Hébant, 1966). A reduced strand may be present or the strand may be completely lacking as in several pleurocarpous mosses which lie prostrate (Frey, 1971). However, no strict correlation exists between the habit and the extent of development of conducting strand. Among the genera which always have a well developed water conducting strand are *Polytrichum, Dawsonia, Tetraphis, Funaria*, and *Bryum*. Reduced strands occur in *Climacium*. Some species of *Grimmia* have a conducting strand, whereas other species lack it. Among the taxa without a strand are *Diphyscium, Orthotrichum*, and the aquatic moss *Fontinalis*.

TABLE 11.1.
Occurrence of hydroids and leptoids in the gametophyte and sporophyte of some representative mosses.

Taxa	Gametophyte (Stem)		Sporophyte (Seta)	
	Hydroids	Leptoids	Hydroids	Leptoids
Polytrichum	+	+	+	+
Funaria	+	−	+	+
Dicranum	+	−	+	−
Buxbaumia	−	−	+	−
Orthotrichum	−	−	−	−

Source. Hébant, 1977.
+ Indicates present.
− Indicates absent.

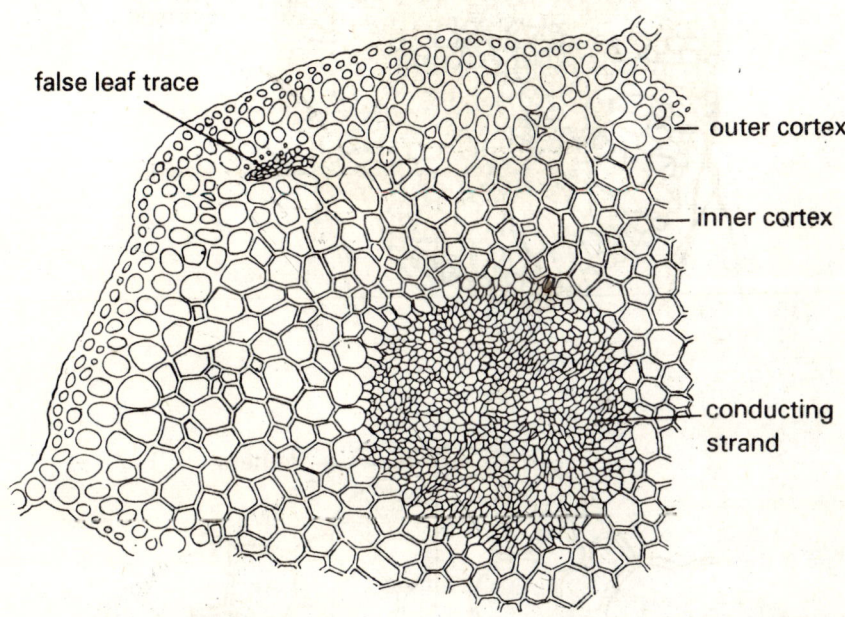

Fig. 11.8. *Mnium undulatum.* Transverse section of robust stem showing a distinct water conducting strand made of hydroids. The inner cortex has thin-walled conducting parenchyma cells, the outer cortex has thick-walled cells. A false leaf trace is also seen in the cortex.
(After Lorentz, 1867-68).

In the gametophytic generation of bryophytes, typical food conducting cells (leptoids) occur only in the members of Polytrichales.

ANATOMY OF THE HORIZONTAL AXIS

The rhizome anatomy of mosses differs from that of the aerial axis. For instance, the creeping axis of *Polytrichum* and *Pogonatum* has a three pronged central solid core of hydroids mixed with stereids; and leptoids mixed with parenchyma occur in the furrows (Fig. 11.9A). The innermost layer of cortex comprises large cells, and these resemble endodermis (Scheirer, 1974; Udar, 1970). At first sight the anatomy of these axes resembles that of dicot roots.

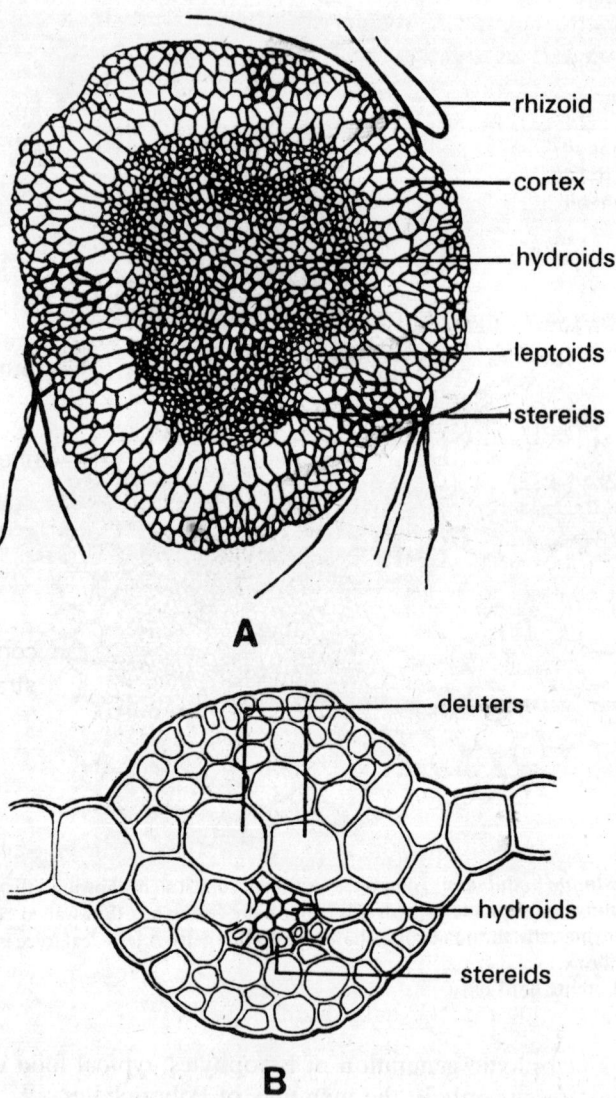

Fig. 11.9 A, B. Internal conduction in gametophyte. A. *Pogonatum* sp. Transverse section of upper region of rhizome. B. *Mnium undulatum*. Vertical section of fully developed green leaf. Food conducting parenchyma (deuters), water conducting cells (hydroids), and supportive cells (stereids) are seen in the midrib. (A. After Udar, 1970; B. After Davy de Virville, 1927, 28).

MIDRIBS AND LEAF TRACES

The most complex midribs of leaves have three types of cells: (1) Deuter cells which conduct food. These are living and have plasmodesmata in their end walls. They are usually wide and elongated (*see* Plate 7.6). (2) Stereids which are thick walled and serve as supporting cells. (3) Hydroids which conduct water (Fig. 11.9B).

The midribs may not be connected to the central strands of the axis by leaf traces. When present the traces are not as complex as the midribs. The traces can be false (not coming in contact with the central strand), or true (which join the central strand). When true, the connection of the trace with the central strand may be loose or firm (Fig. 11.10A-F; Hébant, 1967).

CONDUCTING STRAND IN SETA AND CAPSULE

The most complex conducting strands occur in Polytrichales. The centrally located strand in seta is surrounded by inner cortex in which many intercellular spaces exist (Favali & Bassi, 1974; Hébant, 1975). The hydrom may have two or more layers of leptoids around it as in Polytrichales (Figs 11.11, 11.12) or there may be a single layer of leptoids as in Funariales. The majority of mosses have only hydrom and no leptom. This suggests that capsules of mosses are fairly efficient in photosynthesis, and there is very little transport of organic food from the gametophyte to the sporophyte and vice versa.

In the capsule the conducting strand of seta ends below the columella. In some mosses like *Funaria* it dilates before termination and in this region it comprises short and wide elements.

DEVELOPMENT AND STRUCTURE OF MATURE CONDUCTING TISSUES

The development of hydroids and leptoids has been studied with light microscope as well as electron microscope, and details are available for *Polytrichum*.

Hydroids

The main events in the development of hydroids are as follows (Figs 11.13 A,B, 11.14A,B):
(1) Increase in dimensions, especially in length.
(2) The end walls become oblique.
(3) The lateral walls become thick.
(4) The nuclei show endopolyploidy.
(5) The protoplasts degenerate due to lysosomal activity, so that at maturity these cells become dead and empty.

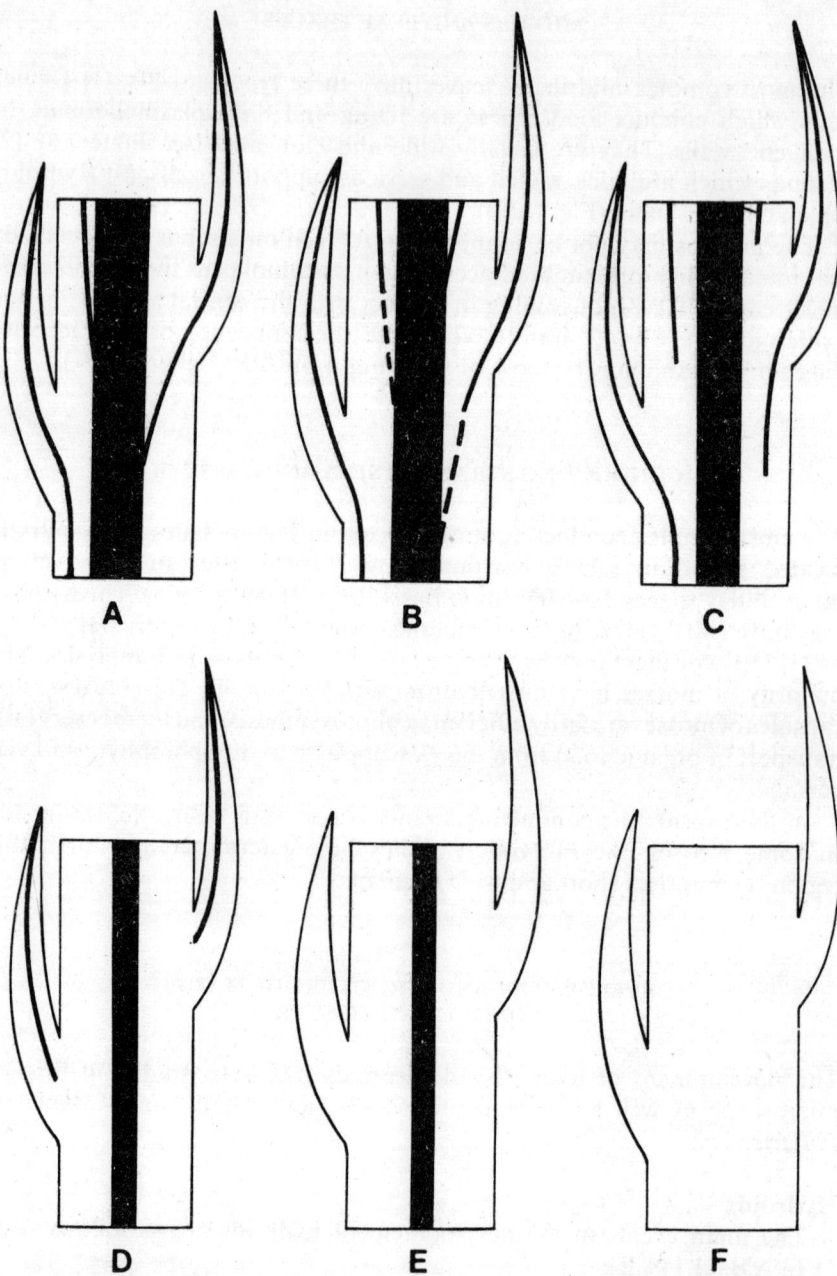

Fig. 11.10 A-F. Diagrams illustrating various types of organization of leaf traces met with in the leafy stems of mosses. A. Leaf traces link up with the central cylinder in a definite way. B. Leaf traces have a loose connection with the central cylinder. C. Leaf traces end blindly in the cortex. D. The hydroids of midrib do not enter the cortex E. The stem has hydroids, but the leaf lacks them. F. Hydroids are lacking from stem as well as leaves.
(After Hébant, 1969).

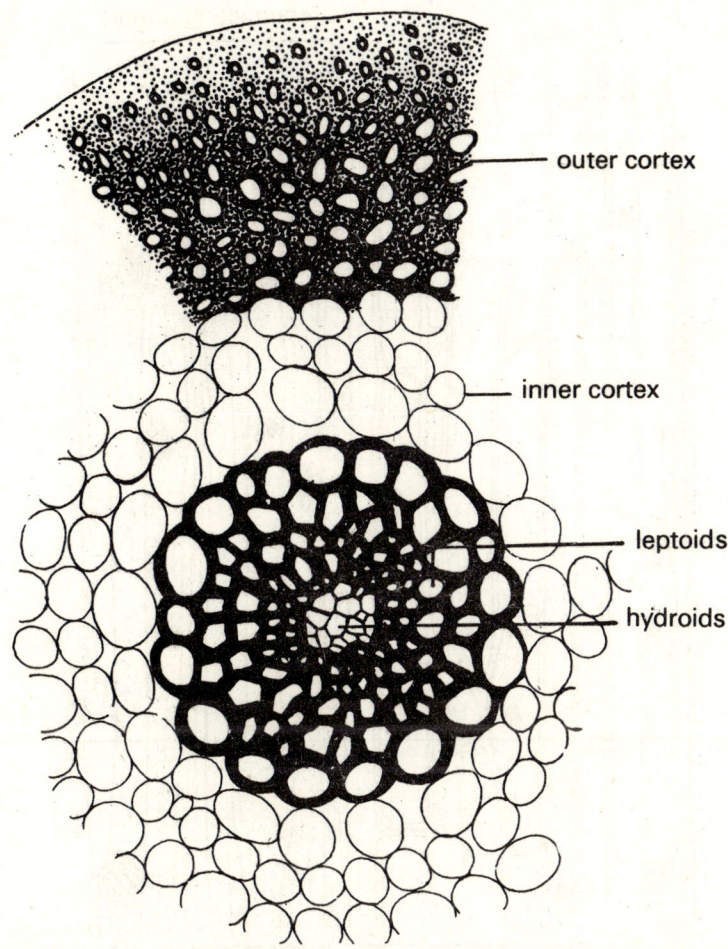

Fig. 11.11. *Polytrichum juniperinum*. Transverse section of mature seta showing the central conducting strand made of thin-walled hydroids surrounded by four or more layers of thick-walled leptoids. The outer cortex has thick-walled cells, the inner cortex is lacunar.
(After Vaisey, 1888).

(6) The end walls get partially hydrolyzed (*see* Plate 7.5). Because of this a network of microfibrils is left and the gaps thus formed result in high permeability (Hallet, 1972; Hébant, 1967, 1974; Hébant & Johnson, 1976; Scheirer, 1975).

Thus, hydroids become preferential pathways for the conduction of water.

The sequence of development in the hydroids of other mosses is comparable to that described above, but in the majority, as in *Funaria*, the lateral walls of hydroids remain thin.

300 *Biology of Bryophytes*

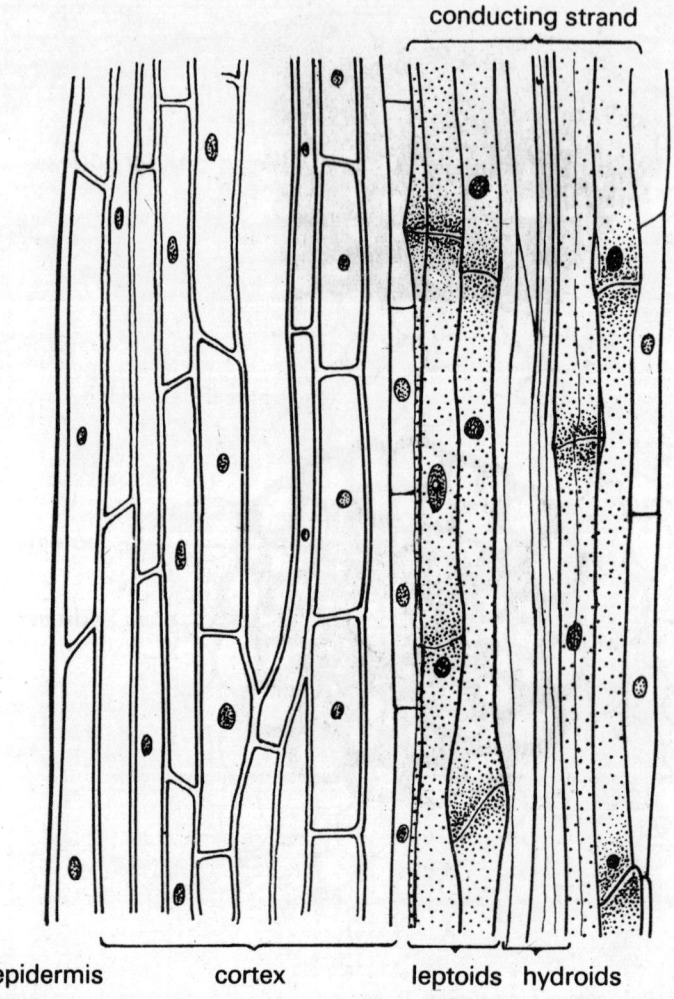

Fig. 11.12. *Atrichum undulatum.* Longitudinal section of seta. The ring of leptoids is mostly two cells thick. Their ends are dilated and the end walls are not usually horizontally placed.
(After Vaisey, 1888).

Eventhough hydroids of Polytrichaceae have several points in common with the tracheary elements of primitive vascular plants (the six features listed above), they show some significant differences. Hydroids lack secondary thickenings in the form of rings, spirals, and reticulum, and the nature of perforations in their end walls is also different. The question of the presence of lignin in bryophytes is discussed elsewhere (*see* page 220).

The hydroids in stem may be associated with one or more types of living cells like stereids, leptoids, internal cortex, and hydrom sheath. The

Conduction in Bryophytes 301

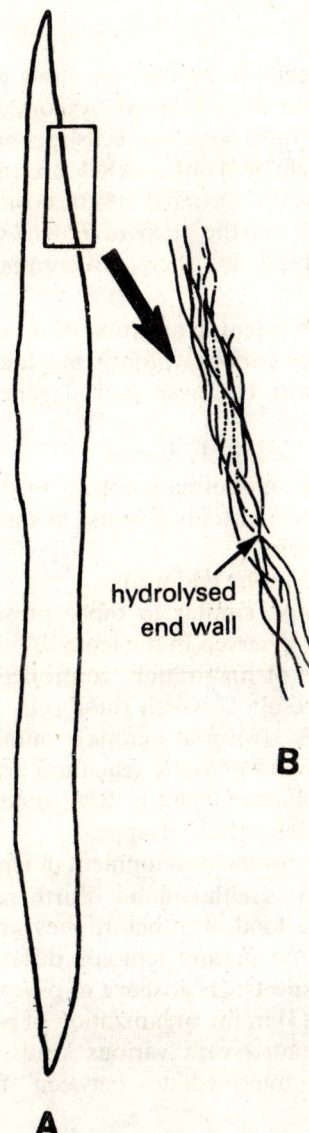

Fig. 11.13 A, B. A water conducting cell of mosses (hydroid). A. Diagrammatic representation. B. Magnified view of the portion marked in A to show partial hydrolysis of the end wall.
(After Hébant, 1977).

cells of hydrom sheath show enhanced enzymatic activity especially during maturation of hydroids.

Leptoids

The food conducting cells in bryophytes show great variability. In the vast majority this function is performed by conducting parenchyma. In many taxa conduction of food occurs in cells less organized than leptoids. Typical leptoids (sieve elements-like cells) are met with only in the members of Polytrichaceae. Detailed studies on the development of leptoids are mostly based on the stem of *Polytrichum* (Hébant, 1973b, 1976). The steps involved in their differentiation are as follows (Fig. 11.14C,D):

(1) Elongation and broadening of extremities.
(2) Nucleus undergoes endopolyploidy, but later shows symptoms of graded degeneration. However, such degenerated nuclei persist in the leptoids.
(3) End walls become obliquely placed.
(4) Well developed plasmodesmata appear in the end walls, which in the highly developed leptoids give rise to small pores.
(5) Lateral walls thicken.
(6) Callose is detected in the end walls.
(7) Refractive spherules, similar to those present in the phloem of pteridophytes, are observed in the leptoids.
(8) In the final stages of maturation, controlled autolysis of contents takes place as a result of which these cells appear clearer. Small, regressed plastids (without grana), mitochondria, and well developed smooth endoplasmic reticulum are among the persisting organelles in the mature leptoids. Ribosomes and dictyosomes are among the organelles which disappear.

Thus, the above features in the development of leptoids closely resemble those of sieve elements in vascular plants. Furthermore, like phloem the leptoids start translocating food even before they are mature (Eschrich & Steiner, 1967). However, the mature leptoids differ from the phloem of vascular plants in three aspects: (1) absence of p-protein, (2) persistence of degenerated nucleus, and (3) in the organization of pores in the end walls.

The leptoids are associated with various kinds of parenchyma cells, which show structures intermediate between typical leptoids and parenchyma cells.

THE INTERPHASE

The zone of contact between the gametophyte and sporophyte is very important and deserves a special mention. Transfer cells have been reported in all the investigated mosses and liverworts. These cells may be present in the foot as well as in the gametophyte as in *Mnium* (Eymé &

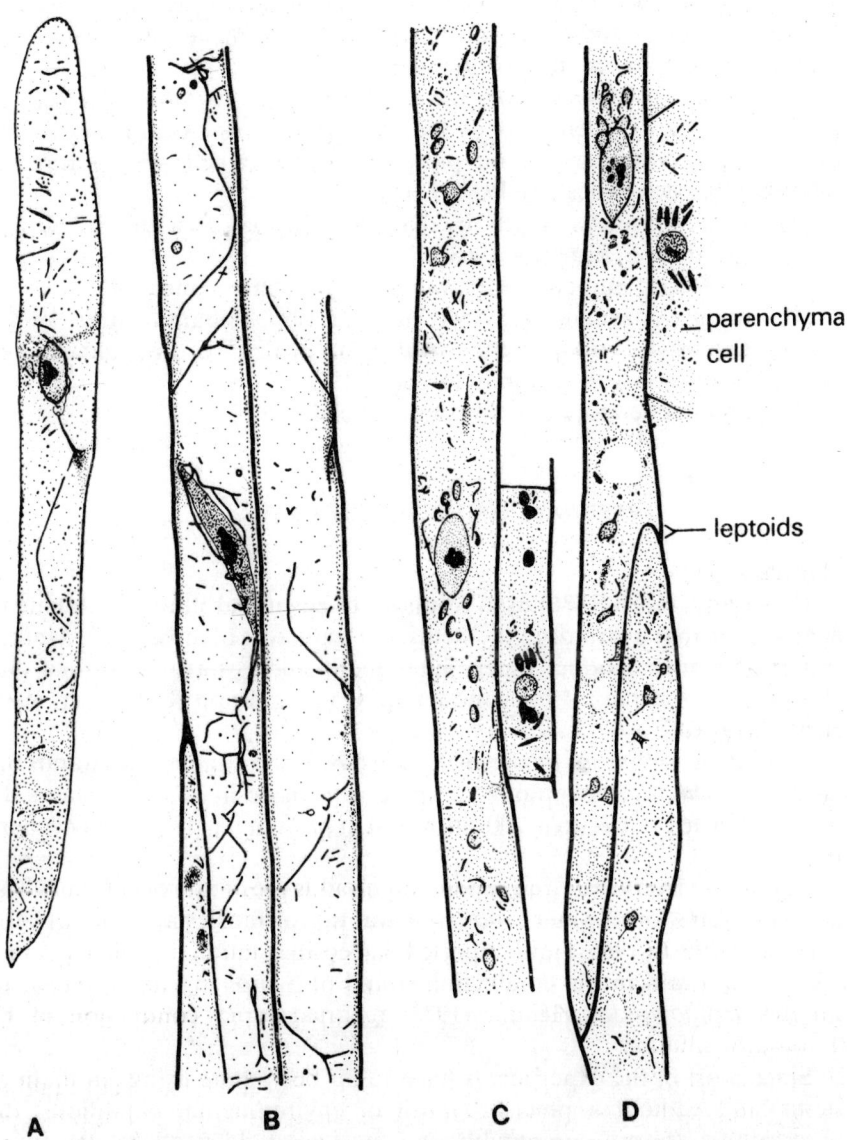

Fig. 11.14 A-D. Conducting tissues in *Polytrichum commune*. A, B. Some stages in the development of hydroids of central strand in the stem. C, D. Some stages in the development of leptoids in the leafy stem. Note difference in size between nuclei of differentiating leptoids and those of associated parenchyma cells. (After Hébant, 1967).

Suire, 1967), *Sphaerocarpos* (Kelley, 1969), and *Physcomitrium* (Lal & Chauhan, 1981). In others like *Polytrichum* (Maier, 1967), *Dawsonia*, and *Dendroligotrichum* (Hébant, 1975) they are present only in the foot. A doubtful report of Gunning and Pate (1969) on *Anthoceros* has indicated their presence only on the gametophytic side.

A noteworthy feature of the contact zone is the absence of plasmodesmata between cells of the two generations. Absorption by the foot is made possible due to the presence of transfer cells which show the following characteristics (*see* Plate 7.16):

(1) Wall ingrowths which increase the surface area of the plasma membrane (Ridgway, 1968).
(2) Abundant mitochondria, which suggest active transport.
(3) Intense enzymatic activity (Chauhan, 1981; Hébant & Suire, 1974).
(4) Extensive endoplasmic reticulum, which is thought to be connected with transport (Hébant, 1975).
(5) Frequent endopolyploidy of the nuclei.

INTERNAL CONDUCTION OF WATER

Gametophyte

Haberlandt (1883, 1886) used solutions of eosin and lithium sulphate to demonstrate that hydroids in the stems and leaf traces of mosses like *Mnium* and *Polytrichum* conduct water. Subsequent studies have confirmed that hydroids are specialized for conduction of water (Bopp & Weniger, 1971; Hébant, 1974).

In the gametophyte of mosses with well developed conducting strands the rates of conduction favourably compare with those in vascular plants. In *Polytrichum* the rate is up to 200 cm/h, and in *Mnium* it is 120 cm/h (Zacherl, 1956).

Among the liverworts, true conducting strands are of rare occurrence. The first demonstration of water conduction was that of Tansley and Chick (1901) in *Pallavicinia*. In 1966, Smith reported that eosin solution travelled up to 1.5 cm in 4 to 5 min in the conducting strand of *Symphyogyna circinata*. In *Takakia lepidozioides* Hébant (1972) reported rapid conduction of K-fluorescein solution.

Since most of the experiments have so far been done using cut thalli or stems, and without a precise control of environmental conditions, the observations are not comparable and may not hold good for the intact plants.

In the gametophytes of mosses, conduction of water is apoplastic as well as symplastic. The former is through the non-living components of cells (cell walls & intercellular space), whereas the latter involves protoplasm. There is evidence that water travels along the cell walls of all tissues. By means of Pb particles produced in situ it was observed that the cell wall material in *Polytrichum* contributes to the radial conduction of water to the hydrom

(Trachtenberg & Zamski, 1979). In fact, in the xeric taxa the walls are quite thick.

Sporophyte

Since the sporophyte has neither capillaries on the surface nor appreciable direct absorption of water, almost the entire water required by the developing capsule is supplied internally. The conducting tissues in the seta are always well developed, irrespective of the presence or absence of conducting strands in the gametophyte. Vaisey (1887) used eosin solution and demonstrated internal conduction in the setae of *Polytrichum* and *Splachnum*. Subsequently, Bopp and Stehle (1957), and Bopp and Weniger (1971) employed fluorescent dyes and observed conduction in the setae of *Funaria* and some members of the Polytrichales.

CONDUCTION OF ORGANIC COMPOUNDS

By using a labelled compound (^{14}C-bicarbonate), Eschrich and Steiner (1967) demonstrated that organic compounds travel in the stem of *Polytrichum* at the rate of 32 cm/h. Trachtenberg and Zamski (1978) observed that leptom of *Polytrichum* stem conducts assimilates, exogenously applied sucrose, and ionic solutes such as sulphate and lead. With the help of ^{14}C-sucrose, Eschrich (1975) demonstrated that translocation occurs in the seta of this moss at the rate of 50 cm/h. Autoradiography was employed to demonstrate that the organic compounds travel through leptoids in both the generations.

Internal transport of organic substances has also been demonstrated in the protonemata of mosses (Bopp & Knoop, 1974; Larpent-Gourgaud, 1974), of labelled sucrose in the thalli of *Marchantia* (Rota & Maravolo, 1975), and of IAA in the rhizoids of *Funaria* (Rose & Bopp, 1983).

Some indirect evidences for the role of leptoids in conduction of organic compounds in *Polytrichum* stem are:
(1) High enzymatic activity in these cells (Hebant, 1973c).
(2) Exudation from leptoids (Hébant, 1975).
(3) Association of aphids, which are also known to feed on the phloem of vascular plants (Müller, 1973).

EVOLUTIONARY TRENDS IN THE CONDUCTING STRANDS

In different taxa the conducting strands may be well developed, reduced, or absent. Hébant (1979) opined that characteristics of conducting tissues can provide additional help in bryophyte systematics. There are three possible views regarding their evolution:

(1) No conducting strand → reduced strand → well developed strand.
(2) Well developed strand → reduced strand → no strand.
(3) Well developed strand ← reduced strand → no strand.

Since all the three types co-exist in many large taxa, it is opined by some workers that this is due to parallel evolution. Some others consider that there has been retrogressive evolution, and the absence of conducting strand is a derived condition (see Hébant, 1977).

REFERENCES

BAYFIELD, N.G., *J. Bryol.*, **7**, 607 (1973).
BOPP, M. and B. KNOOP, *Bull. Soc. bot. Fr., Coll. Bryologie*, **121**, 145 (1974).
BOPP, M. and E. STEHLE, *Z. Bot.*, **45**, 161 (1957).
BOPP, M and H.-P. WENIGER, *Z. PflPhysiol.*, **64**, 190 (1971).
BOWEN, E.J., *Ann. Bot.*, **49**, 844 (1935).
BRUCH, TH., W.PH. SCHIMPER, and TH. GUMBEL, *Bryologia europaea, seu Genera Muscorum europaeorum monographice illustrata;* 6 vols.; E. Scheizerbart, Stuttgart, 1836-55.
BURR, R.J., B.G. BUTTERFIELD, and C. HÉBANT, *Bryologist*, **77**, 612 (1974).
CAMPBELL, E.O., K.R. MARKHAM, and L.J. PORTER, *N.Z.J. Bot.*, **13**, 593 (1975).
CHAUHAN, E., "Ultrastructural, cytochemical, and histoenzymological studies on some aspects of sporophyte development in the moss *Physcomitrium cyathicarpum* Mitt.," Ph.D. Thesis, Univ. Delhi, India, 1981.
CLEE, D.A., *Ann. Bot. N.S.*, **1**, 325 (1937).
CLEE, D.A., *Ann. Bot. N.S.*, **3**, 105 (1939).
DAVY DE VIRVILLE, A., *Rev. gén. Bot.*, **39**, 364-383, 449-457, 515-522, 560-586, 638-662, 711-726, 767-783 (1927).
DAVY DE VIRVILLE, A., *Rev. gén. Bot.*, **40**, 30-44, 95-110, 156-173 (1928).
DELOIRE, A., C. HÉBANT, and JEAN-MICHEL HENON, *J. Hattori bot. Lab.* No., **46**, 61 (1979).
ESCHRICH, W., "Bidirectional transport." In S. Aronoff et al., Eds, *Phloem Transport*, Plenum Press, New York, 1975. Pp. 401-416.
ESCHRICH, W. and M. STEINER, *Planta*, **74**, 330 (1967).
EYMÉ, J. and C. SUIRE, *C.r. Acad. Sci., Paris*, **265D**, 1788 (1967).
FAVALI, M.A. and M. BASSI, *Nova Hedwigia*, **25**, 451 (1974).
FREY, W., *Nova Hedwigia*, **20**, 463 (1971).
GUNNING, B.E.S. and J.S. PATE, *Protoplasma*, **68**, 107 (1969).
HABERLANDT, G., *Ber. dt. bot. Ges.*, **1**, 263 (1883).
HABERLANDT, G., *Jb. wiss. Bot.*, **17**, 359 (1886).
HALLET, J.-N., *Annls. Sci. nat. (Bot.), Sér.* **12**, **13**, 19 (1972).
HÉBANT, C., *C.r. Acad. Sci., Paris*, **263D**, 1065 (1966).
HÉBANT, C., *Naturalia monspeliensia, Sér. Bot.*, **18**, 293 (1967).
HÉBANT, C., *Naturalia monspeliensia, Sér. Bot.*, **19**, 75 (1968).
HÉBANT, C., *Rev. Bryol. Lichénol.*, **36**, 721 (1969).
HÉBANT, C., *C.r. Acad. Sci., Paris*, **275D**, 189 (1972).
HÉBANT, C., *C.r. Acad. Sci., Paris*, **277D**, 1445 (1973a).
HÉBANT, C., *J. Hattori bot. Lab.* No., 37, 211 (1973b).
HÉBANT, C., *Protoplasma*, **77**, 231 (1973c).
HÉBANT, C., *J. Hattori bot. Lab.* No., 38, 565 (1974).
HÉBANT, C., *J. Hattori bot. Lab.* No., 39, 235 (1975).
HÉBANT, C., *Protoplasma*, **87**, 79 (1976).
HÉBANT, C., *The Conducting Tissues of Bryophytes*, J. Cramer, Germany, 1977.
HÉBANT, C., *Protoplasma*, **96**, 205 (1978).

HÉBANT, C., "Conducting Tissues in Bryophyte Systematics," In G.C.S. Clarke and J.G. Duckett, Eds. *Bryophyte Systematics*, Academic Press, New York, 1979. Pp. 365-383.
HÉBANT, C., *J. Hattori bot. Lab.* No., 47, 63 (1980).
HÉBANT, C. and R.P.C. JOHNSON, *Cytobiologie*, 13, 354 (1976).
HÉBANT, C. and C. SUIRE, *Rev. Bryol. Lichénol.*, 40, 171 (1974).
KELLEY, C., *J. Cell Biol.*, 41, 910 (1969).
LAL, M. and E. CHAUHAN, *Protoplasma*, 107, 79 (1981).
LARPENT-GOURGAUD, M., *Bull. Soc. bot. Fr., Coll. Bryologie*, 121, 161 (1974).
LORENTZ, P.G., *Jb. wiss. Bot.*, 6, 363 (1867-68).
MÅGDEFRAU, K., *Z. Bot.*, 29, 337 (1935-36).
MAIER, K., *Planta*, 77, 108 (1967).
McCONAHA, M., *Am. J. Bot.*, 26, 353 (1939).
McCONAHA, M., *Am. J. Bot.*, 28, 301 (1941).
MÜLLER, F.P., *Entomologische Abhandlungen*, 39, 205 (1973).
PAOLILLO, D.J., JR. and J.A. REIGHARD, *Bryologist*, 70, 61 (1967).
PROSKAUER, J., *Phytomorphology*, 10, 1 (1960).
RIDGWAY, J.E., "Ultrastructural features of the sporophytic-gametophytic interphase in bryophytes," *26th Annual EMSA Meeting*, 1968. Pp. 86-87.
ROSE, S. and M. BOPP, *Physiol. Plant.*, 58, 57 (1983).
ROTA, J.A. and N.C. MARAVOLO, *Bot. Gaz.*, 136, 184 (1975).
SCHEIRER, D.C., "The anatomy, histochemistry, and ultrastructure of the gametophyte of *Dendroligotrichum dendroides* (Hedw.) Broth. (Bryopsida: Polytrichaceae)," Ph.D. Thesis, Pennsylvania State University (University Microfilms, Ann. Arbor. no. 75-09836), 1974.
SCHEIRER, D.C., *Bryologist*, 78, 113 (1975).
SCHIMPER, W.PH. Mémoire pour servir à L'histoire naturelle des Sphaignes, Pp., 96 Pl. 1-24, Paris, 1857.
SIEGEL, S.M., *The Plant Cell Wall*, Pergamon Press, New York, 1962.
SMITH, J.L., *Univ. Calif. Publ. Bot.*, 39, 1 (1966).
TANSLEY, A.G. and E. CHICK, *Ann. Bot.*, 15, 1 (1901).
TRACHTENBERG, S. and E. ZAMSKI, *J. exp. Bot.*, 29, 719 (1978).
TRACHTENBERG, S. and E. ZAMSKI, *New Phytol.*, 83, 49 (1979).
UDAR, R., *An Introduction to Bryophyta*, Shashidhr Malaviya Prakashan, Lucknow, India, 1970.
VAISEY, J.R., *Ann. Bot.*, 1, 73 (1887).
VAISEY, J.R., *J. Linn. Soc. (Bot.)*, 24, 262 (1888).
WINKLER, S., *Öst. Bot. Z.*, 117, 348 (1969).
ZACHERL, H., *Z. Bot.*, 44, 409 (1956).

12 Water Relations

BRYOPHYTES have long been used for physio-ecological studies. In the field of water relations the problems which have received most attention are absorption and conduction of water and solutes in the gametophyte and sporophyte, and the remarkable resistance to drought shown by the spores and vegetative cells of many bryophytes.

Bryophytes have very limited power of withdrawing water from substrate. Water required for metabolic processes is therefore mostly derived from water falling on or flowing over the plants.

ABSORPTION AND CONDUCTION OF WATER*

Water movement in bryophytes has been the subject of study at least since the time of Hedwig (1782). Haberlandt (1883, 1886) demonstrated rapid movement of dye solutions in the central strands of the stem of *Polytrichum juniperinum* and *Plagiomnium undulatum*. He observed that there is an upward movement of water from base of stem to leaves, comparable at least in some respects to the transpiration stream in higher plants. Bowen (1931, 1933a,b,c) studied various mosses using dyes, NO_3, Li^+, and other tracers to follow water movement. It has been observed that all bryophytes are capable of absorbing (as well as loosing) water over practically the entire surface of the leafy shoot or thallus. There is a sharp correlation between the occurrence and distribution of many species of mosses and conditions of moisture of the habitat. Buch (1945, 1947) made a significant study on the area of absorption as well as the path of conduction of water in the gametophyte. His observations also emphasized the physiological differences with respect to water relation which exist among bryophytes. He concluded that bryophytes have two major physiological groups—endohydric and ectohydric.

ENDOHYDRIC

Mosses which have a well developed conducting strand, such as *Bryum*

---

* Also see Chapter 11.

capillare, Mnium undulatum, Polytrichum commune, and *P. juniperinum* absorb water mainly by rhizoids at the base and transfer the water reserves from the base to the actively photosynthesizing leaves at the apex. Such mosses have a transpiration stream, and so long as the soil remains moist they are able to maintain turgidity in an atmosphere with a considerable saturation deficit, even when no external conduction of water can take place. The leaves in these mosses are relatively impermeable to water and have a cuticle-like covering e.g. *Bartramia pomiformis*. In some species the cuticle of endohydric mosses has thin spots through which water is lost or absorbed more readily than elsewhere. These can function as hydathodes when the plant is in a saturated atmosphere. Endohydric mosses include most tuft—forming true mosses (acrocarpous bryopsida) for example Funariales and Polytrichales. The leaves of large endohydric mosses like *Polytrichum* and *Dawsonia* have marginal wings which fold over the photosynthetic lamellae in the centre (Sarafis, 1971). These hinge movements are due to the swelling of lamellae (Zanten, 1974) and result in the leaf being reflexed under moist conditions, but on drying the leaves straighten and tend to hug the main axis and help in retardation of water loss. The tips of lamellae on these leaves also have a waxy covering which repels water and facilitates gas exchange (Proctor, 1979a,b).

According to Buch (1947) endohydric bryophytes can be readily recognized by a number of features: (1) they have well developed basal rhizoid system, (2) their young fully expanded leaves are difficult to stain with dilute solutions of basic dyes such as toluidine blue, (3) the external cell walls of the leaves are relatively impermeable to electrolytes, and (4) they occur on loose substrata i.e. soil or humus and do not usually grow on rocks or barks.

ECTOHYDRIC**

Mosses which lack well differentiated conducting strand are termed ectohydric'. These are capable of absorbing water (and dissolved substances) through almost any part of the external surface of the shoot or thallus and are quickly stained by basic dyes. There is no regular internal movement of water within the stem. Such mosses absorb water rapidly into the moss carpet by capillary systems which are of three categories:

(1) Interorgan capillary systems e.g. spaces within sheathing leaves or leaf bases or between a tomentum of rhizoids;
(2) Epiorgan capillary systems e.g. spaces between papillae, ridges, or folds in leaf or stem surfaces;
(3) Intraorgan capillary systems e.g. non-photosynthetic cells in *Sphagnum* leaves and the porous cells of leaf bases in a number of moss families

---

**This group includes a great majority of mosses as *Rhacomitrium, Orthotrichum, Ulota, Cryphaea,* and all the leafy liverworts.

Considerable amount of water is held in the larger capillary spaces between the shoots, within the moss carpet (Proctor, 1979a). Absorption of dew is also very important as a source of water to enable photosynthesis in mosses (Lange, 1969; Kellomäki et al., 1978; Kappen et al., 1979). Retardation of water loss from ectohydric mosses is enhanced by the tendency to form carpets or tufts. In addition many species possess leaves with hair points. These hair points reflect a proportion of the incoming solar radiation so that less water evaporation occurs. Also, they increase the diffusion path length for water vapour to reach the boundary layer and hence the rate of water loss is reduced still further (Proctor, 1979a). By removing hair points of *Grimmia pulvinata* and *Tortula intermedia* Proctor (1980) observed that the hair tips reduce water loss by about 35 percent.

Ectohydric species predominate on hard, impermeable, and often nutrient-poor substrata, notably rock surfaces. The bryophytes of acid peat bogs and corticolous taxa are also very largely ectohydric. In ectohydric types cuticle has not been demonstrated.

For capillary action of mosses morphological features such as the arrangement and distance between leaves, the presence of rhizoids or paraphyllia on the stem surface, and the branching and direction of growth are important. In ectohydric species the leaves often 'revive' in a few seconds, whereas in endohydric species the air dry leaves become turgid very slowly.

MYXOHYDRIC

This group has the features of both ectohydric and endohydric types. As an example may be cited *Funaria hygrometrica*. In such mosses both external and internal conduction take place. Once water is in the stem the central strand is important, but the excess of water moves through the leaves by external channels.

Myxohydric species occur predominantly on moist, porous, and often nutrient-rich substrata. They are prominent in bryophytic vegetation of forests and related shady habitats.

WATER HOLDING CAPACITY AND GROWTH RATE

The importance of water relations in determining the ecological range of bryophyte species was demonstrated by Overbeck and Happack (1957). They investigated the water holding capacity and growth rate in various species of *Sphagnum* occurring in North Germany. In sphagna, differences in the water holding capacity presumably depend on the morphological construction of the shoots. The growth rate (measured as growth in length in the field) varies from 3.3 to 10.7 cm/annum in *S.*

rubellum to over 41.5 cm in *S. cuspidatum* and *S. plumulosum*, but higher growth rates would be attained if the plants always had sufficient water supply. Great differences in water holding capacity between the species were found e.g. *S. rubellum* holds 46 g/100 ml volume, *S. magellanicum* 19 g, and the less densely tufted *S. recurvum* only 10 g/100 ml volume.

Studies on the growth rate of a wide range of bryophytes indicate that many bryophytes grow very slowly and this is one of the important factors affecting competition between them and higher plants. Tallis (1959) studied the growth rate in different habitats and under different conditions of a widely distributed and ecologically important moss *Rhacomitrium lanuginosum*. Under British conditions the growth in length of this moss never exceeds 20 mm/annum and there is strong evidence that in some habitats like calcareous grassland, its growth rate is very slow and that is why it is usually absent in this habitat. On the other hand, Overbeck and Happack (1957) observed that the growth rate of sphagna, measured as yearly production of dry matter, is remarkably high. *S. rubellum*, a relatively slow growing species, grows about twice to four times as fast as mosses like *Camptothecium* and *Hylocomium*.

Hagerup (1935) studied a large number of species in Denmark and came to the conclusion that most European bryophytes had two annual growth periods, one in spring from December to March, and another from June or July to August or September. He noticed that even bryophytes in non-seasonal tropical conditions showed periodic growth, and concluded that the phenomenon depended chiefly on an internal 'autonomic' rhythm and not on the annual cycle of climatic conditions. On the other hand, most of the mosses studied by Lackner (1939) in the continental climate of East Prussia, showed only one annual growth period. He, therefore, suggested that growth rate is dependent on the environmental conditions and is much less controlled by an autonomic rhythm. Jendralski (1955) investigated the species of the oceanic climate of the Rhineland. She found that some species which Lackner had observed to have only one growth period in East Prussia had two such periods in Denmark, under her experimental conditions. She, however, agreed with him in concluding that the periodic growth of mosses is largely controlled by external climatic conditions and is not autonomic. Romose (1940) also observed that *Camptothecium sericeum* if kept moist grew uninterruptedly through the summer and had no autonomically determined resting period. According to Overbeck and Happack (1957) the same is true of sphagna.

Thus, periodic growth may be influenced to some extent by an inherent rhythm, but the most important determining factors are temperature and moisture.

DESICCATION AND REHYDRATION

Bryophytes have little direct control over water loss. They are in general poikilohydric i.e. they become dormant in the dry season by losing most of their water. However, they have responded to dry environments by developing physiological tolerance to desiccation. Keilin (1959) called this phenomenon as anhydrobiosis. On remoistening normal metabolism is resumed more or less rapidly. However, desiccation tolerance does not solve all the water supply problems of bryophytes, and some water movement is essential even for bryophytes in shady, constantly moist habitats. As a moss dries, the water content falls and the water potential (measured in bars) becomes more negative. At low (negative) water potential the cytoplasm within the cells becomes viscous and gel-like.

Dilks and Proctor (1979) observed that mosses growing in shaded woods during summer would have a water content of about 20 percent dry weight and a water potential of about -200 to -400 bars. When the water content falls to 10 percent dry weight, the water potential is around -1000 bars and is observed in mosses during dry but not extremely hot conditions in the open in summer.

Busby and Whitfield (1978) reported that mosses from the circumboreal woodlands of northern Canada, such as *Pleurozium schreberi*, *Hylocomium splendens*, and *Tomenthypnum nitens*, carry on the processes of photosynthesis and respiration until the water potential falls to -55 to -100 bars. Dilks and Proctor (1979) observed that in the tolerant species, *Camptothecium lutescens*, net photosynthesis was positive down to about -150 bars but was so up to -80 bars in the sensitive species, *Hookeria lucens*.

In basal cells of moss leaves the protoplasm condenses at the ends of the cells as the moss dries, leaving the central parts empty. At the same time the chloroplasts become round and compact, and vesicles disappear (Tucker et al., 1975). The cells remain in this condition until remoistened. When a drought tolerant moss is rewetted, the following sequence of events occurs. There is initially a release of carbon dioxide which is thought to be due to displacement of absorbed gas. Pockets of gas are seen to form immediately after rehydration at the distal and proximal ends of leaf cells of the desiccation tolerant moss *Tortula ruralis*. However, these shrink and disappear within 10 to 30 sec as the cytoplasm expands to fill the whole cell (Tucker et al., 1975). After this initial release of gas there is a period of enhanced respiration lasting from one to a few hours. During this time the cell organelles regain their normal appearance (Noailles, 1978). The more rapid the desiccation rate, the longer and more pronounced the period of enhanced respiration (Krochko et al., 1979). During this time the process of photosynthesis commences and accelerates. After 4 to 24 h, normal rates of respiration and photosynthesis are observed (Dilks & Proctor, 1976; Peterson & Mayo, 1975; Proctor, 1981).

Gupta (1977) subjected three leafy liverworts (*Porella platyphylla*, *Plagiochila asplenioides*, & *Scapania undulata*) and one moss (*Plagiothecium undulatum*) to desiccation for different periods (2-82 days). In all these species the rate of leakage of solutes reached the highest value in the first two minutes after rewetting, and then it slowed down. This reduction in leakage may be due to the reassemblage of membrane structure whereby normal semipermeability is regained. It may also be due to the rapid decrease of solutes in the dead or injured tissues after rewetting. The leakage from viable tissues is possibly due to the shock resulting from sudden immersion of desiccated shoots in water, whereas in the dead or injured cells it is because of disruption of membranes. The capacity to retain soluble cell contents was different in the four species. In *Scapania* (a susceptible species) even a very moderate desiccation treatment (at 96.5% RH) for two days resulted in the loss of about 60 percent of the soluble labelled cell constituents within 1 h. The more severe treatment for two days at 50 percent RH resulted in a loss of about 90 percent solutes. However, the loss in *Porella* (a resistant species) after two days of treatment at 50 percent RH was only about one-fifth of that in *Scapania*. It is noteworthy that in spite of leakage of large quantities of solutes from plants they were still alive and capable of photosynthetic activity. In both these liverworts reabsorption of some of the lost solutes by the viable cells was observed on rewetting.

In drought tolerant mosses, mRNA is apparently conserved during drying and polyribosomes are reformed on rehydration without the need for new mRNA synthesis. Thus protein synthesis can begin immediately so that repair can be rapid (Dhindsa & Bewley, 1978).

MECHANISM OF DAMAGE

The loss of viability on drying can be due to several reasons. Effects have been observed on enzymes such as NADP-linked glyceraldehyde phosphate dehydrogenase (Stewart & Lee, 1972), and on the capacity to synthesize proteins (Bewley, 1974, 1979). It has been noticed that the loss of photosynthate and essential nutrients during remoistening is also an additional problem (Gupta, 1977). Krochko et al. (1978) compared the effects of fast and slow drying on the ultrastructure of two mosses: *Cratoneuron filicinum* (drought sensitive) and *Tortula ruralis* (drought tolerant). Fast drying of *Tortula* for 24 h over silica gel resulted in the swelling of chloroplasts and mitochondria. These organelles lost their integrity when the plant was rewetted. However, within 24 h of rewetting these organelles regained their normal appearance. The chloroplasts and mitochondria of *Cratoneuron* showed a similar damage on rewetting after fast drying, but 25 h later the cells showed extreme degradation and loss of cell contents. When the sensitive species was dried more slowly (for 12 h over RH of 75%) only 20 percent of the cells exhibited swollen organelles on rewetting, and after 24 h recovery there were approximately equal numbers of degraded cells and cells with normal-looking nuclei and mitochondria. The fact that *Tortula* can survive swelling of the cell components whereas the sensitive species can not, suggests that sensitive and tolerant species differ in the stability, availability, and capacity for synthesis of the enzymes needed for repair (Dhindsa & Bewley, 1978) Krochko et al. (1979) suggested that for dried mosses it is important to retain the ability to synthesize ATP than to conserve normal cellular levels of this substance. In the desiccation tolerant moss *Tortula ruralis*, control levels of ATP are regained within 30 min even after rapid drying. In the sensitive species *Cratoneuron filicinum*, ATP levels steadily declined following rewetting of fast dried specimens. However, slowly dried material was able to resume ATP synthesis and regain normal levels within 2 h. It is clear, therefore, that although mosses are poikilohydric, it is an advantage for many species to lose water as slowly as possible.

GROWTH-FORMS

Structural features of bryophytes which increase their ability to hold capillary water are important, since they extend the period during which the plant can remain metabolically active, and they also extend the period in which mineral ions can be absorbed from very dilute solutions. Such features are common in tropical bryophytes, e.g. the 'hyalocysts' of the Calymperaceae and Leucobryaceae, and 'water sacs' of the Lejeuneaceae, Jubulaceae, Pleuroziaceae, and some other hepatics. Some species (*Colura, Pleurozia*) have specialized valve-like flaps which close these sacs during periods of dryness. Water holding adaptations are usually absent or are poorly developed in the epiphytic under-growth of the moist shaded places, but are present in many 'cloud forest' bryophytes e.g. *Pleurozia, Trichocolea*, and *Frullania*. According to Biebl (1964) the thick cell walls (in hepatics often collenchymatous) common in 'cloud forest' bryophytes serve as water stores.

Hosakawa and Kubota (1957) studied the osmotic pressure by plasmolyzing the leaf cells with KNO_3 solution. The resistance to desiccation was determined by survival time after exposure to a range of relative humidities controlled by sulphuric acid solutions in petri dishes. They observed a clear correlation between the osmotic values of the leaf cells and the vertical range of the species of the trees, and also between both of these and survival time at low humidities. The osmotic values and drought resistance were lowest in the 'stump' species and highest in *Ulota crispula* which grew on the upper branches. Drought resistance was greater in winter than in summer.

The upper limit on trees of some species is determined mainly by sensitivity to drought and the lower limit of others by inadequate illumination.

Some indications of the eco-physiological characteristics of bryophytes are given by their growth-forms. Some mosses are gregarious and have aggregated shoots, whereas others are solitary and have separated shoots. In the humid tropical forests solitary bryophytes (most of which are mosses and many of large size) are abundant, varied in form, and numerous in species. Solitary bryophytes, especially those of the hanging growth-form, are more exposed to desiccation and extremes of temperature than are gregarious bryophytes. Recently, Richards (1983) has proposed a provisional growth-form classification for tropical forest bryophytes (Table 12.1; Fig. 12.1).

TABLE 12.1.
Classification of Growth-Forms of tropical forest bryophytes. Based on Giesenhagen (1910), Gimingham and Birse (1957), and Iwatsuki (1960), with modifications.

A. Social forms. Leafy shoots or thallus branches aggregated.
(1) Cushions (shoots mainly erect and radiating to form dome-shaped masses).
  (a) Large cushions (> 5 cm diameter)—*Leucobryum*, some *Schistochila* spp.
  (b) Small cushions (< 5 cm diameter—*Octoblepharum*, many species of *Calymperes* and *Syrrhopodon*.
(2) Turfs (shoots upright, more or less parallel, "like the pile of a carpet").
  (a) Tall turfs (> 2 cm high), branches mostly erect—*Leucoloma*, some *Syrrhopodon* spp., *Herbertus*, some *Schistochila* spp.
  (b) Tall turfs with divergent or creeping branches, mostly of limited growth—*Acroporium, Macromitrium, Mastigophora, Sphagnum*.
  (c) Short turfs (< 2 cm high)—*Calymperes* sect. *Macrhimanta, Diphyscium*.
  (d) Open turfs (shoots somewhat separated, often springing from a persistent rhizoidal or protonemal system)—*Fissidens, Drepanophyllum, Epipterygium*.
(3) Mats (primary stems creeping horizontally over the substratum, lateral branches erect or parallel, mostly limited in growth, forming closely interwoven mats).
  (a) Rough mats (main shoots adhering to substratum, with abundant short erect laterals)—*Ectropothecium, Sematophyllum, Trachypodopsis*.
  (b) Smooth mats (branches in same plane as main shoot, often closely interwoven)—*Callicostella* and many other Hookeriaceae, *Isopterygium, Radula*, many Lejeuneaceae, *Frullania* spp.
  (c) Thread-like forms (delicate, creeping, sparingly & irregularly branched—small Lejeuneae, including many epiphyllous species.
  (d) Thallose mats—*Metzgeria, Monoclea forsteri, Symphyogyna* sect. *Repentes, Dumortiera*.
  (e) Wefts—*Thuidium*.

B. Solitary forms. Leafy shoots or thallus branches not closely aggregated.
(1) Protonemal bryophytes (protonema persistent, leafy shoots minute, scattered)—*Ephemeropsis, Micromitrium, Protocephalozia*.
(2) Unbranched dendroid (terrestrial) forms—*Dawsonia, Pogonatum*.
(3) Branched dendroid, not dorsiventral, forms (epiphytic or terrestrial)—*Hypnodendron, Porothamnium, Rhodobryum, Symphyogyna* sect. *Dendroides*.
(4) Feather forms (secondary stems more or less pinnately branched, dorsiventral, epiphytic, or saxicolous)—*Hypopterygium, Neckeropsis, Callicosta bipinnata, Bryopteris*.
(5) Bracket mosses (shoots horizontal, not dorsiventral, epiphytic or saxicolous)—*Leiomela, Spiridens*.
(6) Hanging bryophytes (secondary branches long, pendulous)—Meteoriaceae, *Frullania* sect. *Meteoriopsis, Plagiochila* spp.

316 Biology of Bryophytes

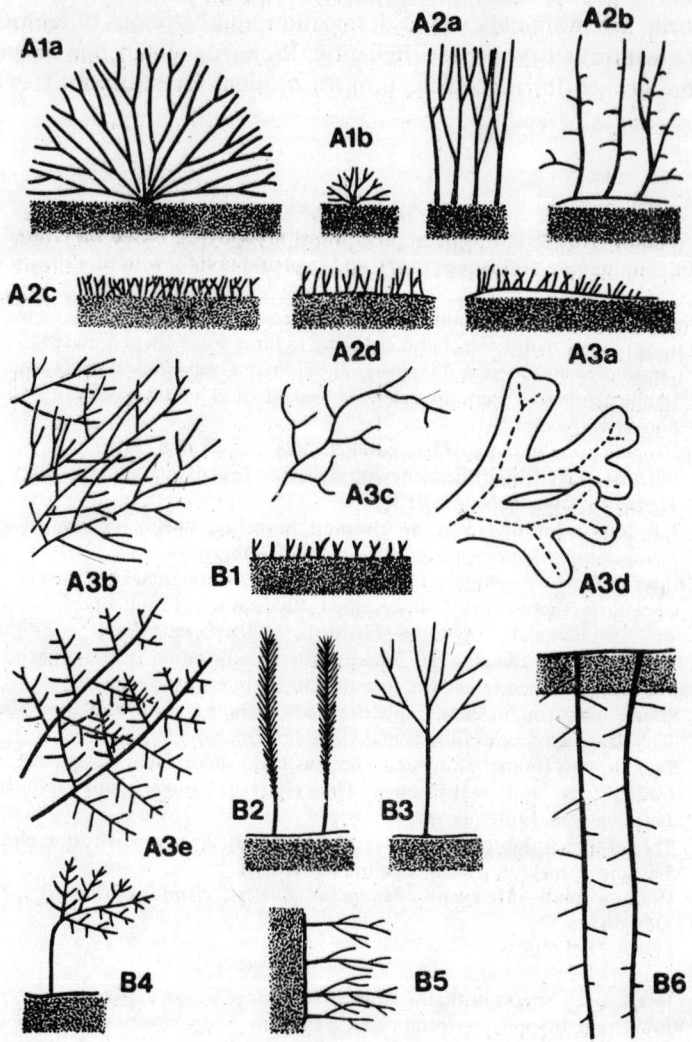

Fig. 12.1. Growth-forms of tropical bryophytes (diagrammatic representation). For details see Table 12.1.

A1 a, Large cushions; A1 b, Small cushions:

A2 a, Tall turfs; A2 b, Tall turfs with divergent or creeping branches; A2 c, Short turfs; A2 d, Open turfs:

A3 a, Rough mats; A3 b, Smooth mats; A3 c, Thread-like forms; A3 d, Thallose mats; A3 e, Wefts.

B1, Protonemal bryophytes; B2, Unbranched dendroid forms; B3, Branched dendroid not dorsiventral forms; B4, Feather forms; B5, Bracket mosses; B6, Hanging bryophytes.

(After Giesenhagen, 1910; Gimingham & Birse, 1957; Iwatsuki, 1960; with modifications).

REFERENCES

BEWLEY, J.D., *Can. J. Bot.*, 52, 423 (1974).
BEWLEY, J.D., *Ann. Rev. Pl. Physiol.*, 30, 195 (1979).
BIEBL, R., *Protoplasma*, 59, 277 (1964).
BOWEN, E.J., *Ann. Bot.*, 45, 175 (1931).
BOWEN, E.J., *Ann. Bot.*, 47, 401 (1933a).
BOWEN, E.J., *Ann. Bot.*, 47, 635 (1933b).
BOWEN, E.J., *Ann. Bot.*, 47, 889 (1933c).
BUCH, H., *Commentat. biol. Soc. Sci. Fenn.*, 9(16), 1 (1945).
BUCH, H., *Commentat. biol. Soc. Sci. Fenn.*, 9(20), 1 (1947).
BUSBY, J.R. and D.W.A. WHITFIELD, *Can. J. Bot.*, 56, 1551 (1978).
DHINDSA, R.S. and J.D. BEWLEY, *Proc. natn. Acad. Sci. (U.S.A.)*, 75, 842 (1978).
DILKS, T.J.K. and M.C.F. PROCTOR, *J. Bryol.*, 9, 249 (1976).
DILKS, T.J.K. and M.C.F. PROCTOR, *New Phytol.*, 82, 97 (1979).
GIESENHAGEN, K., *Ann. Jard. bot. Buitenzorg Suppl.*, 3, 711 (1910).
GIMINGHAM, C.H. and E.M. BIRSE, *J. Ecol.*, 45, 533 (1957).
GUPTA, R.K., *Can. J. Bot.*, 55, 1186 (1977).
HABERLANDT, G., *Ber. dt. bot. Ges.*, 1, 263 (1883).
HABERLANDT, G., *Jb. wiss. Bot.*, 17, 359 (1886).
HAGERUP, O., *Danske Vidensk. Selsk. Biol. Meddel.*, 11, 1 (1935).
HEDWIG, J., *Fundamentum Historiae Naturalis Muscorum Frondosorum*, Leipzig, 1782.
HOSAKAWA, T. and H. KUBOTA, *J. Ecol.*, 45, 579 (1957).
IWATSUKI, Z., *J. Hattori bot. Lab.* No., 2:, 159 (1960).
JENDRALSKI, U., *Decheniana*, 108, 105 (1955).
KAPPEN, L., O.L. LANGE, E.D. SCHULZE, M. EVENARI, and U. BUSCHBOM, *Flora*, 168, 85 (1979).
KEILIN, D., *Proceeding of the Royal Society*, 8150, 149 (1959).
KELLOMÄKI, S., P. HARI, and T. KOPONEN, *Bryophytorum Bibliotheca*, 13, 485 (1978).
KROCHKO, J.E., J.D. BEWLEY, and J. PACEY, *J. exp. Bot.*, 29, 905 (1978).
KROCHKO, J.E., W.E. WINNER, and J.D. BEWLEY, *Pl. Physiol.*, 64, 13 (1979).
LACKNER, L., *Planta*, 29, 534 (1939).
LANGE, D.L., *Planta*, 89, 90 (1969).
NOAILLES, M., *Ann. Sci. Nat. Sér.*, 12, 249 (1978).
OVERBECK, F. and H. HAPPACK, *Flora*, 144, 335 (1957).
PETERSON, W.L. and J.M. MAYO, *Can. J. Bot.*, 53, 2897 (1975).
PROCTOR, M.C.F., "Structure and eco-physiological adaptations in bryophytes." In G.C.S. Clarke and J.G. Duckett, Eds, *Bryophyte Systematics*. Systematics Association Special Volume No. 14, Academic Press, London, 1979a. Pp. 479-509.
PROCTOR, M.C.F., *J. Bryol.*, 10, 531 (1979b).
PROCTOR, M.C.F., "Diffusion resistance in bryophytes." In E.D. Ford and J. Grace, Eds, *Plants and their Atmospheric Environment*. J. Symp. Brit. Ecol. Soc., 1980. Pp. 219-229.
PROCTOR, M.C.F., "Physiological ecology of bryophytes." In W.J. Schultze-Mötel, Ed., *Advances in Bryology* Volume No 1, J. Cramer, Germany, 1981. Pp. 79-166.
RICHARDS, P.W., "The ecology of tropical forest bryophytes." In R.M. Schuster, Ed., *New Manual of Bryology* Volume No. 2. The Hattori Botanical Laboratory, Nichinan, Miyazaki, Japan, 1983. Pp. 1233-1270.
ROMOSE, V., *Dansk. Bot. Ark.*, 10, 1 (1940).
SARAFIS, V., *N.Z. J. Bot.*, 9, 711 (1971).
STEWART, G.R. and J.P. LEE, *New Phytol.*, 71, 461 (1972).
TALLIS, J.H., "A study of the biology and ecology of *Rhacomitrium lanuginosum* Brid.," Ph.D. Thesis, University College of North Wales, Great Britain, 1957.
TUCKER, E.B., J.W. COSTERTON, and J.D. BEWLEY, *Can. J. Bot.*, 53, 94 (1975).
ZANTEN, B.O. AAN, *Bull. Soc. bot. Fr., Coll. Bryologie*, 121, 63 (1974).

Subject Index

Abscisic acid, effect on
  apogamy, 112
  bud formation, 49
  protonemal growth, 112
Acetylcholine, 142, 182, 183
Activated charcoal, 279, 280
Active transport, 168, 304
Acid rain, 259
Acylated flavonoids, 220
Acyclic sesquiterpenoids, 200
Adenosine 3′:5′-cyclic monophosphate (Cyclic AMP)
  effect on
    bud formation, 47
    gametangial induction, 104
    protonemal differentiation, 40
  endogenous, 181, 182
Adenylate cyclase, 182
Adventitious shoots, 64, 66, 81
Agar, effect of concentration on
  apogamy, 108, 111
  gametangial induction, 98
  gemma cup production, 32
  protoplast isolation, 274
  spore germination, 26
Alkanes, listed, 185, 186
Allergenic activities, 233, 240
Alternation of generations, 107, 119, 124
  views about, 125, 126
Alternative pathways in life cycle, 107
Amino acids
  analogue resistant mutant, 283
  effect on formation of
    buds, 47, 48
    gametangia, 100
    gemmae, 34, 35
  presence of, 233, 237
Amphomorphic, 125
Anhydrobiosis, 312
Anthocyanins, 220
Antiauxins, effect on
  bud induction, 43
  caulonema, 43
  gametangial production, 101
  gemma germination, 34
  regeneration, 82
  sporogon development, 117
Antibiotic activity, 175-177
Antifungal property, 177
Antiseptic property, 175
Antithetic theory, 119, 125, 126
Antitumour activities, 240
Aphids, 305
Apical dominance
  effect on regeneration, 77-83
  hormonal mechanism of, 79-82
  in gametophyte, 77-82
  in protonema, 82
  in sporophyte, 77, 108
  influenced by nutrients, 78, 81, 82
  physiological basis for, 83
Apigenin C-glycosides, 213
Apigenin O-glycosides, 211
  listed, 212
Apogamy
  factors controlling
    age of tissue, 108, 114
    casein hydrolysate, 113
    chloral hydrate, 112
    coconut milk, 113
    genetic constitution, 114, 115
    growth hormones, 112
    hydration, 108, 111, 113
    inorganic nutrients, 112, 113
    light, 111, 113, 114
    products of moss metabolism, 114
    sporogon factor, 112, 113
    sugars, 111-113
    yeast extract, 113
  from callus, 111-114
  occurrence in diplophase and haplophase, 108-110
  role of calyptra in sporogon development, 116-118
  spore production in apogamous sporophytes, 109, 110
Apoplastic conduction, 169, 304
Apospory, 54, 107

320   Biology of Bryophytes

   in liverworts, 119
   in mosses, 118
Archesporium, origin in
   Andreaeales, 160
   Bryales, 160
   Sphagnales, 160
Aromadendranes, 197, 198, 203
Arsenates, as pollutants, 254
Aurones, 219
Autolysis, 302
Autonomic rhythm, 311
Autoradiography, 305
Auxins
   effect on
      bud formation, 43
      gametangial production, 101
   endogenous, 178, 179
Auxotrophic mutants, 282, 283
Axoneme 156, 157

Bark discs, 259, 260
Bicarbocyclic diterpenoids, 205
   clerodane-type, 205
   dolabellane-type, 205
   labdane-type, 205
   sacculatane-type, 205
Bicarbocyclic sesquiterpenoids, 200
   acorane-type, 201
   azulene-type, 202
   bicyclogermacrane-type, 202
   bicyclohumulane-type, 202
   cadinane-type, 201
   carotane-type, 201
   caryophyllane-type, 200
   chamigrane-type, 200
   cuparane-type, 200
   drimane-type, 200
   eremophilane-type, 201
   guaiane-type, 201
   himachalane-type, 200
   pinguisane-type, 202
   santalane-type, 202
   seco-aromadendrane-type, 202
   selinane-type, 201
   trichothecane-type, 200
Biflavonyls, 218, 219
Bioassay for cytokinins, 45
Bioindicators of pollution, 250
Bisabolanes, 194, 195, 200
Blepharoplast, 156, 157
Blue-far-red pigment system, 131
Blue light absorbing pigment, 132-134
Blue light, effect on
   bud formation, 42, 143
   callus morphogenesis, 140

   capsule growth, 139, 140
   growth, 134-140
   protonemal growth, 138, 139
   regeneration, 67, 69
   spore germination, 23, 134-136
Bracket mosses, 315, 316
Brood buds, 26
Brood cells, 33, 253
Bryokinin, 114, 179
Budding fission, 84
Bud formation (induction)
   bioassay for cytokinins, 45
   caulonema specific proteins, 46
   directly on spore, 44
   factors affecting, 42
      abscisic acid, 49
      adenine, 47
      adenosine 3':5'-cyclic monophosphate, 47
      amino acids, 47, 48
      auxins, 43
      chelates, 48
      chemical factors, 42
      critical length of filaments, 41
      critical size of protonema, 41, 44
      cytokinins, 43-46
      gibberellins, 46, 47
      kinetin-auxin interaction, 44
      light, 42, 143
      minerals, 48
      organisms, 50
      pH, 49, 50
      physical factors, 42
      protonemal age, 41
      sugars, 49
      temperature, 43
      vitamins, 48, 49
   Funaria type, 41
   induction in dark, 42
   moruloid buds, 44
   on chloronemal cells, 41
   parameters in the study of, 42
   photoreceptors involved in, 134
   Polytrichum type, 41
   protein content during bud formation, 45
   RNA content during bud formation, 45
   target cells, 45

Cadinanes, 195, 196, 201
Cadmium, as pollutant, 252
Callose, 153, 159, 165, 289, 293, 302
Callus,
   differentiation in, 111-114, 123, 140
   factors for induction of, 119, 122, 123
   formation from different parts, 120, 121

Calyptra, role in sporogon development, 116-118
Capillary channels (systems), 285-287, 309
    epiorgan, 309
    interorgan, 309
    intraorgan, 309
Carbohydrates, 225
    carbohydrate-nitrogen ratio, 99
    effect on gametangial induction, 99
    listed, 228, 229
    taxonomic value of, 230
Carotenoids, 225
    listed, 226, 227
Caryophyllanes, 196, 197, 200
Casein hydrolysate, effect on,
    apogamy, 113
    callus formation, 123
    protonemal growth, 28
Cauloids, 74, 75
Caulonema,
    characteristics of, 39
    development from chloronema, 39
Caulonema specific proteins, 46
Cesium, as pollutant, 261, 262
Chalcones, 219
Chelates (chelating agents), effect on
    bud formation, 48
    gametangial induction, 102, 103
    gemma germination, 35
    endogenous levels of iron and copper, 103
Chemical constituents, 175
    amino acids, 233, 237
    antibiotics, 175-178, 240
    carbohydrates, 225, 228-230
    carotenoids, 225-227
    dihydrostilbenes, 232, 233
    enzymes, 233-236
    flavonoids, 211
    growth substances, 178
        non-specified, 183, 184
        specified, 178-183
    inorganic compounds, 233, 239
    lignins, 221-224
    lipids, 185
    miscellaneous, 239
    organic acids, 230, 231
    quinones, 233, 238
    terpenoids, 191-210
Chloral hydrate, effect on
    apogamy, 112
    spore germination, 26
Chlorocysts, 154
Chloronema
    buds on, 41

characteristics of, 39
Chromium, as pollutant, 254
Coconut milk, 113, 122
Colchicine, 26
Composite gemmae, 28
Conducting parenchyma, 289, 293, 302
Conducting strands
    evolutionary trends in, 305, 306
    in liverworts, 289-293, 304.
    in mosses, 293-295, 304, 305
    reduced, 294, 309
    well developed, 289, 292-296, 304, 308, 309
Conduction
    external
        adaptations in gametophyte, 285, 286
        in sporophyte, 286
        methods to determine the rate of, 288
        rate of, 288
        significance of, 286, 288
    internal
        cells involved in hornworts, 289, 290
        cells involved in liverworts, 289-293
        cells involved in mosses, 293-295
        conduction of water, 304, 305
        conduction of organic compounds, 305
        rate of, 304, 305
Contact zone see sporophyte-**gametophyte** junction
Contaminants in cultures, effect **on**
    bud formation, 50
    protonemal growth, 50
    spore germination 26
Copper mosses, 256
Correlative factors, 71
Correlative inhibition (systems), **73, 75, 86**
Critical factor (governing factor), 94
Critical size of protonema, 41, 44, 184
Cryoprotectants, 280
Culture techniques, use in studies on
    pollutants, 252-254, 257, 259, 264
    protoplasts, 275-277
Cuprophilic bryophytes, 256
Cushions, 315
Cuticular components, 191
Cyclic **AMP** see adenosine 3':5'-cyclic monophosphate
Cyclic nucleotides, 104
Cytokinins
    bioassay for, 45
    effect on formation of
        buds, 43-46
        gametangia, 102
        gemmae, 28, 29

endogenous, 179
interaction with IAA, 44, 102
Cytoplasmic polarity, 166

Damage to plants due to
   desiccation, 313
   heavy metals, 250-255
Dark induction of buds, 42
Day-neutral plants, 93, 94
De-differentiation
   in bud primordia, 43, 45
   in caulonema, 40, 43, 61
   in gametophytes 61, 66
   in wall labyrinth of transfer cells, 168
   visible signs at the cellular level, 83
Desiccation
   effect on regeneration, 70
   mechanism of damage, 313, 314
   sensitive species, 313
   tolerant species, 312, 313
Deuter cells, 153, 154, 297
Dew, as a source of water supply, 310
Dictyotene stage, 164
Differentiation
   controls in, 124
   in callus, 111, 112, 123,
   in cultured cells, 140
   of gametophyte, 116
   of protonema, 39
   of sporophyte, 116
Dihydroflavonoids, 218, 219
Dihydrostilbenes, 232, 233
   allergenic properties of, 233
   lunularic acid, 232
   lunularin, 232
Diterpenoids, 204
   classified, 205, 206
Dormancy, 137
Drimanes, 193, 194, 200, 201
Drop culture, 275, 276
Drought *see* desiccation
Dwarf-males, 104, 105
Dyes, use in the study of conduction, 308

Ectohydric mosses, 251, 253, 308-310
Egg
   free suspension of, 158
   isolation of, 158, 159
   nuclear evaginations in, 159
Electrical vector, 144
Elemanes, 197, 200
Embryonization, 55, 77
Endogenous rhythms, 135
Endohydric mosses, 251, 308, 309

Endopolyploidy, in conducting cells, 297, 302
Enzymatic method for protoplast isolation, 269, 270
Enzymes, 233
   for protoplast isolation, 273, 274
   listed, 234-236
   in the interphase cells, 170, 171
Epigenetic changes, 115
Ethylene, 183
Evaporative flux, 289
Exposure factor, 249, 261
External conduction *see* conduction

Factor F and H, 40, 183
Far-red light, effect on
   growth, 136-139
   protonemal growth, 138, 139
   regeneration, 68, 69
   rhizoid formation, 30, 31, 140, 141
   spore germination, 23, 134-136
   vegetative propagation, 140, 141
Fatty acids, 186-191
   change with shoot age, 190
   listed, 188
   influence of temperature and light on, 190
   role in building up membranes, 190, 191
Feather forms, 315
Flavonoids, 211
   acylated, 220
   anthocyanins, 220
   apigenin C-glycosides, 213, 214
   apigenin O-glycosides, 211-213
   aurones, 219
   biflavonyls, 218
   chalcones, 219
   dihydroflavonoids, 218, 219
   flavones, 211
   flavonols, 217, 218
   6-and 8-hydroxylated flavon derivatives, 216, 217
   isoflavones, 217
   luteolin C-glycosides, 214
   luteolin O-glycosides, 214, 215
   proanthocyanidins, 220
   sphagnorubins, 221
   tricetin derivatives, 215, 216
Flavones, 211
Flavonols, 217, 218
Flavoprotein, 144
Fluorescent dyes, 288, 304, 305
Fluorides, as pollutant, 260
Food conducting cells *see* leptoids

Funaria type bud formation, 41
Fusion bodies, 280
Fusogenic agents, 280

Gametangial induction
  factors affecting
    animal sex hormones, 104
    carbohydrates, 99
    chelating agents, 102, 103
    cyclic nucleotides, 104
    growth regulators, 101, 102
    humidity, 98
    hydration of the medium, 98
    light duration, 92-94
    light level, 95
    light quality, 95
    mineral salts, 104
    nitrogenous substances, 99-101
    pH, 103, 104
    temperature, 95-97
    temperature-photoperiod interaction, 97, 98
  day-neutral plants, 94
  long-day plants, 92-94
  short-day plants, 92-94
Gametophytic factor, 116
Gaseous pollutants
  fluorides, 260
  ozone, 260, 261
  sulphur dioxide, 257-260
Gemma/gemma cup production, as influenced by
  amino acids, 34
  humidity, 32
  light, 30
  liquid medium, 32, 34
  nitrate, 34
  temperature, 32
Gemmae
  composite, 28
  liverwort gemmae, effect of
    chelating agent, 35
    gravitation, 30, 31
    growth regulators, 32-34
    humidity, 32
    pH, 35
    light, 30
    nitrogenous substances, 34
    physical treatments, 32
    sucrose, 35
    temperature, 32
  experiments on moss gemmae, 26, 28, 29, 31-34
  in situ germination of, 32, 33, 35, 81
  nature of, 26, 28
  nematogones, 32
  protonemal, 26-29
  secondary, 28
Gemma factor, 29, 184
Gene block hypothesis, 124
Germacranes, 197, 200
Germ rhizoid formation, 21, 22
Gibberellins
  effect on
    bud formation, 46, 47
    gametangial formation, 101, 102
    gemma germination, 34
    growth, 142
    spore germination, 25
  endogenous, 142, 179
Green light
  effect on
    bud formation, 42, 143
    capsule growth, 140
    growth, 138
    protonemal growth, 41, 138
    regeneration, 69
    spore germination, 134, 135
    vegetative propagation, 141
Growth
  forms, 314-316
  orthotropic, 137
  periods, 311
  photoreceptors involved in, 132, 133
  rate, 310, 311
Growth substances (growth hormones/regulators)
  effect on
    apogamy, 112
    callus induction, 122, 123
    gametangial induction, 101, 102
    gemmae, 32-34
    spore germination, 25
  non-specified
    factor F and H, 183
    gemma factor, 184
    sporogon factor, 184
  Specified
    acetylcholine, 182, 183
    auxins, 178, 179
    cyclic AMP, 181, 182
    cytokinins, 179
    ethylene, 183
    gibberellins, 179
    lunularic acid, 180, 181

Hair points, 310
Hanging bryophytes, 315, 316
Hanging drop culture, 276
Haustorium, 167-169

Heavy metals
  as pollutants, 250
    arsenates, 254
    cadmium, 252
    chromium, 254
    lead, 251, 252
    mercury, 253, 254
    nickel, 254, 255
    vanadium, 255
    zinc, 253
  damage due to, 251-255
  getting rid of excess, 251
  metal tolerance, 255, 256
  stability pattern of, 255
  stimulation of growth by, 252, 254
Heterotrichous habit, 16, 40, 138
High-energy reactions, 131, 135, 136
Histoenzymological studies, 170, 171
Homologous theory, 119
Horizontal axis, anatomy of, 295, 296
Humidity, effect on
  gametangial induction, 98
  gemmae, 32
  regeneration, 71
Humulanes, 196, 197, 200
Hydration of medium, effect on
  apogamy, 111
  gametangial induction, 98
Hybrid sporophyte, 74, 108, 113, 140, 142, 179, 182, 184
Hydrogen-ion concentration (pH), effect on
  bud formation, 49, 50
  gametangial induction, 103, 104
  gemmae, 35
  protoplast isolation, 275
  regeneration, 70
  spore germination, 25
Hydroids, 152, 169, 170, 293-297, 299
  comparison with tracheary elements, 300
  development of, 297, 299
  structure at maturity, 300-302
Hydrolysis of cell walls, 153, 289, 299
Hydrom, 293, 297
Hydrom sheath, 300, 302
6- and 8-Hydroxylated flavone derivatives, 216, 217

Incubous, 288
Index of atmospheric purity (IAP), 249, 260
Index of radiosensitivity, 264
Indole-3-acetic acid (IAA), effect on
  bud induction, 43
  gemma germination, 32-34
  protonemal differentiation, 40

regeneration, 79-81
spore germination, 25
Inductive effects of light, 42, 68, 69, 131, 132
Infra-red radiation, 143
Inhibitors, use in studies on regeneration, 82, 87
Inorganic compounds (minerals)
  endogenous, 233, 239
  effect on
    apogamy, 112, 113
    bud formation, 48
    gametangial induction, 100, 101, 104
    spore germination, 24
Intercalary meristem, 118
Intercalation theory, 119, 125
Internal conduction see conduction
Interphase see sporophyte-gametophyte junction
Interpolation theory, 119, 125
Isoflavones, 217

Kinetin (6-furfurylaminopurine)
  deficient mutant, 44
  effect on gemmae, 34
    inhibition by blue light, 42
    regeneration, 80
    spore germination, 25

Lead
  as pollutant, 251, 252
  tolerance as a factor in natural selection, 252
Light duration see photoperiod
Leaf traces, 297, 298
Leptoids, 153, 169, 170, 293-297, 302, 305
  comparison with sieve elements, 302
  development of, 302
  structure at maturity, 302
Leptom, 293, 297
Leucocysts, 154
Life-forms (Growth-forms), 249, 314-316
Light, effect on
  apogamy, 111
  bud formation, 42, 143
  chloroplast replication, 136
  development, 145
  differentiation in callus, 140
  gametangial induction, 92-95
  gemmae, 30, 31, 92
  growth, 136-140
  metabolism, 141, 142
  polarotropism, 132, 134, 144, 145
  protonemal differentiation, 40
  protoplast isolation, 274
  protoplast regeneration, 141

Subject Index 325

regeneration, 67-69, 141
senescence, 142, 143
spore germination, 21-23, 134-136
tropic responses, 144, 145
vegetative propagation, 140, 141
Light level effects, 21, 30, 42, 67, 68, 95, 111, 113, 123, 140, 274
Light quality effects, 22, 23, 30, 31, 42, 67-69, 95, 111, 114, 134-136, 140-145
Lignins, 221-224
Limisphere, 156
Lipids, 185
    alkanes, 185
        listed, 185, 186
    cuticular components, 191
    fatty acids, 186
        change with shoot age, 190
        listed, 188
        influence of temperature & light on, 190, 191
        role in building up membranes, 190
    neutral lipids, 185, 187
    polar lipids, 185
    storage lipids, 190
Liquid medium, 34, 98, 123, 275
Long-day plants, 92-94
Low-energy reaction system, 131
Low temperature requirement, for
    gametangial production, 95, 96
    spore germination, 23
Luminous moss, 14
Lunularic acid
    chemical nature of, 180, 181, 232
    ecological significance of, 142, 181
    levels in different species, 181
    occurrence of, 180
    production of, 141, 142, 180
Lunularin, 178, 232
Luteolin C-glycosides, 214
Luteolin O-glycosides, 214
    listed, 215
Lysosomal activity, 297

Maalianes, 197, 198, 203
Mats, 315, 316
Mechanical method for protoplast isolation, 269
Meiosis in mosses
    postmeiotic changes, 165
    unique features of, 163, 164
Mercury, as pollutant, 253, 254
Metal tolerance, 255, 256
Microanalytic technique, 84
Microchamber culture, 276
Mircodroplet culture, 276

Midrib, anatomy of, 297
Mineral nutrients, 104, 124
Monocarbocyclic sesquiterpenoids
    bisabolane-type, 200
    elemene-type, 200
    germacrane-type, 200
    humulane-type, 200
Monoplastidy, 162
    mechanisms of, 162
Monoterpenoids, 192, 193
Moruloid buds, 28, 44
Moss bag technique, 257
Mucilage hairs, 285
Mucilage plug, 158
Multiple drop array (MDA) technique, 276
Multilayered structure (mls), 156
Mutants
    amino acid analogue resistant, 282, 283
    auxotrophic, 282, 283
    auxin-resistant, 40
    auxin-cytokinin resistant, 282, 283
    induction of, 40, 41, 43, 44, 282
    kinetin deficient, 44
    morphological, 282
    nutritionally deficient, 282
    X-ray induced, 41, 43
Myxohydric mosses, 251, 253, 310

Nematogones, 32
Neurohormone see acetylcholine
Neutral lipids, 185, 187
Nickel, as pollutant, 254, 255
Nitrogenous substances, effect on
    gametangial induction, 99-101
    gemmae, 34
2(N-morpholino) ethane sulfonic acid (MES) buffer, 274
Nutritionally deficient mutants see auxotrophic mutants
Nutrients, as a factor in
    gametangial induction, 99, 104
    regeneration, 78, 79
Nutritive cells, 163

Oogenesis, 157-160
Open chain diterpenoids, 205
Orange light, effect on growth, 138
Organic acids, 230
    listed, 231
Orthotropic growth (orthogeotropic growth) 137, 142, 179
Osmolality, 270, 275
Osmoticum, 269-271
Oxidative polymerization, 222
Ozone, as pollutant, 260, 261

Paraphyllia, 286
Peat adsorbers, 263
Peat mosses, 257
Perine, 148, 161
  origin of, 166, 167
Peristrophe, 83
pH see hydrogen-ion concentration
Photochemical systems, 131
Photomorphogenesis, defined, 131
  involvement in
    bud formation, 143
    growth, 136-140
    metabolism, 141, 142
    senescence, 142, 143
    spore germination, 134-136
    tropic responses, 144, 145
    vegetative propagation, 140, 141
Photoperiod effects, 22, 30, 69, 92-94, 135, 137
Photoreactive systems, 136
Photoreceptors, involved in
  bud induction, 134
  growth, 132, 133
  metabolism, 133
  polarotropism, 134
  senescence, 133
  spore germination, 132
  vegetative propagation, 133
Phototropism, 144, 145
Physiological isolation, 72, 76
Physiological tolerance, 312
Phytochrome, 131-134
  mediated responses
    bud induction, 143
    cell permeability, 142
    chloroplast replication, 68
    germination of propagula, 68, 69, 140
    growth, 136-138
    phototropism, 144, 145
    protonemal ramification, 69, 138, 139
    regeneration, 141
    senescence, 143
    spore germination, 23, 135, 136
    synthesis of hormones, 142, 183
    synthesis of proteins, 87, 142
    synthesis of RNA, 87, 142
Pin cushion moss, 285
Pinguisanes, 198, 199, 202
Placental region see sporophyte-gametophyte junction
Plastids
  in foot, 168
  in sporogenous cells, 162
  in spores, 163, 165
  in regenerating cells, 83-85

  number during sporogenesis, 161, 162
  regression in, 302
Plant indicators, 249-251, 253, 256-258, 261, 262
  reliability of, 263, 264
Plasmodesmata, 156, 157, 160, 169, 170, 289, 293, 297, 302, 304
Plasmotic resistance to SO_2, 259
Poikilohydric, 312, 314
Polar lipids, 185
Polarity
  apico-basal, 82
  dorsiventral, 82
  effect on
    callus induction, 119
    regeneration, 77, 82
Polarotropism, 144, 145
  photoreceptors involved in, 134
Pollutants
  culture technique for study of, 252-255, 257, 259, 264
  defined, 249
  effects of, 250-264
  exposure factor, 249, 261
  factors influencing degree of damage, 249, 258, 263
  herbarium specimens as indices, 250-253
  index of atmospheric purity, 249, 260
  primary, 249
  secondary, 249, 260
Polyploidization
  factor for apogamy, 115
  in conducting cells, 297, 302
Polyploid races, 118
Polytrichum type bud formation, 41
Pre-treatment for spore germination
  high temperature, 23
  low temperature/cold treatment, 23
Primary pit fields, 289
Primary pollutants, 249
Primary protonema, 39
Proanthocyanidins, 220
Progesterone, effect of, 104
Protonema
  caulonema, 39
  chemical control of differentiation, 40
  chloronema, 39
  factor F, 40
  factor H, 40
  growth, 138-140
  heterotrichous habit, 16, 40
  mutants of
    auxin resistant, 40
    kinetin deficient, 44

X-ray induced, 41, 43
primary, 39
secondary, 39
thallose/thalloid, 14
ultrastructure of, 150-152
Protonemal bryophytes, 315
Protonemal differentiation, 39, 40
Protonemal gemmae, 26-29
Protoplast
    defined, 268
    applications of culture, 268, 269
    techniques of culture
        agar plating, 276, 277
        liquid culture, 275, 276
    factors affecting isolation of
        agar, 274
        age of material, 275
        calcium, 274
        enzymes, 273, 274
        light, 274
        MES buffer, 274
        osmolality, 275
        pH, 275
        pretreatment of the tissue, 271, 273
        rinsing of the tissue, 271
        wetting agents, 274, 275
    factors affecting regeneration of, 279, 280
    fusion and somatic hybrids, 280
    induction and isolation of mutants, 282, 283
    isolation of, 269-271
    regeneration of, 141, 277-280
    source material for, 271, 272

Quinones, 233
    listed, 238

Radiations
    callus induction by, 122
    commonly investigated, 263
    effect on regeneration, 70, 263, 264
    factors influencing the degree of effect
        external, 263
        internal, 263
        mode of action of, 70, 263, 264
        tolerance to, 263, 264
Radionucleides
    cesium, 261, 262
    strontium, 262
    uranium, 262, 263
Reactivation, 54, 77, 79, 83
Red/far-red system see phytochrome
Red light, effect on
    bud formation, 42, 143

callus morphogenesis, 140
capsule expansion, 139, 140
growth, 136-140
    nuclear volume, 139
    protein synthesis, 139, 142
    protonemal growth, 42, 138, 139
    regeneration, 67-69
    rhizoid formation, 30, 140, 141
    RNA synthesis, 139, 142
    spore germination, 23, 134-136
    vegetative propagation, 140, 141
Refractive spherules, 302
Regeneration
    changes occurring in regenerating cells
        budding of chloroplasts, 84
        budding of mitochondria, 84, 86
        DNA synthesis, 84, 87
        peristrophe, 83
        protein synthesis, 84, 87
        RNA synthesis, 84, 86, 87
        systrophe, 84, 85
    defined, 54
    factors affecting
        age, 74
        antiauxin, 82
        apical dominance, 77-83
        correlative inhibition, 75, 76
        humidity, 71
        IAA, 79-83
        isolation, 58, 71
        kinetin, 80
        light, 67-69
        location in the plant, 73, 74
        nutritional disruption, 78
        pH, 70
        polarity, 60, 77, 82
        radiation, 70
        reserve food, 73
        season, 71
        size of fragment, 72, 73
        temperature, 72
        time, 58
        wounding, 59, 71, 72
    importance of the study of, 54
    morphology of regenerants, 61-66
    of protoplasts, 277-280
    potentialities of various organs for, 55, 56
        leaves, 56-59, 62-64
        setae, 59-61, 118
Regressed plastids, 302
Rehydration, sequence of events on, 311-313
Reparation, 55

Reproductive biology, 92
Restitution, 54
Rhizoid development
  from spore, 21, 22
  effect of
    auxins, 32, 34
    chelating agents, 35
    gravitation, 31
    light, 30, 31, 140, 141
    photoperiod, 32
    phytochrome, 140, 141
    temperature, 32
Rhizoidal grooves, 286
Rhizome, anatomy of, 295, 296
Rhizonemata, 63
Rhythms,
  autonomic, 311
  circadian, 135
  diurnal and annual in resistance to pollution, 264

Secondary gemmae, 28
Secondary metabolites, 107
Secondary pollutants, 249, 260
Secondary protonema, 39, 56, 58, 61
Self-inhibition, 180
Senescence
  role of light in, 142, 143
  photoreceptors involved in, 133
Seta
  conducting strand of, 297
  regeneration from, 59-61, 118
Sesquiterpene lactones, 199, 200, 240
Sesquiterpenoids, 193
  classified, 200-204
Shoots, direct origin from
  gemma, 29
  leaf, 61, 64
  stem, 61
Short-day plants, 92-94
Sinks for pollutants, 250, 254, 255, 262
Social forms, 315
Solitary forms, 315
Somatic hybrids, 268, 280
Spermatogenesis, 155-157
Sphagnorubins, 221
Spore germination
  factors affecting, 21
    agar, 26
    chloral hydrate, 26
    colchicine, 26
    growth regulators, 25
    H_2O_2, 26
    hydrogen-ion concentration, 25, 26
    light, 21, 22, 134-136
    minerals, 24, 25
    mold fungi, 26
    sugars, 23, 24
    temperature, 23
  mechanism of, 26
  photoreceptors involved in, 132
  stages in, 135
  types in
    Andreaeales, 14
    Anthocerotales, 12
    Bryales, 14
    Jungermanniales acrogynous, 1
    Jungermanniales anacrogynous, 6
    Marchantiales, 10
    Sphaerocarpales, 12
    Sphagnales, 13
    Tetraphidales, 14
Sporeling types in mosses, listed, 18
Spore production, in apogamous sporophytes, 109, 110
Spore sac
  characteristics of, 160
  role in spore development, 161
Spore wall
  layers in, 148-150
  mode of formation, 165-167
  ultrastructure of, 148-150
Sporogenesis, 160-167
Sporogenous cells, 160, 161
Sporogon factor
  nature of, 113, 114
  production of, 113, 114, 184
Sporophyte-gametophyte junction
  histoenzymological studies, 170, 171
  ultrastructure of, 167-169, 302, 304
Stability pattern of metal ions, 255
Steroids (sterols), 206
  listed, 207-210
Storage lipids, 190
Stereids, 153, 154, 294, 296, 297
Stereome, 294
Strontium, as pollutant, 262
Subapical meristem, 117
Succubous, 288
Sugars, effect on
  apogamy, 111, 112
  bud formation, 49
  callus induction, 119, 121
  gemma germination, 35
  spore germination, 23
Sulphur dioxide, as pollutant, 257-260
Sulphur mosses, 256
Symplastic conduction, 304
Synergism of IAA-kinetin, effect on bud initiation, 43, 44

gametangial induction, 102
Systrophe, 84, 85

Tapetum see spore sac
Target cells, 45
Temperature, effect on
   bud formation, 43
   gametangial induction, 95-97
   gemmae, 32
   growth, 311
   protonemal differentiation, 39, 40
   regeneration, 72
   spore germination, 23
Temperature-photoperiod interaction, effect on
      bud formation, 43
      gametangial induction, 97, 98
Tetracarbocyclic diterpenoids, 205
   kaurane-type, 205
   neoverrucosane-type, 206
   verrucosane-type, 206
Terpenoids, 191, 192
   monoterpenoids, 192, 193
   diterpenoids, 204-206
   sesquiterpenoids, 193
      aromadendranes, 197, 198
      bisabolanes, 194
      cadinanes, 195
      caryophyllanes, 196
      drimanes, 193
      elemanes, 197
      germacranes, 197
      humulanes, 196
      maalianes, 197, 198
      pinguisanes, 198, 199
      sesquiterpene lactones, 199, 200
   triterpenoids, 206
      sterols, 206-210
Testosterone, effect of, 104
Tetracarbocyclic sesquiterpenoids, 204
Total resistance to SO_2, 259
Totipotency, 54, 78, 82, 123
Toxiphilous species, 258
Tracers, use in the study of
   auxin transport, 178
   bud formation, 45
   conduction, 305, 308
   metabolism, 179
   regeneration, 83, 84, 86, 87
   spore germination, 136
Transfer cells, 167-170, 302, 304
Transpiration stream, 308
Tricarbocyclic diterpenoids, 205

Tricarbocyclic sesquiterpenoids, 203
   6(5-1)-abeo-aromadendrane-type, 204
   aromadendrane-type, 203
   cedrane-type, 203
   cycloguaiane-type, 204
   gymnomitrane-type, 203
   isolongifolane-type, 203
   longibornane-type, 203
   longicyclane-type, 203
   longifolane-type, 203
   longipinane-type, 203
   maaliane-type, 203
   sativane-type, 203
   ylangane-type, 204
Tricetin derivatives, 215, 216
Triterpenoids, 206
Tropic responses,
   in liverworts, 144
   in mosses, 144, 145
Turfs, 315, 316

Ultrastructure of
   antheridium, 155
   hydroids, 153, 297, 299, 300
   leaf, 153, 154
   leptoids, 152, 153, 302
   oogenesis, 157-160
   protonema, 150-152
   seta, 169, 170
   spermatogenesis, 155-157
   spore, 148-150
   spore sac, 161
   sporogenesis, 160-167
   sporophyte-gametophyte junction, 302, 304
   stem, 152, 153
Uranium, as pollutant, 262, 263

Vaginula, 167-171
Vanadium, as pollutant, 255
Vegetative propagation
   photoreceptors involved in, 133
   role of light in, 140, 141

Water conducting cells see hydroids
Water holding capacity of *Sphagnum*, 310, 311
Water potential, 312
Water sacs, role of, 285
Wetting agents, 274, 275
Wounding, effect on regeneration, 71, 72
Wound reaction phase, 86

X-ray induced mutant, 41, 43

Yeast extract, effect on
  apogamy, 113
  protonemal growth, 28

Yellow light, effect on
  growth, 138
  spore germination, 22

Zinc, as pollutant, 253

Plant Index

(Numbers in bold face indicate plates)

Abietinella, 187, 207, 234
Abietinella abietina, 186
Acroporium, 315
Adelanthus, 208
Aerobacter, 50
Aerobolbus, 228
Aerobryopsis longissima, 254
Agrobacterium, 50
Agrobacterium tumefaciens, 50
Alternaria brassicola, 178
Amblystegium, 59, 118
Amblystegium juratzkanum, 110
Amblystegium riparium, 110
Amblystegium serpens, 27
Anacamptodon latidens, 16
Anastrepta, 208
Anastrepta orcadensis, 198, 205
Andreaea, 16, 38, 39, 207
Andreaea fauriei, 14, 16
Andreaea petrophila, 14, 16
Aneura, 208, 226
Aneura latifrons, 119
Aneura palmata, 119
Aneura pinguis, 198
Anisothecium molliculum, 26, 29, 41, 44, 46, 47, 50
Anisothecium spirale, 26, 46
Anoectangium, 44, 57, 70, 274
Anoectangium stracheyanum, 43, 44, 46, 48
Anoectangium sublaeteviren, 57
Anoectangium thomsonii, 22, 43, 44, 46-48, **2.5**, 120, 132, 134, 139, 143, 270, 272, 274, 275, 277
Anomodon, 187, 207
Anomodon rostratus, 175, 261
Anoplolejeunea, 208
Anthelia, 208
Anthelia julacea, 204
Anthelia juratzkana, 204
Anthoceros, 12, 13, 119, 155, 175, 208, 230, 304
Anthoceros erectus, 177
Anthoceros fusiformis, 66

Anthoceros husnoti, 92, 93, 97
Anthoceros laevis, 92, 93, 97, 155, 171
Anthoceros punctatus, 92, 93, 97
Aphanorrhegma serratum, 19
Apometzgeria, 228
Apometzgeria pubescens, 216
Archidium, 166
Archilejeunea, 208
Aspergillus flavus, 26
Asterella, 24, 38, 187, 188, 208, 226, 228, 235
Asterella angusta, 22-24, 101, 120, 121, 123, **5.11**, 177
Asterella lindenbergiana, 191
Asterella odora, 25
Asterella sanguinea, 176, 177
Asterella tenella, 97, **4.2**
Asterella wallichiana, 120, 122, 123
Athalamia hyalina, 149, **7.1, 7.2**
Athalamia pusilla, 60, 119, 120, 122
Atrichum, 58, 59, 61, 71, 73-75, 80, 123, 153, 166, 176, 234
Atrichum pallidum, 59, 60, 63
Atrichum undulatum, 18, 57-59, 61, 62, 71, 74, 120-122, 162, 183, 250, 256, 300
Aulacomnium, 69, 226, 234
Aulacomnium androgynum, 27
Aulacomnium heterostrichum, 162
Aulacomnium palustre, 56, 133, 141
Aulacopilum, 57, 58, 61
Aulacopilum japonicum, 63, 64
Aulacopilum piliferum, 19
Avena, 144

Bacillus polymyxa, 177
Bacillus subtilis, 176
Balantoipsis, 228
Barbilophozia, 195, 208, 228
Barbilophozia barbata, 205
Barbilophozia floerkei, 205
Barbilophozia lycopodioides, 205
Barbula, 27, 29, 56, 58, 61, 94, 176
Barbula arcuata, 27

Barbula glauca, 27
Barbula gregaria, 27, 29, 41, 46-50, 71, 93-99, 101-104
Barbula indica, 56, 58
Barbula revoluta, 27
Barbula rigidula, 27
Barbula trifaria, 27
Bartramia, 24, 187
Bartramia pomiformis, 23, 162, 309
Bartramidula, 94
Bartramidula bartramioides, 24, 41, 46, 48, 49, 52, 93-99, 101, 103, 104, 106
Bazzania, 208, 228, 231
Bazzania pompeana, 185, 193-195, 239
Bazzania trilobata, 193, 195
Blasia, 76, 81, 119, 208, 228, 231
Blasia pusilla, 72, 75, 77, 81, 119, 120
Blyttia, 78
Botrytis cinerea, 178
Brachiolejeunea, 208
Brachythecium, 57, 207, 234
Brachythecium albicans, 63
Brachythecium campestre, 110
Brachythecium frizidium, 164
Brachythecium procumbens, 176, 177
Brachythecium rivulare, 186
Brachythecium rutabulum, 18, 262, 264, 282
Breutelia, 207
Brothera, 58
Brothera leana, 58, 71, 72
Bruchia, 166
Bryopteris, 208, 315
Bryum, 28, 29, 46, 58, 61, 94, 110, 118, 157, 166, 222, 226, 294, 308
Bryum argenteum, 22, 41, 93-96, 99, 101-105, 4, 1, 182, 183, 258-260
Bryum atrovirens, 43, 44, 46-48
Bryum capillare, 56, 62, 217, 309
Bryum cellulare, 55, 58, 63
Bryum coronatum, 22, 24, 27-29, 41, 46-50, 94-99, 101-104, **4.1**, 120, 122, 123, **5.7**
Bryum cryophilum, 211, 220
Bryum klinggraeffii, 27-29, **1.1**, 41, 44, 47, 49, 50, 133, 141, 142, 184
Bryum micro-erythrocarpum, 27
Bryum pallescens, 44, 46, 48
Bryum radiculosum, 27
Bryum riparium, 27
Bryum rutilans, 220
Bryum rubens, 27
Bryum sauteri, 27
Bryum tenuisetum, 27
Bryum ventricosum, 225
Bryum violaceum, 27
Bryum weigelii, 216, 220

Bucegia romanica, 217
Buxbaumia, 167, 295
Buxbaumia aphylla, 19, 27, 151, **262**

Callicosta bipinnata, 315
Callicostella, 315
Calliergon, 58, 71
Calobryum, 1, 12
Calobryum rotundifolium, 2
Calopodium crispifolium, 164
Calycularia, 208
Calymperes sect. *Macrhimanta*, 315
Calypogeia, 3, 208, 228, 241
Calypogeia granulata, 107, 120-122
Calypogeia integristipula, 185
Calypogeia meylanii, 186
Calypogeia tosana, 2
Camptothecium, 228, 230, 311
Camptothecium lutescens, 312
Camptothecium sericeum, 311
Campylium, 187, 234
Campylopus, 57, 58, 61, 207
Campylopus atrovirens, 57
Campylopus introflexus, 186
Campylopus laetus, 177
Campylopus richardii, 44
Candida albicans, 175, 177
Carrpos, 66, 218
Carrpos sphaerocarpos, 218
Catharinea, 59, 164, 165, 234
Catoscopium, 234
Cavicularia, 208, 231
Cavicularia densa, 11
Cephalozia, 1, 228
Cephalozia hamatiloba, 2
Cephalozia media, 93, 100
Cephaloziella, 120, 123
Cephaloziella phyllacantha, 239, 256
Cephaloziella massalongi, 239, 256
Ceratodon, 22, 24, 58, 59, 72, 136, 166, 187, 234
Ceratodon purpureus, 42-44, 47, 58, 72, 82, 118, 132, 134, 136, 139, 142, 144, 145, 149-151, 214, 250, 252, 258-260
Ceratolejeunea, 1
Chamberlainia, 226
Cheilolejeunea imbricata, 7
Chiloscyphus, 208, 228, 235, 240
Chiloscyphus polyanthus, 196
Cinclidium artichum, 166
Cinclidium stygium, 166
Cinclidium subrotundom, 166
Cladonia subtenuis, 261
Clasmatocolea, 208
Climacium, 57, 187, 207, **294**
Climacium americanum, 120

Colura, 314
Conocephalum, 12, 66, 78, 175, 187, 188, 191, 208, 212, 214, 215, 227, 228, 231, 235, 238, 240, 288, 289
Conocephalum conicum, 92, 93, 95, 97, 101, 175-177, 181, 185, 191-193, 196, 200, 214, 215, 232
Conocephalum supradecompositum, 193, 197, 219
Corsinia, 66, 78
Corsinia coriandrina, 217, 218
Cratoneuron, 187, 234, 313
Cratoneuron filicinum, 313, 314
Cryphaea, 309
Cryptochila, 208
Cryptothallus mirabilis, 21, 23, 25, 93-96
Ctenidium, 207
Ctenidium molluscum, 186, 258, 260

Dalbergia sissoo, 279
Dawsonia, 167, 222, 226, 234, 294, 304, 309
Dawsonia grandis, 222
Dawsonia longiseta, 58, 222
Dawsonia papuana, 170, 222
Dawsonia polytrichoides, 57-59, 222
Dawsonia superba, 57, 58, 151, 169, 222
Dendroceros, 12
Dendroceros crispus, 290
Dendroceros javanicus, 12
Dendroligotrichum, 153, 167, 222, 304
Dendroligotrichum dendroides, 153, **7.5, 7.6**, 222, 307
Desmatodon, 109
Desmatodon bogosicus, 27
Desmatodon randii, 108-110, 115, 125, 126, 165
Desmatodon ucrainicus, 108-111
Dicranella, 47
Dicranella coarctata, 22, 24, 43, 44, 47-49
Dicranella heteromalla, 18, 183, 253
Dicranella jamesonii, 27
Dicranella schreberi, 121
Dicranella staphylina, 27
Dicranella varia, 27, 251, 253
Dicranolejeunea, 208
Dicranoweisia cirrata, 258
Dicranoweisia crispula, 25
Dicranum, 59, 78, 166, 187, 207, 212, 234, 295
Dicranum bergeri, 222
Dicranum flagellare, 58
Dicranum elongatum, 190
Dicranum scoparium, 25, 59, 132, 136, 170, 219, 261
Dicranum undulatum, 25, 58
Dicranum viride, 57
Didymodon michiganensis, 27

Didymodon recurvus, 27, 29, 133, 134, 139, 143
Diphyscium, 61, 167, 294, 315
Diphyscium fulvifolium, 19
Diplophyllum, 187, 188, 228, 237, 240
Diplophyllum albicans, 92, 93, 177
Distichum, 187, 234
Ditrichum, 226, 234
Ditrichum cornubicum, 27
Ditrichum cylindricum, 27
Ditrichum pallidum, 162, 163, 165, 166, **7.15, 7.16**
Ditrichum pusillum, 27
Ditrichum vaginas, 225
Drepanocladus, 58, 71, 187
Drepanolejeunea japonica, 7
Drepanophyllum, 315
Drummondia sinensis, 16, 21
Dryopteris, 144
Dryptodon stratus, 256
Dumortiera, 66, 228, 315
Dumortiera hirsuta, 175, 177, 197

Ectropothecium, 315
Encalypta, 166
Encalypta rhaptocarpa, 20
Encalypta streptocarpa, 287
Entodon, 22, 41, 50, 226
Entodon myurus, 22, 25, 43, 44, 46, 50, **2.5**
Entodon seductrix, 120, 225
Ephemeropsis, 315
Ephemerum, 166
Epipterygium, 315
Equisetum, 222
Eriopus remotifolius, 167
Eucladium verticillatum, 27
Eurhynchium, 166, 234
Eurhynchium hians, 261
Eurhynchium praelongum, 252, 253
Eurhynchium striatum, 162
Exormotheca, 228
Exormotheca tuberifera, 101

Fimbriaria, 288
Fimbriaria bleumeana, 287
Fissidens, 58, 59, 61, 78, 148, 167, 315
Fissidens garberi, 261
Fissidens limbatus, 160-162, 165-167
Fontinalis, 187, 226, 234, 294
Fontinalis antipyretica, 189, 225, 239
Fossombronia, 6, 55, 56, 75, 77, 78, 228
Fossombronia himalayensis, 80, 99, 101, 120, 122
Fossombronia japonica, 8
Fossombronia pusilla, 120, 121, 123

Frullania, 1, 3, 6, 199, 208, 229, 235, 240, 285, 314, 315
Frullania asagrayana, 26
Frullania brittoniae ssp. *truncatifolia*, 233
Frullania dilatata, 199
Frullania pedicellata, 232
Frullania sect. *Meteoriopsis*, 315
Frullania tamarisci, 192, 199
Frullania truncatifolia, 5
Funaria, 22, 24, 26, 42, 43, 45, 46, 49, 50, **2.2, 2.4,** 56, 58, 59, 61, 69, 75, 116-118, 122, 123, 135, 136, 138-140, 144, 151, 167, 183, 222, 234, 274, 279, 288, 294, 295, 297, 299, 305
Funaria hygrometrica, 17, 22, 23, 25, 27, 39-50, 52, **2.2,** 59, 63, 67, 71, 74, 82, 84, 94, 96, 108-118, 120-123, **5.1, 5.4, 5.6, 5.7, 5.9,** 132-134, 136, 139-143, 149, 150, 152, 157, 162, 165, 168, 173, **7.4,** 179, 181-184, 219, 233, 252, 253, 261, 269, 270, 272, 273, 277-279, **10.2,** 310
Fusarium scripi, 26

Gaffkya tetragena, 176
Geothallus, 12, 13, 149
Georgia pellucida, 108, 110, 111, 112, 114, 121, 219
Gigaspora gigantea, 50
Gigaspora margarita, 50
Glomus mosseae, 50
Glyphomitrium humillimum, 16, 21
Gongylanthus, 208
Grimmia, 58, 70, 167, 294
Grimmia doniana, 251
Grimmia orbicularis, 262
Grimmia pulvinata, 60, 110, 118, 260, 310
Grimmia tenera, 255
Gymnocolea, 209, 229
Gymnocolea acutiloba, 239, 256
Gymnocolea inflata, 185, 205, 256
Gymnomitrion, 209
Gymnomitrion obtusum, 195
Gymnostomum, 57, 70
Gyroweisia tenuis, 27

Haplocladium, 71
Haplocladium angustifolium, 60, 75, 118, **5.6**
Haplomitrium, 209, 212, 289, 292
Haplomitrium gibbsiae, 216, 220, 293
Hedwigia, 57, 187
Hedwigia ciliata, 16, 20
Herbertus, 209, 315
Herpetineuron, 57, 61
Herpetineuron toccoae, 62
Heterophyllum haldaneanum, 120

Heteroscyphus, 209, 229
Homaliodendron, 58, 61
Homalium guillaini, 254
Homalothecium, 228, 230, 234
Hookeria, 57, 207
Hookeria lucens, 59, 263, 312
Hydrangea macrophylla, 180
Hygrobiella, 209
Hygrohypnum, 187, 226
Hygrohypnum luridium, 189
Hylocomium, 58, 78, 167, 187, 212, 234, 311
Hylocomium splendens, 187, 251, 253-255, 312
Hymenophyton, 212, 213, 289
Hymenophyton flabellatum, 211, 217
Hymenophyton leptopodum, 217, 219
Hyophila, 57
Hyophila involuta, 24, 27, 29, 44, 48, 52, 106, 120, 122, **5.8,** 176
Hyophila tortula, 27, 28
Hypnodendron, 315
Hypnum, 148, 187, 207, 234, 251, 261
Hypnum circinale, 164
Hypnum cupressiforme, 118, 251-254, 259, 260
Hypnum cupressiforme ssp. *imponens*, 186
Hypnum curvifolium, 262
Hypnum rusciforme, 162, 165
Hypnum serpens, 118
Hypopterygium, 57, 315

Isopterygium, 315
Isotachis, 209
Isotachis japonica, 185, 239

Jackiella, 209
Jamesoniella, 209, 229, 230
Jubula, 209, 229, 285
Jubula javanica, 4
Jungermannia, 1, 175, 192, 209, 222
Jungermannia cordifolia, 192, 196
Jungermannia infusca, 204
Jungermannia lanceolata, 119
Jungermannia obovata, 192
Jungermannia sphaerocarpa, 186, 204
Jungermannia subulata, 120-122
Jungermannia thermarum, 204
Jungermannia torticalyx, 204

Klebsiella pneumoniae, 176, 177

Leiomela, 315
Lejeunea, 1, 6, 209
Lejeunea vaginata, 7

Plant Index   335

Lepicolea, 209
Lepidozia, 4, 209, 229
Leptobryum, 28, 29, **2.2**, 94
Leptobryum pyriforme, 27, 28, 47, 49, **2.2**, 70, 94-96, 101, 102, 183, 222
Lescuraea saxicola, 16
Lethocolea, 229
Leucobryum, 57, 58, 153, 154, 207, 234, 285, 315
Leucobryum albidum, 261
Leucobryum brevicaule, 57
Leucobryum glaucum, 186, 285
Leucodon, 41
Leucolejeunea, 1, 209
Leucoloma, 315
Leucophanes candidum, 154, 169
Liriodendron tulipifera, 261
Lophocolea, 84, 187, 188, 209, 227, 229, 235
Lophocolea bidentata, 84
Lophocolea cuspidata, 92, 93
Lophocolea heterophylla, 92, 93, 119, 120-122, 187, 272
Lopholejeunea, 1, 3
Lophozia, 209
Lunularia, 29, 35, 66, 72, 78, 79, 95, 209, 215, 235, 238, 281, 288
Lunularia cruciata, 30, 32, 33, 35, 92, 93, 95, 96, 132, 133, 137, 138, 140, 141, 178, 180, 181, 214, 232, 255, 270, 272
Lycopodium, 222

Macrodiplophyllum, 209
Macrodiplophyllum plicatum, 186
Macromitrium, 61, 315
Macromitrium gymnostomum, 17
Macromitrium sulcatum, 150
Macrosporiella, 61
Macvicaria, 209
Madotheca, 289
Madotheca platyphylla, 211
Makinoa, 177, 209
Makinoa crispata, 9, 73, 120
Mannia, 21
Mannia dichotoma, 120, 122
Mannia fragrans, 25
Marchantia, 10, 12, 21, 29, 34, 35, 37, 66, 72, 74, 81-83, 95, 127, 156-159, 175, 176, 183, 187, 188, 209, 212, 214, 215, 222, 227, 229, 235, 237, 238, 264, 288, 305
Marchantia alpestris, 256
Marchantia berteroana, 32, 211, 213, 214, 217, 219
Marchantia foliacea, 213-216
Marchantia nepalensis, 30, 32 34, 35, 81, 120, 122, 132, 133, 138, 140

Marchantia paleacea, 176, 177
Marchantia palmata, 177
Marchantia polymorpha, 24, 25, 30-32, 34, 35, 76, 78, 82, 83, 92, 93, 95, 97-99, 101, 119, 120, 122, 132, 133, 137, 140, 142, 159, 171, **7.10**, 175, 178, 179, 181, 183, 187, 194, 197, 211, 214, 219, 222, 232, 252-255, 264, 270, 272, 274, 275, 277-280, 282, **10.1, 10.3**
Marchantia stellata, 175
Marchesinia, 209
Marsupella, 209
Mastigophora, 315
Megaceros, 12, 209
Merceya, 61, 73
Merceya gedeana, 57
Merceya latifolia, 239
Merceya ligulata, 57, 239, 256
Metzgeria, 78, 209, 216, 220, 229, 315
Metzgeria conjugata, 9
Metzgeria furcata, 64
Micrococcus flavus, 175
Micrococcus rubens, 175
Microdus, 41
Microdus brasiliensis, 47, 102-104
Microdus miquelianus, 44
Micromitrium, 315
Microsporum gypseum, 178
Mielichhoferia, 256, 257
Mielichhoferia elongata, 256
Mielichhoferia macrocarpa, 239, 256
Mielichhoferia mielichhoferi, 239, 256
Mnium, 55, 61, 73, 141, 148, 154, 159, 165, 176, 177, 187, 211, 222, 226, 234, 302, 304
Mnium affine, 68, 84, 133, 141, 213, 214, 218, 261
Mnium cuspidatum, 162, 167, 175, 187, 189, 213, 264
Mnium hornum, 25, 141
Mnium medium, 162, 187
Mnium orthorhynchium, 169
Mnium rostratum, 162
Mnium undulatum, 141, 158, 159, 162, 213, 260, 295, 296, 309
Mnium punctatum, 189
Moerckia, 289, 292
Moerckia flotowiana, 73, 81, 120
Molendoa, 57, 70
Monoclea, 66, 229
Monoclea frosteri, 217, 315
Monoselenium, 66
Mucor racemosus, 26
Mycobacterium phlei, 176, 177
Mylia, 187, 188, 209, 229
Mylia taylorii, 186, 198

Myuroclada, 57

Nardia, 1, 3, 16, 229
Nardia sieboldii, 2
Neckera, 71, 187, 207
Neckera crispa, 186
Neckeropsis, 315
Neohattoria, 210
Nicotiana rustica, 179
Nicotiana tabacum, 179
Noteroclada, 81
Noteroclada confluens, 73, 81, 120, 121
Notothylas, 12
Notothylas indica, 66
Notothylas orbicularis, 92, 93, 97
Nowellia, 210
Nowellia curvifolia, 119

Octoblepharum, 22, 56, 58, 59, 72, 74, 136, 315
Octoblepharum albidum, 24, 27, 31, 132, 136
Oligotrichum, 57, 59, 61, 71
Oligotrichum hercynicum, 239
Oligotrichum semilamellatum, 59, 63
Omphalanthus, 210
Oncophorus, 57
Orthotrichum, 58, 294, 295, 309
Orthotrichum obtusifolium, 259, 260
Orthotrichum rupestre, 175, 177
Oxymitra, 66
Oxystegus cylindricus, 177

Pallavicinia, 78, 210, 229, 237, 289, 304
Pallavicinia canaras, 101
Pallavicinia longispina, 8
Pallavicinia lyellii, 119, 291, 294
Pedinophyllum, 210
Pellia, 25, 156, 170, 183, 187, 188, 210, 227, 229, 231, 237, 238, 288
Pellia endiviaefolia, 232
Pellia epiphylla, 92, 93, 95, 97, 98, 119, 186, 198, 232
Pellia fabbroniana, 186, 187, 198
Pellia neesiana, 11, 119
Penicillium martensii, 26
Petalophyllum, 56
Petalophyllum ralfsii, 120
Phaeoceros, 156, 210
Phaeoceros bulbiculosus, 92, 93, 97
Phaeoceros laevis, 92, 93, 97, 171
Phaeoceros laevis subsp. carolinianus, 173
Phascum, 108, 166

Phascum cuspidatum, 108, 110, 111, 115, 116
Philonotis, 102, 226, 234
Philonotis fontana, 253
Philonotis glaucesens, 50
Philonotis turneriana, 96, 97, **4.1**
Physcomitrella, 44, 274
Physcomitrella patens, 23, 24, 37, 43-45, 94, 96, 132, 134, 136, 139, 141, 143, 145, 165, 179, 270-272, 277-279, 282, 283
Physcomitrium, 17, 52, 59, 61, 80, 117, 128, 146, 156, 166, 272, 304
Physcomitrium coorgense, 47, 56, 59, 61, 109-113, 120, 121, 123, 128, **5.2,**157-159, **7.8, 7.11, 7.12**
Physcomitrium cyathicarpum, 56, 59, 61, 118, 159, 160, 169, 171, **7.17,** 306
Physcomitrium pyriforme, 57, 59, 67, 71, 74, 77, 94, 96, 108-116, 118, 121, 123, **5.1, 5.3, 5.5, 5.10,** 133, 140, 142, 165, 179, 182, 184, 282
Physcomitrium sphaericum, 41
Physcomitrium turbinatum, 41, 43, 57, 59, 67, 71, 75, 118, 132, 134, 139, 143, 144
Pinnatella, 57
Pinus longifolia, 196
Pinus verginiana, 261
Plagiochasma, 21, 24, 66, 177, 212, 229-231, 235
Plagiochasma appendiculatum, 23, 25, 132, 138, 176, 177
Plagiochasma articulatum, 101
Plagiochasma intermedium, 22-26
Plagiochasma peruvianum, 219
Plagiochasma rupestre, 219
Plagiochasma tenue, 219
Plagiochila, 198, 210, 229-231, 237, 288, 315
Plagiochila acanthophylla, 197
Plagiochila asplenioides, 197, 211, 214-216, 222, 247, 263, 312
Plagiochila asplenioides var. major, 230
Plagiochila carringtonii, 230
Plagiochila semidecurrens, 197, 198
Plagiochilion, 210
Plagiomnium cuspidatum, 81, 83, 253
Plagiomnium undulatum, 308
Plagiomnium vesicatum, 270, 272
Plagiothecium, 187
Plagiothecium denticulatum, 177
Plagiothecium undulatum, 312
Platyhypnidium, 207, 228
Platyhypnidium riparioides, 230
Plectocolea, 229

Plant Index

Pleurozia, 314
Pleurozium, 187, 212, 235
Pleurozium schreberi, 187, 211, 251, 253, 255, 312
Pogonatum, 41, 57, 59, 61, 71, 156, 169, 235, 286, 296, 315
Pogonatum aloides, 22, 44, 47, 48, 169, 177
Pogonatum microstomum, 59, 63
Pogonatum perichaetiale, 59
Pogonatum urnigerum, 25, 26, 222
Pohlia, 42, 57, 58, 64, 67, 226
Pohlia elongata, 46
Pohlia lescuriana, 27
Pohlia muyldermansii, 66, 68
Pohlia nutans, 42, 46, 47, 49, 99, 132, 134, 138, 139, 143, 183
Polyota, 285
Polytrichadelphus magellanicus, 222
Polytrichum, 59, 61, 63, 65, 71, 72, 74, 81, 88, 117, 123, 136, 140, 148, 153, 156, 166, 175, 176, 187, 207, 222, 226, 235, 286, 289, 294-297, 302, 304, 305, 309
Polytrichum aloides, 94, 96
Polytrichum commune, 17, 47, 59, 70, 81, 119, 121, 122, 150, 152, 154, 161, 162, 165, 169, 170, **7.7**, **7.13**, **7.14**, 175, 177, 179, 191, 222, 225, 259, 263, 289, 303, 309
Polytrichum formosum, 167
Polytrichum juniperinum, 65, 68, 69, 87, 116, 121, 132, 155, 156, **7.9**, 175, 189, 240, 250, 269, 272, 277, 285, 299, 308, 309
Polytrichum piliferum, 168, 170, 250
Polytrichum strictum, 66
Porella, 177, 194, 197, 198, 210, 213, 229, 240, 313
Porella baueri, 213
Porella bolanderii, 213
Porella capensis, 213
Porella cordaeana, 213
Porella densifolia, 194, 198
Porella faurieri, 194
Porella gracillissima, 194, 197
Porella japonica, 200, 205
Porella macroloba, 194
Porella perottetiana, 194
Porella pinnata, 119
Porella platyphylla, 186, 192, 199, 213, 312
Porella platyphylloidea, 213
Porella vernicosa, 194, 197, 198, 205
Porothamnium, 315
Pottia intermedia, 110, 115, 121, 165, 270, 272, 274
Pottia lanceolata, 109, 110, 121
Preissia, 66, 78, 288

Preissia quadrata, 92, 93, 97, 98
Protocephalozia, 315
Pseudomonas aeruginosa, 175-177
Pseudoscleropodium, 187, 207
Pseudoscleropodium purum, 186
Psilotum, 222
Ptilidium, 210
Ptilidium ciliare, 198
Ptilidium pulcherrimum, 119
Ptilium, 235
Ptilium crista-castrensis, 222
Ptychanthus, 210
Ptychanthus striatus, 7, 247
Ptychomitrium, 58
Ptychomitrium sinense, 16, 20
Pylaisiella, 41
Pylaisiella polyantha, 259, 260
Pylaisiella selwynii, 41, 48, 50, 121, 122

Radula, 1, 5, 177, 210, 229, 240, 315
Radula complanata, 119, 192, 232, 258
Reboulia, 21, 66, 210, 212, 215, 229, 288
Reboulia hemisphaerica, 25, 99, 101, 120, 121, 123, 177, 213, 217
Rhacomitrium, 57, 187, 207, 309
Rhacomitrium lanuginosum, 186, 255, 311, 317
Rhizobium, 50,
Rhizogonium, 57, 61
Rhodobryum, 55, 61, 222, 315
Rhodobryum giganteum, 61
Rhodobryum roseum, 61
Rhodotorula, 26, 50
Rhynchostegium, 228, 230
Rhynchostegium serrulatum, 261
Rhytidiadelphus, 78, 186, 187, 207, 235
Rhytidiadelphus squarrosus, 251
Rhytidiadelphus triquetrus, 260
Riccardia, 210, 227, 229-231
Riccardia jackii, 195, 196
Riccardia miyakeana, 8
Riccardia multifida, 92, 93, 120, 121, 123, 196, 197
Riccardia pinguis, 73, 92, 93, 95, 120, 121, 123, 163
Riccardia sinuata, 239
Riccia, 10, 66, 97, 98, 100, 103, **4.3**, 123, 148, 149, 156, 175, 210, 212, 215, 222, 230, 231, 236
Riccia bicarinata, 219
Riccia crystallina, 22, 23, 66, 71, 93-104, **4.3**, 120, 121, 123, 218
Riccia discolor, 133, 138
Riccia duplex, 100, 104

Riccia fluitans, 66, 100, 181, 214, 215, 220
Riccia frostii, 66, 70, 71, 93-95, 98-102, 120, 122, 123, 138
Riccia gangetica, 93-104, **4.3**
Riccia glauca, 66, 93, 97, 159
Riccia plana, 230
Ricciocarpos, 210, 212, 215, 227
Riella, 12, 13, 55, 67, 69, 72-75, 77, 84, 86, 87, 149, 212, 214, 215
Riella helicophylla, 67, 76, 82, 84, 87, 88

Saccogyna, 229
Saelania glaucescens, 204
Salmonella typhi, 176, 177
Sarcina, 175
Sarcina lutea, 175, 176
Scapania, 3, 16, 210, 225, 229, 236, 313
Scapania ligulata, 4
Scapania nemorosa, 35, 119, 120, 197
Scapania parvidens, 120, 121
Scapania parvitexta, 186
Scapania subalpina, 197
Scapania uliginosa, 197
Scapania undulata, 170, 193-197, 205, 222, 230, 312
Schistochila, 315
Schistostega pennata, 14, 17, 27
Scleropodium, 207
Scleropodium toureti, 186
Scopelophila ligulata, 256
Selaginella, 222
Sematophyllum, 315
Semibarbula orientalis, 27
Septoria nodorum, 178
Sewardiella, 55, 56, 84
Solenostoma crenulatum, 181
Solenostoma triste, 181, 204
Sphaerocarpos, 12, 13, 23, 137, 144, 149, 158, 159, 163, 215, 227, 236, 304
Sphaerocarpos cristatus, 132, 135
Sphaerocarpos donnellii, 23, 103, 120, 132-136, 144, 157, 167, 269, 270, 272, 275-278, 280, 282
Sphaerocarpos texanus, 66, 102, 120, 122, 272
Sphagnum, 14, 15, 38, 39, 48, 57, 61, 74, 150, 153, 155, 156, 168, 175, 176, 187, 189, 207, 211, 222-226, 228, 230, 231, 233, 235, 252, 257, 285, 309, 310, 315
Sphagnum angustifolium, 189
Sphagnum balticum, 230
Sphagnum cuspidatum, 177, 186, 191, 311
Sphagnum fimbriatum, 189, 252
Sphagnum fuscum, 186, 189
Sphagnum girgensohnii, 14, 15

Sphagnum imbricatum, 15
Sphagnum lescurii, 7.3
Sphagnum magellanicum, 186, 189, 213, 221, 230, 311
Sphagnum majus, 189
Sphagnum medium, 224
Sphagnum meridense, 14
Sphagnum nemoreum, 189, 252, 254
Sphagnum palustre, 177, 186, 191, 222
Sphagnum papillosum, 189
Sphagnum plumulosum, 94, 98, 311
Sphagnum portoricense, 175, 177
Sphagnum recurvum, 154, 186, 311
Sphagnum rubellum, 186, 311
Sphagnum strictum, 175
Sphagnum teres, 186
Spiridens, 315
Splachnum, 54, 58, 72, 73, 75, 78-80, 83, 84, 110, 187, 305
Splachnum ampullaceum, 44, 46, 71, 74, 75, 83, 85, 86, 178
Splachnum luteum, 112
Splachnum ovatum, 110, 111, 114, 179
Splachnum pedunculatum, 110
Splachnum rubrum, 104, **4.4**, 220
Splachnum sphaericum, 111
Splachnum vasculosum, 220
Spruceanthus, 210
Staphylococcus aureus, 175-177
Stephensoniella, 10
Stictolejeunea, 1
Streptococcus pyogenes, 175
Symbicizidium, 210
Symphyogyna, 289, 292
Symphyogyna circinata, 304
Symphyogyna sect. *Dendroides*, 315
Symphyogyna sect. *Repentes*, 315
Syrrhopodon, 58, 70, 315
Syzygiella, 210

Takakia, 210, 216, 289, 292
Takakia lepidozioides, 290, 304
Targionia, 21, 22, 24
Targionia hypophylla, 22-26, 177
Tayloria, 187
Tetraphis, 57, 61, 63, 69, 294
Tetraphis geniculata, 27
Tetraphis pellucida, 17, 26, 27, 32-34, 65, 110, 121, 122, 133, 141, 178
Tetraplodon, 50
Tetraplodon mnioides, 49
Thamnium, 208
Thamnium alopecurum, 206
Thamnobryum, 187
Thuidiopsis, 226

Thuidium, 187, 208, 226, 235, 315
Thuidium abietina, 186
Thuidium bipinnatulum, 16
Thuidium delicatulum, 121
Thuidium delicatum, 261
Thuidium tamariscifolium, 206
Timmia megapolitana, 162
Timmiella, 42, 176
Timmiella anomala, 41, 44, 47-50, 184, 246
Timmiella barbuloides, 153, 154
Tomenthypnum, 235
Tomenthypnum nitens, 254, 312
Tortella, 57, 58, 208
Tortella caespitosa, 43, 47
Tortula, 57, 169, 187, 228, 235, 313
Tortula amplexa, 27
Tortula bolanderi, 27
Tortula desertorum, 27
Tortula intermedia, 310
Tortula latifolia, 27
Tortula muralis, 27, 169, 258
Tortula papillosa, 27
Tortula princeps, 230
Tortula ruralis, 27, 312-314

Tortula stanfordensis, 27
Trachypodopsis, 315
Trematodon brevicalyx, 27, 29, 44
Trichocolea, 229, 314
Trichocoleopsis, 210
Trichocoleopsis sacculata, 4, 205
Tricholejeunea, 210
Trichophyton mentagrophytes, 178
Trichophyton rubrum, 178
Trocholejeunea sandvicensis, 197, 199
Tuzibeanthus, 210

Ulota, 167, 309
Ulota crispula, 314
Uromyces fabae, 178

Venturiella, 57, 58, 61
Vibrio cholerae, 176, 177

Weissia, 50
Weissia controversa, 50, 255
Wiesnerella, 210, 240
Wiesnerella denudata, 193

Author Index

(Numbers in bold face indicate plates)

Abe, S., 129.
Abel, W.O., 46, 272, 284.
Agnistikova, V.N., 52, 246
Ahlawat, A.S., 145
Ahmad, S.M., 23
Allard, H.A., 92
Allen, C.E., 163-165,
Allsopp, A., 17, 18, 34, 35, 39-42, 44, 46, 47, 49, 52 73, 76, 77, 79, 81, 99, 120-123, 146, 228-230.
Alston, R.E., 213, 214, 218
Anderson, N.H., 194-196, 198, 247
Anderson, L.E., 14.
Anderson, W.H. 115, 189, 243, 247
Andrews, S., 27, 28
Angerman, C., 234.
Anhut, S., 217
Anthony, R.E., 92, 93
Apostolakos, P., 157
Aratani, T., 240.
Armentano, T.V., 49
Asakawa, Y., 177, 192-200, 205, 208-210, 229, 232, 233, 240, 243
Ashton, N.W., 36, 43, 44, 51, 145, 178, 280, 282-284
Asprey, G.F., 25
Aumaître, M.P., **69, 88, 139, 142**
Averett, J.E., 245

Bach, A., 271, 272.
Bandurski, R.S., 241
Banerjee, R.D., 175-177
Barbier, C., 158, 159, 174
Barclay-Estrup, P., 253.
Barkman, J.J., 249
Barthelmess, A., 282
Barz, W., 240.
Basile, D.V., 34, 120.
Bassi, M., 154, 169, 297
Bates, J.W., 251
Batra, A., 272
Bauer, L., 17, 23, 44, 74, 105, **4.4**, 108, 110-115,
121, 135, 136, 140, 146, 179, 182-184
Bayfield, N.G., 286, 289
Bazaz, F.A., 140
Becquerrel, P., 18.
Belkengren, R.O., 42
Belkin, M., 240
Bell, P.R., 156-159
Bendz, G., 196, 211, 220, 225, 226, 239
Benes, I., 204, 208, 241
Benson-Evans, Kathryn, 23, 25, 35, 92-98, 101.
Benesová, V., 185, 186, 188, 195, 198, 208, 209, 226, 227, 232, 239, 241, 244, 247
Berggren, A., 256
Berkley, E.E., 27.
Bernardini, C., 245
Berrie, G.K., 119
Berthier, J., 63, 66-68
Beutelmann, P., 44, 53, 121, 179
Bewley, J.D., 313, 314, 317
Bhandari, N.N., 58-61, 63, 73, 160.
Bhatla, S.C., 22, 40, 41, **2.1**, 92-94, 96, 99, 101-104, **4.1**, 182.
Biebl, R., 263, 264, 314.
Binding, H., 269, 272, 278
Binns, A.N., 34.
Birse, E.M., 315, 316
Bissonette, P., 240.
Bittner, Karolina, 134
Black, W.A.P., 207, 230, 233
Bland, D.E., 222, 223
Blevins, D.G., 53
Boatman, D.J., 48
Böhrs, H.-L., 40
Bonnot, E.J., 156, 157
Bopp, M., 39-41, 43-46, 49, 51, 53, **2.2**, 54, 57, 58, 61, 67, 71-75, 116, 117, 183, 270, 272, 273, 275, 277, 278, 288, 304, 305
Borenhagen, H., 119
Borodin, J., 21, 134
Bostic, S.R., 98, **4.2**.
Bouillant, M.-L., 245

Author Index 341

Bourgeois, G., 192, 198
Bowen, C.C., 172
Bowen, E.J., 287, 288, 308
Bower, F.O., 125
Brandes, H., 41, 45, 49, 101
Brassard, G.R., 27, 256, 262
Braverman, J.B.S., 259
Brehm, B.G., 244, 245
Brenner, K., 53
Briarty, L.G., 172
Briggs, G.R., 252
Brooks, R.R., 263, 266
Brown, D.H., 251, 255, 257, 262
Brown, P.R., 34, 81
Brown, R.C., 150, 161-163, 165, 166
Browning, A.J., 168
Bruch, Th., 287
Brunel, A., 234, 235
Bryan, V.S., 115
Buck, G.W., 262
Buch, H., 308, 309
Bucke, C., 238
Bunning, E., 54, 75, 120, 122
Burdon, R.S., 147, 247
Burgeff, H., 119
Burgess, J., 272
Burkholder, P.R., 48
Burkitt, A., 253
Burr, R.J., 292, 293
Busby, J.R., 312
Buschbom, U., 317
Butler, W.L., 145
Butterfield, B.G., 306
Butterfass, T., 162

Caldicott, A.B., 186, 188, 191
Campbell, D.H., 59
Campbell, E.O., 245, 293
Canonica, L., 177
Capelle, G., 234, 235
Caponetti, J.D., 50
Cardenas, E., 242
Carothers, Z.B., 156, 158, 171, 173
Carpenter, S.B., 265
Casagrande, C., 242
Castaldo, R., 154, 172, 173
Catalano, S., 186, 207
Celakovský, L., 125
Chandra, Prabha, 52, 146
Chandra, S., 235, 236
Châu, H.M., 241
Chauhan, E., 169, 171, 172, 304
Chevallier, D., 161
Chi, S., 265
Chick, E., 304

Chopin, J., 245
Chopra, N., 115
Chopra, R.N., 22, 26-29, 31, 35, 39-44, 46-50, **2.1,**
   2.3, 2.4, 2.5, 54, 56, 63, 70, 81, 92-105, **4.1,**
   4.3, 110-113, 118, 120-122, **5.1, 5.4, 5.7, 5.8,**
   133, 142, 143, 182, 184
Chopra, R.S., 57-61, 63, 71, 73, 74, 125
Clee, D.A., 288
Cocking, E.C., 269, 277
Coker, P.D., 258, 259
Colaiace, J., 92, 93
Cole, J.R., 243
Comeau, G., 74, 258, 261, 266
Connolly, J.D., 195, 199, 204
Coombes, A.J., 253
Corbella, A., 198, 242
Cornhill, W.J., 242
Correns, C., 27, 56
Corrigan, D., 186, 189
Costerton, J.W., 317
Courtice, G.R.M., 283, 284
Costin, C.R., 240
Courtoy, R., 95
Cove, D.J., 23, 24, 43, 44, 50, 53, 132, 134, 136, 139,
   141, 143, 145, 272, 280, 282-284
Craigie, J.S., 225, 229, 230, 231, 233
Crosby, M.R., 14
Crum, H., 115
Cukierski, M., 156
Cutter, E.G., 81, 83
Cybulska, A., 53
Czarnowska, K., 250, 256
Czeczuga, B., 225, 226

Dacknowski, A., 32
Dagar, J.C., 35, 133, 138
Daly, G.T., 249, 258, 259
Das, R.R., 38
Das, V.S.R., 99, 101, 228-231
Datta, P.K., 234, 235
Davidonis, G.H., 76, 81
Davy de Virville, A., 296
DeGreef, J., 132, 133, 137, 143, 183, 227
Deichgräber, G., 170
Deloire, A., 288
DeMaggio, A.E., 151
DeMarco, A., 242
Demaret, F., 265
Demkiv, O.T., 67, 121, 122, 132, 133, 139, 141,
   142
Denizot, J., 149
DeProft, M., 242
Desai, S., 40
De Sloover, J., 249
Devosalle, L., 249

342 Biology of Bryophytes

Dhindsa, R.S., 313, 314
Dhingra-Babbar, Sadhana (Dhingra, Sadhana), 27, 29, 44
Dickson, H., 32, 74, 78
Diers, L., 157-159
Dietert, M.F., 50, 94
Dilks, T.J.K., 312
Dill, F.J., 163, 164
Diller, V.M., 102
Dominguez, X.A., 240
Dornowska, E., 53
Douin, C., 73
Douin, F., 225
Doyle, W.T., 23, 132, 135, 149
Drew, E.A., 228, 229
Drozdova, T.V., 213
Dubovaya, L.P., 52, 246
Duckett, J.G., 155. 173
Ducombs, G., 240
Dufresne, S.J., 89
Dutel, J., 36

Eakin, R.E., 40, 43, 44, 47
Edelbluth, E., 246
Edwards, S.R., 27
Eglinton, G., 186, 188, 191
Egunyomi, A., 22, 24, 27, 31, 56, 58, 59, 72, 74. 132, 136
Ekman, R., 190, 207, 244
Elliot, A.M., 247
Eltz, H.V., 41, 183
Engel, P.P., 96, 282
Erbisch, F.H., 263
Erichsen, U., 46, 51
Erickson, M., 222
Eschrich, W., 152, 302, 305
Etzold, H., 134, 144
Euranto, E.K., 244
Evans, A.W., 78
Evans, P.K., 277
Evensen, K.B., 53
Evenari, M., 317
Eymé, J., 148, 162, 163, 167, 302

Farr, M.E., 174
Favali, M.A., 154, 169, 297
Featherstone, D.R., 50, 284
Fedyk, D. (Y.D.), 127, 142, 146
Felix, M.D., 241
Fell, J., 45
Fellner, F., 66
Ferrari, G., 242
Fitting, H., 17, 18, 29, 31, 32
Fitzgerald, D.B., 241
Foulquier, L., 265

Foussereau, J., 240, 244
Frank, M.L., 261
Franke, H., 244
Franke, W., 234
Frederick, S.E., 121
Fredericq, H., 132, 137, 145, 227, 242
Freeland, R.O., 225, 226
French, J.C., 116-118, 140
Freudenberg, K., 222
Frey, W., 294
Fries, N., 42
Fujiki, H., 241
Fujita, Y., 193
Fulford, M., 1, 16, 25, 64, 105
Furuta, H., 57, 58, 61, 71-73

Gaal, D.J., 83
Galatis, B., 157
Gamborg, O.L., 284
Gams, H., 257
Gambardella, R., 154, 171, 173
Garber, E.D., 244, 284
Gardner, G., 46
Gariboldi, P., 242
Garner, W.W., 92
Gartrell, F.E., 258
Gautheret, R.J., 123
Gay, L., 68, 69, 73, 87, 88, 121, 269, 272, 277, 279
Geissler, G., 233
Geissler, P., 256
Gellerman, J.L., 185, 187-190, 240, 247
Gemmell, A.R., 58, 59, 61, 71, 74, 75, 80
Gemmrich, A.R., 25
Genevès, L., 148, 161, 162, 165
Gianni, F., 169
Gibbs, R.D., 222
Giesenhagen, K., 315, 316
Gilbert, O.L., 249, 258-260
Giles, K.L., 54, 58, 68, 69, 84, 133, 141
Gimingham, C.H., 315, 316
Gleizes, M., 197
Goebel, K., 17, 26, 42, 64, 78, 143, 161, 167
Gordon, A.G., 258
Gorham, E., 258, 261
Gorham, J., 180, 235
Górska-Brylass, A., 156, 159
Gorton, B.S., 40, 43, 44, 47
Goto, H., 245
Gradstein, S.R., 208
Granger, J.M., 260
Grauby, A., 265
Grasmück, I., 146
Grimsley, N.H., 50, 272, 274, 277, 278, 280, 282, 283

Grolle, R., 243
Gross, D., 178, 180
Gross, G.G., 244
Grossman, H.H., 70, 263
Grumbach, K.H., 244
Grusak, M.A., 120
Gullvåg, B.M., 167, 266
Gumbel, Th., 306
Gunckel, J.E., 70, 263
Gunning, B.E.S., 168, 304
Günther, I., 25
Gupta, K.G., 176
Gupta, R.K., 312, 313
Gupta, Urmilla, 22, 42, **2.4,** 143
Gwóźdź, E.A., 270, 272, 275, 277-279, **10.2**

Haberlandt, G., 139, 153, 304, 308
Hackenberg, D., 43, 46
Hackemesser, H., 259
Hagerup, O., 311
Hahn, H., 68
Hahn, M., 45
Halbsguth, W., 31, 34, 82, 133, 140
Hall, P., 241
Hallaway, M., 242
Hallet, J.-N., 299
Hamner, K.C., 24, 30, 34, 92, 93, 98
Hancock, J.A., 27, 262
Handa, A.K., 40, 181
Handszu, A., 45
Happack, H., 310, 311
Harding, A.E., 242
Hari, P., 317
Harrington, A.J., 89
Harrison, M.A., 247
Hartman, E.L., 257
Hartmann, E., 36, 51, 53, 114, 133, 134, 140, 142, 144-146, 182, 183, 233
Hatanaka, A., 235
Hatanaka-Ernst, M., 40, 43, 49
Hartwell, J.L., 175
Hasse, K., 234
Hattori, S., 241
Hausmann, M.K., 155
Hayashi, S., 195, 245
Hayes, L.E., 176
Heald, F.D.F., 21, 23, 57-59, 63, 67, 134
Hébant, C., 153, 157, 167, 169, 170, 222, 234-236, 290, 292-295, 297-299, 301-306
Hebrard, J.P., 262
Heckman, C., 149
Hedwig, J., 1, 308
Heggestad, H.E., 260
Heitholt, J.J., 53
Heitz, E., 17, 25, 73, 84

Helgeson, J., 53
Hendrix, F.F., 50
Henon, Jean-Michel, 306
Hepler, P.K., 53, **2.3,**
Herout, V., 232, 241, 244, 247
Herrmann, R.G., 227
Herzfeld, S., 70
Herzfelder, H., 117
Higinbotham, N., 168
Hillson, C.J., 70, 263
Hirose, Y., 120-122, 128, 129, 247
Hockenhull, Y., 254
Hofer, K., 264
Hoffman, G.R., 23, 24, 261, 262
Hofmeister, Q.E., 125
Hofmeister, W., 1
Holligan, P.M., 228, 229
Hopkins, B.J., 194, 209, 232
Horgan, R., 53
Horner, H.T., 149, 163
Hörster, H., 192
Hosakawa, T., 314
Houget, J., 234, 235
Hughes, J.G., 59, 92-96, 110, 111, 116, 118
Hughes, K.W., 94, 96
Hulbary, R.L., 161-163
Huneck, S., 185, 187, 193, 195, 196, 200, 204-208, 210, 227, 232, 240, 241
Hurel-Py, G., 40

Idzikowska, K., 151
Ihantola, A., 225, 226
Ikan, R., 219
Ilahi, I., 73, 76, 77, 81, 120, 121, 123
Immirzi, A., 242
Inoue, H., 12, 21, 23-25, 135, 241
Iori, A.M., 245
Ishida, Y., 245
Ishikawa, M., 128, 129
Ishikawa, S., 22
Iverson, G.B., 26
Ives, D.A.J., 207
Iwamura, T., 128
Iwasa, K., 44, 48
Iwatsuki, Z., 20, 315, 316

Jacques, R., 69, 90, 146
Jaffe, L.F., 134, 144
Jahn, H., 40, 42, 44, 47, 134
Jain, D., 38
Jamieson, G.R., 189
Janicke, S., 243
Jarvis, L.R., 161, 165
Jaskiewicz-Mroczkowska, B., 53
Jayaswal, R.K., 53

Jendralski, U., 311
Jenkins, G.I., 134, 141, 145, 272
Jensen, K.G., 161-163
Johnsen, I., 259
Johnson, R.P.C., 299
Johri, M.M., 40, 53, 181
Jolad, S.D., 240
Jommi, G., 242
Jones, R.A., 235, 236
Joslyn, M.A., 259
Juniper, B.E., 191

Kachroo, P., 10, 13, 17, 57, 59, 61, 67, 71, 77, 118
Kajiwara, T., 243
Kalaitzakis, J., 228, 230
Kanda, H., 14, 17, 72
Kano, M., 241
Kappen, L., 310
Kapur (née Rekhi), Anita (Rekhi, Anita), 40, 41, 42, 44, 47-49, 51, **2.1**, **2.3**, 120, 122, 184
Karsten, I., 67
Karunen, P., 185, 189-191, 207, 225, 226, 244
Kasai, R., 245
Kass, L.B., 88, 136
Katoh, K., 107, 120-122, 129
Kaufman, P.B., 228, 229, 247
Kaul, A., 22, 24-26, 38, 132, 133, 138, 140
Kaul, K.N., 32, 34, 66, 78, 120, 122
Kaul, R., (Kaul, Ratnum), 47, 181
Kaul, Rita, 132, 133, 138, 140
Kaur, G., 172
Keil, M., 42
Keilin, D., 312
Kelley, C., 167, 304
Kellomäki, S., 310
Kemp, J., 53
Kende, H., 45, 46, 51
Kersten, H., 105
Kessler, B., 134
Khanna, K.R., 115
Kilbinger, H., 133, 142, 183
Kinzel, H., 228, 229, 231
Klebs, G., 42, 98, 99, 138, 143
Klein, B., 40, 41, 44, 183
Klein, E., 185, 193, 196
Klingenberg, B., 146
Kloos, C., 242
Knapp, E., 282
Knoche, H., 199, 240
Knoop, B., 51, 61, 82, 278, 283, 305,
Kockel, H., 183
Kofler, L., 22, 25, 39
Kohl, F.G., 225
Kolattukudy, P.E., 191

Kondaiah, N.M., 246
Konecny, K., 241
Koponen, T., 317
Korcz (Korcz-Zajchert), I., 53
Koskimies, K., 189
Köster, J., 240
Kosuge, T., 50
Kozlowski, A., 211, 214
Kramer, C.M., 240
Kraus, E.J., 99
Krapf, G., 154
Kraybill, H.R., 99.*
Kreh, W., 64, 74, 76
Kreher, R., 247
Kreitner, G.L., 156, 173
Krepinsky, J., 242
Krisko, M.E.P., 132, 140
Kristiansen, J., 234
Krochko, J.E., 312-314
Krupa, J., 136
Krutov, S.M., 198
Kubota, H., 314
Kumar, S.S., 71, 73, 74
Kumra, P.K., 22, 24, 27, 29, 40, 41, 43, 44, 46-51, 54, 56, 93-99, 101-104, **4.1**, 110-112, 118, 120, 123, **5.4**, **5.6**, **5.7**, **5.9**, 133, 141, 142, 184
Kumra, Suman (Dua, Suman), 26, 29, 40, 41, 44, 46, 47, 50, **2.3**, 93-97, 99-104, **4.3**, 120, 122, 123
Kurz, E.H., 26
Kuzin, A.M., 263
Kuzmiakova, N., 208

Laage, A., 134
Lachli, A., 266
Lackner, L., 311
Lal, J., 27
Lal, M., 41, 47, 55, 56, 58, 59, 61, 63, 71, 72, 109-113, 116, 118, 120, 121, 123, **5.2**, **5.5**, 156-160, 169, 184, 304
Lambert, A.M., 163, 165
Lambinon, J., 265
Lang, W.H., 119
Lange, D.L., 310
Lange, O.L., 266, 317
Lark, P.M., 48
Larpent-Gourgaud, M., (Gourgaud, M.), 69, 82, 88, 89, 132, 133, 139, 141, 142, 305
Larpent, J.P., 89, 90, 146
Larson, D.A., 148, 166
LaRue, C.D., 32, 33, 35, 57-59, 61, 71, 73, 74, 78, 79, 118, 178
Laurie, A.E., 250, 251, 253
Lawalree, A., 265
Lawson, J.A., 255

Author Index

Lazarenko, A.S., 27, 109-111, 113, 115, 121, 125, 126, 184
LeBlanche, F., 74, 249, 253, 255, 258-261, 266
Ledén, K., 251, 261
Lee, J., 254
Lee, J.A., 251, 252
Lee, J.P., 313
Leech, R.M., 242
Lehmann, H., 86
Lehnert, B., 43
Leitgeb, H., 21
Lekareva, T.A., 52, 246
Lemmon, B.E., 161-163, 165, 166, 171
Lepp, N.W., 252-255
Lersten, N.R., 172
Leshyak, E.N., 128
Lester, P., 265
Lewis, D.H., 229, 230
Lichtenthaler, H.K., 227, 229, 238
Ligrone, R., 153, 171, 172
Liljenberg, C., 185, 191
Lin, J.C.J., 53
Lindberg, G., 219, 246
Linderman, R.G., 50
Linke, H., 244
Linstead, P.J., 272
Linton, L., 180, 232, 235
Lippincott, B.B., 53, 129
Lippincott, J.A., 53, 129
Listowski, A., 134, 138
Liu, C.-B., 240, 246
Lockwood, L.G., 93, 100
Logan, A., 242
Longendrofer, D.H., 174
Lööf, L.G., 241
Lorentz, P.G., 295
Lounamaa, K.J., 254, 256
Lupova, L.M., 52, 246
Lüttge, U., 154, 168
Lyon, A.G., 35

Maass, W.S.G., 225, 229-231, 233
Machlis, L., 103, 120, 133, 137
Mackenbrock, K., 240
Maćkowiak, T., 53
MacNutt, M.M., 58, 59, 61, 72-75, 83, 85
MacQuarrie, I.G., 46, 54, 71-73, 75, 79, 80
Madsen, G.C., 175-177
Maeda, M., 245, 270, 272, 277, 279
Maeda, T., 245
Mägdefrau, K., 289
Maheu, J., 27
Maier, K., 167, 168, 170, 235, 304
Maier, U., 168, 170, 235
Maleville, J., 244

von Maltzahn, K.E., 44, 46, 54, 58, 59, 61, 68, 71-75, 78-80, 83-86, 133, 141, 178
Mansell, R.L., 234, 235
Manskaja, S.M., 213
Manton, I., 155, 156
Maravolo, N.C., 34, 82, 89, 146, 178, 179, 235, 245, 305
Marby, T.J., 245
Margaris, N.S., 228, 230
Markham, K.R., 211-220, 222, 224, 240, 247, 306
Marrioquin, J., 242
Marsh, B.H., 127
Marshal, ÉL., 59, 118
Marshal, ÉM., 59, 118
Marsili, A., 186, 206-208, 242
Mårtensson, O., 204, 220, 241, 256
Martin, J.T., 191
Martino, V.D., 154
Mason, T.G., 234, 235
Matasov, V.I., 115, 165
Matsuda, R., 241, 243
Matsuo, A., 185-188, 194-198, 204, 208-210, 239, 243
Matsushima, H., 284
Matsuura, T., 243
Mattison, N.L., 235
Matzke, E.B., 72, 75, 119
Mayer, A., 243
Mayo, J.M., 312
McCleary, J.A., 175-177
McClymont, J.W., 148, 166
McConaha 288
Meeuse, B.J.D., 234, 235
Mehra, P.N., 10, 13, 55, 56, 60, 75, 77, 78, 80, 84, 99, 101, 119, 120, 122, 124
Melchert, T.E., 213, 214, 218
Melstrom, C.E., 142, 179
Mèndez, J., 230
Menon, M.K.C., 52, 2.1, 63, 110-113, 115, 116, 121, 123, 5.1, 5.3, 5.5, 5.10, 140, 184
Menshun, M., 242
Metelska, M., 53
Meuche, D., 208
Meyer, D.E., 120, 123
Meyer, M.W., 234, 282
Meyer, S.L., 17, 22, 57, 59, 67, 71, 75, 80, 118, 134
Meyer, Y., 284
Mikola, H., 185, 190, 244
Miksche, G.E., 222
Miller, D.H., 120, 133, 137
Miller, J.H., 68
Miller, M.W., 92, 93, 264
Miller, N.G., 245

Misra, L.P., 52, 146
Mitra, G.C., 17, 18, 36, 39-42, 44, 46, 47, 49, 90, 99, 128, 132, 134, 139, 143
Miyake, K., 128, 129
Miyata, I., 16, 57, 58, 61, 62, 70
Mizuno, T., 16
Mizutani, M., 241
Młodzianowski, F., 151, 152, 272, 279
Moewus, F., 35
Mogensen, G.S., 150, 166
Mohr, H., 23, 131, 132, 134-136
Molisch, H., 211, 213
Mondano, M., 253
Monroe, J.H., 94, 96, 150
Montague, M.J., 66, 120, 122, 243
Moore, A., 240
Moore, N.A., 245
Morelli, I., 206, 208, 242, 245
Mori, K., 272, 277, 278, 284, **10.1**
Morris, Y., 94
Morton, R.A., 242
Moser, J.W., 156
Moutschen, J., 264, 282
Mueller, D.M.J., 148, 151, 160-162, 165, 167
Mueller, M.K., 23
Mueller, T.A., 23
Mues, R., 211, 214-217, 227, 240, 245, 247
Müller, F.P., 305
Mühlethaler, K., 84, 86
Muller, J.C., 240, 241, 246
Müller, H., 17, 27
Mummery, R.S., 46, 49
Munroe, M.H., 76, 81
Muraoka, S., 27, 57, 58, 61-64
Muromtsev, G.S., 47, 179

Nachmony-Bascomb, S., 30, 180
Naef, J., (J.B.), 42, 134, 143
Nagai, J., 72
Naidu, C.S.P., 246
Nair, H., 41, 44
Nakai, K., 245
Nakamoto, T., 245
Nakamura, H., 256
Nakayama, M., 245
Nakosteen, P.C., 94, 96
Narayanaswami, S., 27, 32-35, 55, 56, 58, 59, 61, 63, 71, 72, 79, 118, 178
Nash III, T.H., 252
Naylor, A.W., 40-42, 134, 143
Nebel, B.J., 40-42, 132, 134, 139, 143, 144
Nehira, K., 1, 11, 14-20, 35, 40
Neidhart, H.V., 161, 162, 166, 167
Neish, A.C., 222
Nessel, M., 51

Nester, E.W., 50
Nickless, G., 265
Niekerk-Blom, C.J., 234, 235
Nilsson, E., 204, 212, 213, 216, 220, 230, 241, 244
Ninnemann, H., 133, 140
Nishida, Y., 14, 16-21
Nishimoto, S., 245
Noailles, M., 312
Noda, Y., 245
Noeske, P., 255
Noguchi, A., 14-16, 27, 57, 58, 61-64, 70-73
Novitskii, V.F., 224
Nowak, U., 272, 279
Nurit, F., 36, 166, 171
Nyman, L.P., 81, 83

O'Connor, C.S., 242
Oehlkers, F., 41
O'Hanlon, S.M.E., 12
Ohta, Y., 120-122, 128, 240, 247
Ohusa, T., 22
Ohyama, K., 284
Olarinmoye, S.O., 89
Olesen, P., 150
O'Neil, A.N., 207
Ono, J., 245
Ono, K., 55, 59, 60, 71, 75, 77, 118, 120, 122, **5.6**, 270, 272, 274, 275, 278, **10.3**
Ono, M., 228, 229
Ono, T., 243
Ophus, E.M., 266
Öpik, H., 163
Orsini, F., 242
Osterdahl, B.G., 244, 246
Otero, J.G., 247
Otsuka, H., 16
Otto, K.R., 31, 34, 133, 140
Ourisson, G., 240, 241, 244, 246
Overbeck, F., 310, 311
Overlach, U., 48
Overton, K.H., 205, 208

Pacchiani, M., 242, 245
Pacey, J., 317
Pahwa, M.S., 99, 101, 120, 122
Pakarinen, P., 189, 257
Pande, S.K., 115
Paolillo, D.J., Jr., 88, 116-118, 132, 136, 140, 148, 153, 155, 156, 161-163, 293
Parihar, N.S., 27, 116
Parke, J.L., 50
Pashuk, C.T., 128
Pate, J.S., 168, 172, 304
Pates, A.L., 175-177
Patidar, K.C., 38

Author Index 347

Patschovsky, N., 134
Paul, H., 211
Pauly, G., 243
Pearman, C., 35, 66
Pental, D., 60, 119, 120, 122
Perold, G.W., 194, 209, 232, 240, 244
Persson, H., 239
Pesey, H., 171
Peterson, W.L., 312
Pierik, R.L.M., 123
Plantefol, L., 243
Pokorny, F.A., 50
Porter, L.J., 211-219, 222, 224, 244, 245, 306
Pringsheim, E.G., 27, 42, 49, 132, 134, 138, 143, 144
Pringsheim, N., 118
Pringsheim, O., 42, 49, 132, 134, 138, 143, 144
Prior, P.V., 34, 81
Proctor, M.C.F., 309-311
Proskauer, J., 66, 289, 290
Prusińska, U., 46, 53
Pryce, R.J., 178, 180, 181, 232, 235
Puerner, N., 266
Pullum, P.A., 263

Quillet, M., 228, 229

Raghavan, V., 41, 44
Rahbar, Kavita, 24, 27, 29, 41, 44, 46, 48, 49, 93-96, 98, 99, 101, 103, 104, 120, 122, **5.8**
Ramana, K.V.R., 246
Ramaut, J.L., 175
Rancken, H., 228, 229
Rao, A.N., 35
Rao, D.N., 249, 250, 253, 255, 259, 260, 266
Rao, K.R., 230, 236
Rao, M.P., 99, 101, 228-231
Rashid, A., 22, 24, 41, 43, 44, 46-48, **2.5**, 110-113, 118, 120-123, **5.1**, **5.11**, 132, 134, 139, 143, 270, 272, 274, 275, 277-279
Rasmussen, L., 252, 253
Raudzens, L., 72, 75, 119
Rawat, M.S., 27-29, **1.1**, 40, 41, 47, 49, 50, 94-96, 101, 102, 184
Redfearn, P.L., 27, 28, 59, 67, 71, 75, 118
Reese, W.D., 56
Reeves, R.D., 266
Reid, E.H., 189
Reighard, J.A., 153, 173, 293
Reis-Crepin, D., 174
Rejment-Grochowska, I., 250, 251, 256
Reski, R., 46
Reynolds, A.C., 179
Reznik, H., 217
Reznikov, V.M., 224

Richard, W.H., 261
Richards, P.W., 23, 24, 315
Richardson, D.G., 243
Ridgway, J.E., 92, 93, 97, 98, 304
Rink, W., 119
Rinne, R.J.K., 253
Ripetsky, R.T., 115, 121, 125, 127, 132, 139, 146, 165, 270, 272, 274
Roberts, M.J., 252
Robbins, W.J., 42, 49, 143
Robitaille, G., 253, 258, 259, 266
Rohwer, F., 183
Romose, V., 311
Rose, S., 305
Rota, J.A., 82, 305
Roth, T.F., 145
Rousseau, J., 25, 35
Rudolph, H., 221, 247
Rühling, A., 250, 251, 253, 255, 257

Saarinen, R., 174
Sachar, R.C., 47, 181
Saenko, G.N., 265
Saito, K., 27
Salin, L.M., 189
Samek, Z., 241, 242, 244
Sanz-Cabanilles, F., 230
Sapěhin, A.A., 162
Sarafis, V., 309
Sarla, 44, 46, 48
Sato, S., 245
Saunders, M.J., 53, **2.3**
Sax, K., 70, 263
Saxena, P.K., 48, 120, 270, 272, 274, 275, 277-279
Schaar, I., 86
Schader, E., 35
Schatz, A., 239, 256
Scheirer, D.C., 153, 154, 296, 299
Schieder, O., 270, 272, 275-277, 280, 282
Schier, W., 212, 217, 219-221
Schild, A., 24, 36, 51, 145
Schimper, W. PH. 288, 306
Schlenk, H., 240, 243, 247
Schmidt, C., 53
Schmiedel, G., 152
Schneider, J., 45, 47, 52, 234
Schneider, M.J., 27, 45, 178, 237
Schneider, Z., 234
Schnepf, E., 152, 170
Schofield, W.B., 247
Schostakowitsch, W., 64, 78
Schraudolf, H., 101
Schreiber, K., 210, 227, 243
Schulz, D., 167, 168

Schulz, N., 134
Schulze, A., 241
Schulze, E.D., 317
Schuster, R.M., 55, 186
Schwabe, W.W., 30, 132, 133, 137, 138, 141, 147, 178, 180, 247
Schwarzenbach, M., 119
Scott, G.A.M., 32
Seabury, F.Jr., 161
Sedmera, P., 241
Sekiya, J., 243
Selkirk, P.M., 27, 57-59, 100, 104
Selmer-Olsen, A.R., 254
Sen, S.P., 175-177
Servettaz, C., 40, 42, 49, 134, 143
Shacklette, H.T., 239, 250, 254, 256, 257
Sharma, Poonam, 27, 29, 44, 47, 103, 104
Sharma, P.D., 57, 59, 61, 63, 71
Sharma, S., 40
Sharp, A.J., 27, 178
Shimwell, D.W., 250, 251, 253
Shropshire, W.Jr., 144
Shukla, R.M., 23, 25, 132, 138
Shunk, B., 240
Siegel, B.Z., 266
Siegel, S.M., 222, 253, 289
Silcox, K.R., 270, 272, 281
Siler, M.B., 163
Simola, L.K., 189, 252, 254, 257
Simon, P. (P.E.), 42, 134, 143
Simone, L.D., 119
Singal, Neeta, 105
Singh, B., 176
Singh, V.B., 66
Singh, V.P., 145
Singhivi, U., 23
Sironval, C., 39
Skaar, H., 251
Skoog, F., 53
Smith, G.M., 116
Smith, H., 131
Smith, J.L., 291-294, 304
Smith, W.H., 253
Snyder, C.R., 263
Söchting, U., 259
Solberg, Y., 254
Sood, Sneh, 22, 23, 25, 31, 35, 40, 41, 43, 44, 46-48, 50, 2.5, 81, 93-104, 4.3, 120, 121, 123
Sorm, F., 241
Spaeth, S.C., 235
Sparrow, A.H., 70, 263, 264
Speitel, T., 266
Spencer, K.C., 219
Spiess, L.D., 41, 44, 46, 48-50, 120-122
Springer, E., 108, 110, 111

Spychala, M., 45, 53
Stahl, E., 118
Stange, L., 54, 55, 67, 69, 72-77, 79, 82, 84
Stanton, D.S., 174
Steeman Nielsen, E., 234
Steere, W.C., 261, 263
Stehle, E., 288, 305
Steiner, A.M., 23, 132-135, 144
Steiner, M., 152, 302, 305
Steinkamp, M.P., 149
Steinnes, E., 254
Stephan, J., 134, 138
Sternhell, S., 242
Stetler, D.A., 151
Stewart, G.R., 313
Stockigt, J., 244
Strain, H.H., 226, 227
Stránsky, K., 185, 186, 226
Strassburger, E., 125,
Streibl, M., 241, 247
Stroemer, J.R., 146, 245
Stumm, I., 270-272, 274, 275, 277-279
Sugawara, Y., 272, 277, 279, 284
Sugimura, T., 241
Suire, C., 148, 156, 162, 163, 167, 170, 192, 198, 228, 229, 231, 232, 234-236, 241, 243, 304
Suleiman, A.A.A., 229, 231, 237
Sultan, N., 35
Sun, C.N., 157
Sussman, M.R., 51
Suzuki, Y., 234
Svensson, G.K., 251, 261
Svensson, L., 192, 196, 239
Svitlyk, E.E., 110, 121
Syratt, W.J., 258
Swanson, E.S., 185, 189
Sypherd, P.S., 245
Szweykowska, A., 40, 42-46, 52, 53, 152

Takao, A., 24
Takeda, R., 107, 120-122, 231, 241
Takemoto, T., 199, 241
Takeuchi, M., 270, 272, 275, 280, 284
Tallis, J.H., 251, 311
Tanaka, D., 245
Tanaka, R., 55, 59, 60, 71, 75, 77, 118, **5.6**
Tanikawa, K., 240
Tansley, A.G., 304
Taoda, H., 249, 258, 260
Tarén, Niina, 25, 26, 34
Taylor, A.O., 84
Taylor, I.E.P., 234, 235
Taylor, J. (Taylor, Jane), 66, 120, 122, 227-229, 243, 247, 270-272, 274
Terenius, L., 241

Theander, O., 228, 230
Theodar, R., 216, 220
Thiéry, J.P., 167
Thomas, F.W., 265
Thomas, R.J., 127, 171, 183, 187, 188, 229, 247, 270, 272, 281
Thorarinsson, F., 266
Thornton, I.M.S., 199, 204, 242
Timoney, R.F., 242
Tiwari, S.D., 256
Tjukavkina, N.A., 213
Toivonen, S., 174
Tokunaga, N., 241
Tolonen, K., 257
Touffet, J., 237
Towers, G.H.N., 222
Toyota, M., 240, 241
Trachtenberg, S., 285, 305
Treboux, O., 23, 24, 134
Tripathi, B.K., 36, 90, 120, 122, 128
Triplett, E.W., 50
Troue, W., 84
Troxler, R.F., 246
Tseng, C.-Liw, 240
Tucker, E.B., 312
Tulecke, W., 110
Türk, R., 260
Tutschek, R., 230, 231, 235
Tyler, G., 249-253, 255, 257

Ubisch, G., von, 42, 49, 134, 143
Udar, R., 10, 66, 71, 235, 236, 288, 296
Ueda, T., 243
Ulychna, K.O., 121, 127
Upper, C.D., 50
Url, W., 263
Uto, S., 245

Vaarama, A., 25, 26
Vaisey, J.R., 153, 299, 300, 305
Valadon, L.G.R., 46, 49
Valanne, N., 22-26, 132, 136, 149, 166
Valio, I.F.M., 30, 132, 133, 138, 141, 142, 178, 180
Van Andel, 40
Vandekerkhove, O., 211-215, 219
Vashisht, B.R., 56
Vashistha, B.D., 26, 27, 29, 40, 43, 44, 46-49, 66, 70, 71, 93-95, 98-102, 120, 122, 123, 132-134, 138, 139, 143
Vašičková, S., 242
Veroustraete, F., 242
Vevle, G., 204, 243
Vian, B., 158
Viell, B., 84, 86-88

Villeret, S., 237
Vieweg, G.H., 266
Vitt, D.H., 189
Voth, P.D., 30, 34, 98, 284
Vochting, H., 78
Voth, P.D., 24, 92, 93, 244, 246
Vowinkel, R., 221

Wagner, P.H., 247
Waldner, M., 16
Waliszewska, B., 270, 272, 275, 277, 278, 279, 10.2
Walkington, D.L., 176, 177, 245
Walland, A., 228, 229, 231
Wallin, Th., 254
Walton, T.J., 191
Wang, T.L., 45, 284
Wann, F., 92, 93, 98, 99
Wanstall, P.J., 258
Ward, M., 59, 71, 72, 81, 120, 121, 123
Wareing, P.F., 52, 146
Warncke, E., 256
Wasiek, G., 53
Watson, E.V., 116
Watt, G., 175
Weier, T.E., 163, 165
Weitz, S., 219
Weniger, H.-P., 304, 305
Wenzel, G., 270, 272, 275-277, 280
Westerdijk, J., 78
von Wettstein, D., 26, 120, 122
von Wettstein, F., 55, 59, 74, 77, 108, 110, 111, 118
Whitaker, B.D., 46
Whitfield, D.W.A., 312
Whitehead, M.E., 263
Whitehouse, H.L.K., 27
Wiedhoph, R.M., 243
Wiencke, C., 167, 168
Wiermann, R., 192, 217
Wigglesworth, G., 18
Wilmot-Dear, C.M., 54
Wilson, J.R., 137
Winkler, S., 289
Winner, W.E., 317
Wirth, V., 260
Witkamp, M., 261
Withrow, R.B., 144
Wolters, B., 177, 183
Woodfin, C.M., 100
Woodward, F.N., 242
Worsdell, W.C., 26
Wren, R.W., 175

Yamamoto, K., 41

Yamasaki, K., 245
Yeaple, D.S., 253
Yoshida, K., 41
Yoshimura, H., 228
Yasuda, S., 222

Zacherl, H., 304
Zamski, E., 285, 305

Zanten, B.O. aan, 309
Zenk, M.H., 244
Zepf, E., 74
Ziegler, H., 266
Zielinski, F., 117
Zimmermann, S., 283
Zinsmeister, D.D., 158
Zinsmeister, H.D., 211, 214-216, 240, 245-247

DATE DUE

MAY 2 0 1990			
DEC 1 1993 RETURNED	JAN 0 5 1994		

DEMCO 38-297